VOLUME 46

Advances in
CHROMATOGRAPHY

VOLUME 46

Advances in CHROMATOGRAPHY

EDITORS:

ELI GRUSHKA
Hebrew University of Jerusalem
Jerusalem, Israel

NELU GRINBERG
Boehringer-Ingelheim Pharmaceutical, Inc.
Ridgefield, Connecticut, U.S.A.

CRC Press
Taylor & Francis Group
Boca Raton London New York

CRC Press is an imprint of the
Taylor & Francis Group, an **informa** business

CRC Press
Taylor & Francis Group
6000 Broken Sound Parkway NW, Suite 300
Boca Raton, FL 33487-2742

First issued in paperback 2019

© 2008 by Taylor & Francis Group, LLC
CRC Press is an imprint of Taylor & Francis Group, an Informa business

No claim to original U.S. Government works

ISBN-13: 978-1-4200-6025-6 (hbk)
ISBN-13: 978-0-367-38833-1 (pbk)

Visit the Taylor & Francis Web site at
http://www.taylorandfrancis.com

and the CRC Press Web site at
http://www.crcpress.com

Contents

Contributors

Luis A. Colón
Department of Chemistry
State University of New York
Buffalo, New York, USA

Ilaria D'Acquarica
Dipartimento di Studi di Chimica e
    Tecnologia delle Sostanze
    Biologicamente Attive
Università "La Sapienza"
Rome, Italy

Stacie L. Eldridge
Department of Chemistry
University of California-Riverside
Riverside, California, USA

Francesco Gasparrini
Dipartimento di Studi di Chimica e
    Tecnologia delle Sostanze
    Biologicamente Attive
Università "La Sapienza"
Rome, Italy

Yvan Vander Heyden
Department of Analytical Chemistry and
    Pharmaceutical Technology
Vrije Universiteit Brussel
Brussels, Belgium

Albert K. Korir
Department of Chemistry
University of California-Riverside
Riverside, California, USA

Michael Lämmerhofer
Christian Doppler Laboratory for
    Molecular Recognition Materials
Department of Analytical Chemistry and
    Food Chemistry
University of Vienna
Vienna, Austria

Cynthia K. Larive
Department of Chemistry
University of California-Riverside
Riverside, California, USA

Li Li
Department of Chemistry
State University of New York
Buffalo, New York, USA

Wolfgang Lindner
Christian Doppler Laboratory for
    Molecular Recognition Materials
Department of Analytical Chemistry and
    Food Chemistry
University of Vienna
Vienna, Austria

Katrice A. Lippa
Analytical Chemistry Division
Chemical Sciences and Technology
    Laboratory
National Institute of Standards and
    Technology
Gaithersburg, Maryland, USA

Debby Mangelings
Department of Analytical Chemistry and
    Pharmaceutical Technology
Vrije Universiteit Brussel
Brussels, Belgium

Philip Marriott
Australian Centre for Research on
    Separation Science
School of Applied Sciences
RMIT University
Melbourne, Australia

David V. McCalley
Centre for Research in
  Biomedicine
University of the West of
  England
Frenchay Campus
Bristol, U.K.

Christiana E. Merrywell
Department of Chemistry
University of California-Riverside
Riverside, California, USA

Domenico Misiti
Dipartimento di Studi di Chimica e
  Tecnologia delle Sostanze
  Biologicamente Attive
Università "La Sapienza"
Rome, Italy

Susan V. Olesik
Department of Chemistry
The Ohio State University
Columbus, Ohio, USA

Marco Pierini
Dipartimento di Studi di Chimica e
  Tecnologia delle Sostanze
  Biologicamente Attive
Università "La Sapienza"
Rome, Italy

Catherine A. Rimmer
Analytical Chemistry Division
Chemical Sciences and Technology
  Laboratory
National Institute of Standards and
  Technology
Gaithersburg, Maryland, USA

Danielle Ryan
School of Science and Technology
Charles Sturt University
Wagga, Australia

Lane C. Sander
Analytical Chemistry Division
Chemical Sciences and Technology
  Laboratory
National Institute of Standards and
  Technology
Gaithersburg, Maryland, USA

Claudio Villani
Dipartimento di Studi di Chimica e
  Tecnologia delle Sostanze
  Biologicamente Attive
Università "La Sapienza"
Rome, Italy

Naijun Wu
Merck & Co., Inc.
Rahway, New Jersey, USA

1 Liquid Chromatographic Enantiomer Separation and Chiral Recognition by Cinchona Alkaloid-Derived Enantioselective Separation Materials

Michael Lämmerhofer and Wolfgang Lindner

CONTENTS

1.1 EVOLUTION OF CINCHONA ALKALOID-DERIVED CHIRAL SEPARATION MATERIALS

Cinchona alkaloids that comprise the group of various natural basic chiral compounds including mainly quinine, quinidine, cinchonidine (CD), and cinchonine (CN) as well as their corresponding C_9-epimers (Figure 1.1) have become important "ex chiral pool" materials for chiroscience and chiral technologies. They were first isolated in 1820 by the two scientists, Pelletier and Caventou, as antimalarial agents from the bark of *Cinchona* sp. Although a first stereoselective total synthesis of quinine was achieved recently for the first time by Stork et al. [1] and shortly thereafter also the first catalytic, stereoselective syntheses of quinine and quinidine [2] have been reported, they are still more cheaply obtained from the bark of various *Cinchona* sp., in particular, from that of *Cinchona succirubra*. Since a long time, organic chemists

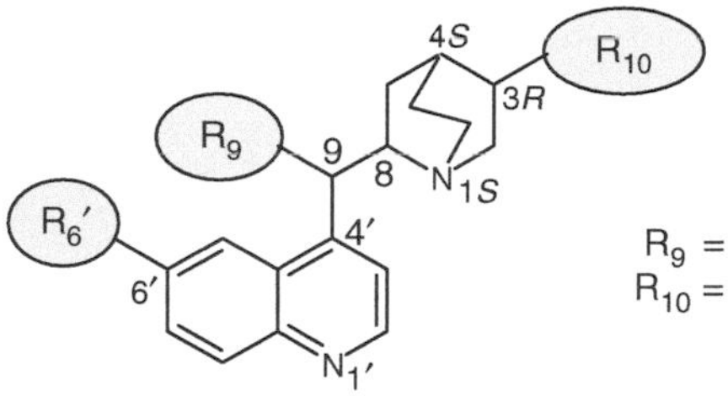

R_9 = OH
R_{10} = vinyl

R_6' = OCH_3			R_6' = H		
QN	(8*S*,9*R*)	Quinine	CD	(8*S*,9*R*)	Cinchonidine
QD	(8*R*,9*S*)	Quinidine	CN	(8*R*,9*S*)	Cinchonine
EQN	(8*S*,9*S*)	Epiquinine	ECD	(8*S*,9*S*)	Epicinchonidine
EQD	(8*R*,9*R*)	Epiquinidine	ECN	(8*R*,9*R*)	Epicinchonine

FIGURE 1.1 Chemistry and stereochemistry of the native cinchona alkaloids quinine, quinidine, cinchonidine, and cinchonine as well as their corresponding C_9-epimeric compounds.

make use of these natural chiral compounds for fractionated crystallization of chiral acids via their diastereomeric salts [3], as chiral shift reagents in NMR spectroscopy [4 7], and not least as chiral catalysts in stereoselective synthesis concepts [8,9].

Also, their first use for chromatographic separation purposes dates back already to the 1950s when Grubhofer and Schleith reported in an early work on quinine-modified resins that enabled the enantiomeric enrichment of mandelic acid by column liquid chromatography (LC) [10,11]. The growing popularity of cinchona alkaloids in separation science was further spurred in the early 1980s by the work accomplished by two groups that focused on the approach of pair chromatography employing quinine as chiral counterion. On the one hand, Izumoto et al. [12–14] could resolve the optical isomers of Cobalt (II) etylenediamine-tetraacetic acid (Co-EDTA) complexes and similar chiral compounds through the use of reversed-phase ion-pair chromatography with cinchona alkaloid cations as the ion-pairing reagents. Better known, on the other hand, is the work by Pettersson et al. [15] who adopted stereoselective ion-pair LC with quinine as counterion for enantiomer separation of chiral acids. Further work on this topic followed [16–18], but it soon became of limited practical importance when various modern bonded chiral stationary phases (CSPs) emerged. However, a kind of renaissance of the ion-pairing capabilities of these chiral auxiliaries has been noticed recently: native cinchona alkaloids and in particular derivatives have received some relevance as chiral additives for the capillary electrophoretic enantiomer separation of chiral acids [19] representing a noticeable alternative to the dominating cyclodextrin-based electrolyte systems.

The first silica-supported CSP with a cinchona alkaloid-derived chromatographic ligand was described by Rosini et al. [20]. The native cinchona alkaloids quinine and quinidine were immobilized via a spacer at the vinyl group of the quinuclidine ring. A number of distinct cinchona alkaloid-based CSPs were subsequently developed by various groups, including derivatives with free C_9-hydroxyl group [17,21–27] or esterified C_9-hydroxyl [28,29]. All of these CSPs suffered from low enantioselectivities, narrow application spectra, and partly limited stability (e.g., acetylated phases).

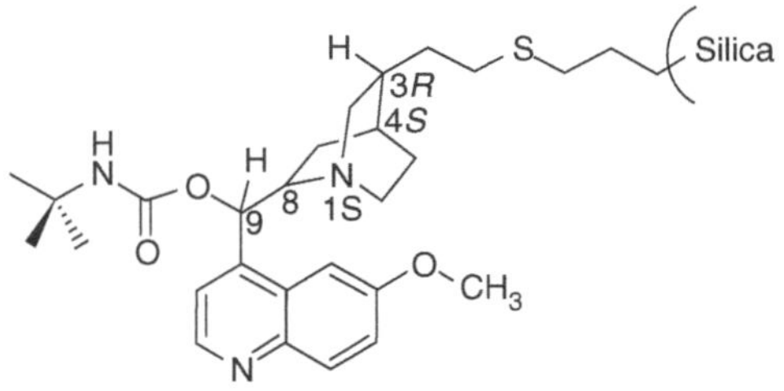

CHIRALPAK ® QN-AX: (8S,9R)   quinine derived
CHIRALPAK ® QD-AX: (8R,9S)   quinidine derived

(Capital letter for technologies and europe)

FIGURE 1.2 Structure and stereochemistry of commercially available cinchona alkaloid CSPs, marketed under trade name CHIRALPAK by chiral technologies europe. QN denotes quinine- and QD quinidine-derived and AX refers to their anion-exchanger capabilities (*vide infra*).

A breakthrough could be accomplished by Lindner and coworkers at the end of the 1980s with the introduction of cinchonan carbamate phases that were obtained through carbamoylation of the secondary alcohol group of the native alkaloids [30]. This intriguingly simple structural modification substantially enhanced the stereodifferentiation capabilities of the cinchona alkaloids' intrinsic chiral recognition characteristics. Later, it turned out that sterically demanding carbamate residues can amplify enantiodistinction power and hence the *tert*-butyl carbamate derivatives of quinine and quinidine evolved as first choice selectors for a variety of applications. They have been introduced into the market in 2002 by Bischoff Chromatography (Leonberg, Germany) and are, since 2005, commercially available from Chiral Technologies Europe (Illkirch Cedex, France) with trade names CHIRALPAK QN-AX and CHIRALPAK QD-AX (Figure 1.2). They are, along with structural analogs, the primary focus of this article.

1.2 GENERAL CHARACTERISTICS OF CARBAMOYLATED CINCHONA ALKALOID-BASED CHIRAL STATIONARY PHASES

Cinchona alkaloids are versatile building blocks for chiral technologies. They offer the advantage of a unique heterogeneous multifunctionality (bulky quinuclidine, planar quinoline ring, secondary alcohol group, and vinyl group) assembled to form a semirigid scaffold with predefined binding clefts for guest insertion. Their stereodifferentiation potential arises from the presence of five stereogenic centers (N_1, C_3, C_4, C_8, and C_9). Amongst the native alkaloids, quinine (QN) and quinidine (QD) turned out to be by far the most powerful, readily available raw materials for chiral auxiliaries such as chiral selectors. In the vast majority, they outperformed their congeneric 6'-demethoxy-analogous cinchonidine and cinchonine as well as the corresponding C_9-epimeric epiquinine (EQN) and epiquinidine (EQD) (absolute configurations as specified in Figure 1.1).

The distinct functional groups mentioned earlier represent effective binding sites or allow for dedicated modifications such as introduction of additional favorable binding sites and immobilization to chromatographic supports. Most importantly, the quinuclidine moiety provides, if the CSP is operated with weakly acidic eluents, a fixed charge center promoting ionic interactions with acidic solutes. Hence, they have most often been adopted as enantioselective weak anion-exchangers for enantiomer separation of structurally diverse chiral acids (the abbreviation AX in the trade name of the commercial CSPs refers to this anion-exchange property). The large planar electron-rich quinoline ring structure promotes directed $\pi-\pi$-stacking with (complementary) functionalities of guest solutes. The secondary hydroxyl group of the native alkaloids, on the contrary, is usually utilized for targeted enantioselectivity modulation through dedicated chemical modifications. Carbamoylation was found to be the most successful one, which allows for the introduction of an H-donor/H-acceptor moiety being composed of a rigid amide partial structure for directed hydrogen bonding and a flexible ester part enabling a minor spatial adaptation of the H-donor/acceptor moiety to the binding prerequisites of the interacting guest solute. Chemically, the carbamate group can be readily obtained through reaction with isocyanates or also by reaction of amines with active ester intermediates of quinine and quinidine, respectively, such as the p-nitrophenyl carbonate. If the carbamate group is combined with a sterically demanding residue, the stereodiscriminating potential can often be profoundly improved, as already pointed out earlier.

By adequate chemical adaptations the chiral recognition properties of the cinchona alkaloid derivatives may be preserved and exploited in various chromatographic and nonchromatographic separation methodologies. For this purpose, the vinyl group at the backside of the quinuclidine ring, which is remote from the active chiral recognition site around the stereogenic center of C_9, can be conveniently utilized because it does usually not interfere with chiral recognition. Thus, these chiral selectors could be successfully employed in high-performance liquid chromatography (HPLC) [30–33], supercritical fluid chromatography (SFC) [34], capillary electro- chromatography (CEC) [35,36], centrifugal partition chromatography (CPC) [37,38], supported liquid membrane (SLM) separations [39], and so forth. For example, for HPLC, straightforward immobilization of the chiral selectors on thiol-modified chromatographic supports like silica particles via radical addition reaction is usually pursued as anchoring strategy and with the tert-butylcarbamates of quinine and quinidine, respectively, yields the CSPs shown in Figure 1.2.

These CSPs are stable under common chromatographic conditions (i.e., within pH range 2–8 and temperatures below 50°C). Since the selectors are covalently anchored to the spherical silica supports (3 and 5 μm for analytical LC, 10–20 μm for preparative LC, or 3 μm for CEC), they tolerate any kind of organic solvents, which gives them a broad choice of elution conditions and great flexibility for selectivity tuning. It is one of the benefits that these columns have multimodal applicability. Indeed, they have been utilized in various operational modes: the reversed-phase (RP) mode, the polar-organic (PO) mode, and the normal-phase (NP) mode. In any case, chiral acids are separated by an anion-exchange retention mechanism (Figure 1.3) [40]. Regardless of which mode (PO, RP, or NP) is adopted, a counterion (competitor acid) is required in that case. Otherwise, the acidic solute will be trapped on

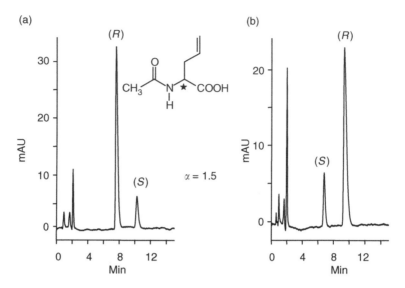

FIGURE 1.3 Enantiomer separation of the chiral acid N-acetyl-α-allyl-glycine on CHIR-ALPAK QN-AX (a) and CHIRALPAK QD-AX (b) by an enantioselective anion-exchange retention process. Chromatographic conditions: Column dimension, 150×4 mm ID; eluent, 1% (v/v) glacial acetic acid in methanol; flow rate, 1 mL min^{-1}; temperature, 25°C; detection, UV 230 nm. (Reproduced from M. Lämmerhofer, et al., *Nachrichten aus der Chemie, 50*: 1037 (2002). With permission.)

the column and does not elute within reasonable time. Neutral and basic compounds can often be separated in the NP mode in accordance with the typical mechanism of donor–acceptor phases (commonly referred to as Pirkle-type phases). However, in some instances also the RP mode turned out to yield successful enantiomer separations if the solute contained hydrophobic patches (*vide infra*).

Although cinchona alkaloid selectors originate from the natural chiral pool, one particular structural feature distinguishes them from other "ex chiral pool" selectors such as proteins, polysaccharides, cyclodextrins, macrocyclic antibiotics, and so forth. It is the fact that they exist in two distinct stereoisomeric forms: The two alkaloids quinine and quinidine, and their corresponding carbamate derivatives likewise, behave often like enantiomers, although they are actually diastereomers. This has to do with the primary stereodirecting centers of chirality at C_8/C_9 in the selectors where QN and QD display opposite configurations. The practical relevance of this pseudoenantiomeric relationship, which in most cases leads to the reversal of the elution order (see Figure 1.3), should not be underestimated (as will be outlined in detail later) and exists in a similar manner for cinchonidine and cinchonine as well as epiquinine and epiquinidine.

It is worthwhile pointing out another issue at this juncture. The low molecular-mass selectors can be grafted onto the surface of mesoporous silica (usually equal to 10 nm pore diameter) in a relatively high molar selector concentration due to their restricted space requirements (yielding typical selector loadings of about 0.4 mmol g^{-1} CSP corresponding to ca. 1.3 μmol m^{-2}). As a consequence, a relatively high number of active binding sites per g CSP should be available, while the individual selector units are still accessible and not too densely packed, which might prohibit the access of the solutes to the binding sites. The result is an elevated sample loading capacity in preparative LC applications (in particular, if compared to CSPs with macromolecular selectors such as protein-type CSPs or macrocyclic selectors such as cyclodextrin or antibiotics CSPs). The low molecular-mass nature of these selectors also greatly facilitates investigations of molecular recognition mechanisms between selector (SO) and selectand (SA) by spectroscopic studies, molecular modeling, X-ray crystal structure analysis, and so forth, and a decent number of papers addressed this issue.

Meanwhile, a wide variety of cinchona alkaloid derivatives have been systematically developed as chiral selectors, which complement each other in their enantiomer discrimination profiles. Considering the variety of derivatives, an overall reasonably broad applicability spectrum, approximating for chiral acids a 100% success rate, is yielded and extreme enantiorecognition levels (α-values above 15) could be realized for some chiral solutes with certain selectors. Moreover, various studies carried out with the CHIRALPAK QD/QN-AX columns in industry and academia clearly document their practical usefulness for solving challenging real-life problems and this should be illustrated by the present review article as well.

1.3 RETENTION MECHANISMS AND METHOD DEVELOPMENT

1.3.1 RETENTION MECHANISM FOR CHIRAL ACIDS

1.3.1.1 Enantioselective Anion Exchange

As mentioned, cinchonan carbamate-type CSPs are broadly applicable for solutes with acidic functionalities. For these analytes, an anion-exchange retention process is primarily at work regardless of the type of mobile phase mode being in use, be it the RP mode with buffered hydro-organic mobile phases (pH 4–7.5) [30], the PO mode [31,41], or NP mode [42]. If operated with (weakly) acidic mobile phases, the tertiary quinuclidine amine (pK_a around 8.9 as calculated for the *tert*-butylcarbamate selector by ACD/Labs 7.00, Advanced Chemistry Development, Toronto, Canada) gets instantly protonated and serves as the fixed ion-exchange site. (*Note*: The quinoline ring has a pK_a of 4.3 and will be largely unprotonated in the usual working pH range of 4–7.5.) Acidic compounds are attracted by electrostatic interactions and it may be argued that the long range but nondirected nature of this interaction is the primary driving force for the solute-CSP adsorption.

A stoichiometric model can conveniently be invoked to explain the ion-exchange retention process [43–46]. As discussed in detail in these cited papers on ion-exchange theory, useful information about the involved ion-exchange process can be deduced from plots of log k vs. the log of the counterion concentration [X], which commonly show linear dependencies according to the stoichiometric displacement model (Equation 1.1)

$$\log k = \log K_z - Z \log[X]. \qquad (1.1)$$

In this equation, the constant K_z, which can be easily inferred from the intercept, represents a system-specific constant that is related to the ion-exchange equilibrium constant K ($L \, mol^{-1}$), the surface area S (in $m^2 \, g^{-1}$), the charge density on the surface, that is, the number of ion-exchange sites q_x available for adsorption (in $mol \, m^{-2}$), and the mobile phase volume V_0 (in L) in the column as described by the following equation:

$$K_z = \frac{K \cdot S \cdot (q_x)^Z}{V_0}. \tag{1.2}$$

From the slope ($Z = m/n$), information can be derived on how many charges are involved in the ion-exchange process. For example, the slope for a given solute will be flatter the higher the charge number n of the employed counterion (e.g., at pH of 8.5, the slope for a given solute is expected to be lower for citrate than phosphate than acetate buffers). Vice versa, the slope of the $\log k$ vs. log [counterion] dependencies is larger for multiply charged solutes. In other words, multiply charged solute species respond stronger to a change in the counterion concentration in terms of their retention behaior.

This should be illustrated by the counterion concentration dependencies of three different model solutes, namely N-3,5-dinitrobenzoylated serine (DNB-Ser), aspartic acid (DNB-Asp), and O-phosphoserine (DNB-PSer), which are structural analogs that differ in the number of nominal charges. The solutes have been analyzed in the RP mode with methanol-phosphate buffer (50:50; v/v) (pH 6.5) under variable phosphate concentrations. The results are shown in Figure 1.4.

At pH 6.5, a high percentage of the phosphate counterion is singly charged ($\sim 84\%$), and therefore the effective charge number of the phosphate counterion (n phosphate) is about -1.2. The three distinct DNB-derivatized solutes differ, as mentioned, also in their effective charge numbers: DNB-Ser has a pK_a-value of 2.91 (± 0.10) (as calculated by ACD/Labs 7.00 software), DNB-Asp pK_a-values of 3.07 (± 0.23) and 4.59 (± 0.19), and DNB-PSer (O-phosphoserine) pK_a-values of 1.76 (± 0.10), 2.61 (± 0.10), and 6.38 (± 0.30). Hence, their effective charge numbers m at pH 6.5 are as follows: -1 for DNB-Ser, -2 for DNB-Asp, and -2.6 for DNB-PSer. Thus, theoretical Z-values of 0.8, 1.6, and 2.1 are calculated from these effective charge numbers for DNB-Ser, DNB-Asp, and DNB-PSer, respectively. Indeed, it can be seen that the experimental $\log k$ vs. $\log [X]$ plots correctly reproduce the expected trends with larger slopes (0.4, 0.9, and 1.1 for DNB-Ser, DNB-Asp, and DNB-PSer, respectively) the higher the charge states of the solutes. The deviation between theoretical and experimental slope values may be due to the super imposed effect of the buffer concentration on other than ionic interactions (e.g. strengthening of hydrophobic interactions by a kind of salting out effect).

From Figure 1.3, it becomes also evident that, while the slopes of the distinct solutes are different, those of the corresponding enantiomers are nearly the same. This means that for both enantiomers an ion-exchange process is at work and both isomers respond almost equally sensitive to the variation of the counterion concentration. In other words, the separation factors are usually almost unaffected by the counterion concentration, which opens up the possibility for a flexible adjustment

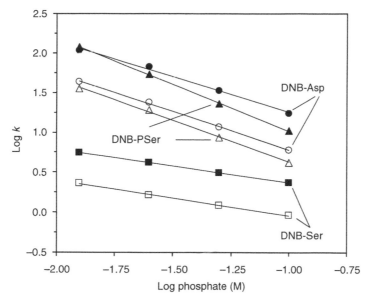

FIGURE 1.4 Dependencies of retention factors k on counterion (i.e., phosphate) concentration [X]. Experimental conditions: Mobile phase, methanol-sodium dihydrogenphosphate buffer (50:50; v/v) (pH_a 6.5; adjusted in the mixture with sodium hydroxide); flow rate, 1 mL min^{-1}; temperature, 25°C; CSP, O-9-[3-(triethoxysilyl)propylcarbamoyl]-quinine bonded to silica [30]; column dimension, 150 × 4 mm ID.

of retention times without significant variation or loss of selectivity. This is a specific feature of these chiral anion-exchangers. For example, DNB-Ser showed the identical separation factor of 2.57 in the buffer concentration range between 20 and 200 mM.

It is also worth noting that the type of counterion is also of prime importance. It strongly affects both retention and enantioselectivities. As a rule of thumb, the elution strength of the counterions in the RP mode decreases roughly in the order citrate > phosphate > formate ~ acetate. This effect on the elution strength can be ascribed to their modulation of K_Z via competitive binding to the anion-exchange site and decreasing affinities of the above counterions in the presented order. (*Remark*: K_Z is actually a combined constant resulting from a competitive adsorption isotherm.) In contrast, α-values concomitantly follow the reversed order, that is, they usually increase in this order.

The stoichiometric displacement model mentioned above provides a simple theoretical basis for describing the effect of the counterion (concentration). In the present case, we have also to consider that the anion-exchanger is a weak base [i.e., we are dealing with a weak chiral anion-exchanger (WAX)]. Acid–base equilibria, for both the ion-exchanger and weakly acidic solutes as well, are superimposed on the previously described ion-exchange process in such a case. Sellergren and Shea [46] presented a theoretical framework that accounts for this issue and is also based simply on the mass action model. Considering the underlying dissociation equilibria for the selector QN and selectand SA, it can be written for the retention factor k (Equation 1.3)

$$k = \phi \cdot \frac{[SA]_s}{[SA]_m} = \phi \cdot \frac{K \cdot [QN^+] \cdot [SA^-]}{[X^-] \cdot ([SA^-] + [SAH])}, \qquad (1.3)$$

wherein ϕ is the phase ratio, K is the ion-exchange equilibrium constant, $[QN^+]$ is the actual concentration of ion-exchange sites, $[X^-]$ is the concentration of counterions in the mobile phase, and the term $([SA^-]+[SAH])$ relates to the free SA in the mobile phase as the sum of ionized and nonionized forms. The equation mentioned above can be simplified in the following equation

$$k = \phi \cdot \frac{1}{[X^-]} \cdot K \cdot \alpha_{SA}^* \cdot \alpha_{QN}^* \cdot [QN]_{tot}, \qquad (1.4)$$

wherein α_{SA}^* and α_{QN}^* are the degree of ionization of solute SA and selector QN, respectively, and $[QN]_{tot}$ is the total ion-exchange capacity (i.e., the surface concentration of selector on the sorbent).

A typical pH profile for a weakly acidic solute such as a carboxylic acid is depicted in Figure 1.5.

It largely obeys the trend predicted by Equation 1.4. Maximal retention may be expected when both selector and analyte are dissociated to a high degree (i.e., where the product of α_{SA}^* and α_{QN}^* reaches a maximum value). For carboxylic acids, this retention maximum is usually found between pH 5 and 6, where the enantiose-lectivity also adopts its optimum (see Figure 1.5). Above this pH, the retention is decreased because (i) the dissociation of the selector becomes increasingly reduced, (ii) the effective counterion concentration may be increased, and (iii) the superim-posed hydrophobic retention increment of the solute on the bonded phase loses its

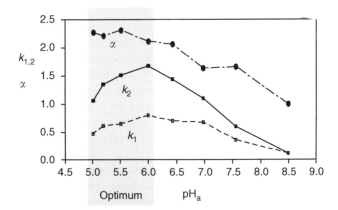

FIGURE 1.5 pH-effect on retention factors k and separation factors α. CSP: O-9-(*tert*-butylcarbamoyl)quinine bonded to silica; column dimension, 150 × 4 mm ID; eluent, methanol-ammonium acetate buffer (80:20, v/v) (adjusted with acetic acid); temperature, 25°C; 1 mL min^{-1}; sample, N-benzoyl-leucine (Bz-Leu). (Reproduced from M. Lämmerhofer et al., *American Laboratory, 30*: 71 (1998). With permission.)

influence when the solute becomes more and more dissociated. In contrast, at lower pH the reduced dissociation of the solute may be made responsible for the lowered retention factors. The asymmetric profile may be explained by (i) the superimposed RP-retention increment, which becomes more important when the pH is lowered and (ii) by a decrease of the effective counterion concentration originating from an increasingly suppressed dissociation of the weakly acidic acetate counterion at lower pH values. For other acids, like phosphonic, sulfonic acids, and acids with additional basic groups, the retention maxima and selectivity optima may be significantly shifted to a different pH.

1.3.1.2   Anion-Exchange Mechanism in PO and NP Modes

The PO mode is a specific elution condition in HPLC enantiomer separation, which has received popularity especially for macrocyclic antibiotics CSPs and cyclodextrin-based CSPs. It is also applicable and often preferred over RP and NP modes for the separation of chiral acids on the cinchonan carbamate–type CSPs. The beneficial characteristics of the PO mode may arise from (i) the offset of nonspecific hydrophobic interactions, (ii) the faster elution speed, (iii) sometimes enhanced enantioselectivities, (iv) favorable peak shapes due to improved diffusive mass transfer in the intraparticulate pores, and last but not least, (v) less stress to the column, which may extend the column lifetime. Hence, it is rational to start separation attempts with such elution conditions. Typical eluents are composed of methanol, acetonitrile (ACN), or methanol–acetonitrile mixtures and to account for the ion-exchange retention mechanism the addition of a competitor acid that acts also as counterion (e.g., 0.5–2% glacial acetic acid or 0.1% formic acid) is required. A good choice for initial tests turned out to be a mobile phase being composed of methanol-glacial acetic acid-ammonium acetate (98:2:0.5; v/v/w).

A thorough study on the ion-exchange mechanism and the effect of distinct counterions in this PO mode was recently presented by Gyimesi–Forras et al. [41]. A large variety of distinct acid additives to methanol, acetonitrile, and tetrahydrofuran (Table 1.1) (without any base added) was investigated in view of the stoichiometric displacement model and their effect on the enantiomer separation of 2-methoxy-2-(1-naphthyl)propionic acid. The stoichiometric displacement model (Equation 1.1) was obeyed also in the PO mode, as revealed by linear plots of log k vs. acid concentration. The slopes and intercepts along with the concentration ranges used with the distinct competitor acids are summarized in Table 1.1.

It can be seen again that the slopes for the enantiomers are, in most cases, very similar, tantamount with a negligible effect of the counterion (competitor acid) concentration in the eluent on α-values. On the other hand, the distinct acid additives can, according to the resultant slopes, be grouped essentially into two clusters: the majority of investigated acids adopted slopes around 0.8–0.9, and solely citric acid, succinic acid, trifluoroacetic acid, and malonic acid showed lower slopes (around 0.5–0.6), which means that they possess apparently a higher effective charge. From a practical viewpoint, more of interest, however, is the relative elution strength of these acid additives in the PO mode. This information can be derived from the intercepts, which are related somehow inversely proportional to the affinity of the

TABLE 1.1

Influence of Acid Additives on Retention Characteristics of 2-Methoxy-2-(1-Naphthyl)Propionic Acid on a O-9-(tert-Butylcarbamoyl)Quinine CSP as Assessed by the Characteristic Parameters of the Stoichiometric Displacement Model (Slopes and Intercepts of log k vs. log Acid Concentration in the Mobile Phase)

Acid (Counterion)	c% (v/v)	Slope[a]	Slope[b]	Intercept[a]	Intercept[b]	pK_a^c
MeOH						
TFA	2.5×10^{-4} to 2×10^{-3}	0.55	0.57	−1.75	−1.51	0.53
Methoxyacetic acid	2.5×10^{-1} to 2	0.90	0.68	−0.56	−0.23	3.55
N-Acetylglycine	2.5×10^{-1} to 2	0.98	0.85	−0.40	−0.01	3.67
Glycolic acid	2.5×10^{-1} to 2	0.88	0.91	−0.45	−0.29	3.74
Formic acid	6.25×10^{-2} to 5×10^{-1}	0.86	0.86	−0.29	0	3.74
Acetic acid	5×10^{-1} to 4	0.75	0.74	0.55	0.80	4.79
Malonic acid	3.14×10^{-2} to 2.5×10^{-1}	0.68	0.73	−1.09	−0.94	2.92; 5.61*
Succinic acid	1.25×10^{-1} to 1	0.53	0.52	0.19	0.43	4.24; 5.52*
o-Phosphoric acid	6.3×10^{-3} to 5×10^{-2}	0.88	0.94	−2.13	−2.05	1.97; 6.42*; 12.53**
Citric acid	1.56×10^{-2} to 2.5×10^{-1}	0.50	0.50	−0.89	−0.69	2.93; 4.23!*; 5.09**
ACN						
Methoxyacetic acid	2.5×10^{-1} to 2	0.74	0.81	−0.38	0.32	
Formic acid	6.25×10^{-2} to 5×10^{-1}	0.85	0.93	−0.29	−0.14	
Acetic acid	5×10^{-1} to 4	0.97	0.99	0.60	0.85	
THF						
Methoxyacetic acid	2.5×10^{-1} to 2	0.84	0.87	−1.05	0.89	
Formic acid	6.25×10^{-2} to 5×10^{-1}	0.53	0.54	−0.39	0	
Acetic acid	5×10^{-1} to 4	0.60	0.60	0.01	0.42	

[a] Data of (−)-(R)-2-methoxy-2-(1-naphthyl)propionic acid.

[b] Data of (+)-(S)-2-methoxy-2-(1-naphthyl)propionic acid.

[c] pK_a values in aqueous solution calculated by ACD Labs/pKa DB 7.0 software version: * pK_{a2}; ** pK_{a3}.

Source: Reproduced from K. Gyimesi-Forras et al., Chirality, 17: S134 (2005). With permission.

counterion (competitor acid) toward the ion-exchange site. The higher the affinity of the acid additive, the lower the intercepts and the higher its elution strength. It is eye-catching in Table 1.1 that trifluoroacetic acid and o-phosphoric acid reveal the smallest intercept values. They effectively eluted the solute even when it was added at concentrations below 0.05%. In other words, these competing agents effectively shielded the charged site of the ion-exchanger and ion-exchange processes were actually more or less suppressed. The situation is similar for citric and malonic acid, which also have a high yet slightly lower affinity toward the ion-exchanger and therefore a strong competing effect and high elution strength as well. In sharp contrast, acetic acid is the counterion with the lowest affinity in the series (largest intercepts) and weakest elution strength. To achieve similar elution strength, it needs to be added in much higher concentrations. The other acids possess a moderate affinity and therefore also a moderate elution strength.

On the other hand, optionally added co-ions of the eluent may also interfere with the ion-exchange process through competitive ion-pairing equilibria in the mobile phase. The effect of various amines added as co-ions to the polar-organic mobile phase was systematically studied by Xiong et al. [47]. While retention factors of 9-fluorenylmethoxycarbonyl (FMOC)-amino acids were indeed affected by the type of co-ion, enantioselectivities α and resolution values R_S remained nearly constant. For example, retention factors k_1 for FMOC-Met decreased from 17.4 to 9.8 in the order

$$\text{ammonia} > \text{diethylamine} > \text{ethylenediamine} > \text{triethanolamine}$$

$$> N,N\text{-diisopropylethylamine} > \text{tripropylamine} > \text{triethylamine}$$

while α-values were always between 1.64 and 1.70. It is conspicuous that the co-ions with closer structural resemblance to the quinuclidine exhibited the higher elution strength. Speaking in practical terms, the separation can be made faster when, for example, ammonia is replaced as co-ion by its stronger congener triethylamine without significant losses in selectivity and resolution.

In a similar manner, NP conditions (chloroform-based eluents) could be adopted for the separation of N-$tert$-butoxycarbonyl-proline (Boc-Pro) enantiomers on a hybrid urea-linked epiquinine-calixarene type CSP and an acidic displacer such as acetic acid promoted elution [42]. Since $\ln k$ vs. $\ln[CH_3COOH]$ dependencies gave straight lines, it may be concluded that this may be attributed to an ion-exchange process still existing in the NP mode. Acids cannot be eluted within reasonable run time without adding an acidic displacer. The practical relevance of the NP mode has to be seen in its much wider solvent choice, which may greatly extend the flexibility in the course of method development.

1.3.1.3 Hydroorganic Elution Mode: Superimposition of RP or HILIC Mechanisms

The cinchona alkaloid-based CSPs are actually mixed-mode RP/weak anion-exchange phases and HILIC/WAX phases, respectively. The surface layer of these

CSPs has, overall, a hydrophobic character (very similar to RP phases with C4–C8 ligands) which stems from contributions of the chiral selectors itself and (capped) linker groups (only a portion of the linkers are utilized for selector attachment) which constitutes a kind of hydrophobic basic layer on the support surface. Hence under typical RP-conditions, hydrophobic interactions between lipophilic residues of the solute and hydrophobic patches of the sorbent may be active and thus a reversed-phase like partition mechanism may be superimposed upon the primary ion-exchange process ($k = k_{RP} + k_{IX}$). This k_{RP}-retention contribution may be especially important for eluents with high aqueous content.

Therefore, it is not further surprising that the organic modifier type and content is an effective variable to adjust retention and to optimize enantioselectivities. Methanol and acetonitrile have been frequently found to be complementary in their selectivity profiles and these two organic modifiers are advised to be tested in a preliminary screening experiment.

Moreover, in various experiments it was found that at a constant total counterion concentration in the eluent the dependence of the retention factors on the organic modifier content φ largely follows the linear solvent strength theory (LSS) (Equation 1.5)

$$\log k = \log k_0 - S \cdot \varphi, \tag{1.5}$$

wherein k_0 represents the (hypothetical) retention factor at 0% organic modifier content (usually obtained by linear extrapolation to the y-intercept), S is a solute-dependent parameter related to its solvent-accessible surface. The larger the hydrophobic contact area, the larger the parameter S and the steeper the curves.

Accordingly, plots of $\log k$ or $\ln k$ vs. percentage of organic modifier content φ often show linearity over a wide range (Figure 1.6a) [48]. The hydrophobic interactions are, of course, weakened with increase of the organic modifier content in the mobile phase, yielding faster elution. The most likely consequences of elevated modifier percentages in the eluent on the observed enantioselectivities may be thereby generally broken down into two scenarios: (i) weakening of nonspecific hydrophobic interactions of the solute enantiomers with the lipophilic basic layer of the sorbent surface or at the chiral selector that concomitantly should lead to an increase of α-values and (ii) weakening of enantioselective hydrophobic interactions at the selector's active site leading to a loss of enantioselectivity with higher modifier content. Indeed, there is also observed a significant effect of the modifier content on the α-values in the presented example of Figure 1.6b. Since the α-values increase with the organic modifier content, it might be argued that nonspecific hydrophobic interactions, which take place nonenantioselectively and may thus lead to lower observed enantioselectivity values, are weakened.

It is, however, also to be noted that deviations from linearity of the aforementioned plots may be readily observed, which may indicate smooth changes in the separation mechanism. If the k vs. percentage of modifier dependency is investigated over the entire or at least a wider range, U-shaped curves may be obtained, in particular with acetonitrile as modifier. While the drop of retention factors with increase of modifier percentage at low modifier contents may follow the described RP-behavior, the trend

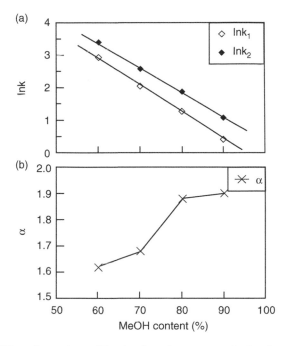

FIGURE 1.6 Effect of organic modifier (methanol) percentage in the eluent on the retention factors (a) and observed enantioselectivities (b) of N-(2,4-dinitrophenyl)-α-(2-chlorobenzyl)-proline employing an O-9-[(2,6-diisopropylphenyl)carbamoyl]quinine-based CSP. Experimental conditions: Eluent, ammonium acetate buffer–methanol (total ionic strength = 25 mM; pH_a = 6.5), methanol content varied between 60 and 90%, while ionic strength and apparent pH were kept constant; temperature, 40°C; flow rate, 0.8 mL min^{-1}. (Reproduced from A. Peter et al., *J. Sep. Sci.*, **26**: 1125 (2003). With permission.)

may just be opposite at high organic modifier contents (around 80% and higher). Thus retention may be raised upon a further increase of the modifier content, which is the typical trend for a hydrophilic interaction chromatography (HILIC) mode. While under RP-regimen acetonitrile exhibits a higher elution strength than methanol, it is also a characteristic feature for a switch to the HILIC mode that the elution strength of acetonitrile-based eluents is lower than that of methanol-based mobile phases [49].

1.3.2 RETENTION MECHANISMS FOR NEUTRAL/BASIC COMPOUNDS

Although the cinchonan carbamate–based CSPs are of primary interest for the separation of chiral acids, it needs to be stressed that the scope of application is, however, not restricted to chiral acids. A few reports in the literature deal with the separation of the enantiomers of neutral and weakly basic chiral compounds, respectively, on quinine carbamate–type CSPs [50–54]. Both RP and NP modes may be applicable.

For example, Gyimesi–Forras et al. [50] performed a detailed investigation on the enantiomer separation of chiral quinazoline derivatives (2-substituted

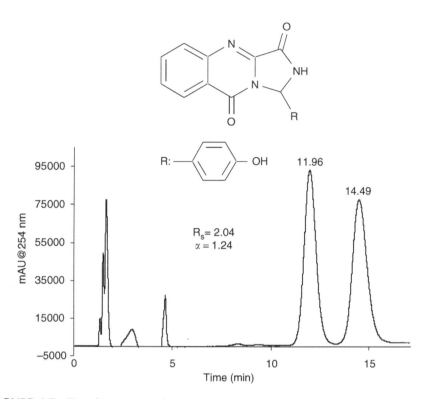

FIGURE 1.7 Enantiomer separation of 2-(4-hydroxyphenyl)-imidazo[1,5-b]-quinazoline-1,5-dione on CHIRALPAK® QN-AX under reversed conditions. CSP: Chiralpak QN-AX (150 × 4 mm ID); mobile phase, 0.1 M ammonium acetate-methanol (60:40; v/v) $pH_a = 6.0$; flow rate, 1 mL min^{-1}; temperature, 25°C; UV detection at 238 nm. (Reproduced from K. Gyimesi-Forrás et al., *J. Chromatogr. A, 1047*: 59 (2004). With permission.)

imidazo[1,5-b]-quinazoline-1,5-diones) on cinchonan carbamate–type CSPs in the RP mode (Figure 1.7). In such a system, pH and buffer type as well as its concentration are of minor relevance, while type and content of organic modifier become the experimental variables of key importance for optimization of the separation. For both methanol and acetonitrile, the LSS theory (see Equation 1.5) adequately described the effect of the organic modifier content, with acetonitrile having stronger elution strength than methanol. It was also found that indeed the more lipophilic quinazoline compounds (as assessed by their log *P* values) were stronger retained on the *O*-9-(*tert*-butylcarbamoyl)quinine-based CSP owing to stronger hydrophobic interactions. However, when strong π–π-interactions came into play and significantly contributed to the retention such as in case of π-acidic substituents, the relative elution orders of distinct analogs were changed (e.g., *o*-chlorophenyl and *p*-chlorophenyl substituents; the latter was stronger retained).

If the hydroorganic elution (RP) mode does fail to separate neutral and basic compounds, respectively, the NP mode may be the method of choice. In this mode, the cinchonan carbamate CSPs appear to function according to principles well known from donor–acceptor phases (commonly referred to as Pirkle-CSPs). This mode of operation has been utilized by Salvadori and coworkers [20,28] and Nesterenko et al. [27] in their early works with quinine-based CSPs having free hydroxyl group. Its scope of application is not fully elucidated for the cinchonan carbamate CSPs yet. However, several papers have been published by Guiochon and coworkers [51–54] describing the principal potential and mechanisms of separation of the enantiomers of neutral chiral arylcarbinols.

In one study, various distinct types of polar modifiers to n-hexane were tested for 3-chloro-1-phenylpropanol (3CPP) and 1-phenylpropanol (1PP) enantiomer separation [53]. Thereby, alcohol modifiers turned out to be more effective displacers of the solutes from the adsorption places on the sorbent surface, yet aprotic polar modifiers provided higher separation factors (with ethyl acetate in n-hexane affording the best separations for these chiral alcohols). It is evident, though, that the optimal choice of polar modifier is strongly solute dependent and can therefore not be generalized.

Retention theory from the work of Lanin and Nikitin [55] (Equation 1.6) was adapted to describe the dependency of retention factors (k) as a function of the mobile phase composition [53]. The concentration of the polar modifier is, besides the type, the primary variable for the optimization of the separation and can be described by competitive adsorption reactions of solute (i.e., sorbate) and polar modifier for which the following relationship can be applied (Equation 1.6)

$$\frac{1}{k} = \frac{1}{\beta \cdot K_s} + \frac{K_m - 1}{\beta \cdot K_s} \cdot X_m^M, \tag{1.6}$$

wherein β is the phase ratio, K_s is the equilibrium constants for the stoichiometric exchange of solvent molecule (diluent) on the active site of the sorbent by solute (sorbate), and K_m is the equilibrium constant for the corresponding competitive adsorption reaction of the modifier molecules on the stationary phase surface. X_m is the mole fraction of the polar modifier in the eluent. According to this stoichiometric displacement model, ethyl acetate plays the role of a competitor for adsorption with respect to the solute and plots of $1/k$ vs. the mole fraction of polar modifier X_m should give straight lines, which was indeed observed for 3CPP [53] (Figure 1.8). n-Hexane, by contrast, acts like a diluent or inactive carrier of the modifier ethyl acetate. Other authors describe the polar modifier effect in terms of the Snyder–Soczewinski model [56].

1.4 STRUCTURAL MODIFICATIONS OF CINCHONAN-TYPE SELECTORS

It is system-intrinsic for highly selective chiral recognition events that the enantiorecognition capability of a chemical receptor or selector system is somehow restricted to a certain group of target selectands. While universal applicability of a CSP may certainly be a desired, frequently postulated property for a powerful chiral phase, it is for sure that it eventually remains a wish because of the contradictive and mutually exclusive nature of "specificity" and "universal applicability." To get around

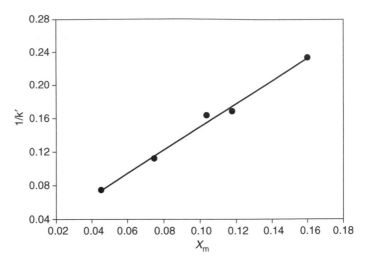

FIGURE 1.8 Effect of the mole fraction of polar modifier (ethyl acetate) in *n*-hexane on the reciprocal of the retention factor for the separation of 3-chloro-1-phenylpropanol enantiomers on a *O*-9-(*tert*-butylcarbamoyl)quinidine CSP. Temperature, 22°C. (Reproduced from L. Asnin, and G. Guiochon, *J. Chromatogr. A, 1091*: 11 (2005). With permission.)

this disturbing dilemma, it has become common among column suppliers to offer a small set of CSPs on the basis of structurally similar yet distinct selector analogs (e.g., cellulose and amylose carbamates and/or esters) having complementary application spectra. With this strategy in mind, systematic selector structure variations by (quantitative) structure–resolution relationships were devised to adjust binding increments of cinchona alkaloid derivatives through dedicated and targeted chemical modifications aiming at fine-tuning, modulation, optimization, and/or expanding the enantioselectivity profiles of cinchonan-based CSPs.

In case of cinchonan-type selectors, the active chiral distinction site is formed by the spaces around the N_1–C_8–C_9–$C_{4'}$ torsion (see Figure 1.1) and functionalities/groups attached to these atoms. Structural modifications within or close to this active site were supposed to considerably impact the enantiorecognition levels opening the avenue for targeted CSP optimization by chemical design. Some strategies that have been pursued are described hereafter [57].

1.4.1 Effect of Carbamoylation and Pendant Carbamate Residue

The favorable effect of the introduction of a carbamate moiety into the cinchonan selectors was already proven by the prototype cinchonan carbamate CSPs (type I and type II) (Figure 1.9) [30], which showed enhanced enantioselectivities and a widened application range as compared to the CSPs with native cinchona alkaloid selectors and those reported earlier in the literature.

To illustrate this beneficial effect of the carbamate-binding increment, Figure 1.10 compares the separation of *N*-(3,5-dinitrobenzoyl)leucine (DNB-Leu) enantiomers on quinine-based and *O*-9-(*tert*-butylcarbamoyl)quinine-based CSPs, both chemically bonded to thiol-modified silica by identical linker and support chemistries, under otherwise identical conditions [32]. It is seen that the enantiomers are hardly baseline separated on the quinine-CSP with free C_9-hydroxyl group (α-value = 1.15). In sharp contrast, the *tert*-butyl carbamate analog easily separates this enantiomeric pair with an exceptional α-value of 15.9. This corresponds to an increase of the differential free energy of binding of *R*- and *S*-enantiomers by a factor of about 20 from 0.35 to 6.85 kJ mol^{-1}. Moreover, it is outstanding that the carbamate

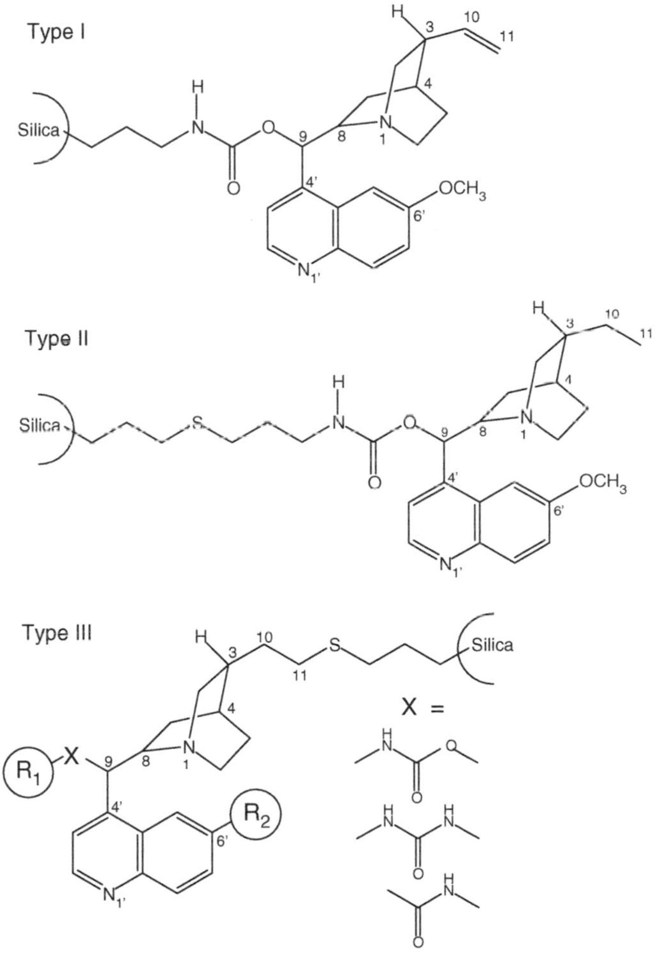

FIGURE 1.9 Selection of cinchonan carbamate CSPs that have been prepared in the course of selector optimization studies (type I: prototype; type II, *O*-9-linked thiol-silica supported prototype; type III, *C*-11-linked thiol-silica supported CSPs; type IV, "dimeric" selectors). (Adapted from M. Lämmerhofer and W. Lindner, *J. Chromatogr. A, 741*: 33 (1996); W. Lindner et al., PCT/EP97/02888, US 6,313,247 B1 (1997); P. Franco et al., *J. Chromatogr. A, 869*: 111 (2000); C. Czerwenka et al., *Anal. Chem., 74*: 5658 (2002).)

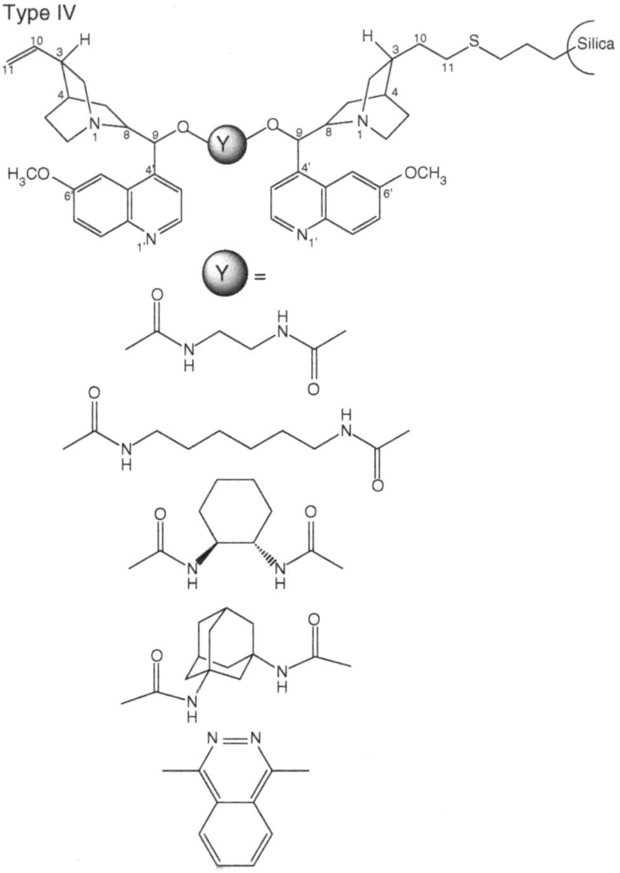

FIGURE 1.9 Continued.

functionality exerts a stereodirecting influence, as can be seen from the reversal of elution order on the quinine carbamate CSP. This is a vivid example for the sensitivity of molecular recognition phenomena to minor structure variations, which clearly emphasizes the effectiveness of dedicated structural changes as a tool to improve the enantiorecognition power of chiral selectors.

Replacement of the carbamate group with isosteric functionalities such as an N-methyl carbamate, urea, or amide group clearly confirmed the favorable qualities of the carbamate group [57]. While the introduction of a urea group, as in case of N-9-(tert-butylcarbamoyl)-9-desoxy-9-aminoquinine selector, instead of carbamate functionality turned out to be virtually equivalent in terms of enantiorecognition capabilities [57,58], the enantiomer separation potential was severely lost on N-methylation of the carbamate group, like in O-9-(N-methyl-N-tert-butylcarbamoyl)quinine [32,58], or its replacement by an amide, such as in case of N-9-(pivaloyl)-9-desoxy-9-aminoquinine selector [57,58]. For example, enantioselectivities dropped for DNB-alanine from 8.1 for the carbamate-type CSP, over 6.6 for the urea-type CSP, to 1.7 for the amide-type CSP, and 1.3 for the N-methyl

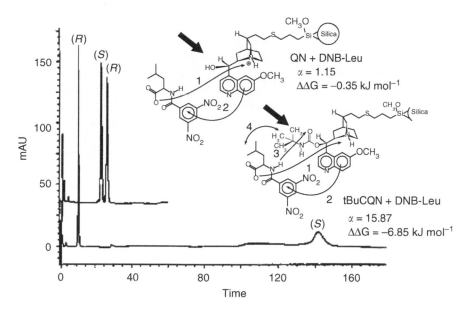

FIGURE 1.10 Comparison of enantiomer separations of DNB-Leu on quinine (QN) based and *O*-9-(*tert*-butylcarbamoyl)quinine (tBuCQN) based CSPs. 1, ionic interaction; 2, π–π-interaction; 3, hydrogen bonding; 4, steric interaction. Experimental conditions: Eluent, methanol-0.1 M ammonium acetate (80:20, v/v) (pH$_a$ = 6.0); flow rate, 1 mL min^{-1}; temperature, 25°C; column dimension, 150 × 4 mm ID; detection, UV 250 nm. Selector loadings, 0.37 and 0.30 mmol g^{-1} for QN- and tBuCQN-based CSPs, respectively. (Reproduced from A. Mandl et al., *J. Chromatogr. A, 858*: 1 (1999). With permission.)

carbamate–type CSP [58] (for structures of CSPs see Figure 1.9, type III with R_1 being *tert*-butyl and R_2 being methoxy). It is clearly evident that the presence of H-acceptor and H-donor site as well as their appropriate spatial orientation is of prime importance. For the amide selector, the H-donor/H-acceptor is spatially differently oriented and thus apparently incapable to form a supportive H-bond simultaneously with other noncovalent bonds such as ionic interaction and/or π–π-interaction. On the other hand, the *N*-methyl carbamate selector lacks the carbamate's H-donor group and thus an optionally established H-bond or steric constraints arising from the methyl-group preclude the formation of a supportive H-bond with the carbonyl of the selector. To conclude, the carbamate-type CSPs generally emerged as favorable structural variants from these studies that were limited to certain solute classes. A generalization must be undertaken with care and it may be worthwhile to screen also the other variants if a baseline resolution cannot be achieved with the standard carbamate-type CSPs. This holds likewise for the CSP-variants presented later.

Moreover, it could be figured out that an effective means to modulate the stereorecognition capabilities of the cinchonan selector motif may be via the carbamate residue. It offers a way of straightforward introduction of bulky alkyl substituents, which may affect the accessibility of active binding sites and/or lead to additional supportive Van der Waals-type interactions.

To verify such a steric effect a quantitative structure–property relationship study (QSPR) on a series of distinct solute-selector pairs, namely various DNB-amino acid/quinine carbamate CSP pairs with different carbamate residues (R_{SO}) and distinct amino acid residues (R_{SA}), has been set up [59]. To provide a quantitative measure of the effect of the steric bulkiness on the separation factors within this solute-selector series, α-values were correlated by multiple linear and nonlinear regression analysis with the Taft's steric parameter E_S that represents a quantitative estimation of the steric bulkiness of a substituent (Note: $E_{S,SA}$ indicates the independent variable describing the bulkiness of the amino acid residue and $E_{S,SO}$ that of the carbamate residue). For example, the steric bulkiness increases in the order methyl < ethyl < n-propyl < n-butyl < i-propyl < cyclohexyl < i-butyl < $sec.$-butyl < t-butyl < 1-adamantyl < phenyl < trityl and simultaneously, the E_S drops from -1.24 to -6.03. In other words, the smaller the E_S, the more bulky is the substituent. The obtained QSPR equation reads as follows:

$$\log \alpha = -1.63(\pm 0.33) - 1.83(\pm 0.32) \cdot E_{S,SA} - 0.40(\pm 0.08) \cdot E^2_{S,SA}$$

$$-0.59(\pm 0.09) \cdot E_{S,SO} - 0.08(\pm 0.01) \cdot E^2_{S,SO} - 0.06(\pm 0.02) \cdot E_{S,SA} \cdot E_{S,SO}$$

$$n = 52, \quad r = 0.9720, \quad s = 0.059, \quad F = 157.15, \quad P\text{-value} = 0.0000.$$

In this equation, n denotes the number of observations (i.e., data points), r is the correlation coefficient, s is the standard deviation of the residuals, F is the value of the Fisher test for the significance of the equation, and the P-value denotes the statistical significance level. The values in parenthesis of this QSPR equation indicate the 95% confidence interval of the estimated coefficients. From the statistical parameters it can be derived that this QSPR model is significant at the 99% confidence level.

This QSPR equation has a physical meaning which is in agreement with the chromatographically derived tentative binding model that was further confirmed by spectroscopic investigations (*vide infra*). The separation factors first increase with larger amino acid and carbamate residues (accounted for by the second and fourth terms of the previous equation), probably due to a growing importance of Van der Waals type and/or dispersive interactions for complex stabilization of the second eluted enantiomer for which a similar QSPR equation is valid [59]. Once the residues become too bulky, steric hindrance or repulsion may lead to a disturbed binding mode and a drop in enantioselectivity (accounted for by the quadratic terms, i.e., the third and fifth terms of the QSPR equation). The last (interrelation) term may indicate that the carbamate and amino acid residues interact with each other and therefore exert a mutual effect upon each other. The response surface given in Figure 1.11 [59] clearly shows that within the investigated series of binding partners, an optimal bulkiness for the selector is obtained with a carbamate residue having $E_{S,SO} \sim -3$, which is featured by substituents like t-butyl ($E_S = -2.78$) or 1-adamantyl ($E_S = -2.90$), and an amino acid residue having $E_{S,SA} \sim -2.2$ (like i-butyl, $E_S = -2.17$). On the contrary, SO–SA pairs having residues with larger and smaller E_S values give lower enantioselectivities.

Overall, the *tert*-butyl carbamates of quinine and quinidine evolved as the most effective chiral selectors among this family because of their broad enantioselectivity spectrum and exceptional degree of selectivity for a wide variety of chiral acids. On

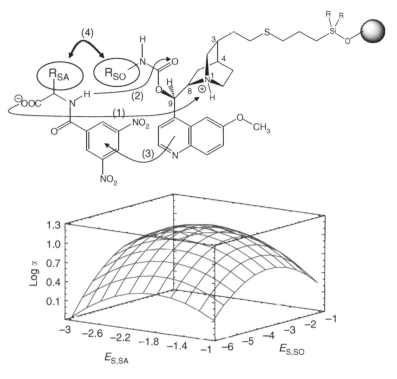

FIGURE 1.11 Systematic optimization of structural binding and stereorecognition increments of amino acid side chain (R_{SA}) and carbamate residue (R_{SO}): Dependency of separation factors on the steric bulkiness of amino acid and carbamate residues as quantified by their corresponding Taft's steric parameters $E_{S,SA}$ and $E_{S,SO}$. (1) ionic interaction; (2) π–π-interaction; (3) hydrogen bonding; (4) steric interaction. (Reproduced from M. Lämmerhofer et al., *J. Sep. Sci.*, 29: 1486 (2006). With permission.)

the other hand, the cinchonan carbamate selector may be decorated via the carbamate residue with additional supportive binding sites such as aromatic functionalities (π-acidic and π-basic aromatic systems, respectively), and so forth. Such modifications showed also great promise and, in particular, complementary selectivity profiles which partly extended or shifted application spectra. A number of aromatic carbamates have been tested including, for example, π-basic ones, such as 2,6-diisopropylphenyl, α- and β-naphthyl, phenyl, trityl as well as π-acidic ones such as 3,4- and 3,5-dichlorophenyl, 3,5-dinitrophenyl, and 3,5-*bis*-(trifluoromethyl)phenyl [31]. Thereby, the 2,6-diisopropylphenyl carbamate CSPs were frequently useful when the tBuCQN-CSP and tBuCQD-CSP displayed unsatisfactory enantioselecitivities (Table 1.2). For example, under the given conditions the enantiomers of atrolactic acid, an aromatic α-hydroxycarboxylic acid, were separated with an α-value of less than 1.1 on the tBuCQN-CSP, while the enantioselectivity was greatly improved on the *O*-9-[(2,6-diisopropylphenyl)carbamoyl]quinine-CSP (DIPPCQN-CSP) and *O*-9-[(triphenylmethyl)carbamoyl]quinine-CSP (TritCQN-CSP) (α-values of 1.37 and 1.27, respectively). Similar trends were observed for other selected test solutes

TABLE 1.2

Separation of Various Chiral Acids on Complementary Cinchonan Carbamate CSPs[a,b]

Selector[c]	k_1	α	R_S	e.o.
	Atrolactic acid			
tBuCQN	49.30	1.04	0.8	$(+) < (-)$
tBuCQD	23.63	1.06	0.9	$(+) < (-)$
DIPPCQN	17.94	1.37	3.5	$(-) < (+)$
DIPPCQD	21.18	1.14	2.9	$(+) < (-)$
TritCQN	12.65	1.27	4.0	$(-) < (+)$
	2-Hydroxybutyric acid			
tBuCQN	12.58	1.12	2.3	n.d.
tBuCQD	8.35	1.11	1.0	n.d.
DIPPCQN	7.15	1.36	4.1	n.d.
DIPPCQD	6.67	1.12	1.3	n.d.
TritCQN	4.83	1.29	1.9	n.d.
	Tetrahydro-2-furoic acid			
tBuCQN	11.03	1.15	2.2	$(R)-(+) < (S)-(-)$
tBuCQD	8.22	1.27	3.8	$(R)-(+) < (S)-(-)$
DIPPCQN	7.89	1.32	2.6	$(S)-(-) < (R)-(+)$
DIPPCQD	6.06	1.09	1.1	$(R)-(+) < (S)-(-)$
TritCQN	6.44	1.24	1.5	$(S)-(-) < (R)-(+)$
	Suprofen			
tBuCQN	5.46	1.11	1.5	$(-) < (+)$
tBuCQD	3.52	1.15	1.7	$(-) < (+)$
DIPPCQN	2.82	1.32	2.7	$(+) < (-)$
DIPPCQD	3.15	1.27	2.7	$(-) < (+)$
TritCQN	2.24	1.23	1.6	$(+) < (-)$

n.d.: not determined.

[a] Column dimension, 150 × 4 mm ID

[b] Mobile phase: ACN-MeOH (50:50, v/v) + 0.5% AcOH; T: 15°C, flow rate:1 mL min^{-1}, 230 nm and ORD.

[c] tBuCQN, *O*-9-(*tert*-butylcarbamoyl)quinine; tBuCQD, *O*-9-(*tert*-butylcarbamoyl)quinidine; DIPPCQN, *O*-9-[(2,6-diisopropylphenyl)carbamoyl]quinine; DIPPCQD, *O*-9-[(2,6-diisopropylphenyl)carbamoyl]quinidine; TritCQN, *O*-9-[(triphenylmethyl)carbamoyl]quinine.

given in Table 1.2 such as arylcarboxylic acids (e.g., suprofen), α-hydroxycarboxylic acids (e.g., 2-hydroxybutyric acid), and tetrahydro-2-furoic acid. More examples of successful enantiomer separations of chiral acids on the DIPPCQN-CSP can be found in the literature [48,60,61].

1.4.2 EFFECT OF C_8/C_9 STEREOCHEMISTRY

The pseudoenantiomeric nature of quinine and quinidine (opposite configurations at C_8 and C_9) as well as of the corresponding carbamates has already been outlined.

In practice, this allows the simple reversal of the chromatographic elution order on exchange of the quinine carbamate CSP by the corresponding quinidine analog (as shown in Figure 1.3 and Table 1.3) [30,31]. This peculiarity may be quite helpful if enantiomeric impurities below 0.5% must be analyzed (elution of the impurity peak in front of the main overloaded peak usually allows to reach lower LOQs). On the other hand, it is also evident that this pseudoenantiomeric behavior is not always obeyed (see Table 1.2) (notice, e.g., the enantiomer elution order for atrolactic acid on the tBuCQN and tBuCQD-CSPs), which is not further surprising because the corresponding quinine and quinidine carbamates are actually diastereomers rather than enantiomers.

In a comprehensive study on the impact of the C_8/C_9 stereochemistry on chiral recognition, the effect of inversion of a single configuration was also elucidated [33]. In addition to the *tert*-butylcarbamate CSPs of quinine and quinidine, their corresponding C_9-epimers were also prepared and chromatographically evaluated with a selected test set of chiral acids (Table 1.3). Consistently, a dramatically lower degree of enantiorecognition was observed for the two epimeric CSPs (tBuCEQN and tBuCEQD CSPs), which makes these CSPs more or less useless for the tested analytes. The relative spatial orientation of the interaction sites, as influenced by the relative C_8/C_9 stereochemistry, prohibits the favorable cooperative multisite binding mechanism in the epimeric CSPs and leads to the loss of stereodifferentiation and chromatographic enantioselectivity.

In conclusion, the C_9-configuration appears to exert a stereocontrolling effect in the active binding pocket and thus determines primarily the elution order, while on the other hand the relative C_8/C_9 configuration has a major impact on the magnitude of the obtained enantioselectivities. Although a generalization has to be undertaken with care, the generally negative results with the epimeric CSPs discouraged their further investigations.

1.4.3 EFFECT OF 6'-QUINOLINE SUBSTITUTION

The derivatives of the 6'-demethoxy alkaloids cinchonine (CN) and cinchonidine (CD), which seem to be—because of their better chemical stability—more popular for enantiomer separation by fractionated crystallization and as catalysts in stereose-lective synthesis applications than quinine and quinidine, showed surprisingly poor chromatographic enantiodifferentiation capabilities [31]. For example, α-values for DNB-Leu dropped from 15.9 on the tBuCQN-CSP to 4.2 on the corresponding tBuCCD-CSP (based on *O*-9-*tert*-butylcarbamoylcinchonidine), from 1.37 to 1.13 for *N*-acetyl-Leu, and from 1.90 to 1.27 for FMOC-Leu. These results led to the assumption that a chemical modification at the quinoline ring could have a major effect and conversely triggered the design of another set of very effective congen-eric CSPs with cinchonan backbone, namely of CSPs with dedicated substitutions at the 6'-position of the quinoline ring (see Figure 1.9) (type III) [62,63]. In fact, for the model solute DNB-Leu, the α-value could be increased from 4.24 ($R_{6'} = H$) to 15.9 ($R_{6'}$ = methoxy), to 20.3 ($R_{6'}$ = isobutoxy), to 31.7 ($R_{6'}$ = neopentoxy) [63]. Similar trends were found for other solutes. Electronic effects (higher electron dens-ity than in quinoline and reinforced π–π-interaction with the π-acidic DNB group)

TABLE 1.3

Enantioseparation Data for N-protected α-Amino Acids and Aryloxycarboxylic Acids on CSPs based on tert-Butylcarbamates of C$_8$/C$_9$ Stereoisomeric Cinchona Alkaloids

Alkaloid Precursor C$_8$/C$_9$ Configurations	tBuCQN-CSP QN (8S;9R)			tBuCEQD-CSP EQD (8R;9R)			tBuCEQN-CSP EQN (8S;9S)			tBuCQD-CSP QD (8R;9S)		
Solute	k_1	α	e.o.	k_1	α	e.o.	k_1	α	e.o.	k_1	α	e.o.
Ac-Leu	1.9	1.73	$R < S$	2.6	1.06	$R < S$	2.5	1.10	$S < R$	1.7	1.41	$S < R$
Bz-Leu	5.6	2.56	$R < S$	5.1	1.08	$R < S$	5.6	1.12	$S < R$	3.5	2.60	$S < R$
DNB-Leu	11.7	15.90	$R < S$	9.4	1.15	$R < S$	11.4	1.16	$S < R$	8.1	12.50	$S < R$
Boc-Leu	3.7	1.21	$R < S$	3.6	1.00	—	3.5	1.07	$S < R$	2.4	1.28	$S < R$
Z-Leu	4.2	1.27	$R < S$	5.4	1.06	$R < S$	6.6	1.07	$S < R$	4.8	1.31	$S < R$
DNZ-Leu	7.7	2.80	$R < S$	9.5	1.05	$R < S$	11.1	1.09	$S < R$	9.1	2.71	$S < R$
FMOC-Leu	9.3	1.90	$R < S$	13.8	1.10	$R < S$	15.3	1.11	$S < R$	11.6	1.85	$S < R$
DNP-Leu	24.0	1.31	$S < R$	17.2	1.00	—	20.9	1.06	$S < R$	19.0	1.09	$R < S$
Dichlorprop	10.9	1.19	$S < R$	10.1	1.03	n.d.	12.4	1.03	n.d.	7.2	1.29	$R < S$

Experimental conditions: Column dimension, 150 × 4 mm ID; mobile phase, methanol-0.1 M ammonium acetate (80:20; v/v) (pH 6); flow rate, 1 mL·min^{-1}; temperature, 25°C; e.o., elution order; n.d. not determined.

tBuCQN, 9-O-(tert-butylcarbamoyl)quinine; tBuCEQD, 9-O-(tert-butylcarbamoyl)epiquinidine; tBuCEQN, 9-O-(tert-butylcarbamoyl)epiquinine; tBuCQD, 9-O-(tert-butylcarbamoyl)quinidine.

Source: Reproduced from N.M. Maier et al., *Chirality, 11*: 522 (1999). With permission.

and/or steric factors may be invoked to explain this selectivity enhancement effect of the larger alkoxy substituents.

1.4.4 OTHER DERIVATIVES

Quinine carbamate C_9-dimers [64] of the general structure QN-X-QN shown in Figure 1.9 (CSPs of type IV) with distinct aliphatic, alicyclic, and aromatic spacers X between the quinine carbamate selector units were elucidated, to figure out whether a second selector moiety in more or less close distance to the primary quinine carbamate selector unit could exert a positively cooperative binding and enantiose-lectivity effect. The major findings of this study can be summarized as follows: (i) The molar selector loadings as determined by elemental analysis were by a factor of about two lower on the dimeric CSPs (mean for nine different CSPs 0.17 ± 0.02 mmol g^{-1}) than on comparable monomeric ones (0.31 ± 0.06 mmol g^{-1}); that is, the molar loading of selector units was about the same. This result is not further surprising, because the dimeric selector has a greater volume extension and covers a larger surface area. (ii) Suspended-state NMR measurements [65] confirmed that the dimeric selectors were immobilized monocovalently rather than *bis*-covalently; the ^{1}H-high-resolution-magic angle spinning NMR (^{1}H-HR-MAS-NMR) spectra showed a nearly equivalent amount of vinyl-protons which should disappear if the dimeric selectors are immobilized by *bis*-covalent anchoring (unpublished results). (iii) Within the series of dimeric selectors with aliphatic di- (C_2), tri- (C_3), tetra- (C_4), and hexamethylene (C_6) spacers, it turned out that the longest spacer was usually more favorable because the two quinine carbamate units could act independently of each other. In contrast, with shorter spacers the enantioselectivity was negatively influenced with α-values getting smaller, the closer the two selector units were linked together (i.e., the shorter the spacer). This trend was consistently observed, for example, for various N-acylated amino acids. (iv) In general, a comparison of dimeric selectors with comparable mono-meric ones showed a stronger retention yet lower separation factors for the dimeric selectors. Hence, in most instances, no positively cooperative effect of the twin select-ors on enantioselectivity, but rather a negatively cooperative influence was revealed by CSPs making use of dimeric selectors of the type QN-X-QN. In a follow-up study, also mixed QN-X-QD dimers have been prepared and were screened by nonaqueous capil-lary electrophoresis (CE) for their enantiomer separation capabilities in comparison to their homomeric counterparts (QN-X-QN and QD-X-QD) [66]. In general, it became quite obvious that the pseudoenantiomeric character is more often violated in case of such dimeric selectors so that the choice of either a quinine or a quinidine-based CSP may be even more critical. A *bis*-quinine carbamate bonded CSP was further tested for the LC enantiomer separation of neutral and basic compounds such as 1,1'-binaphth-2,2'-diol, chlorpheniramine, propranolol, and promethazine, with minor success though [67].

Some other cinchona alkaloid derivatives that have been synthesized and were evaluated as chiral selectors for liquid chromatographic enantiomer separation comprise CSPs based on cinchonan hydrazides [68], cinchonan ureas [42], cin-chonan amides [42], and urea-linked cinchonan-calixarene hybrid selectors [42]

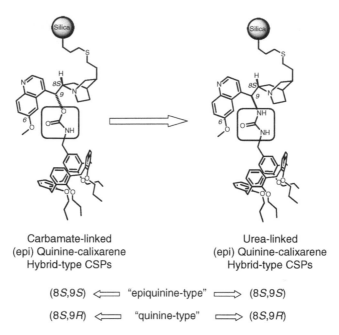

Carbamate-linked
(epi) Quinine-calixarene
Hybrid-type CSPs

Urea-linked
(epi) Quinine-calixarene
Hybrid-type CSPs

(8S,9S) ⟸ "epiquinine-type" ⟹ (8S,9S)

(8S,9R) ⟸ "quinine-type" ⟹ (8S,9R)

FIGURE 1.12 CSPs on the basis carbamate and urea-linked cinchonan-calixarene hybrid selectors as examples for CSPs with complementary enantioselectivity profiles. (Reproduced from K.H. Krawinkler et al., *J. Chromatogr. A, 1053*: 119 (2004). With permission.)

(Figure 1.12). It is beyond the scope of the present review article to discuss all of the findings of these studies in detail. However, in general, it can be stated that also these CSPs may reveal valuable complementary selectivity profiles for particular classes of chiral solutes. This may apply to an even greater extent to CSPs based on Sharpless' phthalazine-derived cinchona alkaloids (see Figure 1.9) (type IV CSPs bottom) [38,59] and mutants thereof (N.M. Maier and W. Lindner, in preparation). They partly showed exceptionally high enantioselectivities for some specific applications (*vide infra*).

1.5 LINKER/SURFACE BONDING STRATEGIES

The "classical" chemical bonding protocol via radical addition reaction of vinyl group containing chromatographic ligands to thiol-modified supports such as 3-mercaptopropyl-silica gel represents a straightforward procedure for the covalent immobilization of cinchona alkaloid derivatives with a controlled loading density of about 0.4 mmol g^{-1} CSP [20,30,32,33]. This is important because if the selector loading density is overdone, as seemed to be the case in a recent study that used supercritical carbon dioxide with shortened reaction time for the preparation of the thiol-functionalized silica gel (0.7 mmol tBuCQN selector g^{-1}) [69], separation factors may be decreased because the binding sites may be blocked by adjacent SO units and not be accessible anymore.

In another study, alternative surface anchoring strategies were investigated [70]. Instead of immobilization via the vinyl-double bond of the quinuclidine at carbon C11, bonding via the carbamate residue as in the prototype cinchona carbamate CSPs (type I and II) was pursued, using, for example, *O*-9-(1,1-dimethylbut-3-enyl) or *O*-9-(3-isopropenyl-α,α-dimethylbenzyl) carbamates to maintain the favorable bulky *tert*-butyl structural element. By this exercise, it was probed whether or not the orientation of the selector unit from the surface and thus the accessibility of the central binding cleft at the quinuclidine moiety would have an effect on the overall enantioselectivity. As the spacer arm was obviously long enough, it made no significant difference corroborating that the participating binding sites are not disturbed by the environment in the CSPs prepared by the "classical" bonding chemistry. This has practical consequence because it would open up flexibility in further chemical modifications at C_{11} position.

It is also worthwhile to outline at this place the immobilization procedure that was used for the preparation of type I CSPs: A bifunctional linker with a terminal isocyanate on one side and a triethoxysilyl group on the other end (3-isocyanatopropyl triethoxysilane) was reacted with the native cinchona alkaloids quinine and quinidine and subsequently the resultant carbamate derivative in a second step with silica [30]. Remaining silanols have been capped with silane reagents, yet, are less detrimental for acidic solutes because of the repulsive nature of such electrostatic interactions. CSPs prepared in such a way lack the hydrophobic basic layer of the thiol-silica-based CSPs mentioned earlier, which may be advantageous for the separation of certain analytes.

A copolymerization approach of *O*-9-[2-(methacryloyloxy)ethylcarbamoyl] cinchonine and cinchonidine with methacryl-modified aminopropylsilica particles was utilized by Lee et al. [71] for the immobilization of the cinchona alkaloid-derived selectors onto silica gel. The CSPs synthesized by this copolymerization procedure exhibited merely a moderate enantiomer separation capability and only toward a few racemates (probably because they were based on less stereodifferentiating cinchonine and cinchonidine). Moreover, the chromatographic efficiencies of these polymer-type CSPs were also disappointing.

Just recently, click chemistry was presented as another viable route for the immobilization of cinchona alkaloid selectors to silica supports [72]. Thus, cinchonan derivatives bearing an alkyne moiety such as *O*-9-*tert*-butylcarbamoyl-10,11-didehydroquinine or *O*-9-(3,5-dinitrophenylcarbamoyl)-10,11-didehydroquinine were allowed to react under mild conditions at room temperature in presence of catalytic amounts of copper(I) iodide (1–5 mol%) in acetonitrile with azido-modified silica gels in a 1,3-dipolar cycloaddition yielding a 1,2,3-triazole linkage. The selector loadings afforded by this new surface modification chemistry was rather comparable to those obtained with the common radical addition reaction strategy or even higher and could be conveniently controlled by the concentration of the chromatographic ligand in the reaction mixture. Also, very comparable enantioselectivity values were obtained for a series of test compounds revealing the insensitivity of the selector bonding chemistry at the C_{11} position on the chiral recognition processes. Although this route is more expensive and appears to be more elaborate due to additional reaction steps, great practical significance of this scientific achievement arises from the fact that it offers a way of immobilization of chiral selectors being incompatible with radical chemistry (i.e., being sensitive to free radicals or capturing free radicals produced by radical initiators) such as was previously observed for the O-9-[(3,5-dinitrophenyl)carbamoyl]quinine selector [31].

1.6 EFFECTS OF SUPPORTS

The selector chemistries described earlier are the major determinants for the thermodynamics of the adsorption process. Since the final goal in chromatographic enantiomer separations will be the baseline resolution of two enantiomers and the separation and distribution process occurs in a column according to a multiple plate model with a multiplicity of mobile-stationary phase distributions, kinetic aspects of these transport processes mostly related to the physical properties of the supporting material have also to be considered in the course of optimization of a chromatographic stationary phase. Recently, various attempts have been undertaken to improve the efficiency, in particular focusing on the fast elution regimen, by using new supports.

In one approach, nonporous 1.5 μm silica particles (MICRA NPS-1.5 μm) were chemically modified with O-9-($tert$-butylcarbamoyl)quinine (via thiol-modified silica intermediate) [73,74]. Due to the small particle size, column length was dictated by the pressure drop to be as low as 30 mm with standard HPLC equipment that was employed (ID used was 4.6 mm). The idea behind the use of 1.5 μm nonporous particles was the expected lower mass transfer resistance term due to the elimination of intraparticulate pore diffusion processes and a low Eddy diffusion term as a result of the smaller particle size. In practice, the achieved plate numbers ranged up to 70,000 m^{-1}, which was less than predicted but still more than typically obtained on a standard porous 5 μm support (about 30,000 plates m^{-1} under similar conditions). Extra-column band broadening and sample overloading effects may be critical for such a system. The nonporous silica has, of course, a much smaller surface area (about 3 m^2 g^{-1} vs. ca. 300 m^2 g^{-1} for porous 100 Å 5 μm particles), which was therefore associated with a selector loading being by a factor of about 100 lower. This fact, along with the short columns used, promoted fast enantioseparations. For example, complete enantiomer separations could be accomplished within 1.5 min, for example, separation of DNB-amino acids, amino acids derivatized with Sanger's reagent (2,4-dinitrophenyl-amino acids), and 1,1'-binaphthyl-2,2'-diyl hydrogenphosphate [73]. This would ideally qualify such a separation system for a two-dimensional (2D) separation of complex mixtures, where a fast enantiomer separation is required in the second dimension (T. Welsch and W. Lindner, *Anal. Bioanal. Chem.* 2007, (in press). If a higher capacity is needed, retention can be easily adjusted via the buffer concentration, and it was shown that a reduction of the total buffer concentration to about 1 mM may yield retention times commonly observed with comparable porous materials [74].

Another recent trend focused on supports in the shape of monolithic columns having the goal to benefit from the high permeability and the improved mass transfer characteristics of such structures. With this goal in mind, Lubda and Lindner [75] prepared enantioselective silica monolith columns with *tert*-butylcarbamoylquinine surface modification. A commercial sol-gel–derived Chromolith Performance Si (100 × 4.6 mm ID) monolith (1.9 μm macropore diameter, 12.5 nm mesopore diameter), encased in a polyetheretherketone (PEEK) plastic tube, was first preactivated with thiol-anchor groups by rinsing with a reaction mixture of 3-mercaptopropyltrimethoxysilane in toluene at a flow rate of 0.2 mL min^{-1} for 1 h at 80°C, followed by a flow-through immobilization of the *tert*-butylcarbamoylquinine selector dissolved in ethanol using AIBN as radical initiator. The 1.9-μm wide macropores surrounding the sponge-like silica skeleton allow the percolation of the mobile phase at a low-pressure drop, also due to the high total porosity of the column. The presence of a narrow distribution of mesopores in the silica skeleton gives access to a large surface area (specific surface area about 300 m^2 g^{-1}) with a favorable adsorption capacity for small molecules. It combines the advantageous properties of a high permeability (described by Tallarek and coworkers [76] by an equivalent sphere dimension for the permeability comparable to that of 15-μm particles) and lower dispersion compared to common 5-μm particles (described by an equivalent sphere dimension for the dispersion [76] comparable to that of 3-μm particles). Both of these characteristic properties, high permeability and enhanced mass transfer, encourage high-throughput separations (i.e., to perform separations faster at high flow velocities). A variety of chiral acids (*N*-derivatized amino acids and profens) could be baseline resolved on such 10 cm long monolithic quinine carbamate columns.

The analysis of the kinetic properties of the monolithic column in comparison to a particulate packed column (5 μm particles, 150 × 4 mm ID) for *N*-acetyl-phenylalanine by adopting the concept of the Van Deemter relationship (H = A + B/u + C × u) clearly showed the benefit of the monolithic support, in particular at higher flow rates. While the plate height minima H_{min} were observed for both monolithic and particulate columns at about the same linear velocity ($u_{min} \sim 0.5$ mm s^{-1}) and having also about the same H_{min} value (ca. 17 μm), the monolithic column outperformed the packed column at higher flow velocity owing to a flatter H/u-curve, that is, due to its smaller mass transfer resistance term (C-term). The C-term was smaller by a factor of about 2.4 for the monolithic column (e.g., 7.3 ms for particulate and 3 ms for monolithic column). In other words, the practical consequence is that if the columns are, for example, operated at a linear flow velocity of 6 mm s^{-1}, the performance of the monolithic column in terms of plate height is by a factor of about 1.5 better than with the particulate column at the same linear flow velocity. At higher flow rate, the gain in efficiency is even more pronounced.

A side aspect was also demonstrated. If required, the column length and thus the peak capacity can be easily amplified by coupling several enantioselective monolithic columns in series [75]. The actual plate count for a single 10-cm-long monolith was about 7,000 for suprofen and it amounted to about 30,000 for six columns in series. Concomitantly, the resolution was improved from 2.1 to 3.5 while the backpressure of the 60-cm-long column was with 45 bar still quite acceptable (opposed to six bar for a single 10-cm-long column).

Zirconia particles have been promoted recently as support replacing silica with the proposed advantage of better chemical, thermal, and mechanical stability. Such zirconia-supported chiral phases have been prepared by Park and coworkers [77] through coating, for example, of *O*-9-[3-(triethoxysilyl)propylcarbamoyl]quinine onto 5 μm zirconia particles (30-nm pore diameter) or carbon-cladded zirconia particles as support [78]. In comparative tests with corresponding silica materials, the separation power of the zirconia-based particles seemed to be impaired by non-enantioselective interaction contributions stemming from the presence of a lot of surface-accessible Lewis acid sites of zirconia. Overall, a real advantage, as claimed, could not be convincingly shown.

Organic supports have been evaluated as well, but remained of less practical relevance for HPLC applications due to some inherent limitations. For example, the *tert*-butylcarbamoylquinine selector was immobilized onto organic polymer beads (Suprema 1000u, from Polymer Standard Services, Mainz) reacting the reactive epoxy groups of the poly(glycidyl methacrylate-coethylene dimethacrylate) beads with butane-1,4-dithiol to obtain thiol-modified organic polymer beads, which were subsequently modified in a second step by grafting the selector through the standard radical addition reaction [79]. Although the quinine carbamate–modified organic polymer beads (10-μm diameter, pore volume 0.9 mL g^{-1}, pore diameter of about 100 nm, specific surface area of ca. 60 m^2 g^{-1}, selector coverage 100 μmol g^{-1}) showed decent enantiomer separations, they were not competitive to the silica-supported analogs mainly because of two reasons: (i) The separation factors were consistently lower by a factor of 1.8 than those for the silica support, presumably because of significant nonspecific interactions of the organic solutes (amino acid derivatives and 2-aryloxypropionic acids) with the lipophilic polymethacrylate backbone of the support. (ii) The chromatographic efficiencies were poor, with plate counts typically by a factor of 5–10 lower, which is supposed to be due to a larger particle diameter (10 vs. 5 μm for silica), a wider dispersity of the particle diameter, and a nonoptimized (broadly distributed) mesopore structure.

Moreover, *in-situ* copolymerization approaches of polymerizable chiral cinchonan carbamate selectors have also been shown to be viable straightforward routes to enantioselective separation media. In one approach, polymethacrylate-type monoliths have been fabricated by copolymerization of functional monomers and crosslinker in presence of porogenic solvents [80–85]. They have been utilized mainly for CEC (and will be described in detail later) while they turned out to be less suitable for HPLC application because of a low crosslinking degree.

On the other hand, Gavioli et al. [86] prepared irregularly shaped polymethacrylate beads by bulk polymerization and *in-situ* incorporation of the chiral selector through copolymerization of the chiral monomer O-9-*tert*-butylcarbamoyl-11-[2-methacrylamidoethyl)thia]-10,11-dihydroquinine, a 3-fold molar excess of *N*-*tert*-butylmethacrylamide, and a 20-fold molar excess of ethylene dimethacrylate (crosslinker) in presence of methanol as pore-forming agent. After grinding and wet-sieving the fraction of particles less than 25 μm were used for chromatographic tests (125 × 4 mm ID stainless steel tubes) and for batch extraction to measure single-component isotherms (i.e., static binding capacities) for *N*-3,5-dichlorobenzoyl-leucine as a test solute. When the same polymeric materials were prepared in presence of templating (*S*)- or (*R*)-3,5-dichlorobenzoyl-leucine solutes (added to the reaction mixture in equimolar amounts related to the chiral monomer), chiral particles with significantly altered binding characteristics could be obtained in comparison to nontemplated polymer. In detail, chromatographic experiments confirmed that (i) no retention and no enantioselectivity was obtained on the polymethacrylate material prepared without chiral monomer, (ii) the enantioselectivity

was the highest ($\alpha = 9.3$) on the polymer templated with the high-affinity analyte (i.e., S-enantiomer of dichlorobenzoyl-leucine), (iii) all other polymethacrylate
CSPs prepared either by analyte templating with the low-affinity enantiomer toward
the chiral monomer (i.e., the weaker binding R-enantiomer) or with the nonchiral
dichlorobenzoyl-glycine, or prepared without analyte templating but otherwise
identical composition of the polymerization mixture yielded lower enantiorecognition ($\alpha = 6.3$ for R-templated polymer, 7.0 for achiral-templated polymer, and
7.1 for nontemplated polymer). The high-affinity analyte-templated polymeric CSP
could even reach the same enantioselectivity level as the silica-supported CSP (5 μm
particles, 100 Å pore diameter) ($\alpha = 8.8$) having immobilized the comparable
tert-butylcarbamoylquinine selector via the thiolpropyl spacer at about the same theoretical surface coverage ($\sim$1 μmol m^{-2}). The measurement of the static binding
capacities provided further insights: It seemed that the increase in enantiorecognition capability of the high-affinity analyte-templated polymer was mostly due to
an increase in the number of accessible binding sites for this enantiomer, here the
S-enantiomer. Apparently, through analyte templating, the conformational binding
space around the active binding site of the chiral monomer was preserved along with
a better access to it, as opposed to the materials that were prepared without templating employing the high-affinity enantiomer. This avoids the binding sites being
buried inside the polymer. The findings of this study are conceptionally certainly
of great scientific interest. The practical application of such polymeric materials for
enantioselective HPLC analysis, however, seems to be prevented by their moderate
chromatographic efficiencies.

1.7 CHIRAL RECOGNITION MECHANISMS

It is a fundamental basis for the specificity of molecular recognition events that
the mechanism is unique for each system. Ultimately, it secures that molecular
information is furnished faithfully and that regulating systems are not disturbed
by exogenic sources. For the present discussion, it has the consequence that only
a few representative model systems can be treated. They have been selected carefully so as to give the readers an illustrative insight into how the cinchona alkaloid
selectors and CSPs work. We may approach such a discussion from two view-
points: first, the intermolecular binding events may be discussed from a macroscopic
point of view in terms of relative binding strengths and binding thermodynamics of
selector–selectand associations, or second, from a microscopic point of view, which
means by a look at the three-dimensional (3D) structures of the formed SA–SO complexes by the help of spectroscopic studies, molecular modeling investigations, and
so forth.

1.7.1 BINDING STRENGTHS AND BINDING THERMODYNAMICS

Various papers have been published that investigated the thermodynamics of
host–guest associations of cinchonan carbamate derivatives either in solution with
the chiral selectors (i.e., the soluble precursors of CSPs) on the one hand [87,88] or

directly with the CSPs on the other hand [51–54,89,90]. As macroscopic entities, such data do not directly provide detailed insight into the chiral recognition mechanisms on a nanoscopic molecular level, but may contribute valuable information in particular if these investigations are carried out in relation to rationally selected structural analogs under identical conditions.

The binding strength of SO–SA associates can be characterized by the binding constant K, which is related to the standard Gibbs free energy of binding $\Delta G°$ according to the following equation:

$$\Delta G° = -R \cdot T \cdot \ln K, \tag{1.7}$$

wherein R is the universal gas constant $(8.3144 \, J \, mol^{-1} \, K^{-1})$ and T is the absolute temperature (in K). Thus, knowledge of the selector–selectand association constant K from, for example, electrophoretic, spectroscopic, or isothermal microcalorimetric methods allows a ready estimation of binding energies.

1.7.1.1 Binding Constants by Microcalorimetric and Spectroscopic Measurements

Isothermal calorimetry (ITC), circular dichroism spectroscopy (CDM), and ultraviolet (UV) spectroscopy have been utilized by Lah et al. [87] as tools for the estimation of binding constants between cinchona alkaloid selectors (native quinine and quinidine as well as their corresponding *tert*-butylcarbamoyl derivatives) with the R- and S-enantiomers of DNB-Leu as model selectand in methanolic solution. The measurements were based on titrations of a certain quantity of selector dissolved in methanol at a constant temperature of 25°C with stepwise incremental addition of either R- or S-enantiomer of the selectand as titrant. After each addition of titrant a specific property was measured that was altered as a response of SO–SA complexation: In ITC, this was the heat generated upon SO–SA interaction (Figure 1.13a), and for the data analysis, the heat effect was given as the enthalpy change in dependence of the molar selectand/selector ratio $[SA]_T/[SO]_T$ (see Figure 1.14). For the CDM spectroscopic determinations, CDM spectra were acquired at various SO/SA ratios (Figure 1.13b) and corrected for by subtraction of the CDM spectra for the pure SO and pure SA to end up with the induced CDM spectra of the SO–SA complex. For the spectrophotometric analysis the corresponding UV spectra were monitored (Figure 1.13c) and treated by the same procedure as outlined for the CDM spectroscopic measurements. For each method, specific equations have been derived from the mass action law, which enabled to readily deduce the binding constants K by curve-fitting procedures (for details see Reference 87). The obtained results for the binding constants of the distinct SO–SA pairs are summarized in Table 1.4.

The calorimetric binding isotherms of the carbamoylated quinine and quinidine selectors clearly reveal that the heats released upon binding are strongly different for S- and R-enantiomers of DNB-Leu, which is commensurate with the remarkable enantioselective molecular recognition capability of these selectors (Figure 1.14a,b). As can be seen from Table 1.4, the binding constants for R- and S-enantiomers differ by about one order of magnitude in case of the carbamate-type selectors. Furthermore,

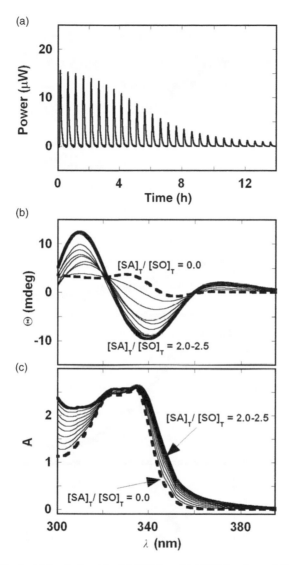

FIGURE 1.13 Typical (a) calorimetric, (b) CDM, and (c) UV titrations of tBuCQD (SO) with (R)-DNB-Leu (SA) in methanol at 25°C. In panel (a) the heat effects measured after successive addition of (R)-DNB-Leu solution are depicted, while in (b) and (c) the CDM and UV spectra, respectively, monitored at different ratios of total SA and SO, $[SA]_T/[SO]_T$ concentrations are shown (note, subscript T stands for total molar amount in the reaction solution). (Reproduced from J. Lah et al., *J. Phys. Chem. B, 105*: 1670 (2001). With permission.)

the ITC titration curves (Figure 1.14a,b) show a high affinity of the *S*-enantiomer of DNB-Leu toward tBuCQN ($K = 3.7 \times 10^3$ M^{-1}) and of the *R*-enantiomer toward tBuCQD ($K = 9 \times 10^3$ M^{-1}) as well as a low affinity for the corresponding opposite enantiomers ($K = 2.2 \times 10^2$ M^{-1} and $K = 1 \times 10^3$ M^{-1}, respectively), reflecting again the pseudoenantiomeric character of these two alkaloid derivatives.

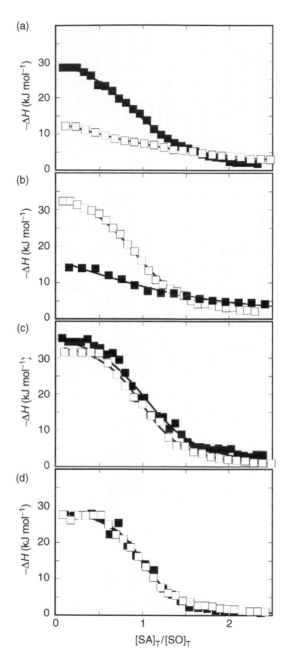

FIGURE 1.14 Calorimetric binding isotherms of (a) tBuCQN, (b) tBuCQD, (c) QN, and (d) QD titrated with (*S*)-DNB-Leu (filled symbols) and (*R*)-DNB-Leu (open symbols) at 25°C. ΔH-values are given in kJ mol^{-1} of added ligand. (Reproduced from J. Lah et al., *J. Phys. Chem. B, 105*: 1670 (2001). With permission.)

TABLE 1.4

Binding Constants K_b for (R)- and (S)-DNB-Leu Binding to the Chiral Selectors tBuCQN, tBuCQD, QN, and QD Determined from the Calorimetric, UV, and CDM Titration Curves in Methanol solutions at 25°C[a]

	$K_b (M^{-1})$						
	ITC			UV		CDM	
	(S)-Leu	(R)-Leu	Gly	(S)-Leu	(R)-Leu	(S)-Leu	(R)-Leu
tBuCQN	3.7×10^3	2.2×10^2	2.2×10^3	2×10^4	2×10^3	5×10^3	
tBuCQD	1.0×10^3	9×10^3	2.3×10^3	3×10^3	2×10^4		1×10^4
QN	1.1×10^4	1.2×10^4	1.6×10^4	3×10^4	3×10^4	3×10^4	1×10^4
QD	1.0×10^4	1.3×10^4	1.9×10^4	1.0×10^5	7×10^4	1.0×10^4	2×10^4

[a] For comparison, the corresponding K_b values for the nonchiral selectand DNB-Gly obtained by ITC are included. A relative error, $\Delta K_b / K_b$, estimated on the basis of the chi-square fitting in which an error of 5% of the measured quantity was assigned for each titration point is about $\pm 30\%$ (ITC) and $\pm 50\%$ (CD, UV).

Source: Reproduced from J. Lah, N.M. Maier, W. Lindner, and G. Vesnaver, *J. Phys. Chem. B, 105*: 1670 (2001).

In contrast, for the native quinine and quinidine selectors, the binding isotherms for R- and S-enantiomers are more or less overlapping and the derived binding constants are therefore virtually identical. In agreement with the chromatographic findings, this points toward a poor enantioselectivity. On the other hand, relatively high heat effects are released upon SO–SA interaction for both SA enantiomers and both native alkaloids suggesting a relatively high affinity in any case (binding constants between 1.0 and 1.3×10^4 M^{-1}). The calorimetric binding isotherms of Figure 1.14 also provide information on the binding stoichiometry and reveal a 1:1 stoichiometry for all investigated SO–SA systems because the inflection point of the curves is located at $[SA]_T/[SO]_T$ ratio of about 1. This was expected for such ion-pairs and has later been confirmed by nuclear magnetic resonance (NMR) spectroscopy (*vide infra*).

The data that had been acquired by CDM and UV spectroscopy complemented the above ITC results. It was found that too strong CDM and UV signals were induced in the spectral range between 200 and 300 nm and therefore the titration curves were constructed by measuring in the wavelength range between 300 and 400 nm (outside the maxima). The CDM difference spectra induced by complexation of (S)-DNB-Leu with tBuCQN were strong, like that of (R)-DNB-Leu with tBuCQD. The fact that the CDM bands in these two complexes had opposite signs is another strong experimental evidence for the pseudoenantiomeric character of these complexes (*vide infra*). In contrast, the corresponding spectra of the opposite complexes, that is, (R)-DNB-Leu with tBuCQN and (S)-DNB-Leu with tBuCQD, were extremely weak so that no binding constants could be derived with trustworthy certainty. Overall, the estimated binding constants from CDM spectroscopy were in reasonable agreement with those of ITC, mutually cross-validating the accuracies of each method (see Table 1.4).

Somewhat larger deviations are seen for the UV spectrophotometrically derived constants. However, the intrinsic binding selectivities as assessed from the ratio of the binding constants for the corresponding enantiomers was a reasonable match between UV and ITC, at least what the carbamate selectors is concerned.

An appreciable advantage of the ITC measurement over other methods is that it gives direct access to thermodynamic data. In fact, besides providing binding constants K and thus free energies of binding $\Delta G°$ (Equation 1.7), enthalpy changes upon binding $\Delta H°$ are directly measurable with acceptable accuracy. Through the well-known Gibbs–Helmholtz equation (Equation 1.8)

$$\Delta G° = \Delta H° - T \cdot \Delta S° \tag{1.8}$$

also entropy changes upon binding $\Delta S°$ are amenable (Table 1.5). It becomes evident from Table 1.5 that the process of SO–SA complexation was driven, in any case, by the enthalpy change on SO–SA interaction, while the entropy change contributed negatively to the binding. The exothermic reactions my be readily explained by the primary H-bond supported ion-pairing interaction in concert with other supportive intermolecular interactions (such as H-bonding and π–π-interactions), of which some are not existing for the low-affinity enantiomer in case of the carbamate-type selector systems or at least are weaker. On the other hand, the unfavorable entropy contributions may arise from lowered translational, rotational, and conformational degrees of freedom in the complexed species.

Although the data mentioned earlier have been acquired under conditions resembling those of typical chromatographic runs (normally methanol-based eluents), the system was devoid of counterions that shield the electrostatic interaction sites and weaken this dominating binding force under typical chromatographic conditions of ion-exchange type systems. Hence, the binding constants mentioned earlier may to some extent overestimate the binding strengths under the real chromatographic situation.

TABLE 1.5
Thermodynamic Profiles for (S)-DNB-Leu and (R)-DNB-Leu Binding to Selectors QN, QD, tBuCQN, and tBuCQD in Methanol Solutions at 25°C Determined from ITC Titration Curves[a]

	ΔG_b^o (kJ mol^{-1})		ΔH_b^o (kJ mol^{-1})		$T\Delta S_b^o$ (kJ mol^{-1})	
	(S)-Leu	(R)-Leu	(S)-Leu	(R)-Leu	(S)-Leu	(R)-Leu
tBUCQN	−20	−13	−33	−27	−13	−14
tBUCQD	−17	−22	−30	−37	−13	−14
QN	−23	−23	−38	−36	−15	−13
QD	−23	−23	−39	−29	−7	−6

[a] The relative errors in ΔG_b^o, ΔH_b^o, and $T\Delta S_b^o$ based on the relative errors in K_b and ΔH_b^o estimated from the ITC titration curves are $\Delta(\Delta G_b^o)/\Delta G_b^o \approx \pm 5\%$, $\Delta(\Delta H_b^o)/\Delta H_b^o \approx \pm 10\%$, and $\Delta(T\Delta S_b^o)/T\Delta S_b^o \approx \pm 30\%$.

1.7.1.2 Measurement of Binding Constants by CE

On the contrary, binding constants in media identical to chromatographic conditions were determined by CE using *tert*-butylcarbamoylquinine as selector and DNB-Leu as selectand with a mixture of methanol—0.1 M ammonium acetate buffer (80:20; v/v) (pH 6) as background electrolyte (BGE) [88]. Various sets of experiments were performed. First, reciprocal affinity CE was adopted in which either (R)-, (S)-, or racemic DNB-Leu were added as additive in variable concentrations to the BGE and, after injection and application of an electric field, the effective electrophoretic mobilities of the cationic chiral selector tBuCQN were measured in dependence of the additive (DNB-Leu) concentration. This reciprocal experimental setup has the decisive advantage that it allows the precise simultaneous determination of both the cationic electroosmotic flow (EOF) and cationic mobilities of the injected tBuCQN selector in the same CE run, which would not be possible if the selector is employed as additive owing to countercurrent mobilities of EOF and anionic DNB-Leu solute. The effective electrophoretic mobilities (u_{eff}) of tBuCQN in presence of the complexing additive (DNB-Leu) will be the result of its mobilities in free (u_A) and complexed form (u_{AB}) averaged over their respective molar fractions (Equation 1.9)

$$u_{\text{eff}} = f_A \cdot u_A + f_{AB} \cdot u_{AB}, \qquad (1.9)$$

wherein f_A represents the molfraction in free form and f_{AB} represents the molfraction in complexed form. Here, A denotes the free cationic tBuCQN selector and AB denotes the ion-paired tBuCQN selector. The latter is a more or less neutral species and shows no or at least much lower electrophoretic mobility. Upon increasing the DNB-Leu concentration in the BGE, a higher fraction of tBuCQN will be complexed (i.e., exist as ion-pair) and hence the effective mobility will be decreased. f_A and f_{AB} can be expressed in terms of the mass action law so that Equation 1.9 can be rewritten by the following equation:

$$K = \frac{u_{\text{eff}} - u_A}{u_{AB} - u_{\text{eff}}} \frac{1}{[B]}, \qquad (1.10)$$

wherein [B] denotes the equilibrium concentration of the free additive, here DNB-Leu. If the binding constant K is small and the additive concentration high relative to the injected concentration of tBuCQN, the equilibrium concentration, which remains unknown, can be approximated by the total additive concentration c_t, which is known. This allows transformation of Equation 1.10 into the linear form (Equation 1.11)

$$u_{\text{eff}} = \frac{1}{K} \left(\frac{u_A - u_{\text{eff}}}{c_t} \right) + u_{AB} \qquad (1.11)$$

This linearized form allows the straightforward determination of binding constants from the slopes of plots of effective mobilities vs. the term $(u_A - u_{\text{eff}})/c_t$. (*Note*: The mobility of the free tBuCQN selector u_A can be easily measured without additive and conversely, u_{eff} of the selector is measured at various DNB-Leu concentrations.) The intercepts of such plots represent the effective mobilities of the complex, a figure

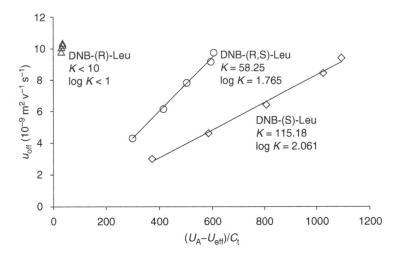

FIGURE 1.15 Dependencies of the effective mobilities of tBuCQN on the concentration of (*R*)-, (*S*)-, and racemic DNB-Leu in the BGE. Experimental conditions: BGE, methanol-0.1 M ammonium acetate (80:20;v/v) (pH = 6.0); T, 25°C. (Reproduced from J. Lah et al., *J. Phys. Chem. B, 105*: 1670 (2001). With permission.)

that is otherwise hardly accessible. Figure 1.15 illustrates the results. The binding constant for the *S*-enantiomer under the selected chromatographic conditions is about $115 \, \mathrm{L\,mol^{-1}}$ and for the *R*-enantiomer lower than $10 \, \mathrm{L\,mol^{-1}}$. (*Note*: A statistical correction for the convex deviation from linearity that may have resulted from factors like altered viscosities and currents upon change of additive concentrations gave slightly distinct binding data with a better statistical fit.)

It becomes obvious from the clustered data points that the binding constant for the *R*-enantiomer is too small to be accurately determined by this method. Hence, indirect affinity CE (resolution method) was utilized to determine the binding constant for the *R*-enantiomer. Indirect affinity CE makes use of the knowledge of the constant for one enantiomer (here, K_S) and in addition of experimental separation data as obtained with the racemate of DNB-Leu in presence of the tBuCQN selector as BGE additive. By use of Equation 1.10, an enantioselectivity factor may be defined as the ratio of the binding constants of *S*- and *R*-enantiomers yielding the following equation:

$$\frac{K_{(S)}}{K_{(R)}} = \frac{(u - u_{\mathrm{eff}(S)})(u_{\mathrm{eff}(R)} - u_{\mathrm{AB}})}{(u - u_{\mathrm{eff}(R)})(u_{\mathrm{eff}(S)} - u_{\mathrm{AB}})}. \tag{1.12}$$

Mobilities of free *S*- and *R*-enantiomers of DNB-Leu (u) are identical (readily measurable in absence of the tBuCQN additive, which gave under given conditions $u = -12.20 \pm 0.02 \times 10^{-9} \, \mathrm{m^2 \, V^{-1} \, s^{-1}}$). Further, *R*- and *S*- complexes are supposed to have zero charge and therefore identical mobilities close to zero (the experimental mobility value for the *S*-complex was available from the reciprocal affinity CE method described earlier and was found to be $u_{\mathrm{AB}} = -0.40 \pm 0.01 \times 10^{-9} \, \mathrm{m^2 \, V^{-1} \, s^{-1}}$).

Finally, the effective mobilities u_{eff} for R- and S-enantiomers of DNB-Leu, at any concentration of tBuCQN as additive, have been determined accurately by the so-called "constant time method" [88]. For example, at 5 mM tBuCQN in the BGE the effective mobilities were $u_{eff}(S) = -6.79 \pm 0.02 \times 10^{-9}$ m^2 V^{-1} s^{-1} and $u_{eff}(R) = -12.15 \pm 0.07 \times 10^{-9}$ m^2 V^{-1} s^{-1}, which after substitution into Equation 1.12 gave $K_S/K_R = 49.8$ and thus a binding constant K_R of 2.5 L mol^{-1}.

There is, however, another way of deriving the binding constant for the weaker binding R-enantiomer, namely, by combination of the resolution method and reciprocal affinity CE with racemate as additive (in this context, it is worthwhile to point out that the combination of these two methods allows the determination of binding constants of individual enantiomers even if only the racemate is available, the only missing information would be the assignment of the configuration). Through knowledge of $K_S/K_R = 49.8$ from the resolution method and $(K_S + K_R)/2 = 58.3$ L mol^{-1} from the reciprocal affinity CE experiment with racemic DNB-Leu as BGE additive, K_R could be calculated as well. (*Note*: The apparent binding constant derived from the slope in Figure 1.15 corresponds to the average of the individual constants of the two enantiomers.) Thus, it follows that $K_S = 114$ and $K_R = 2.3$ L mol^{-1}, being in reasonable agreement with the values determined earlier.

Overall, it is seen that the binding constants in presence of counterions (total buffer concentration of 20 mM) are lower by a factor of about 10 than in a plain methanolic medium without counterions (see Table 1.4). In contrast, the intrinsic enantioselecitivity is still in the same order of magnitude or even higher ($K_S/K_R \sim 17$ by ITC and ca. 50 under CE conditions) because of a more proper balance of the otherwise dominating ionic interaction.

1.7.1.3 Binding Energetics by Chromatography

Investigations of binding energetics with soluble CSP analogs (i.e., the selectors) are unfortunately only rough approximations and simplifications of the real situations that actually exist in the HPLC experiment. Spacer and support surface chemistries can make a significant contribution to the overall binding process as well. If knowledge on the binding mechanism on the sorbent surface under LC conditions is the ultimate goal, a better way to look at binding thermodynamics may be by means of variable termperature HPLC.

Two principal approaches with a different degree of complexity and differentiation have been pursued: Thermodynamics investigations under linear and nonlinear chromatography conditions.

Thermodynamic Studies by Linear Chromatography. The first one is under linear chromatography conditions at variable temperature making use of the van't Hoff equation (Equations 1.13 and 1.14) [48,50,52,53,89],

$$\ln k = -\frac{1}{T} \cdot \frac{\Delta H^\circ}{R} + \frac{\Delta S^\circ}{R} + \ln \phi \qquad (1.13)$$

or written in terms of enantioselectivity

$$\ln k_2/\ln k_1 = \ln \alpha = -\frac{1}{T} \cdot \frac{\Delta\Delta H^\circ}{R} + \frac{\Delta\Delta S^\circ}{R}, \qquad (1.14)$$

wherein ϕ is the phase ratio, k is the chromatographic retention factor, and α is the chromatographic separation factor.

Oberleitner et al. [89] and Török et al. [90] investigated the thermodynamics of N-acyl (DNB, DNZ, Z, Bz, Ac) derivatives of aliphatic (Ala, Leu, α-neopentyl-glycine) and aromatic (α- and β-Phe, α-arylsubstituted Gly) amino acids on the tBuCQN-CSP using hydro-organic eluents (MeOH-acetate buffer pH 6) (denoted in Table 1.6 as ion-exchange process in RP mode). Gyimesi-Forrás et al. [50] studied the temperature effect of a set of imidazo[1,5-b]quinazoline-1,5-diones on the same CSP also with hydro-organic conditions (methanol or acetonitrile mixed with acetate buffer, pH 6) (RP-type separation of neutrals), and Asnin et al. [52,53] described the thermodynamics of arylcarbinols on tBuCQD-CSP under normal-phase conditions, namely 2,2,2-trifluoro-1-(9-anthryl)-ethanol (TFAE) in toluene- acetonitrile (ACN) [52] or 1- and 2-phenylpropanols in hexane-ethyl acetate (EA) [53]) (NP-type separation of neutrals).

The major observations from these thermodynamic studies by linear chromatography approach can be summarized as follows: Although the mechanisms at the molecular level are supposed to be considerably different for amino acid derivatives, quinazolones, and arylcarbinols, it was a common trend that the retention factors decreased with increasing temperature. The negative ΔH-values (heats of adsorption $-\Delta H^\circ$) suggested that for both enantiomers, the solute distribution process was exothermic in any case. The calculated ΔS-values were also always negative (i.e., they showed the same sign). This means that enthalpy and entropy terms had opposite effects on adsorption ("compensation effect"), which may be explained by consumption of the energy released upon binding through the changes in the structural order of bulk solution and surface layer (loss of rotational, translational, and conformational degree of freedom of the binding partners in the bound state relative to the free forms). While the enthalpic contribution promoted adsorption, the entropy of the system reduced the tendency for solute–sorbent interaction. A representative selection of examples of thermodynamic parameters for the separation process is given in Table 1.6. For the investigated N-acyl amino acid derivatives, it was found in terms of separation that $\Delta\Delta H$-values were in the range of 3.2–15.7 kJ mol^{-1} with exception of Z– and Ac-derivatives for which they were <3 kJ mol^{-1} (see Table 1.6). This is a substantial difference for the heats of adsorption for the opposite enantiomers so that it may be argued that this is most likely due to a distinct number of intermolecular interactions for R- and S-enantiomers. In contrast, the $\Delta\Delta H$-values were much smaller for quinazolones (between 1.7 and 3.0 kJ mol^{-1}) and arylcarbinols (typically between 0.4 and 1.4 kJ mol^{-1}) (Table 1.6). These small differences observed in particular for the latter compound class, on the contrary, suggest that the enantiomer distinction process is probably not based on a different number of interactions taking place but rather on slightly different steric factors leading to minor stability

TABLE 1.6
Thermodynamic Parameters Obtained on O-9-($tert$-Butylcarbamoyl)Quinine and Quinidine-based CSPs, Respectively, for Selected Solutes

SA	Conditions	$\Delta\Delta H$ (kJ mol^{-1})	$T\Delta\Delta S$ (kJ mol^{-1})	$\Delta\Delta G$ (kJ mol^{-1})	% $\Delta\Delta S$ (rel. to $\Delta\Delta H$)
Ion-exchange process in RP mode; tBuCQN-CSP[a]					
DNB-Ala	MeOH-0.2 M NH$_4$Ac (80:20; v/v) (pH$_a$ 6)	−11.33	−6.25	−5.08	55
DNB-Leu		−14.09	−7.36	−6.73	52
DNB-Npg		−15.7	−7.97	−7.73	51
DNZ-Leu		−9.47	−6.9	−2.57	73
Z-Leu		−2.17	−1.51	−0.66	70
Bz-Leu		−7.79	−5.41	−2.38	69
Ac-Leu		−2.11	−1.28	−0.83	61
RP-separation of neutrals (imidazo[1,5-b]-quinazoline-1,5-diones); tBuCQN-CSP[b]					
3-Phenyl	MeOH-0.1 M NH$_4$Ac (40:60; v/v) (pH$_a$ 6)	−3	−2.4	−0.6	80
3-(2 Chlorophenyl)	ACN 0.1 M NH$_4$Ac (20:80; v/v) (pH$_a$ 6)	−2.2	−1.8	−0.4	82
3-(4-Chlorophenyl)	ACN-0.1 M NH$_4$Ac (20:80; v/v) (pH$_a$ 6)	−1.7	−1.4	−0.3	82
NP-separation of neutrals; tBuCQD-CSP[a]					
TFAE	1% ACN in toluene	−1.4	−1.09	−0.32	78
TFAE	2% ACN in toluene	−1.2	−0.9	−0.3	75
TFAE	3% ACN in toluene	−1.2	−0.93	−0.27	78
1PP	5% EA in n-hexane	−0.75	−0.55	−0.2	73
2PP	5% EA in n-hexane	0.24	0.2	0.04	83
3CPP	5% EA in n-hexane	−0.4	−0.22	−0.18	55
3CPP	5% EA in n-hexane + 0.02% H$_2$O	−0.08	0.08	−0.16	−100
3CPP	8% EA in n-hexane	−0.62	−0.44	−0.18	71
3CPP	11% EA in n-hexane	−0.58	−0.39	−0.19	67

NH$_4$Ac, ammonium acetate; Npg, α-neopentyl-glycine; DNZ, 3,5-dinitrobenzyloxycarbonyl; Z, benzyloxycarbonyl; Bz, benzoyl; Ac, acetyl; TFAE, 2,2,2-trifluoro-1-(9-anthryl)-ethanol; 1PP, 1-phenyl-propanol; 2PP, 2-phenylpropanol; 3CPP, 3-chloro-1-phenylpropanol; ACN, acetonitrile; EA, ethyl acetate.

[a] $T = 295$ K.
[b] $T = 298$ K.

Data summarized from References 89, 50, 53.

differences for the distereomeric adsorption complexes [52]. Typically, the $\Delta\Delta H$-values were prevailing in magnitude over $\Delta\Delta S$-values, rendering the separations under enthalpic control. However, the entropy term made a significant negative contribution to the separation (i.e., the separations could be better from an enthalpic viewpoint but are impaired by entropic factors). Notably, $\Delta\Delta S$-values reached a percentage relative to $\Delta\Delta H$-values as high as about 50–73% for the investigated amino acid derivatives [89,90], 80–82% for the tested quinazolones [50], and 55–83% for most of the arylcarbinols. Asnin et al. [52] determined also the solvent effect of the thermodynamic quantities for TFAE at distinct acetonitrile content in toluene (1–3%; v/v) and of ethyl acetate in n-hexane for 3CPP [53]. It was found that the heats of adsorption are higher in eluents with low percentage of polar modifier, which was valid for both enantiomers. This evidently points toward a strong solvation effect of interaction sites that competitively weakens the solute–sorbent interactions at these sites.

Two aspects are worth to be mentioned separately because they showed deviating behavior from the outlined general trends: Within the homologous series of arylcarbinols, 2-phenylpropanol (2PP) showed a reversed enantiomer preference toward the tBuCQD-CSP and, hence, opposite elution order (indicated by positive signs for $\Delta\Delta H$ and $\Delta\Delta S$). This is certainly due to a change of the chiral recognition mechanism within this series of homologs, which was supported by the graphical illustration of the compensation effect of the series of arylcarbinols. When the $\Delta\Delta S$-values of the arylcarbinol separations were plotted vs. the corresponding $\Delta\Delta H$-values, all the experimental values for 1-arylcarbinols were lying on a straight line, while the data point for 2PP was not on this compensation line [53]. The other aspect to be pointed out is related to the separation of 3CPP with wet n-hexane-ethyl acetate eluent containing 0.02% water. This separation was of particular interest because enthalpic and entropic terms of the separation, $\Delta\Delta H$ and $\Delta\Delta S$, do possess the same magnitude but opposite signs. The effect was that the separation was more or less independent of the temperature. The small amount of water in the eluent seems to shield the active adsorption sites (strong solvation of H-bonding sites) and thus decrease retention factors (ca. 10%) as well as lower separation factors (ca. 0.2–1%) as compared to the corresponding separation without water traces. The general mechanism, though, appeared to be essentially unchanged because the data point was lying on the compensation effect line.

Adsorption Isotherm Measurements and Site-Selective Thermodynamics. For heterogeneous surfaces like CSPs, the adsorption isotherms are usually composite isotherms and often a Bi-Langmuir model (Equation 1.15) describes reasonably well the adsorption behavior [54].

$$q = \frac{q_{ns} \cdot b_{ns} \cdot C}{(1 + b_{ns} \cdot C)} + \frac{q_s \cdot b_s \cdot C}{(1 + b_s \cdot C)}. \tag{1.15}$$

In Equation 1.15, q represents the adsorbed amount of solute, q_{ns} and q_s are the saturation capacities (number of accessible binding sites) for site 1 (nonstereoselective, subscript ns) and site 2 (stereoselective, subscript s), and b_{ns} and b_s are the equilibrium constants for adsorption at the respective sites [54]. It is obvious that only the second term in this equation is supposed to be different for two enantiomers. Expressed in terms of linear chromatography conditions (under infinite dilution where the retention factor is independent of the loaded amount of solute) it follows that the retention factor k is composed of at least two distinct major binding increments corresponding to nonstereoselective and stereoselective sites according to the following equation [91]:

$$k = \phi \cdot \sum_{1}^{n} K_i = \phi \cdot (K_{ns} + K_s) = \phi \cdot (q_{ns} \cdot b_{ns} + q_s \cdot b_s), \qquad (1.16)$$

wherein K_{ns} and K_s are the Henry constants for sites 1 and 2, respectively. The physical meaning is that the retention factor of two distinct solute species is different due to either a different equilibrium constant or a different number of accessible binding sites, or both and in case of enantiomers owing to the figures of the stereoselective sites only.

From this discussion, it becomes clearly evident that the shortcoming of the approach mentioned earlier to derive thermodynamic parameters by linear chromatography is the neglectance of the surface heterogeneity of the CSP with its distinct adsorption sites. Contributions arising from adsorption at nonenantioselective binding sites (site 1) and enantioselective sites (site 2) are lumped together and are not deconvoluted, which would give a higher degree of information. Hence, by this macroscopic contemplation valuable information is lost.

On the contrary, a more advanced methodology makes use of nonlinear chromatography experiments: If the adsorption isotherms are measured under variable temperatures, the corresponding thermodynamic parameters for each site can be obtained in view of the van't Hoff dependency (site-selective thermodynamics measurements) [51,54]. Thus, the adsorption equilibrium constants of the distinct sites b_i ($i =$ ns, s) are related to the enthalpy (ΔH_i) and entropy (ΔS_i) according to the following equation [54]:

$$b_i = \exp\left(\frac{-\Delta H_i}{RT}\right) \exp\left(\frac{-\Delta S_i}{R}\right). \qquad (1.17)$$

Using this methodology via measurement of adsorption isotherms, Guiochon and coworkers investigated site-selectively the thermodynamics of TFAE [51] and 3CPP [54] on a tBuCQD-CSP under NP conditions using the pulse method [51], the inverse method with the equilibrium-dispersive model [51, 54], and frontal analysis [54].

For the sake of illustration, Table 1.7 gives here the numerical coefficients of the adsorption isotherms for TFAE [51]. The experimental data, which were acquired separately for the two enantiomers (single-component isotherms), fitted well to the bi-Langmuir model, which is indeed in accordance with the presence of two distinct types of adsorption sites, nonstereoselective and stereoselective ones, as previously outlined. The numerical values for q_{ns} and b_{ns} were virtually identical for the two enantiomers, as expected, because the nonstereoselective sites are indifferent for enantiomeric species. Moreover, the number of accessible stereoselective sites q_s turned out to be much lower than that of nonstereoselective ones q_{ns} (by a factor of about 25–40). The contrary relationship was found for the equilibrium constants, which were higher by a factor of about 8–16, that is, about one order of magnitude, on the stereoselective sites (high-energy sites) than on the nonstereoselective sites (low-energy sites). Of course, the binding strength at the stereoselective sites

TABLE 1.7
Numerical Coefficients of the Isotherm

$T(°C)$	q_{ns} (mmol L^{-1})	b_{ns} (L mmol^{-1})	q_s (mmol L^{-1})	b_s (L mmol^{-1})
(R)-TFAE				
15	262	0.0601	6.7	0.955
22	299	0.0431	6.5	0.662
30	318	0.0317	9.9	0.342
40	355	0.0203	13.4	0.233
(S)-TFAE				
15	282	0.0614	8.00	0.843
22	293	0.0483	8.17	0.542
30	217	0.0312	15.8	0.264
40	359	0.0224	13.6	0.191

Source: Reproduced from G. Götmar, L. Asnin, and G. Guiochon, *J. Chromatogr. A,*
1059: 43 (2004). With permission.

was different for the *S*- and *R*-enantiomers. The significant nonstereoselective interaction contribution deteriorated, however, the observed separation factors, which
was therefore lower than it could be with regards to the intrinsic enantiorecognition
capability of the chiral selector per se. As a consequence, the nonenantioselective
interaction sites need to be kept at a minimum as far as possible or such interactions should be set off by the experimental conditions without impairing the
stereoselective ones.

Moreover, it is seen from Table 1.7 that the equilibrium constants at both the
nonstereoselective site (b_{ns}) and the stereoselective site (b_s) were decreased when
the temperature was raised, while concomitantly more nonstereoselective as well as
stereoselective adsorption sites became available or accessible at elevated temperatures. The larger number of accessible binding sites at elevated temperatures has been
explained by solvation effects: The sorbent surface and thus the interaction sites are
solvated to a lesser degree so that they are better available for solute interactions [51].

Considerations putting experimentally found saturation capacities in relation to
the molar selector densities ($\sim$450 μmol g^{-1} equivalent to $\sim$1.5 μmol m^{-2}) revealed
that for the given solute there was a mismatch between accessible stereoselective
binding sites and totally available ligand moieties, which means that the selector
utilization rate was quite low. Only about 2% of the immobilized chromatographic
ligands could be utilized and saturated with TFAE [51]. The ligands appeared to
be too densely packed on the surface so that only a low fraction was available for
adsorption.

The site-selectively derived thermodynamic parameters obtained by adaptation
of Equation 1.17 (Table 1.8) clearly revealed that the heat of adsorptions are exothermic on both enantioselective and nonenantioselective sites, and the difference in
the adsorption enthalpies on enantioselective and nonenantioselective sites is about
10 and 15 kJ mol^{-1} for *R*- and *S*-enantiomers, respectively. The differential enthalpy
change upon adsorption of *R*- and *S*-enantiomers at the enantioselective site $\Delta\Delta H_s$

TABLE 1.8
Thermodynamic Characteristics of the Adsorption Sites According to the Bi-Langmuir Model

	(R)-TFAE	(S)-TFAE
$-\Delta H_s$ (kJ mol^{-1})	43.9	46.4
$-\Delta H_{ns}$ (kJ mol^{-1})	32.3	31.2
ΔS_s (J mol^{-1} K^{-1})	−152	−163
ΔS_{ns} (J mol^{-1} K^{-1})	−136	−131

Source: Reproduced from G. Götmar, L. Asnin, and G. Guiochon, *J. Chromatogr. A, 1059*: 43 (2004). With permission.

amounts to 2.5 kJ mol^{-1}, but is partly consumed by the entropy that counteracted the enthalpy-driven adsorptions at either site. It is probably noteworthy that this compensation effect was larger for the enantioselective site that might be explained by a stronger multisite binding process that may lead to an overall lower translational, rotational, conformational, and vibrational degree of freedom.

Very similar characteristics (bi-Langmuir model) and overall trends have been also found for 3CPP on the tBuCQD-CSP in the NPLC mode, yet the saturation capacities at the enantioselective sites q_s were much larger than in case of TFAE [54]. This effect was ascribed to the smaller space requirements of 3CPP compared to TFAE with its extended anthryl-ring system. On the contrary, the equilibrium constants at the enantioselective sites were 10–30 times smaller for 3CPP than TFAE. It is also notable that b_{ns} was extremely low, yet their number was extremely high (one order of magnitude higher than the high-energy binding sites).

The data mentioned earlier might pretend that saturation capacities and thus column loadability in preparative separations are, in general, comparably low. However, it must be emphasized that the results mentioned earlier with regards to saturation capacities and sample loading capacities, respectively, may be not necessarily representative for the situation of chiral acidic solutes, which follow an ion-exchange adsorption process that is known for enhanced loading capacities. For several reasons, the establishment of adsorption isotherms for such a system is more complicated and the adsorption behavior of chiral acids on cinchonan carbamate CSPs needs yet to be elucidated. Preliminary experiments in the batch chromatography mode (pulse method) with N-FMOC-α-allyl-glycine and tBuCQN-CSP indeed point toward a high sample loading capacity of such an ion-exchange system.

1.7.1.4 Estimation of Binding Strengths by Transfer NOE Measurement

Another, yet completely different access to macroscopic binding strengths of selectands on CSPs has been described by Hellriegel et al. [65] employing suspended-state NMR spectroscopy. Thus, HR-MAS 2D transfer-nuclear Overhauser effect spectroscopy (NOESY) was utilized to distinguish solutes strongly binding to the

CSP, which was suspended in the mobile phase, from weakly binding ones. This technology makes use of the observation of the so-called transferred nuclear Overhauser effect (trNOE). The basis is that the weak positive intramolecular NOE usually observed for low molecular mass ligands becomes a strong negative NOE, if the ligand is binding to a macromolecular species such as a CSP. The strong negative NOE can be detected, quasi as memory of binding, in the free ligand after its dissociation from the CSP. Consequently, this NMR technique allows straightforward and rapid distinction between effective binders (negative cross-peaks) and weak binders (positive cross-peaks). When HR/MAS 2D trNOESY spectroscopy experiments were carried out with tBuCQN-CSP suspended in a 0.1 mol L^{-1} solution of either enantiomer of DNB-Leu or Ac-Phe in deuterated methanol, the intensities for the negative cross-peaks were always higher for the stronger binding S-enantiomers (factor of 7.4 for DNB-Leu and 2.4 for Ac-Phe). Thus, the trNOESY experiments correctly reflected the situation found in chromatography both in terms of enantiomer preference as well as in relative magnitudes of enantioselectivities for DNB-Leu and Ac-Phe. This approach could serve as fast screening methodology of relative binding strengths of CSPs and their enantioselective molecular recognition capability in a microscale format.

1.7.2 Structure of SO–SA Complexes by Spectroscopic, X-Ray Diffraction, and Molecular Modeling Investigations

Various endeavors have been undertaken to get insight into the 3D selector–selectand complex structures and to elucidate chiral recognition mechanisms of cinchonan carbamate selectors for a few model selectands (in particular, DNB-Leu). Such studies comprised NMR [92–94], FT-IR [94–96], X-ray diffraction [33,59,92,94], and molecular modeling investigations (the latter focusing on molecular dynamics [92,93,97], and 3D-QSAR CoMFA studies [98]).

1.7.2.1 NMR Spectroscopy

NMR spectroscopy proved to be a useful tool for investigating selector–selectand complex structures. Various sets of information can be derived from NMR spectroscopic studies including the following: (i) Continuous SO–SA titration experiments may unveil the binding stoichiometries. (ii) Information of the conformational preferences of the binding partners may be obtained from dihedral coupling constants and intramolecular NOEs. (iii) Self-association phenomena of selector and selectands may be deduced from shifts of particular NMR signals upon continuous dilution of SO, SA, and/or SO–SA complexes. (iv) Knowledge about the intermolecular interactions and 3D-arrangement of the host–guest complex may also be gained from significant complexation-induced chemical up- or downfield shifts and intermolecular NOEs.

Since the cinchona alkaloid-derived selectors do possess two basic sites, the quinuclidine and quinoline nitrogens with the former being more basic, ion-pairs of 1:1 or 2:1 stoichiometry could be formed. To shed light on this issue, the binding stoichiometry was investigated by Maier et al. [92] and Czerwenka et al. [93]

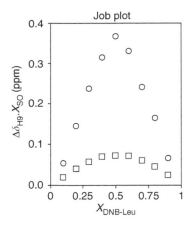

FIGURE 1.16 Job plot for the complexes of O-9-$tert$-butylcarbamoyl-6′-neopentoxy-cinchonidine selector and DNB-Leu enantiomers. Symbols: R-enantiomer (□), S-enantiomer (○). The shifts of the most diagnostic H_9 proton was monitored as function of increasing molar fractions X of DNB-Leu enantiomers and the total concentrations were 10 mM in methanol-d_4. (Reproduced from N.M. Maier et al., *J. Am. Chem. Soc., 124*: 8611 (2002). With permission.)

by continuous variation NMR titration experiments in which the selector, O-9-$tert$-butylcarbamoyl-6′-neopentoxy-cinchonidine, was complexed with either enantiomer of the selectand, namely, DNB-Leu [92], DNB-Ala-Ala [93], or DNB-Ala-Ala-Ala [93] at various molar selector–selectand ratios and the change of the chemical shift of the benzylic proton of the methylene bridge at the carbon C_9 ($\Delta\delta_{H9}$) was monitored. Straightforward interpretation of the results was made possible by construction of a Job plot (Figure 1.16). The maxima were for both of the DNB-Leu enantiomers observed at a 1:1 SO–SA ratio ($X_{DNB-Leu} = 0.5$), indicating a 1:1 complexation stoichiometry, which is well in accordance with the assumed 1:1 ion-pair formation at the most basic tertiary quinuclidine nitrogen and was later also found for the stronger binding S-enantiomers of DNB-Ala-Ala and DNB-Ala-Ala-Ala [93]. These results confirmed the 1:1 binding stoichiometry inferred from the ITC titrations (*vide supra*).

Various studies were focusing on the conformational behavior of the cinchonan carbamate selectors in free and complexed form, which could readily be derived from the dihedral coupling constant of the H_8–H_9 protons ($^3J_{H8H9}$) and intramolecular NOEs as measured by 2D-NOESY [92,93] or two-dimensional rotating frame Overhauser effect spectroscopy (2D-ROESY) [65] spectra.

It has been shown by NMR and molecular modeling that native cinchona alkaloids [99–101] as well as derivatives thereof such as the carbamates can adopt a number of energetically favorable conformations. The backbone is relatively rigid and conformational flexibility is mainly associated with changes of the dihedral angle $C_{4'}$—C_9—C_8—N_1 (torsional angle T_1) along with that of the $C_{3'}$—$C_{4'}$—C_9—C_8 torsion (torsional angle T_2). The profound experimental data in the literature point toward the fact that, as for native cinchona alkaloids [99–101], also for their carbamates such as the O-9-($tert$-butylcarbamoyl)quinine as well as the O-9-($tert$-butylcarbamoyl)-6′-neopentoxy-cinchonidine selectors, mainly three conformational states are important

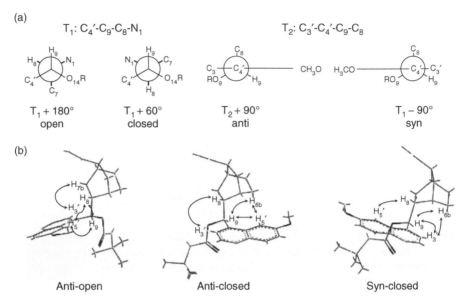

FIGURE 1.17 Preferential conformational states of cinchonan carbamate selectors shown in Newman projection (a) and line models of 3D images (b) as exemplified by *O*-9-(*tert*-butylcarbamoyl)quinine (arrows indicate intramolecular NOEs). (Reproduced from K. Akasaka et al., *Chirality, 17*: 544 (2005). With permission.)

that may coexist in solution concurrently: the *anti*-open, *anti*-closed, and *syn*-closed conformers (Figure 1.17), a terminology that has been coined by Dijkstra et al. [99].

The torsional angle T_1 determines whether *open* or *closed* conformational state is dominating and T_2 decides whether the *anti* or *syn*- arrangement exists (see Figure 1.17, upper panel) [92,99–101]. Experimentally, inter-ring NOEs (i.e., intramolecular NOEs between quinuclidine and quinoline ring) as well as the dihedral coupling constant $^3J_{H8H9}$ allow pinpointing which of the three conformers are existing under the given conditions. In the *anti*-open conformation, the quinuclidine nitrogen points away from the quinoline plane, which potentially could give rise to intramolecular NOEs between $H_{5'}$—H_8, $H_{5'}$—H_9, H_8—H_9 (H_8H_9 in *gauche* orientation), and $H_{3'}$—H_{7b} (see Figure 1.17, bottom panel). On the other hand, it points toward the quinoline plane in the *closed* conformation (H_8H_9 in *trans* orientation), for which the T_2 (which rotates the quinoline ring around the $C_{4'}$—C_9 axis by about 180°) may adopt either the *anti*-orientation (NOEs between $H_{5'}$—H_{6b}, $H_{5'}$—H_9, H_9—H_{6b} and $H_{3'}$—H_8) or *syn*-arrangement (NOEs between $H_{3'}$—H_{6b}, $H_{3'}$—H_9, H_9—H_{6b} and $H_{5'}$—H_8) (*syn* refers to the orientation in which the $C_{6'}$ methoxy and *O*-9-substituents are on the same side of the rotation axis and *anti* to the orientation in which they are on opposite sides; Figure 1.17, upper panel) [92,99–101]. Which one of the conformers is higher populated and thus dominating depends on several factors including the type of derivative (substituents) [99], solvent [99,100], protonation [92–94,99], and complexation with guest molecules [92–94,99].

Although the various cinchonan carbamate selectors that were investigated had distinct substituents and were in complex with different selectands, common conformational preferences were found quite consistently in all of these studies [92–94].

Most characteristic for the overall conformation of the cinchonan carbamates in solution is the dihedral angle of the torsion H_8–C_8–C_9–H_9, which can be inferred from the vicinal coupling constant of the NMR proton signal of H_9. The experimentally measured coupling constant $^3J_{H8H9}$ represents actually the average over the populations $P_{(i)}$ of the different conformers in solution according to [100]

$$^3J_{H8H9} = \sum P_{(i)} \cdot {^3J_{H8H9(i)}}. \tag{1.18}$$

On the basis of calculations and a modified Karplus equation, the coupling constants for the individual major conformers—*anti*-closed, *syn*-closed, and *anti*-open—have been reliably determined to be 9.4, 9.6, and 1.7 Hz, respectively, by Bürgi et al. [100]. With these values in hand and Equation 1.18 approximate distributions of conformers have been calculated from the measured coupling constants of H_9 for free selectors as well as the corresponding solutions of the selector in complex with selectand enantiomers. Solutions must be appropriately diluted as to avoid the measurements being biased by autoassociation phenomena of the selector, which are quite common for native cinchona alkaloids [102] as well as carbamate derivatives [95], in particular in apolar and aprotic solvents such as chloroform and acetonitrile. To resemble chromatographic conditions, the NMR measurements were usually performed with methanol d_4 at concentrations less than 20 mM, for which autoassociation was less an issue and, for example, found to be $3.7(\pm 1.8) \times 10^1$ M^{-1} for tBuCQN.HCl and $<1 \times 10^{-1}$ M^{-1} for tBuCQN base at 25°C [94].

Consistently, in all NMR studies that reported on the conformational states of the carbamoylated cinchonan selectors the free selector (*O*-9-*tert*-butylcarbamoyl-6′-neopentoxy-cinchonidine [92,93] or *O*-9-*tert*-butylcarbamoyl-quinine [94]) existed in methanolic solution always as a mixture of open and closed conformers as indicated by a vicinal $^3J_{H8H9}$ coupling constant of about 4.5 Hz (open/closed conformer ratio calculated to be about 2:1). This finding was also supported by the presence of characteristic inter-ring NOEs such as $H_{3'}$–H_{7b} indicative for *anti*-open conformer, $H_{3'}$–H_8 characteristic for the *anti*-closed conformer, besides other NOEs. In any case, upon protonation of the cinchonan carbamate selectors, for example, with hydrochloric acid (to form mono-HCl salt), a conformational transition of the selector to the exclusive existence of the *anti*-open conformation took place, as displayed by an alteration of the $^3J_{H8H9}$ coupling constant from about 4.5 Hz to about 1.5 Hz (*O*-9-*tert*-butylcarbamoyl-6′-neopentoxy-cinchonidine) [92,93] and 1.8 Hz (*O*-9-*tert*-butylcarbamoyl-quinine) [94], respectively. As a general trend, on complexation of the cinchonan carbamate selectors with acidic selectands such as DNB-Leu [92], DNB-Ala-Ala [93], DNB-Ala-Ala-Ala [93], and 2-methoxy-2-(1-naphthyl)propionic acid [94], the conformational preference of the selector resembled that of the HCl salt. In case of the more stable complexes (always the *S*-enantiomer for the selectands mentioned earlier), the vicinal $^3J_{H8H9}$ coupling constant indicated open conformer populations greater than 90%, while the less stable complexes (always the *R*-enantiomer) showed a little higher percentage of closed conformers (10–20% depending on the type of selectand). These results derived from the vicinal $^3J_{H8H9}$ coupling constant were also supported by the existence of characteristic intramolecular NOEs, which suggested with high probability the coexistence of all three conformers in case of the less stable complexes and a strong preference for the *anti*-open conformer for the more stable complexes.

Besides, information on intermolecular interactions has been derived in these studies from complexation-induced shifts (CIS). The chemical shift is an indicator for the shielding of a nucleus and thus for the electronic state of a specific proton. Since the electronic environment may change on complexation, CIS can be used to monitor where host–guest contacts may take place. If these interactions occur stereoselectively, the CIS will be different for the two guest enantiomers ($\Delta\delta$ distinct from 0) giving possibly some insight into the chiral recognition mechanism.

When 1:1 complexes of selector and selectand enantiomers were investigated, the NMR signals for the quinuclidine protons especially those being adjacent to the N_1 nitrogen were significantly shifted indicating ion-pairing interactions [92–94]. Such CIS were, however, also observed for the HCl salt of the selectors and were always relatively similar for both R- and S-complexes suggesting that ion-pair formation occurs largely nonstereoselectively; that is, this primary interaction does exist for both enantiomers [92–94]. A pronounced CIS was also always observed for H_9, that is, the proton at the prominent C_9 stereogenic center around which the distinct intermolecular interaction sites are assembled. In case of the complexation of S-DNB-Leu with the O-9-*tert*-butylcarbamoyl-6'-neopentoxy-cinchonidine selector, a strong downfield shift of $\Delta\delta = 1.02$ ppm was detected [92]. This was believed to be a net effect of various influential factors: First, it was assumed that upon complex formation and concomitantly induced conformational change of the selector (closed–open transition), H_9 may get into the influence of the deshielding portion of the quinoline or carbamate groups (ring current fields). The effect was, however, more pronounced for the stronger binding S-enantiomer than for the weaker binding R-enantiomer ($\Delta\delta = 0.26$ ppm) [92]. Hence, it was argued that, second, the stereoselective CIS is most likely partly also a result of stereoselective interactions in the surrounding environment of H_9. Similar observations were made by Czerwenka et al. [93] for complexes of the same selector with DNB-alanine di- and tripeptides as selectands. It is also interesting to note that, on the contrary, such stereoselective CIS of H_9 could not be observed for the complexes of O-9-*tert*-butylcarbamoyl-quinine selector with 2-methoxy-2-(1-naphthyl)propionic acid enantiomers [94] being subjected to the same closed–open transition in the course of complexation and thus also indirectly supporting the argumentation that the strong CIS mentioned earlier for H_9 is not solely originating from the influence of ring current fields.

As much as the quinoline ring is concerned useful information on the existence or absence of π–π-interactions with corresponding aromatic moieties in the selectands could indeed be derived by help of CIS of aromatic protons. Substantial upfield CIS in the range of $\Delta\delta = -0.24$ to -0.37 ppm have been detected in the S-complex of DNB-Leu with the O-9-*tert*-butylcarbamoyl-6'-neopentoxy-cinchonidine selector for the quinoline protons and the proton in para-position of the DNB group, while the corresponding R-complex was devoid of this effect [92]. Essentially the same observation was made for the DNB-Ala-Ala selectand, but not for DNB-Ala-Ala-Ala, for which such CIS was absent in the stronger binding *S*-complex. The stereose-lective occurrence of a strong CIS in case of DNB-Leu and DNB-Ala-Ala was regarded as an experimental evidence for stereoselective π–π-interactions, which was missing for the DNB-Ala-Ala-Ala tripeptide probably because of a spatial mis-match of quinoline and DNB group as a result of the elongated peptide chain [93]. The absence of a pronounced CIS in the aromatic regions for the complexes of 9-*O*-*tert*-butylcarbamoyl-quinine and 2-methoxy-2-(1-naphthyl)propionic acid suggested that π–π-interactions do not make a significant contribution as complex stabilization forces and for chiral recognition [94].

Overall, NMR allowed to derive valuable information on the solution-state con-formational preferences and existence or absence of π–π-information. Moreover, it became clearly evident that the SO–SA complexation is in case of cinchona alkal-oid selectors a dynamic event that involves conformational changes and induced fit phenomena for an optimal binding. Tentative SO–SA binding models could be derived on the basis of the NMR results and other complementary findings (*vide infra*). Intermolecular NOEs that have been detected between specific nuclei of selector and selectand in case of complexes of the 9-*O*-*tert*-butylcarbamoyl-6'-neopentoxy-cinchonidine selector and the stronger binding DNB-(*S*)-Leu and DNB-(*S*)-Ala-(*S*)-Ala enantiomers, being indicative for a close spatial approxim-ation in the complex, could be partly explained in view of these tentative binding models [92,93]. On the contrary, intermolecular NOEs were reported to be absent for the SO–SA system consisting of 9-*O*-*tert*-butylcarbamoyl-quinine and 2-methoxy-2-(1-naphthyl)propionic acid, which is in consent with an overall weaker binding as a result of a more flexible arrangement of the SA enantiomers at the selector's active site and the absence of strong simultaneous multisite interactions.

1.7.2.2 FT-IR Spectroscopic Investigations

Fourier transform-infrared (FT-IR) spectroscopic studies on SO–SA complexa-tion provides information that may be complementary to that of NMR and other techniques, namely, in particular, on the involvement of functional groups in inter-molecular and intramolecular interactions. Attenuated total reflectance (ATR) IR spectroscopy has been used for the study of binding modes of cinchona alkal-oid selectors either in solution [95] or in solid state [94], or directly on the CSP [96].

For the study in solution, the ATR technique with a cylindrical internal reflection cell and acetonitrile-based solutions of *O*-9-allylcarbamoyl-10,11-dihydroquinidine as selector and *S*- and *R*-enantiomers of DNB-Leu, DNP-Leu, as well their *N*-methylated congeners were employed for the measurement of the IR vibrations. Equimolar SO–SA mixtures were analyzed for spectral differences between *R*- and *S*-complexes to extract information on the intermolecular binding forces and chiral recognition mechanisms [95]. Besides nonstereoselective ion-pair formation that was evidenced by disappearance of the COOH stretching band of the free SA for all invest-igated amino acid derivatives (between 1744 and 1750 cm^{-1}), particular attention was paid to the most characteristic carbonyl stretching vibration of the selector's car-bamate group (amide I band in the 1720–1740 cm^{-1} region). This band is shifted to lower frequencies if the carbonyl is hydrogen bonded because of a weakening of the force constant for the stretching vibration (the stronger the H-bond the lower the frequency). In the acetonitrile solution, the C=O stretching band was only slightly shifted from 1725 cm (uncomplexed selector) to 1721 cm^{-1} in the stronger bound R-complex of DNB-Leu, while it was strongly shifted in the corresponding weak S-complex to 1739 cm^{-1}. (To avoid confusions, it must be emphasized at this place that we discuss here a quinidine-derived selector, while for the NMR studies we referred to quinine-derived selectors with reversed enantiomer affinity relationships.) These shifts indicate that the free selector tends to aggregate in acetonitrile solutions driven by H-bonds and that this H-bond from autoassociation of the selector is replaced in the R-complex by a stronger intermolecular H-bond with the selectand's NH (see Figures 1.10 and 1.11). In contrast, in the S-complex the H-bond from autoassociation of the selector is disrupted and not replaced by an intermolecular H-bond with the selectand featuring the carbonyl stretch at a frequency, which may be expected for a free carbamate carbonyl. The frequencies of the C=O stretch vibrations of the free selector and weaker S-complex were shifted to lower wavenumbers on addition of water, indicating strong solvation effects of the carbonyl group by water. In contrast, the strong H-bond and thus the C=O stretching band in the stronger bound R-complex was less affected by the addition of water, showing that this stereoselective H-bond is stable even in polar protic solvents and is not disrupted by solvation. It may be concluded that this H-bond is an important contribution to intermolecular complex formation and due to its stereoselective occurrence to chiral recognition as well.

This interpretation was also supported by the spectra of the corresponding N-methyl-leucine derivative in which the H-donor of the selectand was substituted by a methyl group and therefore not available for hydrogen bonding. Both complexes showed a similar spectral behavior as the weak S-complex of DNB-Leu: The C=O stretch was always shifted from 1725 (uncomplexed autoassociated selector) to 1739 cm^{-1} (indicative for disrupted H-bonds) in the S-complex and R-complex as well. These FT-IR data may be regarded as an unequivocal proof for the existence of a stereoselective H-bond between the NH of DNB-Leu and the selector's carbonyl group (Figures 1.10 and 1.11).

The same set of measurements for DNP-Leu as selectand and O-9-allylcarbamoyl-10,11-dihydroquinidine as selector displayed a considerably different picture. While no indication for H-bond formation between the aromatic α-amino group and the carbamate of the selector could be deduced, the IR spectra of free and respective complexed forms suggested the occurrence of π–π-interactions of the DNP group as evidenced by a shifted C=C stretching vibration from 1594 to 1588 cm^{-1} in the stronger bound S-complex (force constant weakened due to delocalized electrons). This type of shift was not found in the weak R-complex, but in the corresponding S-complex of DNP-N-methyl-Leu as well. The spectral data nicely reflect the binding relationships in HPLC where both DNP-Leu and DNP-N-methyl-Leu are well separated with comparable separation factors, but with an elution order that is opposite compared to DNB-Leu.

Self-association of the selector and stereoselective yet intercomplex H-bond formation could be identified in an ATR FT-IR study of Akasaka et al. [94] in which the SO–SA complexation of O-9-($tert$-butylcarbamoyl)-quinine and 2-methoxy-2-(1-naphthyl)propionic acid was investigated in solid state.

Wirz et al. [96] suggested a more elegant methodology, namely, ATR IR spectroscopy in combination with modulation spectroscopy and phase-sensitive detection as a powerful tool to investigate the different adsorption of enantiomers (DNB-Leu) at chiral liquid–solid interfaces, which were in the presented study, the O-9-($tert$-butylcarbamoyl)quinine-based CSP and a corresponding cinchonidine-based CSP. For such type of measurements, the CSP was fixed on the internal surface of an internal reflection element that was mounted in a flow-through cell. The experimental setup allowed to periodically alter the concentration of the adsorbate flushed through the detection cell (*concentration modulation*) as well as the type of adsorbate (which refers here to the configuration, that is, either R- or S-enantiomer) (*absolute configuration modulation*): hence, modulation of R-enantiomer against solvent, modulation of S-enantiomer against solvent, and modulation of R-enantiomer against S-enantiomer was possible with the set-up proposed in Reference 96. During the modulation experiments, time-resolved IR spectra were monitored, and by subsequent phase-sensitive data analysis phase resolved, that is, demodulated spectra have been obtained. If a system rapidly responds to the stimulation, the demodulated spectra are difference spectra between the two distinct states of the system.

The major advantages of such modulation spectroscopy experiments are that (i) only the signals that change periodically (with the alteration of the external parameter) will give a signal in the demodulated spectra whereas the signals that do not change periodically with the stimulation are cancelled out and (ii) since the noise is efficiently filtered out, the phase-sensitive data analysis generates high quality spectra with a good signal-to-noise ratio, much better than conventional difference spectra. For the present case of absolute configuration modulation, only signals were detected that resulted from different interactions of the two opposite enantiomers. This greatly facilitated the interpretation of the spectra.

Figure 1.18 depicts the demodulated spectra of (a) DNB-(R)-Leu vs. solvent, (b) DNB-(S)-Leu vs. solvent, and (c) and (d) DNB-(R)-Leu vs. DNB-(S)-Leu with modulation period of T $=$ 299 s and 523 s, respectively, employing O-9-($tert$-butylcarbamoyl)quinine immobilized on silica as support. The major differences in the adsorption complexes of R- and S-enantiomers of DNB-Leu on the tBuCQN-CSP can be easily derived by comparison of these demodulated spectra. A number of bands change in intensity, for example, the symmetrical stretching $\nu_s(NO_2)$ at 1346 (1345) cm^{-1} or the amide I band of the SA at 1670 (1668) cm^{-1}, while their frequencies remain nearly unchanged. Since the intensities of the absorbencies in the spectra of the S-enantiomer are higher, it was assumed that the S-enantiomer is stronger adsorbed to the CSP, which is in accordance with the liquid chromatographic observations. More notably, the amide I band of the selector (stretching vibration stemming from C$=$O of carbamate group) is shifted significantly from 1725 up to 1745 cm^{-1} on adsorption of the R-enantiomer of DNB-Leu (weak complex) (Figure 1.18a), which was regarded as an indication for the loss of a hydrogen bond or the formation of a weaker one upon complexation on the surface.

On the other hand, more bands were shifted in the corresponding stronger binding S-adsorbate (Figure 1.18b). For example, the amide I band of the selector was shifted

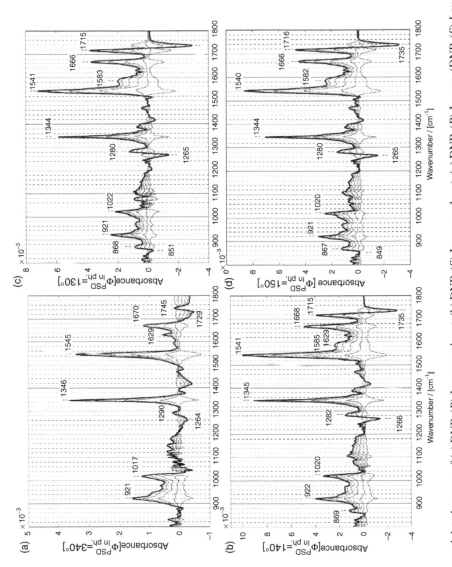

FIGURE 1.18　Demodulated spectra of (a) DNB-(*R*)-Leu vs. solvent, (b) DNB-(*S*)-Leu vs. solvent, (c) DNB-(*S*)-Leu vs. (DNB-(*R*)-Leu vs. (DNB-(*S*)-Leu with modulation period of 299 s, and (d) DNB-(*R*)-Leu vs. (DNB-(*S*)-Leu with modulation period of 523 s on *O*-9-(*tert*-butylcarbamoyl)quinine-based CSP. The boldface spectra are the so-called in-phase spectra. (Reprinted from R. Wirz et al., *Anal. Chem.*, 76: 5319 (2004). With permission.)

from 1735 to 1715 cm^{-1} (the band at 1735 cm^{-1} was decreasing and the absorbance at 1715 cm^{-1} was growing with addition of S-adsorbate). This finding is in accordance with the previously described solution-phase ATR IR spectroscopy study and points toward the existence of a strong H-bond that was not found in the weak R-complex. Concomitantly, the amide II band of the SA (N–H deformation band), which did not show up in the R-complex, was shifted in the S-complex from the region between 1500 and 1550 up to 1585 cm^{-1}, and, additionally, a relatively broad N–H stretching vibration (at 3341 cm^{-1}) appeared in the demodulated spectra of the S-adsorbate. It may be assigned to a H-bonded amide (H-donor group for the stereoselective hydrogen bonding interaction between SA's amide and SO's carbamate carbonyl as claimed earlier). As another dissimilarity in the R- and S-adsorption complexes, the shift of the asymmetric stretching vibration of the nitro group ν_{as} (NO$_2$) from 1545 cm^{-1} in the R-adsorbate to 1541 cm^{-1} in the S-adsorbate (as could be explained by a weakening of the force constant due to delocalized electrons) may be regarded as an evidence for a stereoselective π–π-interaction (since it was only found for the S-complex). The demodulated spectra of R- vs. S-DNB-Leu in Figure 1.18 c,d look similar to the demodulated spectra of the S-enantiomer of DNB-Leu vs. solvent and are a confirmation of the discussed findings.

Also control experiments were performed in which the transmission spectra of free selector and in complex with S- and R-enantiomers were measured in solution and the difference spectra confirmed similar binding states in solution and on the surface of the CSP.

1.7.2.3 Single-Crystal X-Ray Diffraction Analysis

One particularly nice feature of the present low-molecular ion-exchange selectors derives from the feasibility of straightforward growing of single crystals from SO–SA complexes through cocrystallization of the basic selectors and an acidic solute enantiomer. The salty character of the SO–SA complexes thereby greatly facilitates the cocrystallization process. Single-crystal X-ray diffraction analysis then allows to generate, with a certain resolution and with uncertainty of atomic positions, a 3D-image of the investigated SO–SA complex structures and thus directly provides comprehensive information of structural features such as complete sets of atomic distances, angles, dihedral angles, spatial distances, as well as to some extent intra- and intermolecular interactions. The technique has been exploited to determine the absolute configurations of individual stereoisomers that were obtained by chromatographic racemate resolution on preparative scale (W. Bicker et al., submitted). However, it was also very helpful for mechanistic investigations of molecular recognition, whereby it must be kept in mind that X-ray crystal structures represent solid-state structures that may deviate from solution situations. In any case, X-ray structures should represent energetically favorable structures in which bond lengths are usually constrained (and not at their equilibrium state) yet the torsional angles that are most decisive for the overall conformational arrangement are very often, if not at their global, at least at a local optimum. Hence, valuable insight into SO–SA binding modes could be obtained from X-ray structures of SO–SA cocrystals [33,59,92,94] (Figures 1.19 and 1.20).

It is seen from Figures 1.19 and 1.20 that the selectors, in agreement with solution-state NMR measurements, displayed in these X-ray structures in any case the *anti*-open conformation. In this *anti*-open conformation, the selector forms a spatially well-defined, preorganized binding cleft, which is opened for the guest selectand and allows for its close approximation. The polar anion-exchanger and hydrogen bonding sites are located close to the center of this active enantiorecognidion site and are

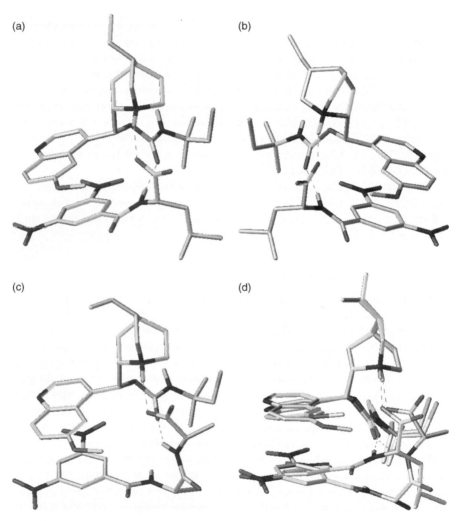

FIGURE 1.19 X-ray crystal structures of selector–selectand complexes (ion-pairs): (a) *O*-9-(β-chloro-*tert*-butylcarbamoyl)quinine with *N*-(3,5-dinitrobenzoyl)-(*S*)-leucine, (b) the pseudoenantiomeric complex of *O*-9-(β-chloro-*tert*-butylcarbamoyl)quinidine with *N*-(3,5-dinitrobenzoyl)-(*R*)-leucine, (c) *O*-9-(β-chloro-*tert*-butylcarbamoyl)quinine with *N*-(3,5-dinitrobenzoyl)-(*S*)-alanyl-(*S*)-alanine, and (d) comparison of the complexes of (a) and (c). Most hydrogens have been omitted for the purpose of clarity. (Reprinted from C. Czerwenka et al., *Anal. Chem., 74*: 5658 (2002). With permission.)

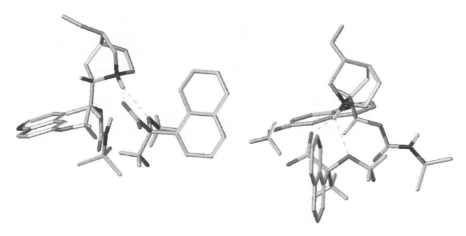

FIGURE 1.20 X-ray crystal structure of the more stable complex between *O*-9-(*tert*-butylcarbamoyl)quinine and (*S*)-2-methoxy-2-(1-naphthyl)propionic acid (two distinct perspectives). (Reprinted from K. Akasaka et al., *Chirality, 17*: 544 (2005). With permission.)

surrounded by hydrophobic bulky groups that may act as hydrophobic shields for the polar interactions and as steric barriers. As can be seen from the presented X-ray structures, the selectands bind into this opened binding pocket, being more or less embraced by the methoxy (or neopentoxy) quinoline and *tert*-butyl carbamate arms. As has been elucidated by NMR (*vide supra*), an induced fit phenomenon appears to be involved in the course of complex formation and the first SO–SA contact seems to be driven by long range but nondirected electrostatic interactions, which are further stabilized by a directed hydrogen bond-mediated ionic interaction (salt bridge) once the contact is close enough. Further, it can be seen that the carbamate group always adopts the preferential *s-cis* conformation in the ester part and a *transoid* orientation in the amide part, as expected.

As far as the DNB-amino acids and DNB-dipeptides in complex with the cinchonan carbamate selectors [33,59,92] is concerned, the X-ray structures confirm the existence of the intermolecular interactions suggested by the spectroscopic studies described above (*vide supra*). The stronger bound complexes shown in Figure 1.19a through 1.19c all feature the hydrogen bond between the SA's amide N–H and the SO's carbamate C=O. In addition, the electron-rich quinoline ring (π-base) and the electron-deficient DNB group (π-acid) adopt a favorable coplanar arrangement to each other indicative for complex stabilization by face-to-face π–π-interactions in case of the high-affinity enantiomer.

If both of the complexes with quinine configurations (Figure 1.19a,c) are superimposed on each other by aligning the molecules on the rigid quinuclidine body, it becomes nicely visible that the conformational arrangements of the selectors are virtually identical despite a distinct guest molecule in the binding pocket (Figure 1.19d). Conformational adaptations to account for the structural peculiarities of the binding partners mainly concern the quinoline plane and the carbamate group (including its residue): The quinoline plane slightly flips around via rotation along the C_9–$C_{4'}$

axis (Figure 1.19d) as to allow for an optimal π–π-interaction with the corresponding aromatic DNB group in the amino acid and peptide derivative, respectively.
On the other hand, the carbamate group exploits its flexible ester part to adjust for
its optimal alignment as to enable H-bonding with the amide group of the guest
molecules.

Another issue is validated by the presented X-ray structures: This is related to the
pseudoenantiomeric character of the *tert*-butylcarbamates of quinine and quinidine
(Figure 1.19a,b). Except for the vinyl on the backside of the quinuclidine ring, both
the complexes that are actually diastereomeric to each other actually look like mirror
images with regard to conformations and intermolecular interactions as well so that
the pseudoenantiomeric experimental chromatographic behavior for DNB-Leu can be
rationalized also on the basis of their X-ray crystal structures.

Figure 1.20 depicts the X-ray crystal structure of the complex between
O-9-(*tert*-butylcarbamoyl)quinine and the (*S*)-enantiomer of 2-methoxy-2-(1-
naphthyl)propionic acid that forms the stronger complex. While the selector adopts
essentially the same conformation (*anti*-open), the overall binding mode seems less
favorable because of lack of additional strong supportive interactions besides the
ionic H-bond. Most notable, a π–π-interaction that was expected is not featured
by the structure of the cocrystallized SO–SA complex. Thus, the observed enantiomer separation, which is characterized by a comparably small chromatographic
α-value (α = 1.79 [41]) is most probably due to different steric environments for the
opposite SA enantiomers or slightly distinctive weak dispersive interactions rather
than a different number of strong noncovalent interactions (such as H-bonding or
π–π-interactions). This example clearly demonstrates that it may not always be so
straightforward to decipher and rationalize the decisive stereodifferentiating forces in
selector–selectand complexes with moderate chromatographic α-values, even on the
basis of X-ray crystal structures. On the contrary, a set of complementary techniques
including various spectroscopic methods and molecular modeling may support each
other and amplify the knowledge of complex structures and molecular recognition
mechanisms.

1.7.2.4 Computational Methods

Cinchona alkaloids and derivatives have been subject of several molecular modeling
studies, mostly for the sake of getting knowledge of conformational states, making
use of *molecular mechanics* [99], *semiempirical methods* (e.g., PM3) [101], *ab-initio*
(with advanced basis sets such as 6-31G**) [96,100] or *density functional theory methods* [96,100]. The major results of these conformational studies were already indicated
earlier in the context of the NMR discussion. Also, a 3D quantitative structure–activity
relationship study using comparative molecular field analysis (3D-QSAR CoMFA)
was set up to elucidate the regions around a set of chiral quinine carbamate selectors
where changes in the calculated electrostatic and steric fields can be correlated to
increases or decreases of the enantiomer discrimination capability of the cinchonan
carbamate selectors [98]. Herein, we restrict ourselves, however, to a brief discussion of computational studies that aimed at the elucidation of intermolecular SO–SA
binding events by *nanosecond stochastic molecular dynamics (MD) simulations*

on the diastereomeric complexes of *O*-9-(*tert*-butylcarbamoyl)quinine/DNB-Leu [97], *O*-9-(*tert*-butylcarbamoyl)-6′-neopentoxy-cinchonidine/DNB-Leu [92], and *O*-9-(*tert*-butylcarbamoyl)-6′-neopentoxy-cinchonidine/DNB-Ala-Ala as well as *O*-9-(*tert*-butylcarbamoyl)-6′-neopentoxy-cinchonidine/DNB-Ala-Ala-Ala [93].

The calculations were performed using continuum models for the solvent (water and chloroform to account for a high-polarity and a low-polarity medium, respectively). After conformational analysis (grid search) and energy minimization of the binding partners using the AMBER* force field implemented in MacroModel 5.5 computation software, a motif-based SO–SA docking was performed; that is, the SAs were docked into the SO's active binding site in such a way that the corresponding potential interactions sites such as quinuclidinium and carboxylate functionalities were fitted close to each other. Subsequently, the docked complexes were energy minimized and thereafter MD simulations of the diastereomeric complexes were run over an extended period of time (1–15 ns) following a procedure as detailed in Reference 97. The MD simulation results were analyzed by various approaches to figure out the desired binding and chiral distinction increments (*vide infra*). The results can be briefly summarized as follows.

The complex structures of the X-ray study as well as the relative enantiomer affinities (i.e., chromatographic elution orders), as shown by the averaged total potential energies of the corresponding diastereomeric complexes obtained by the stochastic dynamics (SD) simulations, could be reasonably well reproduced by the computations. In most cases, the computed energy differences for the *R*- and *S*-complexes (ΔE) were reported to agree within reasonable errors with the chromatographic experiments (methanol-buffer, 80:20): for example, the aqueous simulation ($\Delta E = 5.85$ kJ mol^{-1}) only slightly underestimated the experimental value ($\Delta E = 8.78$ kJ mol^{-1}) for the *O*-9-(*tert*-butylcarbamoyl)-6′-neopentoxy-cinchonidine/DNB-Leu SO–SA pair, while the chloroform simulation significantly overestimated the experimental value ($\Delta E = 15.15$ kJ mL^{-1}). (*Note*: Chloroform less closely matches the experimental chromatographic solvent and may overemphasize electrostatics.) Besides the total potential energies, the component energy terms (stretch, bend, torsion, Van der Waals, and Coulomb) were also discussed for both the diastereomeric complexes. More of interest were, however, the intermolecular energies (i.e., the nonbonded interactions, which are the atom–atom interactions that are 1,3 or greater in distance), which were extracted in such a way to have the intramolecular contributions removed. In any case, the complexes were held together mainly by forces from the electrostatic term and this contribution was increased in chloroform [92]. It was found that the most enantiodiscriminating contribution arises from the long-range electrostatic interactions rather than short-range Van der Waals contributions [92,93,97], except for the relatively poorly resolved SO–SA system composed of *O*-9-(*tert*-butylcarbamoyl)-6′-neopentoxy-cinchonidine/DNB-Ala-Ala-Ala [93], which showed an approximately equal enantiomer distinction contribution of both the terms.

During the SD simulations, it was found that the selectands were spatially constrained in their motion to a very confined region around the selector, yet were still moving (fluctuating) around enabling its rearrangement on the SO's binding site. This has been visualized by the MD trajectories (Figures 1.21 and 1.22) by

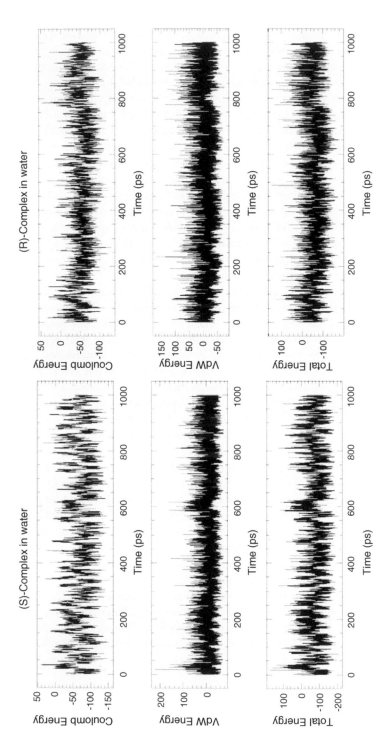

FIGURE 1.21 Trajectories of the intermolecular energies for (a) the strong S-complex and (b) the weak R-complex between O-9-(tert-butylcarbamoyl)-6'-neopentoxy-cinchonidine and DNB-Leu over the 1 ns simulation period in the polar (water) medium. (Top) Coulomb energy; (middle) Van der Waals energy; (bottom) total energy. (Reprinted from N.M. Maier et al., *J. Am. Chem. Soc., 124:* 8611 (2002). With permission.)

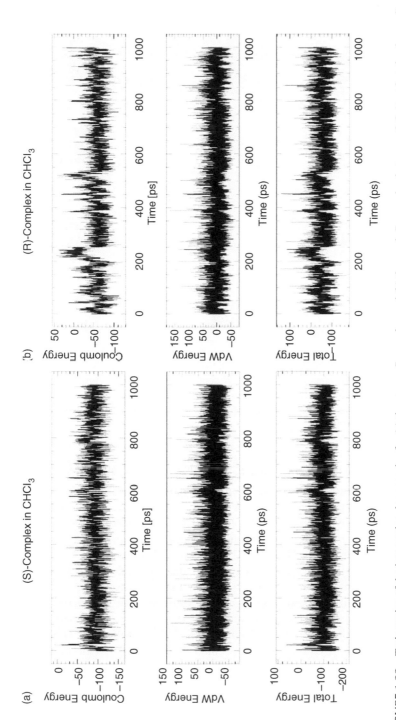

FIGURE 1.22 Trajectories of the intermolecular energies for (a) the strong *S*-complex and (b) the weak *R*-complex between *O*-9-(*tert*-butylcarbamoyl)-6′-neopentoxy-cinchonidine and DNB-Leu over the 1 ns simulation period in the nonpolar (chloroform) medium. (Top) Coulomb energy; (middle) Van der Waals energy; (bottom) total energy. (Reprinted from N.M. Maier et al., *J. Am. Chem. Soc.*, *124*: 8611 (2002). With permission.)

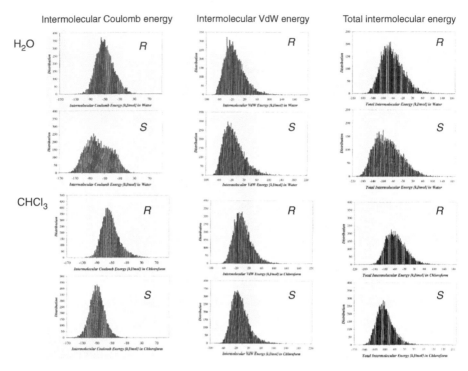

FIGURE 1.23 Distributions of (a) electrostatic energies, (b) Van der Waals energies, and (c) total intermolecular energies for *S*- and *R*-complexes of *O*-9-(*tert*-butylcarbamoyl)-6′-neopentoxy-cinchonidine and DNB-Leu sampled during the simulation period of 1 ns in water (top) and chloroform (bottom). (Reprinted from N.M. Maier et al., *J. Am. Chem. Soc., 124*: 8611 (2002). With permission.)

plotting the computed energies, total intermolecular energies (bottom), as well as intermolecular electrostatic energies (top) and Van der Waals energies (middle), as a function of time for the polar water environment (Figure 1.21) and the non-polar chloroform environment (Figure 1.22). It was evident from the trajectories that there exist huge fluctuations and thus in intermolecular energies as the system evolved in time (Figures 1.21 and 1.22). This was also confirmed by histogram plots of the energy distributions (Figure 1.23) that differed in shape (unimodal or slightly bimodal) and width, depending on the investigated SA–SO system [92,93,97]. In any case, there were significant differences in the energy distributions between *R*- and *S*-enantiomers and the distributions were shifted to lower energy values for the more stable *S*-complexes, for example, the *S*-complex of *O*-9-(*tert*-butylcarbamoyl)-6′-neopentoxy-cinchonidine and DNB-Leu (Figure 1.23).

To get a deeper insight on which structural increments were mainly responsible for the differential energies in the diastereomeric complexes, a further differentiation was undertaken in these computational studies by partitioning the chiral selectors into fragments (quinuclidine ring, carbamate group with residue, and quinoline) and extracting the corresponding energy contributions individually [92,93,97]. The outcome of this approach appeared to be quite inconsistent or, in other words, seemed to be solute specific, not very surprisingly. While for the *O*-9-(*tert*-butylcarbamoyl)-quinine/DNB-Leu [97] as well as *O*-9-(*tert*-butylcarbamoyl)-6′-neopentoxy-cinchonidine/DNB-alanine di- and tripeptides [93] the quinuclidine increment was the most important group contribution to hold the complexes together, in case of the *O*-9-(*tert*-butylcarbamoyl)-6′-neopentoxy-cinchonidine/DNB-Leu SO–SA system [92], this role was attributed to the carbamate fragment. In any case, the quinoline was found to be the second most stabilizing contribution [92,93,97]. The computed energy differences for the quinoline fragment of the diastereomeric complexes of DNB-Leu and *O*-9-(*tert*-butylcarbamoyl)-6′-neopentoxy-cinchonidine selector as well as histogram plots (Figure 1.24) of the intermolecular distances of dummy atoms that were placed in the centroids of the quinoline ring and the DNB group, which were monitored during the simulations, both clearly suggested the occurrence of a stereoselective π–π-interaction for this SO–SA system: It is clearly seen from Figure 1.24 that the

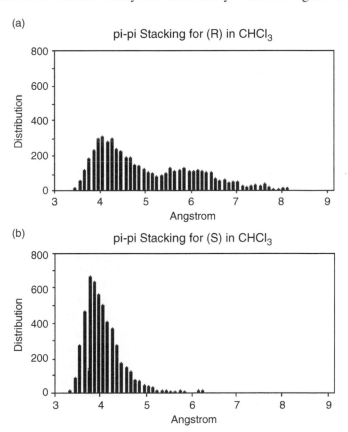

FIGURE 1.24 Histogram plots of intermolecular distances between dummy atoms placed at ring centroids of the DNB group of DNB-Leu and the quinoline of the *O*-9-(*tert*-butylcarbamoyl)-6′-neopentoxy-cinchonidine selector (chloroform environment). (a) *R*-complex and (b) *S*-complex. (Reprinted from N.M. Maier et al., *J. Am. Chem. Soc.,* *124*: 8611 (2002). With permission.)

distances between the aromatic planes are closer for the stronger binding S-complex [92]. A central role in enantiodiscrimination of the cinchonan carbamate selectors could also be ascribed to the carbamate group on the basis of the performed computations [92,93,97]. This structural fragment turned out to be a major stereodiscriminating group mainly by long-range electrostatic contributions.

All the chromatographic, spectroscopic, and molecular modeling investigations as well as X-ray crystallography together gave a quite consistent picture in terms of binding and chiral recognition mechanisms for the investigated SO–SA model systems. Some of the information may be partly expanded to structurally similar selectands. Further, information regarding stereoselective molecular SO–SA association phenomena in the transition phase (from solution to gas phase) or gas phase has lately been compiled by various experiments employing electrospray ionization mass spectrometry [103–107], which may to some extent complement the data obtained by solution-phase and solid-state techniques. Overall, the present cinchonan carbamate selector class probably belongs to the currently best-investigated chromatographic selector systems regarding its molecular and chiral recognition mechanisms.

1.8 CHROMATOGRAPHIC APPLICABILITY SPECTRUM

1.8.1 ANALYSIS OF AMINO ACIDS AND DERIVATIVES

The chiral distinction capability of cinchonan carbamate CSPs for underivatized amino acids has not been fully elucidated yet, in contrast to the large embodiment of N-acylated and N-arylated amino acid derivatives (*vide infra*). However, it seems that chiral amino acids can be successfully resolved into enantiomers if the amino acid side chain (R residue) contains a functionality that represents a strongly interactive binding site with the selector such as an extended aromatic ring system like in thyroxin (T_4).

T_4 as well as its monodeiodine analog triiodothyronine (T_3) are amino acid–type thyroid hormones secreted by the pituary gland. They play an important biological role as they are amongst others required for the stimulation of the metabolic rate and the regulation of the growth and the development in infants. From a pharmaceutical viewpoint, the L-enantiomer of T_4 (levothyroxin) is of major interest, because it is used for replacement therapy of hypothyroidism disorders. Pharmacological investigations clearly revealed that the D-enantiomer does possess a completely different activity profile. It does not enhance the metabolic rate, but besides some other adverse effects may alter lipogenic enzyme levels. Thereby it influences the lipid metabolism. The D-enantiomer has therefore also been used as antihyperlipidemic agent. Hence, highly pure levothyroxin should be administered in substitution therapy of thyroid disorders treatment, which raises the demand for a fast and robust quality control method of levothyroxin tablets. Such an analytical assay for monitoring the enantiomeric purity of levothyroxin tablets was recently proposed by Gika et al. [108]. The method makes use of a O-9-(*tert*-butylcarbamoyl)quinine-CSP (Chiralpak QN-AX) and allows the simultaneous separation of T_4 enantiomers besides T_3 enantiomers (Figure 1.25). The analysis of pharmaceutical formulations from the market showed that significant amounts of D-T_4 can be present in such tablets. For example, the

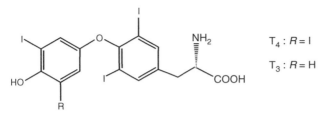

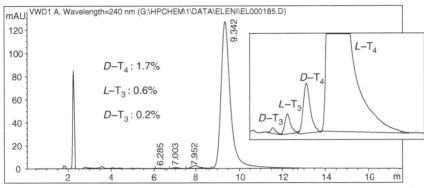

FIGURE 1.25   HPLC determination of impurities in a levothyroxin (L-T_4) formulation. Experimental conditions: Column, Chiralpak QN-AX (150 mm × 4 mm ID); mobile phase, acetonitrile-50 mM ammonium acetate (60:40, v/v) (pH$_a$ 4.5); flow rate, 0.7 mL min^{-1}; UV detection, 240 nm; temperature, 25°C. Sample, T4-200 tablets (Uni-Pharma, Greece) containing 0.2 mg L-T_4 sodium per tablet; the tablet was pulverized, suspended in methanol-10 mM sodium hydroxide (1:1; v/v) and after ultrasonication for 5 min the residues were removed by filtration. An aliquot of 10 μL of the filtrate was directly injected. (Reproduced from H. Gika et al., *J. Chromatogr. B, 800:* 193 (2004). With permission.)

chromatogram of levothyroxin tablets depicted in Figure 1.25 shows that the formulation contains about 1.7% of D-T_4. Aside of this substantial enantiomeric impurity, both enantiomers of T_3 were also present in detectable amounts (0.6% and 0.2% of L- and D-T_3, respectively).

More thoroughly investigated than the applicability of cinchonan carbamate CSPs for enantiomer separations of underivatized amino acid has been their application profile for *N*-derivatives [30–32,47,48,74,89,90,109–115]. Thereby, the derivatization of the amino group allows for the introduction of chromophoric or fluorophoric labels with favorable detection properties but also of interactive sites to support chiral distinction. Moreover, the amphoteric nature of amino acids gets eliminated by such chemical modifications that render the derivatives acidic, being best suitable for the establishment of an anion-exchange process. Figure 1.26 gives a representative overview of amino acid derivatives that have been resolved into enantiomers on cinchonan carbamate–type CSPs. Standard derivatization protocols have been reported, for example, by Czerwenka et al. [116]. Tables 1.3 and 1.9 summarize selected chromatographic enantiomer separation data for amino acid derivatives obtained on *O*-9-(*tert*-butylcarbamoyl)quinine- and quinidine-based CSPs, respectively, with a hydro-organic mobile phase.

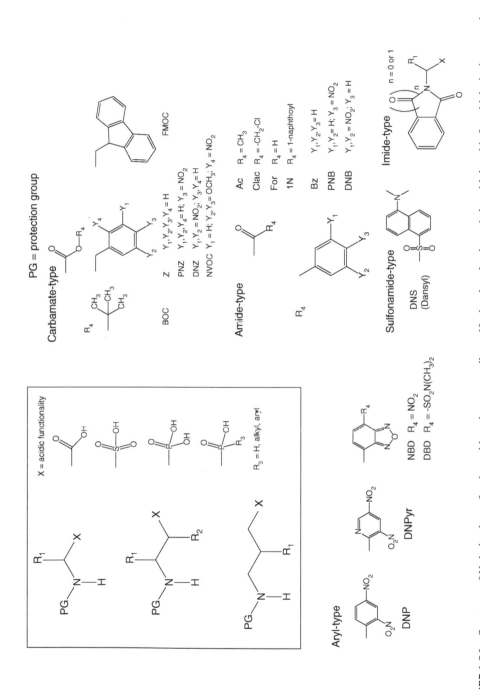

FIGURE 1.26 Spectrum of *N*-derivatives of amino acids and corresponding sulfonic, phosphonic, and phosphinic acids for which cinchonan carbamate-based CSPs showed broad enantiomer separation capabilities.

TABLE 1.9
Selected Examples of Enantiomer Separation Data of N-Derivatized Amino Acids (See Also Table 1.3), Amino Sulfonic, Phosphonic and Phosphinic Acids on tBuCQN-Based CSP[a]

Compound	k_1	α	e.o.
N-acyl (carbamate type)			
NVOC-Leu	7.16	3.59	D < L
FMOC-Leu	9.26	1.9	D < L
FMOC-Gln	4.13	1.48	D < L
FMOC-Cys	15.514	1.622	D < L
FMOC-Cys-S-trityl	55.893	1.425	D < L
FMOC-Arg	1.143	1.675	D < L
FMOC-Arg-PMC	21.4	1.604	D < L
FMOC-Phe	15.65	1.56	D < L
PNZ-Leu	6.11	1.67	D < L
Z-Tle	4.22	1.54	D < L
Z-Tyr	12	1.24	D < L
Z-Ser	6.1	1.21	D < L
Boc-Val	3.55	1.4	D < L
BOC-Tyr	6.27	1.25	D < L
DNZ-α-Aminobutyric acid	6.95	2.46	D < L
DNZ-Val	7.14	3.26	D < L
DNZ-Phg	12.24	1.83	D < L
DNZ-2-Aminopropanesulfonic acid	7.12	1.88	$(R) < (S)$
DNZ-2-Aminobutanesulfonic acid	7.48	2.21	$(R) < (S)$
DNZ-2-Amino-3,3-dimethylbutanesulfonic acid	6.97	3.99	$(R) < (S)$
DNZ-1-amino-1-phenylmethane phosphonic acid	47.3	2.73	
DNZ-1-amino-1-(2-chlorophenyl)methane phosphonic acid	50.16	2.27	
DNZ-1-amino-1-(2-bromophenyl)methane phosphonic acid	58.06	2.2	
DNZ-1-amino-1-phenylmethane P-methyl phosphinic acid	14.31	2.26	
DNZ-1-amino-1-phenylmethane P-ethyl phosphinic acid	15.69	1.75	
N-acyl (amide type)			
DNB-Leu	11.74	15.88	D < L
DNB-N-Me-Leu	10.91	1.07	n.d.
DNB-α-Me-Leu	16.06	1.33	n.d.
DNB-Asp [b]	18.92	1.4	L < D
DNB-Glu [b]	4.38	4.9	D < L
DNB-α-aminoadipic acid [b]	5.52	4.31	D < L
DNB-cysteic acid [b]	19.81	1.12	L < D
PNB-Leu	7.08	4.1	$(R) < (S)$
Bz-Phe	8.9	1.91	D < L
Bz-β-Phe	4.8	1.62	
1N-Leu	9.83	1.43	$(R) < (S)$
Ac-Phe	3.87	1.41	D < L
Clac-Phe	6.75	1.29	

Continued

TABLE 1.9
(Continued)

Compound	k_1	α	e.o.
For-Phe	5.32	1.24	
For-Tle	2.67	1.25	D < L
Phthalyl-Val	8.48	1.17	
N-acyl (sulfonamide type)			
DNS-Phe	35.99	1.44	D < L
DNS-Ser	17.45	1.31	
N-aryl			
DNP-Leu	24.01	1.31	L < D
DNP-*N*-Me-Leu	23.31	1.39	
DNP-Pro	12.04	1.36	L < D
DNP-2-amino-3,3-dimethylbutanesulfonic acid	13.98	1.26	(R) < (S)
DNP-1-amino-1-phenylmethane phosphonic acid	111.91	1.32	
DNP-1-amino-1-phenylmethane P-methyl phosphinic acid	23.89	1.13	
DBD-Leu	15.41	1.39	
DBD-Phe	24.25	1.31	
DNPyr-Phe	31.62	1.18	

[a] *Experimental conditions*: Mobile Phase: MeOH/0.1M ammonium acetate (80:20), $pH_a = 6.0$, T: 25°C, flow rate: 1 mL min^{-1}; e.o., elution order; for structures of protection groups see Figure 1.26.
[b] Same conditions as above, but 1 M ammonium acetate buffer.

While the distinct amino acid residues have mostly only a modulating effect (see Table 1.9) (e.g., FMOC-protected amino acids), the type of protection group or derivative formed decides on the molecular and chiral recognition mechanism and hence on the obtained elution order as well as the level of enantiomer recognition (i.e., magnitudes of α-values) that can be afforded. From a practical point of view, we may distinguish between two groups of *N*-derivatives:

1. *N*-acylated amino acid derivatives such as of the carbamate type (BOC, Z, FMOC, PNZ, DNZ, NVOC, etc.), of the amide type (Ac, Clac, For, Bz, DNB, etc.), and of the sulfonamide type (DNS) obtained by derivatization with corresponding chloroformates, acid chlorides, or *N*-hydroxysuccinimide esters as reagents. FMOC, Z, BOC-amino acids, and also more uncommon analogs such as NVOC (photocleavable protection group), and so forth, are important building blocks for peptide synthesis and need to be highly enantiomerically pure for this purpose to avoid significant amounts of diastereomeric peptide impurities in the final products. Acetyl and benzoyl-amino acid derivatives are important substrates in biocatalytic production processes of non-proteinogenic amino acid enantiomers. In contrast, DNB-derivatives do not have a noteworthy practical relevance in amino acid and peptide chemistry but have been mainly utilized for mechanistic investigations. DNS-labeling on the other hand was formerly a prominent derivatization strategy for sensitive amino acid analysis owing to the fluorescent dimethylaminonaphthalin label. All these *N*-acylated amino acids

possess the hydrogen donor–acceptor system at the stereogenic center, which may give rise to H-bond formation with the SO's carbamate C=O, as discussed for DNB-Leu (*vide supra*). In these systems, the expected elution order for α-amino acids is D before L on quinine-derived CSPs and L before D on quinidine-derived CSPs. This is valid for amino carboxylic, phosphonic, phosphinic, and sulfonic acids likewise (note altered priorities in CIP terminology, i.e., *R/S* terminology may pretend erroneously a reversal of elution order and change of chiral recognition mechanism). However, there exist a few exemptions to this rule (e.g., for α-amino acids with acidic side chain). For example, DNB-Asp and DNB-cysteic acid (both possessing one methylene group between acidic side chain functionality and stereogenic center) show reversed elution orders compared to the other amino acid congeners, while, on the contrary, the acidic amino acid analogs DNB-Glu and DNB-α-amino adipic acid both having the acidic group farther distant from the stereogenic center (i.e., having two and three methylene groups, respectively) adhere to the common elution orders. Obviously the second acidic group in the amino acid residue of DNB-Asp and DNB-cysteic acid plays a more dominant role in terms of ion-pair formation with the basic selector rather than the α-carboxylic functionality. Thus, the common molecular recognition mechanism that exists for the other DNB-amino acids gets disturbed, which leads to the inversion of the elution order. The situation is different for DNB-Glu and DNB-α-amino adipic acid probably due to an improper spacing of the acidic group in the amino acid side chain so that the elution order remains the same as for the other amino acid derivatives without acidic group in the residue. In general, for selectands having an amide-type or carbamate-type H-bond interaction site at the stereogenic center enantioselectivity levels with α-values of about 1.2–1.3 can commonly be reached (e.g., Boc-, Z, Ac-, Clac-, For-protected aliphatic and aromatic amino acids), which can be substantially exceeded if additionally a supportive π–π-interaction can be established (like in DNB, Bz, DNZ, and FMOC derivatized amino acids) (see Table 1.9). Since the H-donor is not available in *N*-acylated derivatives of secondary amino acids (such as proline or pipecolinic acid), enantioselectivity may significantly drop compared to primary amino acids or may be even completely lost (Table 1.9) (see DNB-Leu and DNB-*N*-methyl-Leu).

In a recent study, Gyimesi-Forras et al. [112] extended the enantioseparation spectrum of cinchonan carbamate CSPs to imide-type alpha-amino acid derivatives. The α-amino acids having a primary amino group were derivatized with ortho-phthaldialdehyde, naphthalene-2,3-dicarboxaldehyde or anthracene-2,3-dicarboxaldehyde as the reagents in acetonitrile (nonaqueous conditions) to form isoindoline-1-one derivatives or the corresponding benzo- and naphtho-condensed analogs. Enantioselectivities ranged typically between 1.1 and 1.4 depending on the reagent, the CSPs investigated (tBuCQN, tBuCQD, and DIPPCQN) and the eluent. The elution order followed that of amide-type amino acid derivatives. It is evident that this methodology is restricted to primary amino acids because the isoindoline-1-one derivatives are not formed with secondary amino acid analogs.

2. *N*-aryl amino acids (such as DNP, DNPyr, DBD, and NBD labeled amino acids): Some of these derivatives are the preferred choice for many analytical purposes because they can be conveniently introduced by straightforward and smooth derivatization procedures. This holds in particular for the Sanger's reagent

(i.e., 2,4-dinitrofluorobenzene DNFB) yielding strongly chromophoric DNP-derivatives (UV $\lambda_{max} \sim 360$ nm) and Imai's 4-fluoro-7-nitro-2,1,3-benzoxadiazole as well as 4-fluoro-7-dimethylaminosulfonyl-2,1,3-benzoxadiazole, respectively, yielding fluorescent NBD ($\lambda_{ex} \sim 470$ nm, $\lambda_{em} \sim 530$ nm) and DBD derivatives ($\lambda_{ex} \sim 450$ nm, $\lambda_{em} \sim 560$ nm), respectively. For these N-aryl derivatives, the molecular recognition mechanism may be altered, as can be simply deduced from changed elution orders in comparison to the N-acyl amino acids. Such a scenario, for instance, is noticed for DNP-derivatives (L before D on quinine-derived CSP and D before L on quinidine-derived CSP). It must, however, be stressed that the chiral recognition mechanism is less consistent within this subclass of amino acid derivatives because NBD and DBD derivatives do show a reversed elution order compared to DNP-derivatives (*vide infra*). A prime advantage of the N-aryl derivatization protocols is their equal applicability for enantiomer separations of primary and secondary amino acids: both can be separated with nearly the same enantioselectivities (cf. Table 1.9, DNP-Leu and DNP-N-methyl-Leu). It may be safely argued that a H-donor at the α-amino group is not involved in chiral recognition and therefore not a required precondition for a successful enantiomer separation.

Derivatization with Sanger's reagent (2,4,-dinitrofluorobenzene) has often been utilized as the preferential strategy for the sensitive stereoselective analysis of various proteinogenic and nonnatural amino acids likewise. For example, the accurate analysis of low amounts of enantiomeric impurities in proline samples, which was facilitated by selection of the column that eluted the enantiomeric impurity in front of the main peak; that is, the quinidine carbamate column for the S-enantiomer and the quinine carbamate column for the R-enantiomer, was described by Kleidernigg et al. [110] (R_S was 4.9 for the quinine-derived CSP and 6.6 for the corresponding quinidine CSP). Enantiomeric impurities, as low as 0.039% and 0.013%, could be reliably determined for (S)-proline and (R)-proline, respectively. In another study, a complex mixture of natural amino acids was successfully separated into enantiomers after precolumn derivatization with Sanger's reagent as their DNP-derivatives in a single run by a step gradient elution methodology, and in the course of optimization of the elution conditions and gradient profiles the authors made use of the computer software DRYLAB [109].

A number of nonnatural amino acids were resolved into individual enantiomers on O-9-(2,6-diisopropylphenylcarbamoyl)quinine-based CSP by Peter and coworkers [48,90,113,114] after derivatization with Sanger's reagent, chloroformates (DNZ-Cl, FMOC-Cl, Z-Cl), Boc-anhydride, or acyl chlorides (DNB-Cl, Ac-Cl, Bz-Cl). For example, the four stereoisomers of β-methylphenylalanine, β-methyltyrosine, β-methyltryptophan, and β-methyl-1,2,3,4-tetrahydroisoquinoline-3-carboxylic acid could be conveniently resolved as various N-derivatives [113]. The applicability spectrum of cinchonan carbamate CSPs comprises also β-amino carboxylic acid derivatives, which were, for example, investigated by Peter et al. [114]. A common trend in terms of elution order of DNP-derivatized β-amino acids was obeyed in the latter study: On the utilized quinine carbamate-based CSP, the elution order was S before R for 2-aminobutyric acid, while it was R before S for the β-amino acids having branched R_1 substituents such as *iso*-butyl, *sec*-butyl, *tert*-butyl, cyclohexyl, or phenyl residues.

This reversal of elution order is due to the changed CIP priorities, but not a result of an altered binding and chiral recognition mechanism. Moreover, in yet another study, BOC and DNP-protected α-substituted proline derivatives have been resolved into enantiomers and elution orders were determined [48]. The method allowed the sensitive and accurate analysis of samples with regards to their enantiomeric purities.

Mitulovic et al. [111] presented a procedure for the preparation of α-carbon deuterium-labelled α-amino acids from native amino acids via a Schiff-base racemization protocol in deuterated acetic acid involving a preparative chromatography step of the obtained Z-protected deuterium-labeled amino acid derivatives on the tBuCQN-CSP [111]. The analytical control of the enantiomeric products after DNP, Z, or DNZ derivatization showed a high enantiomeric excess (97–98%) and also a high isotopic purity (99%) by MS.

For the purpose of sensitive analysis of hydrophobic D-amino acids in tissues (cerebrum, cerebellum, liver, kidney) and physiological fluids (plasma, urine) of rats Hamase et al. [117] have set up a 2D column switching HPLC system with fluorescence detection which had implemented a multiloop device (400 μL each) for multiple peak fraction collection, a reversed-phase column in the first dimension and a cinchona alkaloid-derived chiral anion-exchanger column in the second dimension (Figure 1.27a). Homogenized tissue, plasma, and urine samples were diluted with methanol, centrifuged, and an aliquot of the supernatants used for pre-column derivatization of the amino acids with 4-fluoro-7-nitro-2,1,3-benzoxadiazole (NBD-F) yielding highly fluorescent NBD-amino acid derivatives (λ_{ex} 470 nm, λ_{em} 530 nm). The resultant samples were injected into the HPLC system where the individual amino acids were resolved in the first dimension on the RP18 narrow bore column (1 mm ID), which was operated at a flow rate of 75 $\mu L\,min^{-1}$. As can be seen from Figure 1.27b, the investigated hydrophobic amino acids (Val, *allo*-Ile, Ile, Leu) could be nicely separated under isocratic elution conditions and were quantitated separately. Fractions of each peak (between 200 and 300 μL), as indicated in Figure 1.27b by gray bars, were collected into individual loops of the multiloop device and then transferred one after the other into the second chromatographic dimension, which was equipped with the enantioselective column, either a Chiralpak QN-AX column (Figure 1.27c) or a Chiralpak QD-AX column (Figure 1.27d). The columns in the second dimension had a wider diameter (4 mm ID) and thus were operated isocratically at a higher volumetric flow rate (1.5 mL min^{-1}) in order to avoid volume-overloading effects. The analysis in the enantioselective column (second dimension) was rapid with a run time less than 10 min each while, on the contrary, the separation in the first dimension was adjusted to be slow to provide enough time for the separation in the second dimension. Overall, the total analysis time per run was with 60 min quite acceptable. In any case, each sample was analyzed twice, once with the Chiralpak QN-AX in the second dimension (Figure 1.27c) and once with the corresponding pseudoenantiomeric Chiralpak QD-AX column in the second dimension (Figure 1.27d) on which the NBD-amino acid enantiomers showed reversed elution orders. Thereby, it could be assured that the minor peaks were correctly assigned to be the D-amino acids. By this confirmation, false positive results for traces of D-amino acids in the biological samples could be avoided.

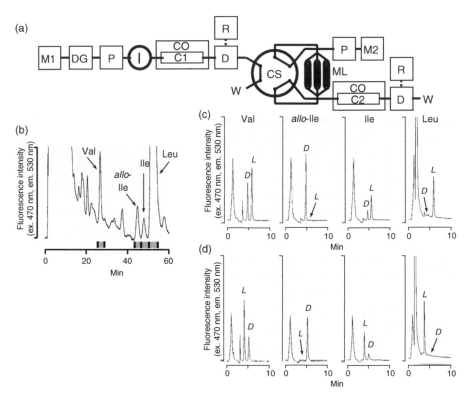

FIGURE 1.27 Analysis of hydrophobic D-amino acids as NBD-derivatives (obtained by pre-column derivatization with 4-fluoro-7-nitro-2,1,3-benzoxadiazole NBD-F) in rat tissue using an online 2-dimensional HPLC system combining RP18 and chiral anion-exchanger columns with fluorescence detection. (a) Experimental setup, (b) first dimension. M1: THF-TFA-H$_2$O (25:0.05:75; v/v); flow rate, 75 μL min^{-1}; C1: Capcellpak C18 MG II (150 mm × 1.0 mm ID, 40°C); (c) second dimension: M2: acetonitrile–methanol (50:50; v/v) containing 10 mM citric acid; flow rate, 1.5 mL min^{-1}; C2: Chiralpak QN-AX (150 mm × 4.0 mm ID, 40°C); (d) same as (c) but C2: Chiralpak QD-AX (150 mm × 4.0 mm ID, 40°C). Fluorescence detection, λ_{ex} 470 nm, λ_{em} 530 nm. Legend: M1 and M2, mobile phase 1 and 2; C1 and C2, column 1 and 2; DG, degasser; P, pump; I, injector; CO, column oven; D, detector; R, integrator; W, waste; ML, multiloop device; CS, column selection valve. (Reproduced with permission from K. Hamase et al., *J. Chromatogr. A*, *1143*: 105–111 (2007).)

This 2D-method was validated for the concentration range between 0.005 and 0.5 pmol for D-amino acids and 0.05–5 pmol for L-amino acids. Within-day and interday precisions were always better than 8% relative standard deviation (RSD) and the accuracies for spiked rat plasma samples were between 95.5% and 100.2%. Limit of detections (LODs) and limit of quantitations (LOQs) were reported to be as low as 3 fmol (S/N = 3–5; corresponding to 0.15 nmol g^{-1} wet tissue) and 5 fmol (corresponding to 0.25 nmol g^{-1} wet tissue). It was concluded that this assay is supposed to be one of the most sensitive analysis method for amino acid enantiomers in mammalian samples.

TABLE 1.10
Amounts of the Enantiomers of Branched Aliphatic Amino Acids in Rat Tissues and Physiological Fluids

	Val		allo-Ile		Ile		Leu	
	D	L	D	L	D	L	D	L
Cerebrum	n.d.	72.33 ± 6.85	n.d	0.60 ± 0.09	n.d.	39.91 ± 3.73	0.46 ± 0.11 (0.40 ± 0.12)	80.98 ± 8.83
Cerebellum	n.d.	66.14 ± 2.39	n.d	0.44 ± 0.03	n.d.	42.88 ± 1.58	trace (trace)	93.66 ± 0.78
Liver	n.d.	247.96 ± 22.27	n.d	1.12 ± 0.06	n.d.	156.73 ± 14.28	0.71 ± 0.10 (0.53 ± 0.05)	286.52 ± 14.37
Kidney	trace (trace)	230.56 ± 15.03	0.37 ± 0.03 (0.38 ± 0.09)	1.09 = 0.06	n.d.	123.83 ± 9.41	0.43 ± 0.08 (0.34 ± 0.09)	235.72 ± 15.16
Plasma	n.d.	191.51 ± 16.82	trace (trace)	1.55 ± 0.12	n.d.	87.14 ± 8.78	1.31 ± 0.20 (1.45 ± 0.26)	146.04 ± 13.80
Urine	8.42 ± 1.53 (8.47 ± 1.53)	14.10 ± 0.48	22.20 ± 5.66 (22.58 ± 6.65)	trace	2.63 ± 0.22 (2.60 ± 0.24)	7.71 ± 0.19	0.40 ± 0.09 (0.41 ± 0.13)	12.47 ± 0.54

Values represent mean ± standard error (SE) (nmol g^{-1} wet tissue) of three rats (Wistar, male, 9 weeks of age, SPF) determind by chiralpak QN-AX; n.d.: lower than the limit of detection (0.15 nmol g^{-1} wet tissue); trace: below the limit of quantification (0.25 nmol g^{-1} wet tissue) Vaules in parentheses are those determined by Chiralpak QD-AX.

Reproduced from K. Hamase et al., *J. Chromatogr. A, 1143*: 105–111 (2007). With permission.

Finally, this powerful validated 2D-HPLC assay was applied for the sensitive quantitative determination of these hydrophobic amino acids in various rat tissues (cerebrum, cerebellum, liver, and kidney) as well as rat plasma and urine samples. The results are summarized in Table 1.10. It is remarkable that significant amounts of various D-amino acids could be found in various tissues and biological fluids of healthy rats. In particular, in urine the D-enantiomer content in relation to the L-enantiomer was unexpectedly high (e.g., up to 60% for Val). It is also worth pointing out that substantial amounts of *allo*-Ile were found in the rat urine and close to 100% of this nonproteinogenic amino acid could be assigned to have D-configuration. The metabolic origins and the biological consequences of these findings will need to be further explored in the future.

1.8.2 Amino Sulfonic, Phosphonic, Phosphinic Acids

Sulfur and especially phosphorous-analogs of amino acids such as chiral amino sulfonic, phosphonic, and phosphinic acids have recently received noticeable attention in pharmaceutical and medicinal chemistry. Since the replacement of the carboxylic acid functionality by a sulfonic, phosphonic, or phosphinic acid groups does obviously not perturb the basic molecular recognition and separation mechanism, these chemical entities may be readily resolved into enantiomers by cinchonan carbamate CSPs as *N*-derivatives with basically the same elution orders as the amino acid congeners (note the changed priorities in CIP terminology) (see Table 1.9 and Figure 1.28). Hence, this chromatographic system has provided a methodology to indirectly assign absolute configurations of enantiomeric amino phosphonic acids [118] or aminophosphinic acids (Figure 1.28) [61] on the basis of the elution order.

For example, in the study by Zarbl et al. [118], four distinct quinine carbamate CSPs were exploited for the stereoselective analysis of α- and β-amino phosphonic acids as well as α-methyl-α-aminophosphonic acids congeners after derivatization to DNP and DNZ derivatives. α-values were typically in the range between 1.1 and 2.8, enabling convenient baseline resolutions of the enantiomers and the accurate enantiomeric excess determination. The chromatographically assigned absolute configurations for a series of enantiomeric α- and β-aminophosphonic acids were cross-validated with an independent methodology, making use of the Mosher method and ^{31}P-NMR spectroscopy to determine the configurations of the hydroxyphosphonic acid precursors. Complete agreement between the results obtained by these distinct complementary methods was found.

In yet another study, the simultaneous separation of the four stereoisomers of α-amino-β-hydroxy propane phosphonic acids as well as the corresponding β-amino-α-hydroxy propane phosphonic acids could be successfully accomplished after labeling with Sanger's reagent [119].

The applicability of cinchonan carbamate CSPs for bioanalytical investigations using HPLC-ESI-MS/MS has been demonstrated by Fakt et al. [120]. The goal was the stereoselective bioanalysis of (*R*)-3-amino-2-fluoropropylphosphinic acid, a γ-aminobutyric acid (GABA) receptor agonist, in blood plasma in order to determine whether this active enantiomer is *in vivo* converted to the *S*-enantiomer. In this enantioselective HPLC-MS/MS bioassay, sample preparation consisted of

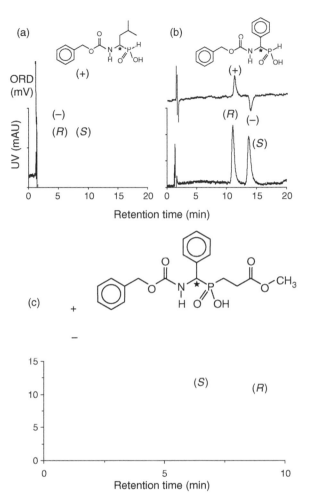

FIGURE 1.28 Chromatograms of the HPLC enantiomer separation of Z-protected α-aminophosphinic acids (a,b) and phosphinic acid-ψ-dipeptide (c) on (a and b) a *O*-9-(*tert*-butylcarbamoyl)quinidine CSP and (c) corresponding *O*-9-(*tert*-butylcarbamoyl)quinine-CSP, respectively. Experimental conditions: Column dimensions, 150 mm × 4 mm ID; mobile phase, methanol-50 mM sodium phosphate buffer (80:20; v/v) (pH$_a$ 5.6); temperature, 40°C; flow rate, 1 mL min^{-1}; detection, UV at 250 nm and optical rotation detection (ORD). (Reproduced from M. Lämmerhofer et al., *Tetrahedron Asymmetry*, *14*: 2557 (2003). With permission.)

ultrafiltration of the plasma samples (for protein removal) and derivatization with 2,4-dinitrofluorobenzene yielding the corresponding DNP-aminophosphinic acids. After centrifugation, aliquots of the supernatants were directly injected into the HPLC system equipped with a 7-μm CN-guard column, a tBuCQN-CSP (5 μm) analytical column, and a PE Sciex API 365 triple quadrupol mass spectrometer for detection (MRM transition of *m/z* 308/262 for the target fluoro-phosphinic acid enantiomers and m/z 290/244 for the internal standard 3-aminopropylphosphinic acid; note, for both analytes, the product ions were obtained by loss of nitro group). The enantiomers were baseline resolved with the MS-compatible eluent and the S-enantiomer eluted before the target R-enantiomer (retention times of 5.1 min for S-enantiomer, 6.8 min for R-enantiomer, and 6.4 min for the internal standard) (Figure 1.29). The method was validated and favorably complied with the requirements for the intended bioanalytical study. The LOQ was 0.5 μmol L^{-1} and linearity could be obtained for plasma over the concentration range between 0.5 and 150 μmol L^{-1}. Intraday precision was about 6.3% and 6.1% at 1.7 μmol L^{-1} for each enantiomer ($n = 10$). Absolute recoveries ranged between 97% and 102%. Application of the validated bioassay to authentic rat and dog plasma samples after administration of (R)-3-amino-2-fluoropropylphosphinic acid showed that *in vivo* racemization does not occur, neither in rat nor in dog. Overall, the general applicability of the tBuCQN-CSP column in the RP mode (hydro-organic eluent) turned out to be advantageous for such bioanalytical studies and the ESI-MS (turbo-ionspray) detection.

1.8.3 PEPTIDE STEREOISOMER SEPARATIONS

In a series of papers, the methodologies applied for amino acids were extended to peptide stereoisomer (enantiomer and diastereomer) separations [59,116,121–123].

In general, $2^{(n-m+o)}$ stereoisomers do exist wherein n is the number of amino acid residues in the peptide, m is the number of achiral residues like Gly, and o is the number of residues with two stereogenic centers like Thr and Ile. It becomes evident that the number of possible stereoisomers is rapidly increasing with peptide chain length. Owing to the huge variability of peptides and possible stereoisomeric forms, it is self-explaining that only model separations could be performed. A few highlights are presented at this place to illustrate what can be achieved in this field on cinchona alkaloid-derived CSPs.

From a practical point of view, the necessity to separate peptide enantiomers is less of an issue when the number of amino acid residues exceed 2 or 3, while diastereoisomer separations remain still of paramount practical importance due to configurational instability of individual residues in certain processes (e.g., racemization in the course of peptide synthesis). Although peptide diastereomers should be principally separable on achiral RP-type separation materials, there is no guarantee that this is actually true for all peptide sequences and all possible diastereomers. CSPs may therefore be a valuable alternative as they may often provide higher levels of diastereoselectivity. For small peptides (e.g., $n = 2$ or 3) the simultaneous separation of all peptide stereoisomers may be of practical relevance and this can be accomplished, for instance, on the present cinchonan-derived CSPs [121] (Figure 1.30). As in the given case of the alanyl-alanine dipeptide, the critical factor for a full separation of all isomers seems to be often the diastereoselectivity, while enantioselectivities are often much better. What the corresponding tri- and tetraalanine peptides was concerned, the peak capacity and resolution of the CSP tended to be insufficient for several diastereomeric peak pairs for a simultaneous separation of all stereoisomers and hence a more powerful 2D approach implementing a RP-type separation of the diastereomers in the first dimension followed by the separation of

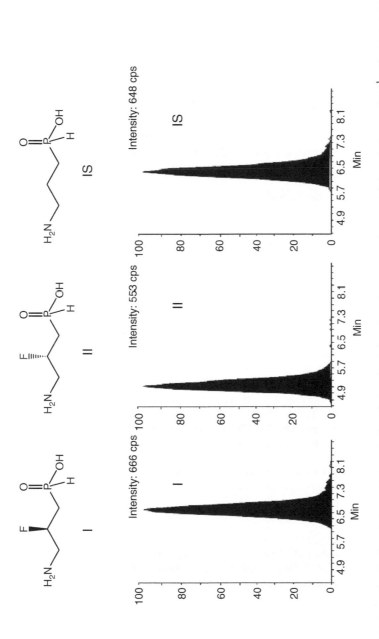

FIGURE 1.29 Enantioselective LC-MS/MS analysis of a spiked human plasma sample, containing 6.70 $\mu mol\,L^{-1}$ of **I** (*R*)-3-amino-2-fluoropropylphosphinic acid, 5.91 $\mu mol\,L^{-1}$ of **II** (*S*)-3-amino-2-fluoropropylphosphinic acid and 7.01 $\mu mol\,L^{-1}$ of IS 3-aminopropylphosphinic acid, determined after derivatization with Sanger's reagent as DNP-derivatives. Experimental conditions: Analytical column, tBuCQN-CSP (5 μm) (150 mm × 4 mm ID); guard-column, Brownlee CN (7 μm) (15 mm × 3.2 mm ID); mobile phase, 0.2 M ammonium acetate (pH 5)-methanol (10:90; v/v); flow rate, 1.3 mL min^{-1}. (Reproduced from C. Fakt et al., *Anal. Chim. Acta*, *492*: 261 (2003). With permission.)

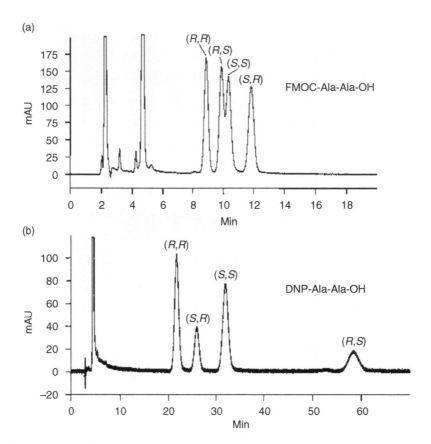

FIGURE 1.30 Micro-HPLC separation of all 4 stereoisomers of the dipeptide alanyl-alanine as FMOC derivatives (a) and DNP-derivatives (b), respectively, on a *O*-9-(*tert*-butylcarbamoyl)quinine-based CSP. Experimental conditions: Column dimension, 150×0.5 mm ID; mobile phase (a) acetonitrile–methanol (80:20; v/v) containing 400 mM acetic acid and 4 mM triethylamine, and (b) methanol-0.5 M ammonium acetate buffer (80:20; v/v) (pH$_a$ 6.0); flow rate, 10 μL min^{-1}; temperature, 25°C; injection volume, 250 nL; detection, UV at 250 nm. (Reproduced from C. Czerwenka et al., *J. Pharm. Biomed. Anal., 30*: 1789 (2003). With permission.)

the enantiomer pairs on the chiral anion-exchange type stationary phase in the second dimension was utilized [122]. Thus, all eight stereoisomers of Ala-Ala-Ala and 9 out of 10 stereoisomers of Ala-Ala-Ala-Ala as DNB-derivatives could be successfully resolved.

In a follow-up study, the effect of introduction of achiral Gly residues into the Ala-peptide chain (di- and tripeptides) on the afforded enantiomer separations was thoroughly investigated [123]. One of the major findings was that the introduction of a Gly residue compromised the enantioselectivity in particular if the Gly residue was located at the *N*-terminus. For example, α-values (for all-*S*/all-*R* enantiomers) decreased in the order Ala-Ala > Ala-Gly > Gly-Ala for the DNB-protected dipeptide series and dropped in the order Ala-Ala-Ala > Ala-Ala-Gly > Ala-Gly-Ala > Gly-Ala-Ala as well as Ala-Ala-Ala > Ala-Gly-Gly > Gly-Ala-Gly (no separation) < Gly-Gly-Ala (reversed elution order for the latter) for the DNB-protected dipeptide series.

To demonstrate the feasibilities of cinchona alkaloid-derived CSPs for chiral recognition of higher order peptides, alanine peptides with up to 10 residues were selected as model peptides and analyzed as N-derivatives—11 distinct protection groups namely, Ac, pivaloyl, Bz, DNB, CC (carbazole-9-carbonyl), DNS, BOC, Z, DNZ, FMOC, DNP—on a number of distinct cinchonan-based CSPs including O-9-($tert$-butylcarbamoyl)-quinine, O-9-($tert$-butylcarbamoyl)-6$'$-neopentoxy-cinchonidine, and 1,4-bis(9-O-quinidinyl)phthalazine-based CSPs, amongst others [59,116]. Besides highlighting the huge structural variability of the cinchona alkaloid-derived selector and CSP family having been proposed for enantiomer separations, the study demonstrated also the exceptional enantiomer separation capability of this class of CSPs for peptides. For example, the O-9-($tert$-butylcarbamoyl)-6$'$-neopentoxy-cinchonidine was able to distinguish between the all-(S) and all-(R) enantiomers of DNB-alanyl-alanine dipeptide with a chromatographic α-value of about 20 (corresponding to a $\Delta\Delta G$ value of 7.43 kJ mol^{-1}). For all carbamate-type CSPs and also the phthalazine-type CSP, a strong drop of α-values was observed within the homologs (Ala)$_n$-peptide series on change from di- to tripeptide, because the supportive $\pi-\pi$-interaction contribution that is simultaneously active with the C-terminal ionic interaction and hydrogen bonding at the α-amide of the C-terminus in case of amino acid and dipeptide derivatives gets lost due to steric reasons. Most notably, the 1,4-bis(9-O-quinidinyl)phthalazine-based CSP enabled the separation of all all-(S)/all-(R) enantiomeric pairs of the DNB-Ala$_n$ ($n = 1$–10) series (Figure 1.31), with a strong drop of enantioselectivity from $n = 2$ to 3, slightly decreasing α-values between $n = 3$ and 6 ($\alpha = 4.4$–2.1), and nearly constant α-values of around 1.8–1.9 when the peptide becomes longer ($n = 7$–10). It is remarkable that peptide enantiomers of this length can be distinguished by a low molecular-mass selector (Figure 1.31).

In another study, the effect of the protection groups mentioned earlier on the stereorecognition levels was investigated [116]. Thereby it became evident that the stereodirecting nature of the protection group, as reported for amino acids, is lost for the peptides. This means also for DNP-protected peptides, the elution order with regards to all-(S) all-(R) enantiomers is the same as for the amide- and carbamate-type protected peptides (all-R before all-S on cinchonan carbamate CSPs) due to the presence of an α-amide group at the C-terminal amino acid residue, which gets stereodirecting character. The practical consequence is that enantioselectivities between all-(S) and all-(R)-enantiomers are relatively similar for the longer peptides regardless of the type of N-protection group.

1.8.4 Enantiomer Separations of Miscellaneous Acids

The enantiomer separation capability of the cinchonan carbamate selectors and CSPs, respectively, is extremely broad what acidic compounds is concerned (success rate close to 100% if the acidic functional group is close to the chiral center). Hence

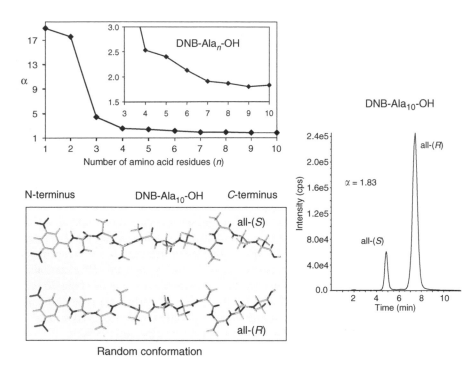

FIGURE 1.31 Separation factors for all-(S)/all-(R)-enantiomers of N-3,5-dinitrobenzoylated alanine peptides (DNB-Ala$_n$-OH) and chromatogram for the DNB-protected decaalanine peptide, DNB-Ala$_{10}$-OH, on a 1,4-*bis*(9-O-quinidinyl)phthalazine-based CSP (5 μm) with LC-ESI-MS. Experimental conditions: Column dimension, 150 mm × 4 mm ID; mobile phase, methanol-0.5 M ammonium acetate buffer (80:20; v/v) (pH$_a$ 6.0); flow rate, 1 mL min^{-1}; temperature, 25°C; injection volume, 50 μL; MS: negative mode, SIM (C. Czerwenka et al., *Anal. Chem.*, *74*: 5658 (2002).)

it is not restricted to the discussed amino acid and peptide derivatives as well as the corresponding sulfonic, phosphonic, phosphinic acid analogs. Table 1.2 already showed the enantioresolution capabilities of quinine and quinidine carbamate CSPs for aliphatic acids such as tetrahydro-2-furoic acid, aliphatic hydroxy acids (e.g., lactic acid, 2-hydroxybutyric acid [40]), and aromatic hydroxy acids (e.g., atrolactic acid, β-phenyllactic acid), which are all important chiral building blocks. Moreover, chromatographic separation data for aryl carboxylic acids (e.g., suprofen) having pharmaceutical importance as nonsteroidal anti-inflammatory drugs (NSAIDs), and for aryloxycarboxylic acids such as dichlorprop, mecoprop, and fenoprop, which are important agrochemicals with stereoselective herbicidal activity profiles, were already shown in Tables 1.2 and 1.3, respectively.

In another study, an enantioselective assay for the analysis of the enantiomer composition of monomethyl 4-(2′,3′-dichlorophenyl)-2,6-dimethyl-1,4-dihydropyridine-3,5-dicarboxylic acid on the tBuCQN-CSP was proposed ($R_S = 3.4$) [49]. This chiral carboxylic acid is the synthesis intermediate and the primary metabolite of the short-acting calcium-channel antagonist clevidipine, butyroxymethyl methyl 4-(2′,3′-dichlorophenyl)-2,6-dimethyl-1,4-dihydropyridine-3,5-dicarboxylate, which was shown to exhibit different pharmacokinetic profiles for the individual enantiomers. The method turned out to be suitable for the enantiomeric excess determination and stereoselective pharmacokinetic studies.

In a recent study, chiral separations for pyrethroic acids, which are the chiral building blocks of synthetic pyrethroids and the primary metabolites of the acid part of these potent ester insecticides, have been developed [62]. For example, a polar-organic mobile phase allowed the complete baseline resolution of all four stereoisomers of chrysanthemic acid (2,2-dimethyl-3-(2-methylprop-1-enyl)-cyclopropanecarboxylic acid) on a O-9-($tert$-butylcarbamoyl)quinine-based CSP ($\alpha_{cis} = 1.20$, $\alpha_{trans} = 1.35$, critical $R_s = 3.03$) (Figure 1.32a). This chiral acid is the precursor of pyrethroids like allethrin, phenothrin, resmethrin, and tetramethrin but not excreted as metabolite. The primary acid metabolite of these pyrethroids is chrysanthemum dicarboxylic acid (3-[(1E)-2-carboxyprop-1-enyl]-2,2-dimethylcyclopropanecarboxylic acid) the stereoisomers of which could also be resolved with a reversed-phase eluent (acetonitrile—30-mM ammonium acetate buffer 90:10, v/v; pH$_a$ = 6.0) and employing an O-9-(2,6-diisopropylphenylcarbamoyl)quinine-based CSP ($\alpha_{cis} = 1.09$, $\alpha_{trans} = 1.50$,

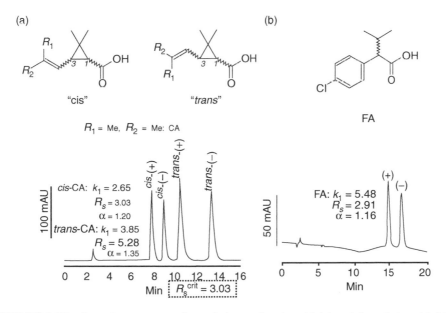

FIGURE 1.32 Stereoisomer separations of chrysanthemic acid (a) and fenvaleric acid (b) employing an O-9-($tert$-butylcarbamoyl)quinine-based CSP in the polar-organic mode (a) and an O-9-(2,6-diisopropylphenylcarbamoyl)quinine based CSP (b). Experimental conditions: Column dimensions, 150 mm × 4 mm ID; eluents (a) 0.06% acetic acid in acetonitrile–methanol (95:5; v/v); (b) acetonitrile-0.3 M ammonium acetate buffer (90:10; v/v) (pHa 6.0); flow rate, 0.65 mL min^{-1}; temperature, 25°C; detection, UV at 230 nm. (Reproduced from W. Bicker et al., *J. Chromatogr. A, 1035*: 37 (2004). With permission.)

critical $R_s = 1.43$). This separation method is supposed to be valuable for stereoselective metabolite studies by HPLC-MS/MS, because no other successful LC separation method has been reported previously for this compound. Further, the applicability range of cinchonan carbamate CSPs covers also permethrinic acid (3-(2,2-dichlorovinyl)-2,2-dimethylcyclopropanecarboxylic acid) and fenvaleric acid (2-(4-chlorophenyl)-3-methylbutanoic acid), which are both precursors of important pyrethroid insecticides (permethrinic acid of permethrin, cyfluthrin, and cypermethrin and fenvaleric acid of fenvalerate) as well as their primary metabolites and thus suitable biomarkers of exposure. All four stereoisomers of permethrinic acid were successfully separated on O-9-($tert$-butylcarbamoyl)quinine-based CSP in a single run ($\alpha_{cis} = 1.20$, $\alpha_{trans} = 1.26$, critical $R_s = 1.65$) with a polar-organic eluent (0.1% acetic acid in acetonitrile—methanol 95:5, v/v) and the enantiomers of fenvaleric acid could be baseline separated on O-9-(2,6-diisopropylphenylcarbamoyl)quinine-based CSP employing the eluent specified earlier for chrysanthemum dicarboxylic acid ($\alpha = 1.16$, $R_s = 2.91$) (Figure 1.32b). Since these pyrethroid pesticides are gaining more and more importance due to their low mammalian toxicity and low bioaccumulation potential, also corresponding assays for stereoselective bioanalytical investigations are badly needed. The excellent compatibility of the separation methods described earlier with MS-detection and the aqueous-based biological samples is expected to be one of the *pros* of the current cinchonan carbamate CSPs in this field of application.

The broad and nearly universal applicability of the cinchonan carbamate CSPs for chiral acid separations is further corroborated by successful enantiomer separations of acidic solutes having axial and planar chirality, respectively. For example, Tobler et al. [124] could separate the enantiomers of atropisomeric axially chiral 2'-dodecyloxy-6-nitrobiphenyl-2-carboxylic acid on an O-9-($tert$-butylcarbamoyl)quinine-based CSP in the PO mode with α-value of 1.8 and R_S of 9.1. This compound is stereolabile and hence at elevated temperatures the two enantiomers were interconverted during the separation process on-column revealing characteristic plateau regions between the separated enantiomer peaks. A stopped-flow method was utilized to determine the kinetic rate constants and apparent rotational energy barriers for the interconversion process in the presence of the CSP. Apparent activation energies (i.e., energy barriers for interconversion) were found to be 93.0 and 94.6 kJ mol^{-1} for the ($-$)- and ($+$)-enantiomers, respectively.

Further, [2.2]paracyclophane-4-acetic acid, a potential drug candidate tested for its anti-inflammatory activity (NSAID could be resolved on the O-9-($tert$-butylcarbamoyl)quinine-based CSP with $\alpha = 1.12$ ($R_S = 2.6$) and elution order (R)-($-$)- before (S)-($+$)-enantiomer [125]. Samples that were assessed to be enantiomerically pure by an enantioselective ^{1}H-NMR spectroscopic method contained 6% and 8% enantiomeric impurity in S- and R-enantiomers, respectively. This clearly reveals that enantioselective HPLC is a more powerful technique than NMR-methods to assess stereoisomeric purity, in particular, if the enantiomeric impurity amounts to less than 10%.

Other examples of successful enantiomer separations of miscellaneous chiral acids on cinchonan carbamate CSPs are collected in Table 1.11.

TABLE 1.11
Applicability Spectrum of Cinchona Alkaloid-Derived CSPs for Miscellaneous Chiral Acids

Solute	CSP	Eluent	Reference
6,6'-Dimethylbiphenyl-2,2'-dicarboxylic acid, 1,1'-binaphthyl-2,2'-diyl hydrogen phosphate, camphor-10-sulfonic acid, acenocoumarol, proglumide, N-succinyl-1-phenylethylamine, 3,4-dihydro-4-(2-naphthyl)-1,3,6-trimethyl pyrimidine-2(1H)-one-5-carboxylic acid	O-9-[3-(3-Silica-propylthio)propyl)-carbamoyl]10,11-dihydroquinidine	Methanol-0.1M ammonium acetate (80:20; v/v) (pH$_a$ = 6.0)	30
Ibuprofen, naproxen, etodolac 2-phenoxypropionic acid, dichlorprop, 2-(2,4-dinitrophenoxy)propionic acid, 2-(2,4-dinitrophenylthio)propionic acid, 3,4-dihydro-2H-pyran-2-carboxylic acid	O-9-[3-(3-Silica-propylthio)propyl)carbamoyl]-10,11-dihydroquinine quinidine, O-9-[3-(silica)-propyl)-carbamoyl]-quinine and quinidine	Methanol-0.1M ammonium acetate and (80:20; v/v) (pH$_a$ = 6.0)	30
2-Hydroxy-2-phenyl butyric acid, 3-(4-hydroxyphenyl)-lactic acid, Mosher acid, 4-(2,3-dichlorophenyl)-1,4-dihydro-2,6-dimethyl 3,5 pyridine dicarboxylic acid monomethyl ester (felodipine precursor)	TritCQN-CSP	Acetonitrile-0.1M ammonium acetate (65:35; v/v) (pH$_a$ = 6.0)	31
3-(α-Naphthoxy)lactic acid	DNPCQN-CSP		
Atrolactic acid, flurbiprofen, naproxen	DIPPCQN-CSP		
2,3,4,5-Tetrahydro-2-furoic acid	DIPPCQN-CSP		
3-Phenyllactic acid	tBuCQN-CSP	Acetonitrile containing 40 mM acetic acid and 10 mM triethylamine	31
2-(tert-Butylsulfonylmethyl)-3-phenyl propionic acid	DNPCQD-CSP	Methanol-0.1M ammonium acetate (80:20; v/v) (pH$_a$ = 6.0)	31

Continued

TABLE 1.11
(Continued)

Solute	CSP	Eluent	Reference
Heptelidic acid	TritCQN-CSP	Methanol-0.1M ammonium acetate (80:20; v/v) (pH$_a$ = 6.0)	31
Camphor-10-sulfonic acid	tBuCQD-CSP	Methanol-0.1M ammonium acetate (80:20; v/v) (pH$_a$ = 6.0)	31
Omeprazole	DNPCQD-CSP	Acetonitrile-0.1M ammonium acetate (65:35; v/v) (pH$_a$ = 6.0)	31
2-Methoxy-2-(1-naphthyl)propionic acid, 2-methoxy-2-(1-naphthyl)acetic acid	tBuCQN-CSP, tBuCQD-CSP and various other cinchonan carbamate CSPs	Methanol-acetic acid (96:4; v/v) and various other conditions	41
trans-2-Phenyl-cyclopropane-1-carboxylic acid	tBuCQN-CSP	Methanol-0.2 M ammonium acetate (80:20; v/v) (pH$_a$ = 6.0)	126
3-Oxo-indane-1-carboxylic acid	tBuCQD-CSP	Acetonitrile-acetic acid (97:3; v/v)	126
1,1'-Binaphthyl-2,2'-diyl hydrogenphosphate	tBuCQN-CSP	Methanol-acetic acid-ammonium acetate (98:2:0.5; v/v/w)	126
3-(4-Fluorophenylsulfonyl)-2-hydroxy-2-methyl-propanoic acid (acid side product of bicalutamide)	tBuCQN-CSP	Methanol-0.2 M ammonium acetate buffer (pH 6) (90:10, v/v)	127

1.9   IMPLEMENTATION OF CINCHONA ALKALOID SELECTORS IN OTHER SEPARATION METHODOLOGIES

1.9.1   SUPERCRITICAL FLUID CHROMATOGRAPHY

A variety of modern instrumental analytical techniques have attracted considerable attention in the last decades as alternative separation and analysis methods with respect to HPLC. This includes, in particular, supercritical fluid chromatography (SFC), which utilizes condensed carbon dioxide (above or near its critical temperature of

31.1°C and pressure of 73.8 bar) as fluid percolating the column. SFC has lately been and is currently strongly promoted by the chemical industry that is involved in chiral separations. It is because of some general advantages such as faster separations owing to the lower viscosities of the supercritical fluids that allow rapid separations, enhanced diffusivities of solutes in such media that may lead to improved mass transfer, simple work-up of collected fractions by depressurization in preparative work, and its better overall assessment in terms of organic waste reduction, safety of solvents, and energy efficiency. Typically, cellulose and amylose carbamate CSPs, Pirkle-type phases (Whelk O1) as well as tartramide CSPs were frequently proposed and tested as stationary phases and enantioselective columns, respectively, for SFC enantiomer separations. In preliminary studies, it could also be shown that quinine and quinidine carbamate CSPs can be utilized in this chromatographic mode [34]. Methanol and citric acid were employed as polar modifier and additive, respectively, in the carbon dioxide based eluent, which enabled adjustment of retention times and selectivity. While the principal feasibility could be proven, the full potential of the cinchona alkaloid-derived CSPs in this chromatographic mode needs yet to be more deeply explored.

1.9.2 CAPILLARY ELECTROPHORESIS

With capillary electrophoresis (CE), another modern primarily analytically oriented separation methodology has recently found its way into routine and research laboratories of the pharmaceutical industries. As the most beneficial characteristics over HPLC separations the extremely high efficiency leading to enhanced peak capacities and often better detectability of minor impurities, complementary selectivity profiles to HPLC due to a different separation mechanism as well as the capability to perform separations faster than by HPLC are frequently encountered as the most prominent advantages. On the negative side, there have to be mentioned detection sensitivity limitations due to the short path length of on-capillary UV detection, less robust methods, and occasionally problems with run-to-run repeatability. Nevertheless, CE assays have now been adopted by industrial labs as well and this holds in particular for enantiomer separations of chiral pharmaceuticals. While native cyclodextrins and their derivatives, respectively, are commonly employed as chiral additives to the BGEs to create mobility differences for the distinct enantiomers in the electric field, it could be demonstrated that cinchona alkaloids [128–130] and in particular their derivatives are applicable selectors for CE enantiomer separation of chiral acids [19,66,119,131–136].

The separation mechanism is based on stereoselective ion-pair formation of oppositely charged cationic selector and anionic solutes, which leads to a difference of net migration velocities of the both enantiomers in the electric field. Thus, the basic cinchona alkaloid derivative is added as chiral counterion to the BGE. Under the chosen acidic conditions of the BGE, the positively charged counterion associates with the acidic chiral analytes usually with 1:1 stoichiometry to form electrically neutral ion-pairs, which do not show self-electrophoretic mobility but migrate with the same velocity as the EOF that is in fused-silica capillaries directed to the cathode. The result is a countercurrent-like migration of free (anodic direction) and complexed solute species (cathodic migration of ion-pairs with EOF), which is favorable in terms of separation, with the stronger bound enantiomer migrating slower to the detector (located at the anodic side) than the weaker bound enantiomer.

Ion-pair CE with cinchona alkaloid type chiral counterions has preferentially been carried out in nonaqueous media (methanol or methanol–ethanol mixtures containing organic acids such as acetic acid or octanoic acid and bases such as ammonia and triethylamine as electrolytes) due to the better solubility of the selectors and especially of the ion-pairs in PO solvents and because of strengthened ion-pair formation in such media. To circumvent the loss of detection sensitivity that might arise in the conventional experimental setup, in which the selector is present in both electrolyte vessels, due to the strong chromophoric quinoline moiety of the cinchonan derivatives, a countercurrent methodology [137–143] and partial-filling technique (PFT) [119,131], respectively, have been normally utilized. In the former, the capillary is first equilibrated with the selector-BGE solution and the CE run is carried out with the electrolyte vessel at the detection end being void of the selector. Thus, upon application of the electric field, the selector zone migrates (with EOF and self-electrophoretic migration) to the injector side clearing the detector before the analyte enantiomers are detected. Separation of the enantiomers takes place, by the mechanism briefly outlined earlier, as long as the solute enantiomers migrate in the selector zone while the enantiomers migrate with same velocity when they have left the selector zone. This means that the solute enantiomers migrate with the same velocity through the detection window. In the PFT-experiment, the capillary is only partially filled with the selector-BGE solution to a certain point before the detection window and the electrophoretic run is carried out with plain BGE, that is, both electrolyte vessels at the inlet and outlet end are devoid of chiral selector. Because the selector zone migrates toward the inlet end, the detector gets never contaminated with UV-absorbing selector. This guarantees a satisfactory detection sensitivity. The advantage of the latter technique is that much less selector solution is consumed (nL volumes per run only). More details on the theory and practical aspects of ion-pair CE with cinchona alkaloid-derived selectors can be derived from recent reviews [19,144].

The PFT technique has been employed, for instance, for the analysis of samples obtained by aminolysis of fosfomycin from fosfomycin biosynthesis studies [119]. For these investigations, all eight isomers that might be formed by aminolysis of fosfomycin needed to be separated (Figure 1.33) to allow to pinpoint unequivocally which stereoisomeric forms were present in the samples. It can be seen that NACE employing O-9-(tert-butylcarbamoyl)quinine and quinidine as chiral counterions allowed the simultaneous separation of all eight components after derivatization with Sanger's reagent to yield the strongly chromophoric N-2,4-dinitrophenyl derivatives using the partial-filling mode (Figure 1.33). This example clearly demonstrates the separation power of CE in combination with the use of cinchonan carbamate selectors.

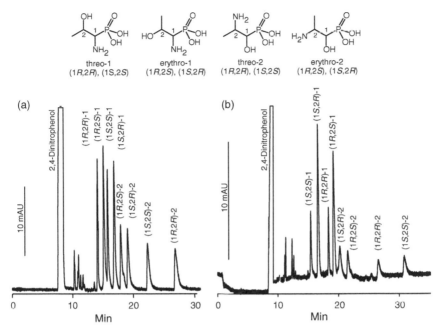

FIGURE 1.33 Separation of the stereoisomers of 1-amino-2-hydroxypropane phosphonic acid **1** and 2-amino-1-hydroxypropane phosphonic acid **2** after derivatization with Sanger's reagent as *N*-2,4-dinitrophenyl derivatives by nonaqueous CE with *O*-9-(*tert*-butylcarbamoyl)quinine (a) and *O*-9-(*tert*-butylcarbamoyl) quinidine (b) as counterions illustrating the reversal of elution orders of enantiomers that can be obtained with the pseudoenantiomeric counterions. Experimental conditions: Fused-silica capillary, 50 μm i.d., 45.5 cm total length, 37 cm to detection window; background electrolyte, 100 mM acetic acid and 12.5 mM triethylamine in ethanol-methanol (60:40, v/v); selector solution, 10 mM counterion in background electrolyte; partial-filling technique, filling of the selector solution with 50 mbar for 5 min (corresponds to ca. 30 cm selector plug length); injection, 50 mbar for 5 s; applied voltage, −25 kV (plain background electrolyte at both inlet and outlet electrode vessels); temperature, 15°C. (Reprinted (with modifications) from M. Lämmerhofer et al., *Electrophoresis*, 22: 1182 (2001).)

1.9.3 CAPILLARY ELECTROCHROMATOGRAPHY

Another emerging high-performance electrokinetic separation technique, namely, CEC, was also evaluated in combination with cinchona alkaloid–derived separation materials for its potential in stereoselective analysis concepts [145,146]. CEC is performed with capillary columns of 75–100 μm ID that contain the CSP (typically 3.5 μm modified silica beads, which are slurry packed into the capillaries and maintained in the capillary by end-frits that are fabricated by sintering of the stationary phase). The peculiarity is that instead of a pressure gradient an electric field is applied across the capillary column. Thus, the solutes are driven through the column by the EOF (ν_{eo}) that is generated on the surface of the charged packing material and, in case of ionized solutes, their electrophoretic migration (ν_{ep}) as well. Owing to their interaction with the CSP the solutes experience a retardation (k_{LC}) so that a truly hybrid separation technique between CE and HPLC will result in which the observed migration velocity (ν_{CEC}) is made up by the individual separation contributions as can be described as follows:

$$\nu_{CEC} = (\nu_{eo} + \nu_{ep}) \cdot \left(\frac{1}{1 + k_{LC}} \right). \tag{1.19}$$

It is evident that the chromatographic term is the only source for enantioselectivity because the retention factors may differ for the distinct enantiomers, while electrophoretic mobilities are identical for enantiomeric species. In other words, electrophoretic mobilities, like ν_{eo}, are nonselective contributions in view of generating chiral separations, but may positively contribute to the selectivity between distinct compounds (such as, for example, chemical impurities) but also of diastereomeric species.

As compared to common HPLC, the aspired advantage of CEC arises mainly from enhanced kinetic properties in the column. The EOF profile is sharper (plug-like) as compared to parabolic flow profiles of HPLC. The absence of pressure-drop limitations enables the use of small particle diameters even at column length of, for example, 25 cm. This provides higher plate counts due to smaller A-term (flow maldistribution) and C-term (resistance to mass transfer) contributions to band broadening. Last but not least, intraparticulate mass transfer by convection in the pores due to the EOF that originates on the stationary phase surface may, if the mesopore diameters are wide enough and double layer overlap is avoided, further enhance the mass transfer leading to less band spreading. An increase of the chromatographic efficiencies; that is, the plate numbers, by a factor of 2–10 as compared to HPLC is common in CEC. Compared to CE, the primary gain would be a higher sample loading capacity that is expected to facilitate the detection of low enantiomeric impurities and the preclusion of the selector additive from the electrolyte vessels, which would avoid contamination of the detection cell (UV detection) or ion source (ESI MS), both usually deteriorating detection sensitivity.

The EOF of most bonded phases, and thus most CSPs likewise, is based on the dissociation of residual silanols, which may lead to a weak and unstable cathodic EOF. In contrast, the cinchona alkaloid phases turned out to be particularly suitable for electro-drive in CEC enantiomer separation due to the presence of the ionic selector on the surface of the CSP. The concentration of these ionic sites determines the ζ-potential and thus the EOF of the CSP, which stand in direct proportional relationship. For example, silica beads modified with the O-9-($tert$-butylcarbamoyl)quinine selector displayed ζ-potentials of -20 to -30 mV in the pH range of 8.5–9.5 with a 10 mM KCl electrolyte solution and thus a cathodic EOF at such high pH values because of the negative net charge of the surface as a result of the excess of dissociated silanols over dissociated quinine selectors [147]. At pH values around 7–8 (depending on the selector loading and surface coverage) the isoelectric point is reached for these modified silica particles. In this pH region, the CEC system is instable and CEC runs are poorly reproducible. In the pH between 6.5 and 3.5, a ζ-potential between $+20$ and $+30$ mV was obtained with a 10 mM KCl electrolyte solution due to excess of positive charges on the surface arising mainly from the protonated quinuclidine ring of the selector. An anodic EOF is the result. It allows a favorable, co-directional separation in which both EOF and electrophoretic migration of the acidic solutes (e.g., N-derivatized amino acids) are directed to the anode that is located at the detector side in such a separation system (negative polarity mode): (being faster than a counterdirectional separation with cathodic EOF and electrophoretic migration towards anode).

Various column technologies have been adapted: (i) packed capillaries [35,36,124,148,149], (ii) organic polymer monolith columns [80–85] and (iii) silica monolith capillaries [150]. In case of the packed capillary column technology, 3.5-μm silica particles (100 Å) modified with the O-9-(tert-butylcarbamoyl)quinine selector have been packed into 100-μm ID-fused silica capillary tubes to a length of 25 cm. The frits were directly sintered on the CSP, which required harsh thermal conditions (at about 550°C for ca. 30 s [36]). This may have, to some extent, negatively impacted the efficiencies that were achievable with such columns, yet secured a satisfactory column stability. Under nonaqueous eluent conditions using acetonitrile–methanol mixtures (60–80% acetonitrile) with 100–400 mM acetic acid and 1–4 mM triethylamine, appealing CEC enantiomer separations for a variety of N-derivatized amino acids and some other chiral acids could be achieved with plate numbers between 70,000 and 120,000 m^{-1}, about a factor of 2–4 higher efficiency than in HPLC. Since 25-cm long columns could be used due to absence of backpressure limitations in CEC, this amounted to about 20,000–25,000 plates/column, which is more efficient by a factor of about 5 than typically observed in corresponding HPLC experiments with standard 15-cm columns and 5-μm particles. The practical applicability to solve real life problems could be demonstrated by a comparative study of HPLC and CEC, which dealt with the determination of the interconversion barriers of stereolabile atropisomeric 2'-dodecyloxy-6-nitrobiphenyl-2-carboxylic acid mentioned earlier [124]. The major shortcoming of the packed column technology, however, turned out to be the limited column longevity because the frits easily broke or caused a poor peak performance. This stimulated the development of monolithic columns, which have certainly advantages in this respect.

Monolithic columns with the chiral anion exchange–type selectors incorporated into the polymer matrix obtained through *in situ* copolymerization process of a chiral monomer (*in situ* approach) [80–83,85] or attached to the surface of a reactive monolith in a subsequent derivatization step (postmodification strategy) [84], both turned out to be viable routes to enantioselective macroporous monolithic columns devoid of the limitations of packed columns mentioned earlier.

In one approach, polymethacrylate-type monoliths have been fabricated by copolymerization of the chiral monomer O-9-[2-(methacryloyloxy)ethylcarbamoyl]-10,11-dihydroquinidine **1** or O-9-(tert-butylcarbamoyl)-11-[2-(methacryloyloxy)ethylthio]-10,11-dihydroquinine **2** (see Figure 1.34a), the comonomer 2-hydroxyethylmethacrylate (HEMA), the crosslinker ethylenedimethacrylate (EDMA) in presence of the binary porogenic solvent mixture cyclohexanol and 1-dodecanol, directly in a single step within fused-silica capillaries. Initiation of the polymerization by either thermal treatment or UV irradiation yielded microglobular polymer morphologies, such as those well known from their corresponding nonchiral

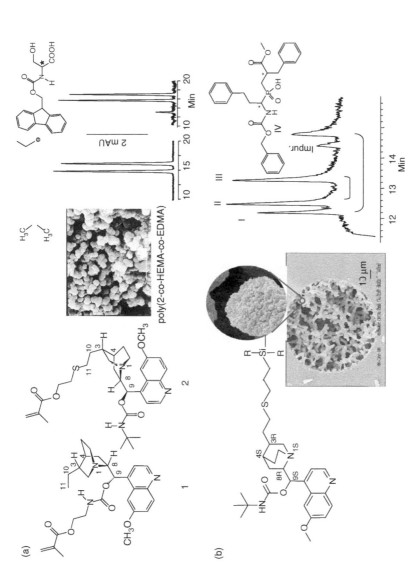

FIGURE 1.34 Monolithic capillary column technologies on the basis of cinchonan carbamate selectors and application examples. (a) Copolymerization of chiral monomer **1** or **2**, HEMA and EDMA in presence of porogenic solvents (cyclohexanol and 1-dodecanol) yields organic polymer monoliths with a macroporous microglobular polymer morphology. N-derivatized amino acids such as DBD-Leu or FMOC-Ser have been separated by CEC with high efficiencies (ca. 100,000 plates m^{-1}) [83]. (b) Silica monolith capillary column (Chromolith CapRod®, Merck, Darmstadt, Germany) derivatized with O-9-($tert$-butylcarbamoyl)quinidine and its CEC application to the stereo:somer separation of N-benzyloxycarbonyl phosphinic pseudo-dipeptide methyl ester Z-hPheψ(PO$_2$HCH$_2$)Phe-OCH$_3$ (note correspondent enantiomers are indicated by brackets) [150]. (Reproduced (with modifications) from M. Lämmerhofer et al., *Electrophoresis, 24:* 2986 (2003). and B. Preinerstorfer et al., *Electrophoresis, 27:* 4312 (2006). With permission.)

congeners, as illustrated in Figure 1.34a. Through systematic studies, the polymerization mixture could be optimized. It was found that the polymerization mixture should contain about 20% (w/w) (related to total monomers) of chiral monomer, 10–20% crosslinker (EDMA), a polar HEMA as comonomer, and a porogenic solvent mixture so as to adjust a pore size of ca. 1 μm.

These enantioselective capillary columns showed extremely good performance in the CEC mode. Plate numbers in excess of 100,000 m^{-1} could be easily achieved for a variety of amino acid derivatives (with chromophoric and fluorophoric labels) (Figure 1.34a) as well as other chiral acids such as 2-aryloxycarboxylic acids.

The postmodification strategy, in which a poly(glycydyl methacrylate-co-ethylene dimethacrylate) monolith was activated with hydrogen sulfide to a thiol-modified monolith and subsequently derivatized with an *O*-9-(*tert*-butylcarbamoyl)quinine selector by radical addition reaction, yielded slightly less efficient capillary columns. However, this procedure has the advantage that only minute amount of chiral selector are needed to end-up with a useful enantioselective capillary column [84].

Likewise promising is the approach that is based on silica monolith technology. Thus, silica monoliths were first prepared in a fused-silica capillary column using sol-gel technology (Chromolith CapRod®, Merck, Darmstadt, Germany) (Figure 1.34b). These capillary columns have an open pore structure and a high total porosity (90–95%). The thin silica skeletons (ca. 1–3 μm) are crosslinked by co-condensation with the fused-silica wall and provide, due to a mesoporous structure, a large active surface area for adsorption. This active surface was chemically modified with mercaptopropyl-silane and finally derivatized with a *O*-9-(*tert*-butylcarbamoyl)quinidine selector. The successful surface modification could be proven by the applicability of the resultant enantioselective capillary column, for example, for the stereoisomer separation of *N*-benzyloxycarbonyl phosphinic pseudo-dipeptide methyl ester Z-hPheψ (PO_2HCH_2)Phe-OCH$_3$ by CEC (Figure 1.34b). This phosphinic acid-ψ-dipeptide is a potent leucine aminopeptidase inhibitor. It contains two stereogenic centers and thus consists of four stereoisomers, which could all be separated from each other with extremely high chromatographic efficiencies (200,000–600,000 plates m^{-1} as calculated by USP methodology). In contrast, HPLC did not allow to simultaneously separate all four stereoisomers because either the peak efficiency was too low or the diastereoselectivity for the critical peak pair was insufficient. CEC clearly proved its benefits for this separation problem. Further, it is worthwhile to mention that the quinidine carbamate-modified monolithic silica capillary column could be favorably applied for nano-HPLC enantiomer separations as well. The particularly nice feature of these capillary columns is their excellent permeability. This enabled the use of a 50-cm long column that delivered an exceptional 45,000 plates per column. Such a long column could be especially helpful for enantiomer separations of complex mixtures where a high peak capacity is required.

1.9.4 LIQUID–SOLID BATCH EXTRACTION

In a series of papers, cinchona alkaloid selectors were in the focus of preparative enantiomer separation methodologies including crystallization [57], countercurrent chromatography (CCC) or CPC [37,38], liquid–solid extraction [64], liquid–liquid extraction (LLE) [151], and SLM processes [39].

In simple experiments, particulate silica-supported CSPs having various cinchonan carbamate selectors immobilized to the surface were employed in an enantioselective liquid–solid batch extraction process for the enantioselective enrichment of the weak binding enantiomer of amino acid derivatives in the liquid phase (methanol-0.1 M ammonium acetate buffer pH 6) and the stronger binding enantiomer in the solid phase [64]. For example, when a CSP with the O-9-(tert-butylcarbamoyl)-6′-neopentoxy-cinchonidine selector was employed at an about 10-fold molar excess as related to the DNB-Leu selectand which was dissolved as a racemate in the liquid phase specified earlier, an enantiomeric excess of 89% could be measured in the supernatant after a single extraction step (i.e., a single equilibration step). This corresponds to an enantioselectivity factor of 17.7 (α-value in HPLC amounted to 31.7). Such a batch extraction method could serve as enrichment technique in hybrid processes such as in combination with, for example, crystallization. In the presented study, it was however used for screening of the enantiomer separation power of a series of CSPs.

1.9.5 LIQUID–LIQUID EXTRACTION

Kellner et al. [151] utilized two-phase LLE in a conventional shake-flask setup to investigate the enantioselective transport of N-derivatized amino acids from the aqueous into the organic phase employing O-9-(1-adamantylcarbamoyl)-10,11-dihydro-11-octadecylsulfinylquinine as a chiral ion-pairing type carrier. For the solvent extraction experiments, the racemic mixture of the amino acid derivatives (e.g., DNB-Leu) was dissolved in the aqueous donor phase (feed phase) that had preferentially a pH between 5 and 6 (compromise for high enantiomeric excess in extract and good yield of extracted S-enantiomer). Dodecane or cyclohexane was a favorable solvent for the organic acceptor phase (extract phase) in which the lipophilic chiral carrier that is insoluble in the feed phase was readily soluble. In the course of the extraction step, the enantiomers of the amino acid derivatives formed lipophilic ion-pairs at the liquid–liquid interface, which owing to their diastereomeric nature with respect to each other did possess differential solubilities in the organic extract phase and/or aqueous feed phase, respectively. This ultimately has led to enantioselective transport. From the organic extract phase, the amino acid derivative could be recovered in a single back-extraction step, preferentially at strongly acidic conditions (pH 1 with phosphoric acid). Under optimized conditions, (S)-DNB-Leu could be obtained from the racemate with an enantiomeric excess exceeding 95% and an overall yield of 70% in a single extraction and back-extraction step with the carrier specified earlier.

1.9.6 SUPPORTED LIQUID MEMBRANE PROCESS

The above LLE experiments actually served as preliminary studies for a supported liquid membrane (SLM) process that have later been described by Maximini et al. [39]. The basic principle of the SLM process is based on LLE yet it has integrated a reextraction step as outlined in Figure 1.35a. Aqueous feed solution and permeate (stripping) solution are separated by a liquid membrane phase that is supported by a microporous hydrophobic membrane (polysulfone with a molecular cutoff of 30 kDa) and solely serves as a septum in this case. The organic liquid wets the pores of the supporting membrane. The membrane was impregnated with the organic solution of the chiral carrier, either O-9-(1-adamantylcarbamoyl)-10,11-dihydro-11-octadecylsulfinylquinine or the corresponding quinidine derivative. At the pore opening interface between the donor solution, which contained the racemate of the selectand (DNB-Leu) and the membrane, the selectand enantiomers undergo stereoselectively ion-pair formation. The SA species being present as ion-pair may, owing to its enhanced lipophilicity, diffuse through the pore to the permeate side where it is transported into the stripping solution because of favorable conditions such as a pH gradient (e.g., pH 6 at the donor and permeate phase). Besides this specific (stereoselective) carrier-mediated transport a minor amount of selectand may also, depending on the conditions, be transported through the membrane by a (nonstereoselective) unspecific diffusional transport process. Conditions must be selected such that this nonstereoselective process is minimized as much as possible.

A prototype of a continuously operating SLM unit (Figure 1.35b) consisted of two coupled hollow fiber membrane modules wherein each membrane module contained 250 polysulfone hollow fibres with a total membrane surface of 0.1 m^2, a molecular cutoff of 30 kDa, an inner diameter of 500 μm and a thickness of 80 μm. Peristaltic pumps maintained a constant inflow and outflow of feed, bleed, and permeate solutions and was controlled via a series of flow and pressure control elements (i.e., via flow controllers and manometers) (see Figure 1.35b). Also, the temperature in the plant could be thermostated and 25°C was selected as a good compromise between faster transport at higher temperatures and higher enantioselectivities at lower temperatures. In the characteristic setup, one hollow fiber module was filled with a 50 mmol L^{-1} solution of the quinine-derived carrier in 1-decanol/pentadecane (30:70; v/v) while the other module contained the corresponding quinidine solution. In order to avoid crystallization/precipitation in the membrane, a 5 vol.% solution of the corresponding carrier covalently immobilized on mercaptopropyl-polysiloxane. The feed solution had dissolved the racemate of DNB-Leu in 50 mmol L^{-1} KH$_2$PO$_4$/K$_2$HPO$_4$ pH 6.5 at a concentration of 50 mmol L^{-1}. The permeate phase was kept at a pH greater than 9. With this setup, DNB-Leu enantiomers could be separated from the racemate in a continuous process at a transmembrane flux J of 29 mmol h^{-1} m^{-2}). The permeate of the quinidine module was enriched with R-enantiomer (80% (R)-DNB-Leu) while the quinine module produced a permeate rich in S-enantiomer (80% (S)-DNB-Leu) corresponding to ee-values of 60% and a separation factor of 4.

In order to end-up with higher ee-values (e.g., 98%) of the products, a multistage SLM process was devised as well and it was calculated that five stages were required to achieve this goal [39]. The general setup is outlined in Figure 1.35c. Stage 1 operates as described earlier and the permeate of the quinidine (QD) module (80% D, that is, R-enantiomer) was after pH adjustment transferred into the next stage. While stage 2 was still performed with two membrane modules (one quinidine- and one quinine-based membrane), stages 3–5 were carried out with only one module each

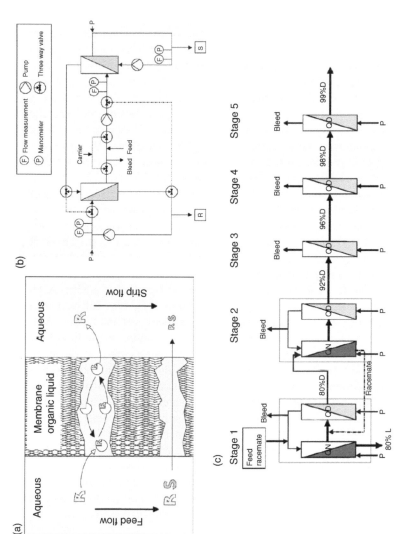

FIGURE 1.35 SLM process using *O*-9-(1-adamantyl-carbamoyl)-10,11-dihydro-11-octadecylsulfinylquinine and corresponding quinidine derivative as chiral carriers for the preparative separation of enantiomers of *N*-derivatized amino acids (e.g., DNB-Leu). (a) Principle of the carrier SLM process with carrier-mediated transport (top) and (nonstereoselective) nonspecific transport processes (bottom). (b) General experimental setup of the SLM production unit with two membrane modules. (c) Multistage SLM purification process. P, permeate; QD/QN, membrane modules supported with quinidine-derived and quinine-derived chiral carriers. R, S, D, L refers to the respective enantiomers of the selectand (DNB-Leu). (Reproduced from A. Maximini et al., *J. Membr. Sci., 276*: 221 (2006). With permission.)

(always quinidine-derived membrane). In the higher stages, the quantity of permeate added was reduced from 300 to 150 mL h^{-1} to obtain higher concentrations in the permeate. By maximizing the yield, as tradeoff, the separation factors afforded by the higher SLM stages were clearly lower (e.g., 4 and 3.4 in stages 1 and 2, respectively, and only about 2 in stages 3–5) (see Table 1.12). Table 1.12 also shows the transmembrane mass fluxes that were obtained in the individual stages as well as corresponding ee-values and yields. The overall yield was 40% of highly pure enantiomer (99% D-enantiomer, that is, 98% ee) (*Note*: Same number of SLM stages would be required on the quinine (QN) module side to end-up with a similar enantiomeric purity of the L-enantiomer (*S*-enantiomer); for the sake of simplicity, only the ultrapurification of the D-enantiomer was experimentally tested.)

The productivity of the plant was assessed to amount to about 5 kg DNB-D-Leu and DNB-L-Leu per mol of carrier with a purity of 99% (i.e., 98% ee). If a similar amount of selector is immobilized on silica gel and the CSP operated in a simulated moving bed (SMB) process, it was assumed to achieve roughly productivities between 1 and 7 kg pure enantiomer per day. Hence, it was concluded that a SLM process could be quite competitive to a SMB process in the production of pure enantiomers.

1.9.7 CENTRIFUGAL PARTITION CHROMATOGRAPHY

The same carriers [37] and also [mono-11-octadecylthio-*bis*-10,11-dihydroquinidinyl)]-1,4-phthalazine [38] have been tested in countercurrent chromatography (CCC) adopting the CPC mode for the preparative enantiomer separation of *N*-derivatized amino acids and aryloxycarboxylic acids such as dichlorprop. CPC actually represents a truly liquid–liquid chromatographic separation method in which the solutes are distributed between two immiscible liquid phases by multiple extraction equilibria. To make the system enantioselective, a chiral selector, that is, the carriers mentioned earlier, were introduced into the liquid stationary phase. In the studies mentioned earlier, the chiral lipophilic selectors were dissolved in the less dense solvent systems that served as stationary phases and hence the more dense mobile phase was pumped through the rotor (CPC column), which contained the stationary phase, in the descending mode (see Figure 1.36). Due to its insolubility in the aqueous phase, the selector leaching into the mobile phase was minimized. The solute, on the contrary, should be distributed between both phases and thus a distribution coefficient of about 1 would be optimal. Favorable solvents for the selectors and as stationary phases, respectively, could be figured out in a series of experiments and it was found that 1-butanol/heptane mixtures or binary or ternary mixtures of isobutyl methyl ketone, diisopropylether, or methyl *tert*-butylether in combination with heptane give favorable results. During operation, the column was rotated at about 1000–1500 rpm superimposing a centrifugal force to the column channels. After initial displacement of a certain amount of stationary phase, which is equal to the void volume, the system reaches an equilibrium state in which there is no leaching of stationary phase anymore. Then a certain amount of racemic sample is injected and the separation of the enantiomers takes place by partition between the two liquid phases. Since diastereomeric ion-pairs are formed between the acidic selectands and the cinchonan-derived selectors, which differ in their solubilities in the distinct phases, enantiomer separation is resulting with

TABLE 1.12
Ultrapurification of 50 mmol L^{-1} DNB-D,L-Leucine in a Cascade of Five Stages with Two Modules and Two Enantiomeric Carrier (Quinine and Quinidine Derivative with 5 vol% Polysiloxane-supported Carrier): Transmembrane Material Stream J, Enantioselectivity α, Enantiomer Excess ee, Purity, and Yield of DNB-D-Leucine

Number of Stage	Addition of Permeate (mL h^{-1})	J (mmol h^{-1} m^{-2})	α$_{number\ of\ stage}$	ee (%)	Purity of DNB-D-Leucine	Yield of DNB-D-Leucine (% mmol)
Racemate					50	
First stage	300	29	4	60	80	72
Second stage	150	26	3.4	83	92	59
Third stage	150	23	2.1	92	96.1	49
Fourth stage	150	23	1.9	96	97.9	44
Fifth stage	150	22	2	98	99	40

Reproduced A. Maximini et al., *J. Membr. Sci.*, 276: 221 (2006). With permission.

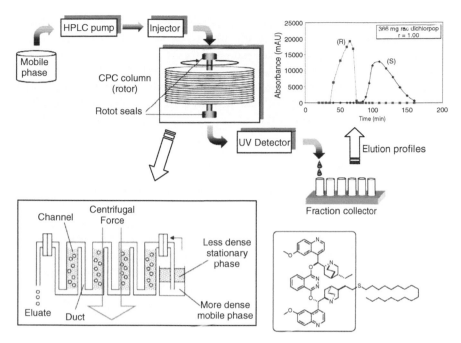

FIGURE 1.36 General scheme for the process of CPC and mobile phase flow regime for the descending mode of CPC (insert bottom left) used for the CPC separation of dichlorprop with [mono-11-octadecylthio-*bis*-10,11-dihydroquinidinyl)]-1,4-phthalazine as chiral selector. Elution profiles for dichlorprop after injection of 366 mg racemate, a molar ratio $r = 1$ of loaded dichlorprop to total selector present in the rotor, and a rotor speed of 1100 rpm. Stationary phase, 10 mM selector in methyl *tert*-butylether; mobile phase, 100 mM sodium phosphate buffer (pH 8); flow rate 3 mL min^{-1}; temperature, 25°C. (Reproduced from E. Gavioli et al., *Anal. Chem.*, 76: 5837 (2004). With permission.)

the enantiomer forming the more stable diastereomeric associate coming out of the column later. The column outlet of the rotor was connected to a UV detector so that a chromatogram can be monitored. Fractions were collected and the enantiomeric composition in these fractions analyzed by enantioselective HPLC to reconstitute the elution profiles such as depicted in Figure 1.36 for dichlorprop. Thus, about 370 mg of dichlorprop—2-(2,4-dichlorophenoxy)propionic acid—could be completely separated into enantiomers which were of about the same molar amount as the chiral carrier [mono-11-octadecylthio-*bis*-(10,11-dihydroquinidinyl)]-1,4-phthalazine) in the rotor.

For the given systems (ionizable selectors and solutes), a modified form of CPC was usually more favorable: The so-called *pH-zone-refining CPC mode*, which is a kind of displacement type of chromatography. In this mode, the column is filled with the acidified stationary phase (e.g., using TFA as retainer), then injection of the sample takes place before the rotor is switched on and elution is started with a basic mobile phase (e.g., using ammonia as displacer in the aqueous mobile phase). Apparent pH and enantiomeric composition were determined for every fraction. It appeared that the enantiomers eluted in refined

sharpened zones so that the productivity could be improved. For example, using the
O-9-(1-adamantylcarbamoyl)-10, 11-dihydro-11-octadecylsulfinylquinine   selector
(10 mM) in isobutylmethylketone containing 10 mM TFA and 20 mM ammonia as
mobile phase at a flow rate of 3 mL min^{-1} (at a rotor speed of 1200 rpm), the loading
capacity and productivity (while maintaining the selectivity) could be improved for
DNB-Leu from about 300 mg with classical CPC mode to about 600 mg with the
pH-zone refining experimental setup.

1.10 CONCLUSIONS

Cinchona alkaloids and derivatives thereof have demonstrated an amazing potential and utility as chiral auxiliaries in diverse stereoselective methodologies. The
present review article was supposed to give a more or less comprehensive, at least
representative, overview of the achievements that have been accomplished with cinchona alkaloid derivatives as chiral selectors in various separation methodologies,
in particular, focusing on liquid chromatographic enantiomer separations but briefly
summarizing also other chromatographic and nonchromatographic separation techniques. Specific attention was also paid to the various investigations of molecular
and chiral recognition mechanisms of these selector systems, especially for chiral
acids such as N-derivatized amino acids. Taking all the presented various mechanistic studies together, which complemented each other in information on molecular
recognition modes and stereodifferentiation forces, it can be concluded that these
cinchonan-derived chiral selector structures certainly belong to the best-investigated
chiral recognition systems. Besides, the practical usefulness of commercial (Chiralpak QN-AX and Chiralpak QD-AX columns) and laboratory-made CSPs based
on quinine and quinidine carbamates and phthalazine derivatives could also be well
documented by a number of applications. More will be shown in the future in terms
of applicability of these cinchonan carbamate-type CSPs and it is safe to state that
they will be especially helpful for enantiomer separations of all kind of chiral acids,
although the application spectrum is certainly not restricted to this class of solutes as
could be convincingly shown. So far, relatively little has been published about the
preparative LC and SFC enantiomer separation power of these chiral anion-exchange
type CSPs, especially for chiral acids. Preliminary investigations, however, have
shown that these CSPs may exhibit exceptionally high sample loading capacities
for chiral acids in the anion-exchange separation mode [152], a property that seems
to be to some extent intrinsic to ion-exchange processes. More work needs to be
directed toward the preparative capabilities of these CSPs in batch LC and continuous SMB processes in order to be able to assess their full enantiomer separation
power. Yet, overall, it can be concluded that the cinchona alkaloid-derived CSPs fill
a niche in the spectrum of the broad range of CSPs that are now (commercially)
available.

ACKNOWLEDGMENTS

Over more than 10 years, there have been a rather large number of people involved in
the development and testing of the various cinchona alkaloid derivatives and CSPs,

whom we would like to gratefully acknowledge at this place. It applies to coworkers in our group and to external partners who are too many to be mentioned individually. Some of them appear as coauthors on the publications but equally valuable work of others has never been published yet, although their impact has been important for the success of the project. We would like to specifically thank also the companies Bischoff Chromatography and Chiral Technologies Europe for their engagement in commercializing some of our developments to make them available to many users worldwide.

ALPHABETICAL LIST OF ABBREVIATIONS

Ac:	acetyl
ACN:	acetonitrile
AcOH:	acetic acid
ATR FT-IR:	Attenuated total reflectance infrared spectroscopy
AX:	anion-exchange(r)
BGE:	background electrolyte
Boc:	*tert*-butoxycarbonyl
Bz:	benzoyl
CC:	carbazole-9-carbonyl
CCC:	countercurrent chromatography
CD.	cinchonidine
CDM:	circular dichroism spectroscopy
CE:	capillary electrophoresis
CEC:	capillary electrochromatography
CIS:	complexation-induced chemical shift
Clac:	chloroacetyl
CN:	cinchonine
CoMFA:	comparative molecular field analysis
CPC:	centrifugal partition chromatography
3CPP:	3-chloro-1-phenylpropanol
CSP:	chiral stationary phase
DBD:	4-dimethylaminosulfonyl-2,1,3-benzoxadiazol-7-yl
DIPPCQD:	*O*-9-[(2,6-diisopropylphenyl)carbamoyl]quinidine
DIPPCQN:	*O*-9-[(2,6-diisopropylphenyl)carbamoyl]quinine
DNB:	3,5-dinitrobenzoyl
DNP:	2,4-dinitrophenyl
DNPCQD:	*O*-9-[(3,5-dinitrophenyl)carbamoyl]quinidine
DNPCQN:	*O*-9-[(3,5-dinitrophenyl)carbamoyl]quinine
DNPyr:	3,5-dinitropyridyl
DNS:	dansyl (5-dimethylaminonaphthalin-1-sulfonyl)
DNZ:	3,5-dinitrobenzyloxycarbonyl
3D-QSAR:	3-dimensional quantitative structure–activity relationship
EA:	ethyl acetate
ECD:	epicinchonidine
ECN:	epicinchonine
EDMA:	ethylene dimethacrylate

ee:	enantiomeric excess
e.o.:	elution order
EOF:	electroosmotic flow
EQD:	epiquinidine
EQN:	epiquinine
ESI:	electrospray ionization
FMOC:	9-fluorenylmethoxycarbonyl
For:	formyl
HEMA:	2-hydroxethyl methacrylate
HILIC:	hydrophilic interaction chromatography
HPLC:	high-performance liquid chromatography
HR/MAS:	high-resolution magic angle spinning
ID:	inner diameter
ITC:	isothermal calorimetry
LLE:	liquid–liquid extraction
LOD:	limit of detection
LOQ:	limit of quantitation
LSS:	linear solvent strength theory
MD:	molecular dynamics
MeOH:	methanol
MRM:	multiple reaction monitoring
MS:	mass spectrometry
1N:	1-naphthoyl
NACE:	nonaqueous capiallary electrophoresis
NBD:	4-nitro-2,1,3-benzoxadiazol-7-yl
4-NBz:	4-nitrobenzoyl
NH_4Ac:	ammonium acetate
NOE:	nuclear Overhauser effect
NOESY:	Nuclear Overhauser Effect Spectroscopy
NP:	normal-phase (mode)
Npg:	α-neopentyl-glycine
NVOC:	6-nitro-veratryl (2-nitro-4,5-dimethoxy-benzyloxycarbonyl)
ORD:	optical rotation detection
PFT:	partial-filling technique
pH_a:	apparent pH (pH measured with a glass electrode in a hydro-organic mixture or organic solvent)
PNB:	p-nitrobenzoyl
PNZ:	para-nitrobenzyloxycarbonyl
PO:	polar organic (mode)
1PP:	1-phenyl-propanol
2PP:	2-phenylpropanol
PSer:	O-phosphoserine
QD:	quinidine
QN:	quinine
RP:	reversed-phase (mode)
RSD:	relative standard deviation

SA:	selectand
SFC:	supercritical fluid chromatography
SLM:	supported liquid membrane
SMB:	simulated moving bed
SO:	selector
T_3:	triiodothyronine
T_4:	thyroxin
tBuCCD:	O-9-($tert$-butylcarbamoyl)cinchonidine
tBuCEQD:	O-9-($tert$-butylcarbamoyl)epiquinidine
tBuCEQN:	O-9-($tert$-butylcarbamoyl)epiquinine
tBuCQD:	O-9-($tert$-butylcarbamoyl)quinidine
tBuCQN:	O-9-($tert$-butylcarbamoyl)quinine
TFA:	trifluoroacetic acid
TFAE:	2,2,2-trifluoro-1-(9-anthryl)-ethanol
TritCQN:	O-9-[(triphenylmethyl)carbamoyl]quinine
trNOE:	transferred nuclear Overhauser effect
WAX:	weak anion-exchanger
Z:	benzyloxycarbonyl

REFERENCES

1. G. Stork, D. Niu, A. Fujimoto, E.R. Koft, J.M. Balkovec, J.R. Tata, and G.R. Dake, *J. Am. Chem. Soc., 123*: 3239 (2001).
2. I.T. Raheem, S.N. Goodman, and E.N. Jacobsen, *J. Am. Chem. Soc., 126*: 706 (2004).
3. P. Newman. *Optical Resolution Procedures for Chemical Compounds*, 1978 (vol. 1), 1981 (vol. 2), 1984 (vol. 3) New York: Optical Resolution Information Center, Manhattan College; 1978–84.
4. C. Rosini, G. Uccello-Barretta, D. Pini, C. Abete, and P. Salvadori, *J. Org. Chem., 53*: 4579 (1988).
5. G. Uccello-Barretta, F. Balzano, and P. Salvadori, *Chirality, 17*: S243 (2005).
6. G. Uccello-Barretta, F. Mirabella, F. Balzano, and P. Salvadori, *Tetrahedron Asymmetry, 14*: 1511 (2003).
7. A. Maly, B. Lejczak, and P. Kafarski, *Tetrahedron Asymmetry, 14*: 1019 (2003).
8. K. Kacprzak and J. Gawronski, *Synthesis*, 961 (2001).
9. T.P. Yoon and E.N. Jacobsen, *Science, 299*: 1691 (2003).
10. N. Grubhofer and L. Schleith, *Naturwiss, 40*: 508 (1953).
11. N. Grubhofer and L. Schleith, *Z. Phys. Chem., 296*: 262 (1954).
12. S. Izumoto, U. Sakaguchi, and H. Yoneda, *Bull. Chem. Soc. Jpn., 56*: 1646 (1983).
13. K. Miyoshi, M. Natsubori, N. Dohmoto, S. Izumoto, and H. Yoneda, *Bull. Chem. Soc. Jpn., 58*: 1529 (1985).
14. K. Miyoshi, N. Dohmoto, and H. Yoneda, *Inorg. Chem., 24*: 210 (1985).
15. C. Pettersson, *J. Chromatogr., 316*: 553 (1984).
16. C. Pettersson and G. Schill, *J. Liq. Chromatogr., 9*: 269 (1986).
17. C. Pettersson and C. Gioeli, *J. Chromatogr., 435*: 225 (1988).
18. A. Karlsson and C. Pettersson, *Chirality, 4*: 323 (1992).
19. M. Lämmerhofer and W. Lindner, in: G. Gübitz and M.G. Schmid (eds.), *Chiral Separations: Methods and Protocols (Book Series: Methods in Molecular Biology—Volume 243)*, Humana Press, Totowa, NJ, USA, pp. 323 (2004).

20. C. Rosini, C. Bertucci, D. Pini, P. Altemura, and P. Salvadori, *Tetrahedron Lett.*, *26*: 3361 (1985).
21. C. Rosini, C. Bertucci, D. Pini, P. Altemura, and P. Salvadori, *Chromatographia, 24*: 671 (1987).
22. P. Salvadori, C. Rosini, D. Pini, C. Bertucci, P. Altemura, G. Uccello-Barretta, and A. Raffaelli, *Tetrahedron, 43*: 4969 (1987).
23. C. Bertucci, C. Rosini, D. Pini, and P. Salvadori, *J. Pharm. Biomed. Anal., 5*: 171 (1987).
24. H.W. Stuurman, J. Köhler, and G. Schomburg, *Chromatographia, 25*: 265 (1988).
25. P. Salvadori, C. Rosini, D. Pini, C. Bertucci, and G. Uccello-Barretta, *Chirality, 1*: 161 (1989).
26. P. Salvadori, D. Pini, C. Rosini, C. Bertucci, and G. Ucello-Baretta, *Chirality, 4*: 43 (1992).
27. P.N. Nesterenko, V.V. Krotov, and S.M. Staroverov, *J. Chromatogr. A, 667*: 19 (1994).
28. C. Rosini, P. Altemura, D. Pini, C. Bertucci, G. Zullino, and P. Salvadori, *J. Chromatogr., 348*: 79 (1985).
29. C. Pettersson and C. Gioeli, *J. Chromatogr., 398*: 247 (1987).
30. M. Lämmerhofer and W. Lindner, *J. Chromatogr. A, 741*: 33 (1996).
31. M. Lämmerhofer, N.M. Maier, and W. Lindner, *American Laboratory, 30*: 71 (1998).
32. A. Mandl, L. Nicoletti, M. Lämmerhofer, and W. Lindner, *J. Chromatogr. A, 858*: 1 (1999).
33. N.M. Maier, L. Nicoletti, M. Lämmerhofer, and W. Lindner, *Chirality, 11*: 522 (1999).
34. S. Andersson, and O. Gyllenhaal, unpublished results.
35. M. Lämmerhofer and W. Lindner, *J. Chromatogr. A, 829*: 115 (1998).
36. E. Tobler, M. Lämmerhofer, and W. Lindner, *J. Chromatogr. A, 875*: 341 (2000).
37. P. Franco, J. Blanc, W.R. Oberleitner, N.M. Maier, W. Lindner, and C. Minguillon, *Anal. Chem., 74*: 4175 (2002).
38. E. Gavioli, N.M. Maier, C. Minguillon, and W. Lindner, *Anal. Chem., 76*: 5837 (2004).
39. A. Maximini, H. Chmiel, H. Holdik, and N.M. Maier, *J. Membr. Sci., 276*: 221 (2006).
40. M. Lämmerhofer, N.M. Maier, and W. Lindner, *Nachrichten aus der Chemie, 50*: 1037 (2002).
41. K. Gyimesi-Forras, K. Akasaka, M. Laemmerhofer, N.M. Maier, T. Fujita, M. Watanabe, N. Harada, and W. Lindner, *Chirality, 17*: S134 (2005).
42. K.H. Krawinkler, N.M. Maier, E. Sajovic, and W. Lindner, *J. Chromatogr. A, 1053*: 119 (2004).
43. J. Stahlberg, *J. Chromatogr. A, 855*: 3 (1999).
44. W. Kopaciewicz, M.A. Rounds, F. Fausnaugh, and F.E. Regnier, *J. Chromatogr., 266*: 3 (1983).
45. M.-C. Millot, T. Debranche, A. Pantazaki, I. Gherghi, B. Sebille, and C. Vidal-Madjar, *Chromatographia, 58*: 365 (2003).
46. B. Sellergren and K.J. Shea, *J. Chromatogr. A, 654*: 17 (1993).
47. X. Xiong, W.R.G. Bayens, H.Y. Aboul-Enein, J.R. Delanghe, T. Tu, and J. Ouyang, *Talanta*, available online June 21, 2006 (2006).
48. A. Peter, E. Vekes, A. Arki, D. Tourwe, and W. Lindner, *J. Sep. Sci., 26*: 1125 (2003).
49. M. Lämmerhofer, O. Gyllenhaal, and W. Lindner, *J. Pharm. Biomed. Anal., 35*: 259 (2004).
50. K. Gyimesi-Forras, J. Kökösi, G. Szasz, A. Gergely, and W. Lindner, *J. Chromatogr. A, 1047*: 59 (2004).
51. G. Götmar, L. Asnin, and G. Guiochon, *J. Chromatogr. A, 1059*: 43 (2004).

52. L. Asnin, G. Götmar, and G. Guiochon, *J. Chromatogr. A, 1091*: 183 (2005).
53. L. Asnin, and G. Guiochon, *J. Chromatogr. A, 1091*: 11 (2005).
54. L. Asnin, K. Kaczmarski, A. Felinger, F. Gritti, and G. Guiochon, *J. Chromatogr. A, 1101*: 158 (2006).
55. S.N. Lanin and Y.S. Nikitin, *Z. Anal. Khim., 46*: 1493 (1991).
56. K. Gyimesi-Forras, N.M. Maier, J. Kökösi, A. Gergely, and W. Lindner, *in preparation* (2006).
57. W. Lindner, M. Lämmerhofer, and N.M. Maier, PCT/EP97/02888; US 6,313,247 B1 (1997).
58. C. Czerwenka, M. Lämmerhofer, N.M. Maier, K. Rissanen, and W. Lindner, *Anal. Chem., 74*: 5658 (2002).
59. M. Lämmerhofer, P. Franco, and W. Lindner, *J. Sep. Sci., 29*: 1486 (2006).
60. M. Lämmerhofer, D. Hebenstreit, E. Gavioli, W. Lindner, A. Mucha, P. Kafarski, and P. Wieczorek, *Tetrahedron Asymmetry, 14*: 2557 (2003).
61. W. Bicker, M. Lämmerhofer, and W. Lindner, *J. Chromatogr. A, 1035*: 37 (2004).
62. W. Oberleitner, Ph.D. thesis, University of Vienna, Austria, 2000.
63. E. Tobler, M. Lämmerhofer, W. Oberleitner, N.M. Maier, and W. Lindner, *Chromatographia, 51*: 65 (2000).
64. P. Franco, M. Lämmerhofer, P.M. Klaus, and W. Lindner, *J. Chromatogr. A, 869*: 111 (2000).
65. C. Hellriegel, U. Skogsberg, K. Albert, M. Lämmerhofer, N.M. Maier, and W. Lindner, *J. Am. Chem. Soc., 126*: 3809 (2004).
66. P. Franco, P.M. Klaus, C. Minguillón, and W. Lindner, *Chirality, 13*: 177 (2001).
67. Z.-Y. Du and R.-T. Xiao, *Yingyong Huaxue, 22*: 1372 (2005).
68. P. Franco, M. Lämmerhofer, P.M. Klaus, and W. Lindner, *Chromatographia, 51*: 139 (2000).
69. N.M. Scully, G.P. O'Sullivan, L.O. Healy, J.D. Glennon, B. Dietrich, and K. Albert, *J. Chromatogr. A, 1156*: 13 July 2007, 68–74 (2007).
70. K.H. Krawinkler, E. Gavioli, N.M. Maier, and W. Lindner, *Chromatographia, 58*: 555 (2003).
71. Y.-K. Lee, K. Yamashita, M. Eto, K. Onimura, H. Tsutsumi, and T. Oishi, *Polymer, 43*: 7539 (2002).
72. K.M. Kacprzak, N.M. Maier, and W. Lindner, *Tetrahedron Lett., 47*: 8721 (2006).
73. M. Lämmerhofer and W. Lindner, *GIT Special—Chromatography International 96*, 16 (1996).
74. V. Piette, M. Lämmerhofer, K. Bischoff, and W. Lindner, *Chirality, 9*: 157 (1997).
75. D. Lubda and W. Lindner, *J. Chromatogr. A, 1036*: 135 (2004).
76. F.C. Leinweber, D. Lubda, K. Cabrera, and U. Tallarek, *Anal. Chem., 74*: 2470 (2002).
77. J.H. Park, J.W. Lee, Y.T. Song, C.S. Ra, J.S. Cha, J.J. Ryoo, W. Lee, I.W. Kim, and M.D. Jang, *J. Sep. Sci., 27*: 977 (2004).
78. J.H. Park, J.W. Lee, S.H. Kwon, J.S. Cha, P.W. Carr, and C.V. McNeff, *J. Chromatogr. A, 1050*: 151 (2004).
79. E. Tobler, Ph.D. thesis, University of Vienna, Austria (2001).
80. M. Lämmerhofer, E.C. Peters, C. Yu, F. Svec, and J.M. Fréchet, *Anal. Chem., 72*: 4614 (2000).
81. M. Lämmerhofer, F. Svec, and J.M. Fréchet, *Anal. Chem., 72*: 4623 (2000).
82. M. Lämmerhofer, F. Svec, J.M.J. Fréchet, and W. Lindner, *J. Microcol. Sep., 12*: 597 (2000).
83. M. Lämmerhofer, E. Tobler, E. Zarbl, W. Lindner, F. Svec, and J.M.J. Fréchet, *Electrophoresis, 24*: 2986 (2003).

84. B. Preinerstorfer, W. Bicker, W. Lindner, and M. Lämmerhofer, *J. Chromatogr. A,* *1044*: 187 (2004).
85. M. Lämmerhofer, *Anal. Bioanal. Chem.,* *382*: 873 (2005).
86. E. Gavioli, N.M. Maier, K. Haupt, K. Mosbach, and W. Lindner, *Anal. Chem.,* *77*: 5009 (2005).
87. J. Lah, N.M. Maier, W. Lindner, and G. Vesnaver, *J. Phys. Chem. B, 105*: 1670 (2001).
88. P. Bartak, P. Bednar, L. Kubacek, M. Lämmerhofer, W. Lindner, and Z. Stransky, *Anal. Chim. Acta, 506*: 105 (2004).
89. W.R. Oberleitner, N.M. Maier, and W. Lindner, *J. Chromatogr. A, 960*: 97 (2002).
90. R. Török, R. Berkecz, and A. Peter, *J. Chromatogr. A, 1120*: 61 (2006).
91. G. Götmar, T. Fornstedt, and G. Guiochon, *Chirality, 12*: 558 (2000).
92. N.M. Maier, S. Schefzick, G.M. Lombardo, M. Feliz, K. Rissanen, W. Lindner, and K.B. Lipkowitz, *J. Am. Chem. Soc., 124*: 8611 (2002).
93. C. Czerwenka, M.M. Zhang, H. Kaehlig, N.M. Maier, K.B. Lipkowitz, and W. Lindner, *J. Org. Chem., 68*: 8315 (2003).
94. K. Akasaka, K. Gyimesi-Forras, M. Lammerhofer, T. Fujita, M. Watanabe, N. Harada, and W. Lindner, *Chirality, 17*: 544 (2005).
95. J. Lesnik, M. Lämmerhofer, and W. Lindner, *Anal. Chim. Acta, 401*: 3 (1999).
96. R. Wirz, T. Buergi, W. Lindner, and A. Baiker, *Anal. Chem., 76*: 5319 (2004).
97. S. Schefzick, W. Lindner, and K.B. Lipkowitz, *Chirality, 12*: 7 (2000).
98. S. Schefzick, M. Lämmerhofer, W. Lindner, K.B. Lipkowitz, and M. Jalaie, *Chirality, 12*: 742 (2000).
99. G.D. Dijkstra, R.M. Kellogg, H. Wynberg, J.S. Svendsen, I. Marko, and K.B. Sharpless, *J. Am. Chem. Soc., 111*. 8070 (1989)
100. T. Buergi and A. Baiker, *J. Am. Chem. Soc., 120*: 12920 (1998).
101. H. Caner, P.U. Biedermann, and I. Agranat, *Chirality, 15*: 637 (2003).
102. G. Uccello-Barretta, L. Di Bari, and P. Salvadori, *Magn. Res. Chem., 30*: 1054 (1992).
103. C. Czerwenka and W. Lindner, *Rapid Commun. Mass Spectrom., 18*: 2713 (2004).
104. C. Czerwenka, N.M. Maier, and W. Lindner, *Anal. Bioanal. Chem., 379*: 1039 (2004).
105. K. Schug, P. Frycak, N.M. Maier, and W. Lindner, *Anal. Chem., 77*: 3660 (2005).
106. K.A. Schug, N.M. Maier, and W. Lindner, *J. Mass Spectrom., 41*: 157 (2006).
107. K.A. Schug, N.M. Maier, and W. Lindner, *Chem. Comm.*, 414 (2006).
108. H. Gika, M. Lämmerhofer, I. Papadoyannis, and W. Lindner, *J. Chromatogr. B, 800*: 193 (2004).
109. M. Lämmerhofer, P.D. Eugenio, I. Molnar, and W. Lindner, *J. Chromatogr. B, 689*: 123 (1997).
110. O.P. Kleidernigg, M. Lämmerhofer, and W. Lindner, *Enantiomer, 1*: 387 (1996).
111. G. Mitulovic, M. Lämmerhofer, N.M. Maier, and W. Lindner, *J. Labelled Cpd. Radiopharm, 43*: 449–461 (1999).
112. K. Gyimesi-Forras, A. Leitner, K. Akasaka, and W. Lindner, *J. Chromatogr. A, 1083*: 80 (2005).
113. A. Peter, G. Török, G. Toth, and W. Lindner, *J. High Resol. Chromatogr., 23*: 628 (2000).
114. A. Peter, *J. Chromatogr. A, 955*: 141 (2002).
115. R. Török, R. Berkecz, and A. Peter, *J. Sep. Sci., 29*: 2523 (2006).
116. C. Czerwenka, M. Lämmerhofer, and W. Lindner, *J. Sep. Sci., 26*: 1499 (2003).
117. K. Hamase, A. Morikawa, T. Ohgusu, W. Lindner, and K. Zaitsu, *J. Chromatogr. A, 1143*: 105–111 (2007).
118. E. Zarbl, M. Lämmerhofer, F. Hammerschmidt, F. Wuggenig, M. Hanbauer, N.M. Maier, L. Sajovic, and W. Lindner, *Anal. Chim. Acta, 404*: 169 (2000).

119. M. Lämmerhofer, E. Zarbl, W. Lindner, B. Peric Simov, and F. Hammerschmidt, *Electrophoresis, 22*: 1182 (2001).

120. C. Fakt, B.-M. Jacobson, S. Leandersson, B.-M. Olsson, and B.-A. Persson, *Anal. Chim. Acta, 492*: 261 (2003).

121. C. Czerwenka, M. Lämmerhofer, and W. Lindner, *J. Pharm. Biomed. Anal., 30*: 1789 (2003).

122. C. Czerwenka, N.M. Maier, and W. Lindner, *J. Chromatogr. A, 1038*: 85 (2004).

123. C. Czerwenka, P. Polaskova, and W. Lindner, *J. Chromatogr. A, 1093*: 81 (2005).

124. E. Tobler, M. Lämmerhofer, G. Mancini, and W. Lindner, *Chirality, 13*: 641 (2001).

125. M. Lämmerhofer, P. Imming, and W. Lindner, *Chromatographia, 59*: 1 (2004).

126. Chiral Technologies, Europe, *Application note (CTE Flyer)* (2005).

127. R. Török, A. Bor, G. Orosz, F. Lukacs, D.W. Armstrong, and A. Peter, *J. Chromatogr. A, 1098*: 75 (2005).

128. A.M. Stalcup and K.H. Gahm, *J. Microcol. Sep., 8*: 145 (1996).

129. A. Bunke and T. Jira, *Pharmazie, 51*: 479 (1996).

130. T. Jira, A. Bunke, and A. Karbaum, *J. Chromatogr. A, 798*: 281 (1998).

131. M. Lämmerhofer, E. Zarbl, and W. Lindner, *J. Chromatogr. A, 892*: 509 (2000).

132. F. Hammerschmidt, W. Lindner, F. Wuggenig, and E. Zarbl, *Tetrahedron Asymmetry, 11*: 2955 (2000).

133. M. Lämmerhofer, E. Zarbl, V. Piette, J. Crommen, and W. Lindner, *J. Sep. Sci., 24*: 706 (2001).

134. C. Czerwenka, M. Lämmerhofer, and W. Lindner, *Electrophoresis, 23*: 1887 (2002).

135. B. Peric Simov, F. Wuggenig, M. Lämmerhofer, W. Lindner, E. Zarbl, and F. Hammerschmidt, *Eur. J. Org. Chem.*, 1139 (2002).

136. P. Hinsmann, L. Arce, P. Svasek, M. Lämmerhofer, and B. Lendl, *Appl. Spectroscop., 58*: 662 (2004).

137. V. Piette, M. Lämmerhofer, W. Lindner, and J. Crommen, *Chirality, 11*: 622 (1999).

138. V. Piette, M. Lämmerhofer, W. Lindner, and J. Crommen, *Chirality, 11*: 622 (1999).

139. V. Piette, W. Lindner, and J. Crommen, *J. Chromatogr. A, 894*: 63 (2000).

140. V. Piette, M. Fillet, W. Lindner, and J. Crommen, *Biomed. Chromatogr., 14*: 19 (2000).

141. V. Piette, M. Fillet, W. Lindner, and J. Crommen, *J. Chromatogr. A, 875*: 353 (2000).

142. V. Piette, W. Lindner, and J. Crommen, *J. Chromatogr. A, 948*: 295 (2002).

143. V. Piette, M. Lämmerhofer, W. Lindner, and J. Crommen, *J. Chromatogr. A, 987*: 421 (2003).

144. M. Lämmerhofer, *J. Chromatogr. A, 1068*: 3 (2005).

145. M. Lämmerhofer, F. Svec, J.M.J. Fréchet, and W. Lindner, *Trends Anal. Chem., 19*: 676 (2000).

146. M. Lämmerhofer, *J. Chromatogr. A, 1068*: 31 (2005).

147. O.L. Sánchez Muñoz, E. Pérez Hernandez, M. Lämmerhofer, W. Lindner, and E. Kenndler, *Electrophoresis, 24*: 390 (2003).

148. M. Lämmerhofer and W. Lindner, *J. Chromatogr. A, 839*: 167 (1999).

149. M. Lämmerhofer, E. Tobler, and W. Lindner, *J. Chromatogr. A, 887*: 421 (2000).

150. B. Preinerstorfer, D. Lubda, A. Mucha, P. Kafarski, W. Lindner, and M. Lämmerhofer, *Electrophoresis, 27*: 4312 (2006).

151. K.-H. Kellner, A. Blasch, H. Chmiel, M. Lämmerhofer, and W. Lindner, *Chirality, 9*: 268 (1997).

152. M. Lämmerhofer, and W. Lindner, in: K. Valko (ed.), *Separation Methods in Drug Synthesis and Purification (Book Series: Handbook of Analytical Separations—Volume 1)*, Elsevier, Amsterdam, Netherlands, 2000, pp. 337.

2 HPLC Chiral Stationary Phases Containing Macrocyclic Antibiotics: Practical Aspects and Recognition Mechanism

Ilaria D'Acquarica, Francesco Gasparrini, Domenico Misiti, Marco Pierini, and Claudio Villani

CONTENTS

2.1 MACROCYCLIC ANTIBIOTICS: STRUCTURE, BIOLOGICAL ACTIVITY, AND PHYSICOCHEMICAL PROPERTIES

Macrocyclic antibiotics represent a class of antibiotics used in therapy against infections caused by gram-positive and gram-negative bacteria and having ring structures with at least 10 members. Despite their common features, macrocyclic antibiotics differ in several of their physicochemical properties and in biological activity. There are hundreds of natural and semisynthetic macrocyclic antibiotics, which comprise a large variety of structural types, including polyene-polyols, *ansa* compounds (e.g., aliphatic-bridged aromatic ring systems), glycopeptides, peptides, and peptide-heterocycle conjugates. In general, these compounds have molecular masses ranging between 600 and 2200 and a variety of functional groups. There are acidic, basic, and neutral types. Some macrocyclic antibiotics absorb quite weakly in the ultraviolet (UV) and visible spectrum while others possess large conjugated groups and absorb intensely [1–3].

This review provides an overview of the literature published to date on macrocyclic antibiotics exploited for enantioselective separations in high-performance liquid chromatography (HPLC). It was not intended as a comprehensive issue on the applications of such antibiotics in sub- and supercritical fluid chromatography (SFC), thin layer chromatography (TLC), capillary electrophoresis (CE), and capillary electrochromatography (CEC). A number of structural properties of the most important macrocyclic antibiotics applied in HPLC enantioseparations are listed in Table 2.1.

2.1.1 ANSAMYCINS

The ansamycins (rifamycin B and rifamycin SV) are produced by certain strains of *Nocardia mediterranei* and act by inhibition of messenger RNA (m-RNA) synthesis, and consequently of bacterial protein synthesis. Structurally, they present a characteristic *ansa* structure, made of a ring containing a naphthohydroquinone system spanned by an aliphatic chain. The two ansamycins differ in the type of substituent on the

TABLE 2.1
Structural Properties of the Most Important Macrocyclic Antibiotics Applied in HPLC Enantioseparations

Macrocyclic Antibiotic	Formula	Molecular Mass	Stereogenic Centers	Sugar Moieties	Ionizable Groups[a]	Publication Year[b]
Rifamycin B	$C_{39}H_{50}NO_{14}$	756	9	—	1	1994
Vancomycin	$C_{66}H_{75}Cl_2N_9O_{24}$	1447	18	2	3	1994
Thiostrepton	$C_{72}H_{85}N_{19}O_{18}S_5$	1663	17	—	1	1994
Teicoplanin A_2-2	$C_{88}H_{97}Cl_2N_9O_{33}$	1878	23	3	4	1995
Ristocetin A	$C_{95}H_{110}N_8O_{44}$	2066	38	6	2	1998
Avoparcin (α)	$C_{89}H_{104}ClN_9O_{36}$	1909	32	5	4	1998
Avoparcin (β)	$C_{89}H_{103}Cl_2N_9O_{36}$	1944	32	5	4	1998
A-40,926 (MDL 62,476)	$C_{83}H_{88}Cl_2N_8O_{29}$	1731	18	2	4	2000
TAG	$C_{58}H_{45}Cl_2N_7O_{18}$	1197	8	—	2	2000
VAG	$C_{53}H_{52}Cl_2N_8O_{17}$	1142	9	—	2	2002
MDL 63,246	$C_{88}H_{100}Cl_2N_{10}O_{28}$	1815	18	2	3	2003
CDP-1	$C_{66}H_{74}Cl_2N_8O_{25}$	1448	18	2	4	2005
A82846A (eremomycin)	$C_{73}H_{89}ClN_{10}O_{20}$	1460	22	3	4	2006

[a] Carboxylic and amine groups.
[b] Year of first introduction in the literature as chiral selector for HPLC.

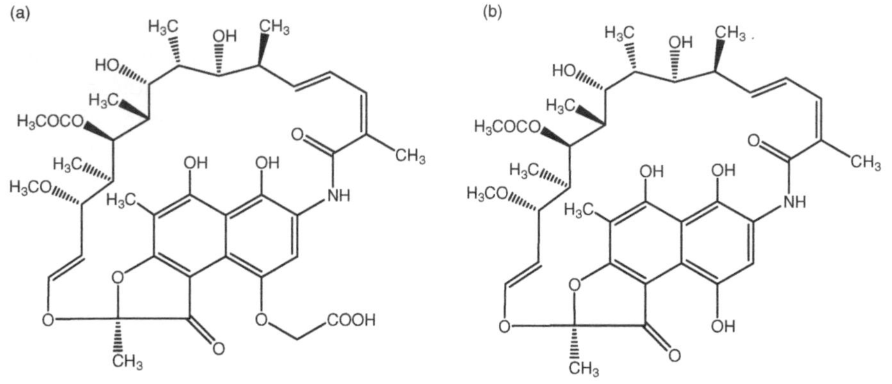

(a)

(b)

FIGURE 2.1   Chemical structures of (a) rifamycin B and (b) rifamycin SV.

aromatic nucleus (Figure 2.1). Rifamycin B ($C_{39}H_{50}NO_{14}$) contains nine stereogenic centers and one amide bond, and has a molecular mass of 756. Ansamycins strongly absorb in the UV and visible spectral regions because of the naphthohydroquinone ring. Each compound has absorption maxima at approximately 220, 304, and 425 nm and minima at approximately 275 and 350 nm [4]. Because of their strong absorbance, CE separations carried out with rifamycins are usually monitored via indirect detection, at relatively high concentrations (20–25 mM). It has also been shown that the ansamycins give aggregation phenomena, which tremendously affect their behavior in solution [5,6]. The ansamycins (in particular, rifamycin B) are only of historical significance in the field of enantioselective separations, in that they were the first chiral selectors used exclusively in CE prior to their use in HPLC [7], and their importance has nowadays declined.

2.1.2   GLYCOPEPTIDES

Glycopeptide antibiotics present a large, cyclic heptapeptide scaffold that is rich in aromatic fragments, surrounded by polar and ionizable groups and carrying carbohydrate moieties at the macrocycle periphery. These sugar units are attached through glycosidic bonds to phenolic or secondary hydroxyl groups of the aglycone. They play important roles in delivering the antibiotic to its target by enhancing its water solubility. The cyclic peptide backbone has a conformationally rigid cup-shaped architecture, with the aromatic fragments rigidly interlocked in a well-defined stereochemical disposition [8]. It is noteworthy the presence of one *cis* peptide bond, which is essential to keep the structures in a rigid macrocyclic form. Glycopeptides are soluble in water, buffers, and acidic aqueous solutions and less soluble at neutral pH. They are moderately soluble in polar aprotic solvents such as DMF and DMSO and are relatively insoluble in most organic solvents [4]. The chemistry, biology, and medicine of the glycopeptide antibiotics are discussed in detail in an attractive comprehensive review [9].

It is now well understood that all glycopeptide antibiotics exert antibiotic activity against Gram-positive bacteria because they stereospecifically bind to the precursor peptidoglycan peptide terminus *N*-acyl-D-alanyl-D-alanine produced during bacterial cell wall biosynthesis, thereby inhibiting the action of bacterial enzymes that would otherwise use these termini to form new cross-links in peptidoglycan [10].

The most successful and most extensively used glycopeptides for enantioselective HPLC [11] include naturally occurring structures (vancomycin and its analogs teicoplanin, TE; ristocetin A; avoparcin; glycopeptide A-40,926) and a series of semisynthetic derivatives (MDL 63,246; teicoplanin aglycone, TAG and vancomycin aglycone, VAG; derivatized vancomycin; crystalline degradation products (CDPs) from vancomycin; methylated teicoplanin aglycone, Me-TAG).

2.1.2.1 Vancomycin

Vancomycin is an amphoteric glycopeptide produced by the soil bacteria *Streptomyces orientalis* (Figure 2.2) [12, 13]. It is an effective and selective antibiotic active mainly against Gram-positive bacteria and some spirochetes; it is especially useful for treating serious infections caused by methicillin-resistant *Staphylococcus aureus* (MRSA). The aglycone portion of the molecule consists of three fused macrocyclic rings, containing five aromatic rings (lettered A through E), and two of the aromatic rings have chlorine substituents. There are also a secondary amine group ($pK_a = 7.2$), a carboxylic acid group ($pK_a = 2.9$), and three ionizable phenolic moieties ($pK_a = 9.6$, 10.5, and 11.7) [14]. The aglycone bears a disaccharide unit, namely L-vancosaminyl-D-glucose, attached at residue 4 of the heptapeptide scaffold. Native vancomycin

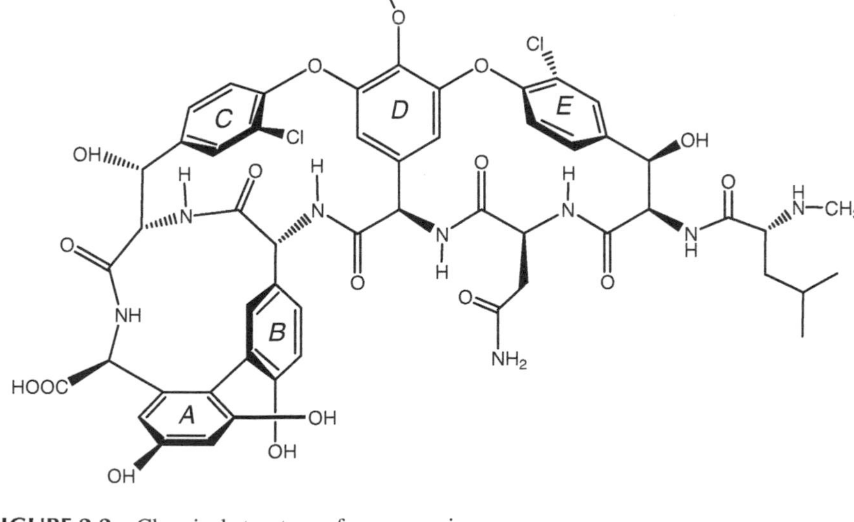

FIGURE 2.2 Chemical structure of vancomycin.

($C_{66}H_{75}Cl_2N_9O_{24}$) has a molecular mass of 1447 and contains 18 stereogenic centers.

Vancomycin was the first macrocyclic antibiotic evaluated as selector for the synthesis of HPLC chiral stationary phases (CSPs) [7], along with rifamycin B (among ansamycins) and thiostrepton (among polypeptides).

2.1.2.2 Vancomycin Analogs

In the vancomycin analog norvancomycin (NVC) [15], produced by the same microorganism, the methyl group connected to the secondary amine nitrogen is lost, that is, leucine rather than N-methyl leucine is the N-terminal amino acid residue. It has a molecular mass of 1433, the loss of the N-methyl group meaning that one primary amino group is present in the aglycone basket of the moiety. NVC was more recently introduced as chiral selector for HPLC, either bonded to silica microparticles [16, 17], or added to the mobile phase [18].

Other vancomycin analogs used as chiral selectors are A82846, LY307599, LY333328, and A35512B. *Amycolatopsis orientalis* NRRL 18090 (formerly designated *Nocardia orientalis*) produces glycopeptide antibiotics A82846 factors A, B, and C. Glycopeptide A82846 differs significantly from vancomycin in having a L-4-*epi*-vancosamine sugar substituted for L-vancosamine in the disaccharide attached at residue 4 and an additional L-4-*epi*-vancosamine attached at residue 6 of the linear heptapeptide. Apart from these differences, factor B (A82846B, also designated LY264826) is identical to vancomycin for the chlorine-substitutions on aromatic rings, while factor A (A82846A, also known as eremomycin) has only one Cl-atom in the N-terminal part of the aglycone, and factor C (A82846C) has none.

LY307599 and LY333328 (also known as oritavancin) are semisynthetic derivatives of A82846B [19]; they differ from A82846B in that an additional biphenyl group was added on the primary amine located on the L-4-*epi*-vancosamine attached at residue 4 (N-alkyl-linked derivatives). The glycopeptide antibiotic A35512B was isolated from *Streptomyces candidus* NRRL 8156 as the major active factor. Chemical degradation studies showed that mild hydrolysis resulted in the release of four neutral sugars: rhamnose, fucose, glucose, and mannose, as well as the liberation of a complex peptide core that retained all the amino acids and from which 3-amino-2,3,6-trideoxy-3-C-methyl-L-*xylo*-hexopyranose (or L-3-*epi*-vancosamine) was isolated [20, 21].

A82846A, also known as eremomycin, was only just introduced as selector for HPLC in 2006 [22, 23]. It has a molecular mass of 1460 ($C_{73}H_{89}ClN_{10}O_{20}$) and contains 22 stereogenic centers (Figure 2.3).

A82846B [24] and LY307599 [25] were developed as chiral selectors for CE; LY333328 [26] and A35512B [27] were applied as chiral mobile phase additives in narrow-bore HPLC.

2.1.2.3 Teicoplanin

Teicoplanin (TE) is produced by certain strains of *Actinoplanes teichomyceticus* [28]. It is applied in the treatment of severe hospital-acquired infections caused by

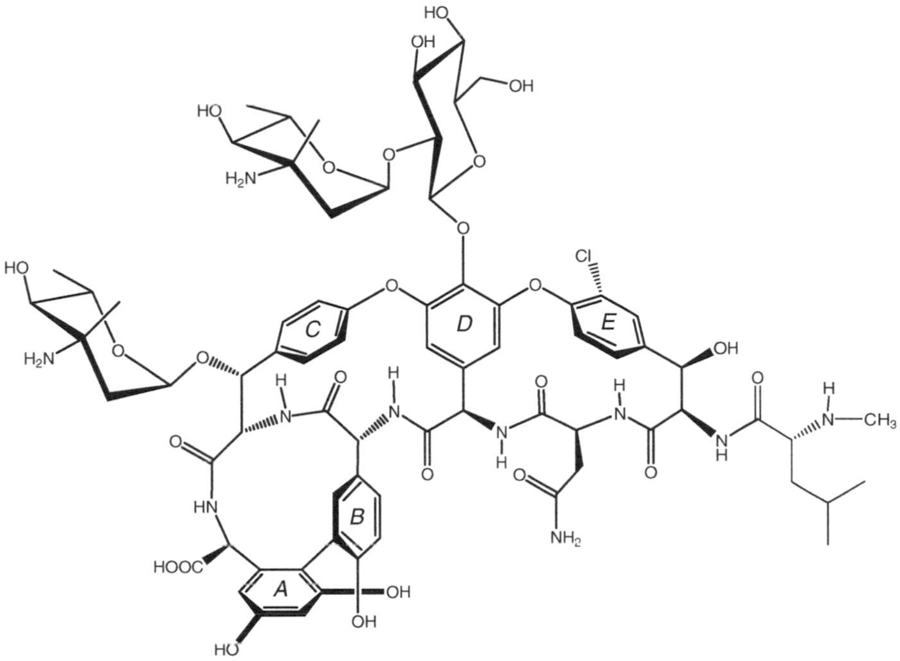

FIGURE 2.3 Chemical structure of A82846A (also known as eremomycin).

Gram-positive bacteria. Structurally, TE contains a heptapeptide aglycone that bears three sugar units. The aglycone consists of four fused medium-size rings, which form a semirigid basket. The basket contains seven aromatic rings (lettered A through G), two of which are chlorosubstituted and four have ionizable phenolic moieties. In the aglycone moiety, there are also a primary amine ($pK_a = 9.2$) and a carboxylic acid group ($pK_a = 2.5$). The three sugar units are D-mannose, D-N-acetylglucosamine, and a residue of D-N-acylglucosamine. For the presence of the latter hydrophobic residue, TE is considerably more surface-active than other related glycopeptides [29]. Five main components of the TE complex have been identified (designated from A_2-1 to A_2-5), differing from each other in the nature and length of the hydrocarbon chain of the N-acylglucosamine moiety. TE A_2-2 ($C_{88}H_{97}Cl_2N_9O_{33}$) is the prevalent component (>85%) of the TE complex (Figure 2.4), with a molecular mass of 1878 (acyl = 8-methyl-nonanoyl), and contains 23 stereogenic centers.

TE was the second macrocyclic antibiotic evaluated as selector for the synthesis of HPLC CSPs, one year later than vancomycin [30].

2.1.2.4 Ristocetins

Nocardia lurida produces glycopeptides ristocetin A and B, differing in the number of sugar moieties attached to the heptapeptide backbone [31]. Ristocetin A consists of an aglycone portion with four joined macrocyclic rings, which contains seven aromatic rings (lettered A through G), none of which are chlorosubstituted. There are also a primary amine group, one methyl ester and four ionizable phenolic moieties.

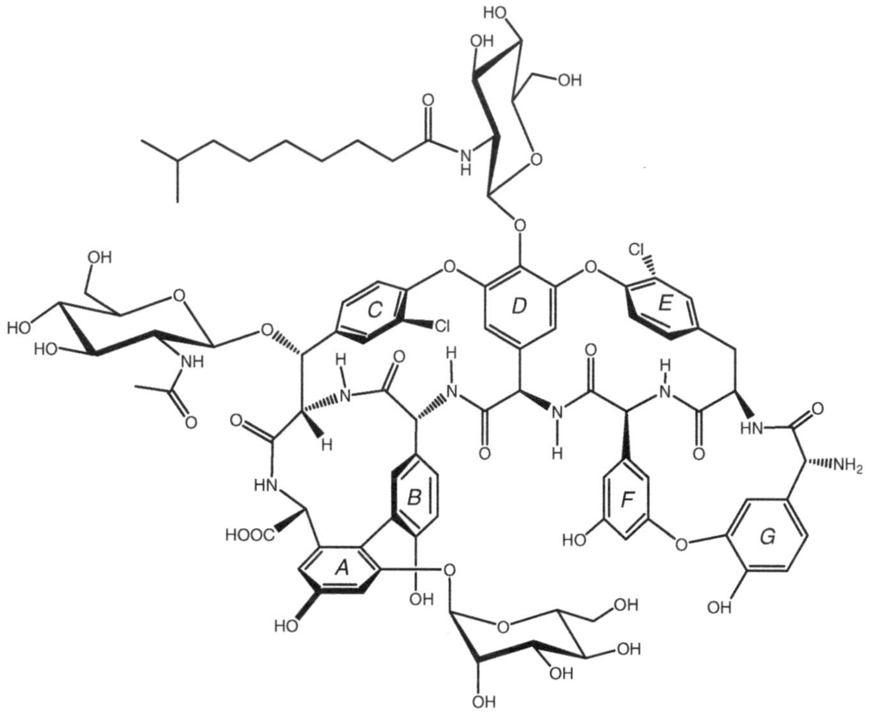

FIGURE 2.4   Chemical structure of teicoplanin A_2-2.

The sugar units are D-mannose, the amino sugar L-ristosamine, and a heterotetrasac-charide made of D-arabinose, D-mannose, D-glucose, and L-rhamnose (Figure 2.5), attached to the peptide nucleus via D-glucose. Ristocetin A ($C_{95}H_{110}N_8O_{44}$) has a molecular mass of 2066 and contains 38 stereogenic centers. Ristocetin B lacks arabinose and mannose of the tetrasaccharide fragment, while the mannose residue directly attached to the heptapeptide scaffold is still in place [32].

Ristocetin A was the third macrocyclic antibiotic evaluated as selector for the synthesis of HPLC CSPs [33].

2.1.2.5   Avoparcin

Avoparcin complex is produced as a fermentation product by a strain of *Streptomyces candidus* [34]. It exists in two major forms, α and β, differing only in one chlorine atom on the aromatic ring F at residue 3 (Figure 2.6). The β component ($C_{89}H_{103}Cl_2N_9O_{36}$) predominates about 3:1 over the α form [35], has a molecular mass of 1944, and contains 32 stereogenic centers. The aglycone portion of the molecule consists of three connected semirigid macrocyclic rings, containing seven aromatic rings (lettered A through G), four of which have ionizable phenolic moieties. In the aglycone moiety, there are also a secondary amine and a carboxylic acid group. There are five sugar units, namely D-mannose, L-rhamnose, the amino sugar L-ristosamine, and a disaccharide unit, L-ristosaminyl-β-D-glucose. An important

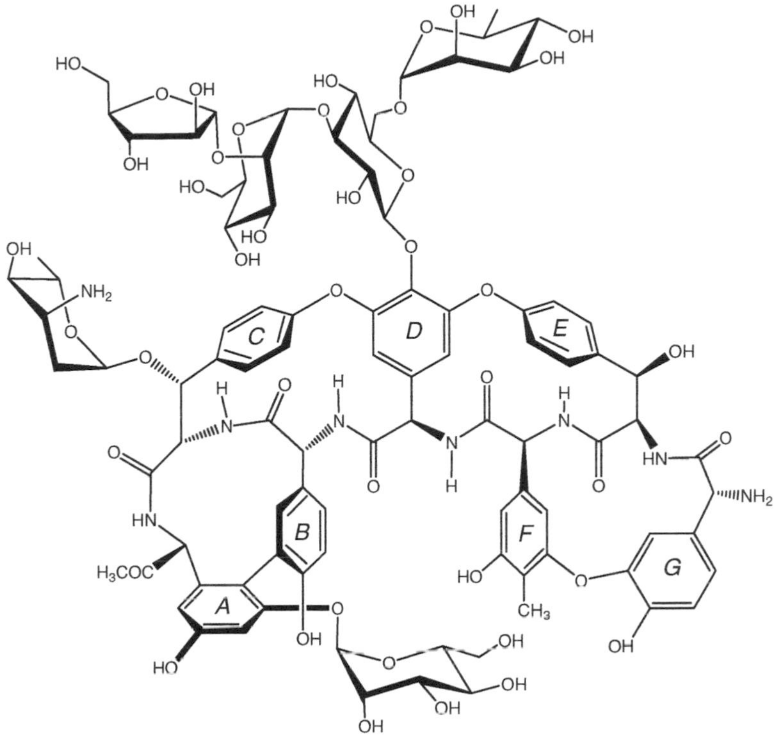

FIGURE 2.5   Chemical structure of ristocetin A.

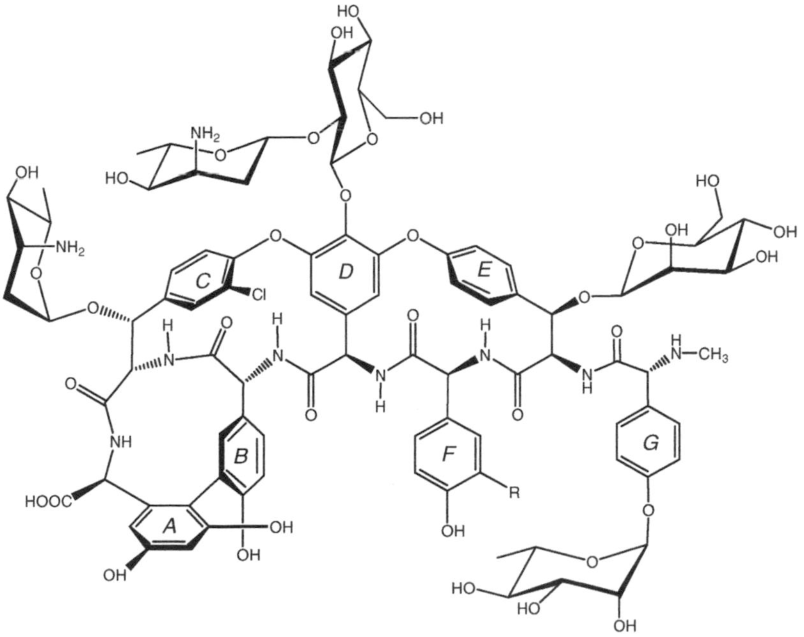

FIGURE 2.6   Chemical structure of avoparcin α (R = H) and avoparcin β (R = Cl).

feature of the avoparcin chemistry is that aqueous solutions of α- and β-avoparcin lose antibacterial activity when they are heated at 80°C for 16 h in the pH range 5–8. When monitored by analytical HPLC, an equilibrium mixture is observed consisting of ca. 30% of starting antibiotic and 70% of a diastereomer, which is called *epi*-avoparcin and which has a longer retention time in the HPLC profile. Equilibration of purified *epi*-β-avoparcin yields the same ratio of β-avoparcin and *epi*-β-avoparcin as observed above. It is especially noteworthy that *epi*-β-avoparcin is 10- to 100-fold less active against gram-positive bacteria than avoparcin [36].

Avoparcin was the fourth macrocyclic antibiotic evaluated as selector for the synthesis of HPLC CSPs [37].

2.1.2.6 Glycopeptide A-40,926 (MDL 62,476)

A-40,926 (or MDL 62,476) is a natural complex of glycopeptides produced by an actinomycete of the *Actinomadura* strain ATCC 39726, which, like the TE complex, contains a fatty acid as part of a glycolipid attached to the peptide backbone [38, 39]. The chemical structure of the prevalent component of A-40,926 complex (factor B, >70%) is reported in Figure 2.7. It differs from TE in the following features: (i) an additional carboxylic group on the *N*-acyl-glucosamine is present, to give a *N*-acylamino-glucuronic acid; (ii) the D-*N*-acetyl-glucosamine residue is missing;

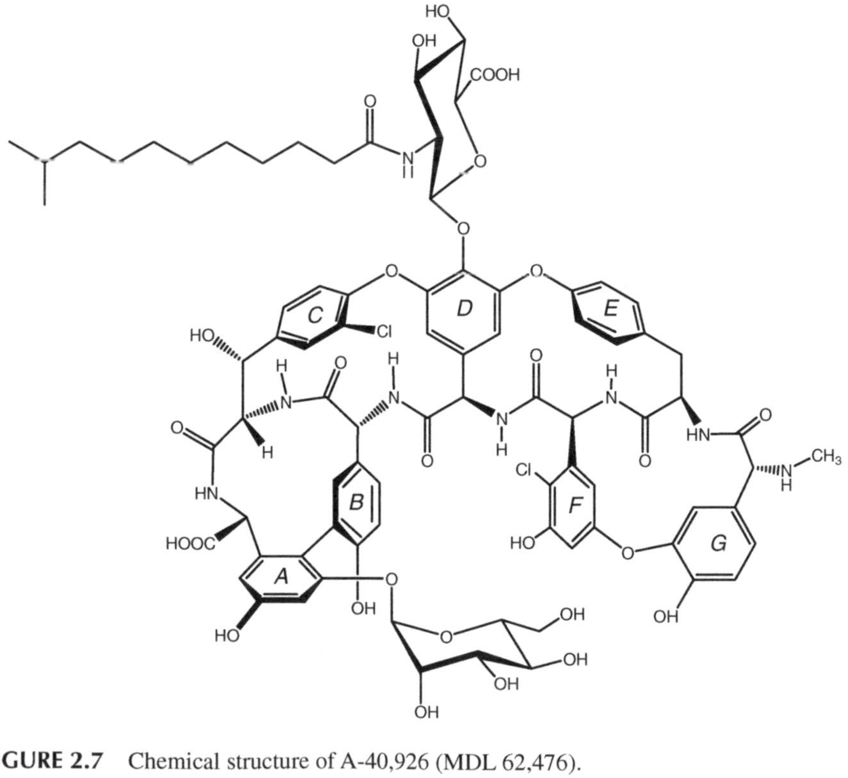

FIGURE 2.7   Chemical structure of A-40,926 (MDL 62,476).

(iii) the primary amine group on the aglycone portion of TE is a methyl-substituted secondary amine; (iv) the chlorine substituent on phenyl ring E in TE is not present in A-40,926, which in turn presents a chlorine atom on its phenyl ring F; (iv) the 9-carbon chain of TE A_2-2 is replaced by an 11-carbon chain in A-40,926 (acyl = 10-methyl-undecanoyl). A-40,926 factor B ($C_{83}H_{88}Cl_2N_8O_{29}$) has a molecular mass of 1731. The presence of two carboxylic acid functions in the structure gives this molecule a unique net negative charge at neutral pH.

A-40,926 was evaluated as selector for the synthesis of HPLC CSPs in 2000 [40, 41].

2.1.2.7 Chemically Modified Glycopeptides

Prior to 1984, the glycopeptide class included few members beyond vancomycin (Figure 2.2), TE (Figure 2.4), ristocetin A (Figure 2.5), and avoparcin (Figure 2.6). With the boost from the threat posed by antibiotic resistance, the class swelled to include thousands of natural and semisynthetic compounds [9]. Structural studies on these compounds have clarified the biological mode of action and serve as a basis for reasonable predictions regarding structure-activity relationships. Although the complexity of these structures first limited the variety of possible chemical modifications aimed to improve biological activity, some excellent results were obtained by certain amide modifications of the terminal carboxylic acid group. Among these, MDL 63,246 was obtained using A-40,926 (MDL 62,476) as starting material by condensation with the proper amine. Moreover, removal of the carbohydrate units could be effected, as well as the attachment of new sugars, as modifications of the outer core of glycopeptides. Acylation or reductive amination at the amino sugar site allowed the synthesis of several new vancomycin analogs (see Section 2.1.2.2). Modifications involving degradation and reassembly of the cyclopeptide core with eventual incorporation of new amino acid components could produce nonbiologically active compounds, such as CDP-1 (see Section 2.1.2.7.5).

All these chemically modified glycopeptides have been helpful in understanding the mechanism of the chiral recognition process mediated by these macrocycles.

2.1.2.7.1 MDL 63,246
MDL 63,246 (Figure 2.8) is a semisynthetic amide derivative of the naturally occurring glycopeptide antibiotic A-40,926 (MDL 62,476), and differs from the latter only for the amide substitution on the terminal carboxylic acid group [42]. Thus, it has the same number of stereogenic centers, and a molecular mass of 1815 ($C_{88}H_{100}Cl_2N_{10}O_{28}$). MDL 63,246 was recently used in the preparation of a CSP for capillary liquid chromatography (LC) and CEC. Enantiomeric separation of selected acidic compounds belonging to the class of nonsteroidal antiinflammatory drugs (NSAIDs) [43] and of some α-hydroxycarboxylic acids [44] were obtained and compared with those achieved with a vancomycin-based CSP.

2.1.2.7.2 Teicoplanin Aglycone
Teicoplanin aglycone (TAG) is the aglycone portion of glycopeptide antibiotic TE A_2-2, that is, it contains the same heptapeptide backbone of TE, but it lacks the sugar

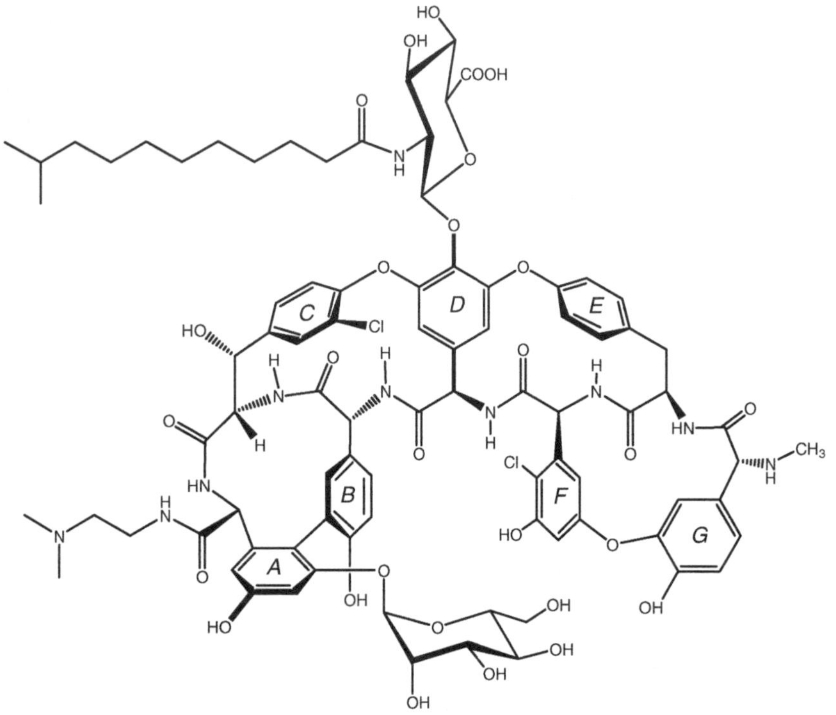

FIGURE 2.8   Chemical structure of MDL 63,246.

units ($C_{58}H_{45}Cl_2N_7O_{18}$, molecular mass of 1197). With the aim of investigating the role of the carbohydrate moieties in the chiral recognition mechanism, TAG was isolated by acidic hydrolysis (H_2SO_4 80% in DMSO) and purified from native TE [45]. Two CSPs were prepared afterward in an identical way, one with the native TE molecule and the other with the TAG residue. Twenty-six compounds were evaluated on the two CSPs under reversed-phase (RP) and polar-organic mode (POM) conditions. It appeared clearly from this study that the carbohydrate units are not necessary for the enantioresolution of common α-amino acids. The cleft near the ureido-terminal of the aglycone basket (see Section 2.2.1.2.1) is an important part of the receptor site for amino acid recognition. Amino acids have an easier access to this site on the TAG CSP, which produces much higher enantioselectivity and resolution factors for these compounds compared to those obtained on the native TE CSP. Nonamino acid compounds that have carboxylate groups also can associate with the aglycone amino acid binding site. Sometimes, these analytes are better resolved by the aglycone CSP. However, many nonamino acid compounds were enantioseparated by a combination of interactions involving both the aglycone basket and its attached sugar units. These compounds are better resolved, or resolved only, on the native TE CSP.

2.1.2.7.3   Vancomycin Aglycone

Vancomycin aglycone (VAG) contains the same peptide scaffold as vancomycin, but it lacks the disaccharide unit ($C_{53}H_{52}Cl_2N_8O_{17}$, molecular mass of 1142). The basic idea behind the VAG project was the same as for teicoplanin; in particular, vancomycin was hydrolyzed under acidic conditions to answer whether the carbohydrate moieties were useful for the enantiorecognition of chiral sulfur-containing compounds [46]. It was found that a few more sulfur-containing compounds could be resolved on the carbohydrate containing CSP than on the aglycone counterpart. However, for the compounds that were separated on both CSPs with identical mobile phases, most had higher enantioselectivity and resolution factors on the VAG CSP. For this class of compounds, the normal-phase (NP) mode was the most effective, followed by the RP mode with methanol–water mobile phases.

2.1.2.7.4 Derivatized Vancomycin

Vancomycin was derivatized with 3,5-dimethylphenylisocyanate (DMP) and evaluated as selector for the synthesis of a CSP for HPLC [7]. The DMP-vancomycin bonded phase [47] was able to resolve several compounds that the vancomycin CSP could not, in particular hydroxyzine and althiazide, under NP conditions.

In an attempt to change and broaden the capabilities of the vancomycin CSP, the glycopeptide was derivatized with (R)- and (S)-(1-naphthylethyl) isocyanate (NEIC) and then bonded to a silica-gel support [48]. A variety of chiral compounds was tested on the two composite stationary phases and the results were compared with the ones obtained using the underivatized vancomycin CSP. The advantages of the NEIC derivatization were not as obvious or substantial as they were in the case of cyclodextrin phases [49]. Moreover, the exact chemical structures of the synthesized NEIC derivatives of vancomycin were not reported.

2.1.2.7.5 Crystalline Degradation Product

Degradation studies of vancomycin produced a nonbiologically active CDP-1 (Figure 2.9). X-ray crystallographic analysis of CDP-1 ($C_{66}H_{74}Cl_2N_8O_{25}$, molecular mass of 1448) revealed that it contained an expanded D-O-E ring, which was originally thought to be representative of the class [50]. It was quickly established, however, that CDP-1 actually arose from an unusual aspartic-isoaspartic rearrangement [51, 52]. Furthermore, the expanded 17-membered D-O-E ring in CDP-1 allowed for equilibration between the two atropoisomers (CDP-1-M and CDP-1-m) of this ring system [52]. Interestingly, the more stable atropoisomer (CDP-1-M) has the chlorine atom in the opposite stereochemistry to that of vancomycin. CDP-1 atropoisomers were formed by the hydrolytic loss of ammonia from vancomycin [53, 54], and then covalently linked to silica gel [55] via the two carboxylic acid groups (see Section 2.2.1.2), leading to a higher surface coverage than for vancomycin, which has only one carboxylic acid group. The CSP obtained was applied in the enantioselective separation of six chiral compounds (three α-amino acids, carbidopa, ibuprofen, and 2-methoxy-2-phenyl ethanol).

2.1.2.7.6 Methylated Teicoplanin Aglycone

Removal of the three saccharides moieties from native TE yielding the aglycone basket TAG (see Section 2.1.2.7.2) introduces three new OH-groups in place of sugars. As a consequence, the TAG CSP could undergo poorer mass transfer, owing to stronger interactions between exposed functional groups and some solutes, in comparison with

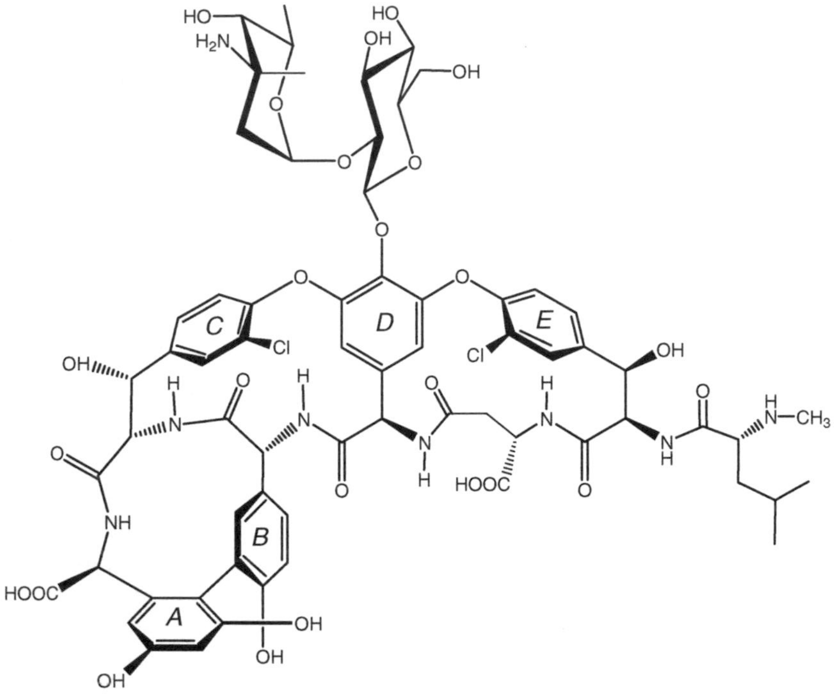

FIGURE 2.9 Chemical structure of CDP-1-M.

native TE CSP. As a result, improved selectivity but sometimes-reduced separation efficiencies of certain kinds of amino acids and their derivatives have been observed [56]. These findings recently led to the preparation of a Me-TAG (Figure 2.10), with the aim to reduce the hydrogen-bonding interactions [57]. Exhaustive methylation was realized using diazomethane (CH_2N_2), to yield a dominant product with five methyl groups, confirmed by mass spectrometry. Alternatively, a reaction was described for TAG bonded to silica gel, using methyl triflate ($CH_3CF_3SO_3$), which preferentially reacts with amines and alcohols. The final result of both the methylation procedures was an average of four to five methyl groups on the CSP structure, and, of course, the esterification of the terminal carboxylic group. HPLC enantiomeric separations of a wide variety of racemic analytes were compared using the three related CSPs (teicoplanin, TAG, and Me-TAG) in two different mobile phase modes (i.e., RP and POM) [58].

2.1.3 POLYPEPTIDES

Macrocyclic antibiotics also include a family of thiopeptides, of which thiostrepton (Figure 2.11) is the parent compound and the most complex member. Produced by *Streptomyces azureus* [59], thiostrepton includes 10 rings, 11 peptide bonds, extensive unsaturation, an imine functionality, a secondary amine, and 17 stereogenic centers [60]; it also contains five thiazole rings and one quinoline nucleus.

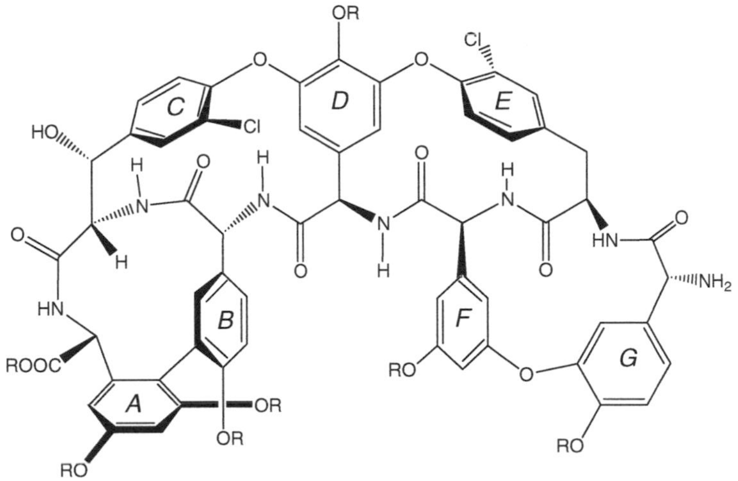

FIGURE 2.10 Chemical structure of Me-TAG (R = H or CH$_3$): the methylation procedures yielded an average of four to five methyl groups.

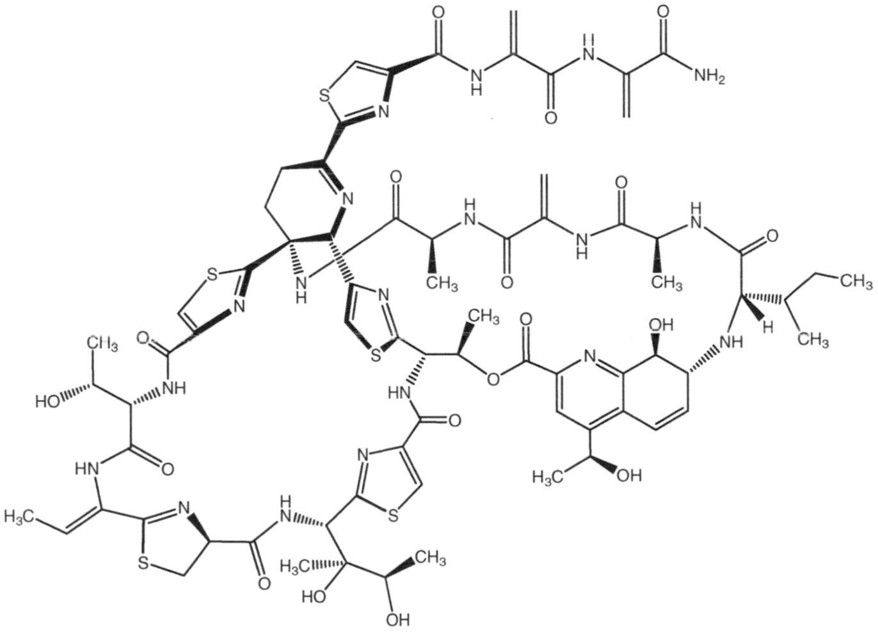

FIGURE 2.11 Chemical structure of thiostrepton.

Thiostrepton (C$_{72}$H$_{85}$N$_{19}$O$_{18}$S$_5$, molecular mass of 1663) has an unusual type of biological activity, in that it binds to ribosomal RNA and to one of its associated proteins, thus disabling protein biosynthesis. It was the first macrocyclic antibiotic evaluated as selector for the synthesis of HPLC CSPs, together with vancomycin and rifamycin B [7], but the application of a thiostrepton-based CSP has currently declined significantly.

2.2 SYNTHETIC STRATEGIES FOR THE PREPARATION OF GLYCOPEPTIDES CONTAINING CSPs FOR HPLC

Any grafting procedure used to connect a chiral molecule to the silica surface should fulfill some basic requirements. (i) The process should give a stable linkage between the chiral selector and the matrix to ensure full mobile phase compatibility and low column bleeding. (ii) Reaction conditions should preserve the stereochemical integrity of the chiral fragment. (iii) Geometrical arrangement of the chiral fragment relative to the silica surface should be planned and realized in a way that it maximizes enantioselectivity. (iv) Chemistry of binding to the solid matrix should be easily scalable for large-scale production.

Two conceptually different binding processes have been exploited in the literature for the grafting of glycopeptides antibiotics: (1) the stepwise assemblage of the target chiral selector on the silica surface, and (2) the direct attachment of a silyl derivative of the chiral selector on unmodified silica particles. The first approach is rather simple, but, obviously, a more homogenous functionalization of the silica surface is achieved with the second approach. Surface coverage data reported in the literature for the different synthetic strategies are summarized in Table 2.2.

2.2.1 SYNTHETIC STRATEGY 1

This strategy consists in the initial modification of the silica surface with organosilanes having suitable anchoring groups, which are either reactive themselves or can be additionally activated for the final attachment of the chiral selector. The choice of the proper silane will depend on the presence of suitable functional groups on the chiral entity to be fixed to the matrix. As macrocyclic antibiotics contain hydroxyl, amine, and carboxylic acid functionalities, they can be linked to the silica surface in a variety of different ways [7, 55]. The obvious drawback of the stepwise assemblage of chiral selectors on the silica surface is the eventual formation of additional polar or ionizable sites on the matrix, which may cause unselective retention of chiral analytes.

2.2.1.1 Carboxylic Acid-Terminated Organosilanes

Carboxylic acid-terminated organosilanes were used in the early studies on chemically bonded glycopeptides to immobilize vancomycin and thiostrepton via their amino groups, leading to the formation of stable amide bonds between antibiotics and modified silica [7]. In a typical reaction, 4 g of dry silica gel is slurried on 50 mL of dry toluene. Two grams of [1-(carbomethoxy)ethyl]methyldichlorosilane or [2-(carbomethoxy)ethyl]trichlorosilane is dissolved in ~15 mL of dry toluene contained in a dropping flask. The organosilane solution is added dropwise over ~30 min

TABLE 2.2
Synthetic Strategies for the Immobilization of Macrocyclic Antibiotics: Surface Coverage Data

Section	Antibiotic	Surface Coverage μmol/g	Surface Coverage μmol/m²	Tradename	Starting Silica Morphology Surface Area (m²/g)	Starting Silica Morphology Pore Diameter (Å)	References
2.2.1.1	Vancomycin	—	—	—	—	—	7
2.2.1.1	Thiostrepton	—	—	—	—	—	7
2.2.1.2	Rifamycin B	—	—	—	—	—	7
2.2.1.2	CDP-1[a]	—	—	—	—	—	55
2.2.1.2.1	Teicoplanin A$_2$-2	165	0.55	LiChrosorb	300	100	61
2.2.1.2.1	TAG	151	0.38	LiChrospher	400	100	45
2.2.1.2.1	A-40,926	147	0.49	LiChrosorb	300	100	40
2.2.1.3	Eremomycin	50	0.17	Diaspher-110-Epoxy	300	110	22
2.2.1.3	Ristocetin A	84	0.28	Diaspher-110-Epoxy	300	110	22
2.2.1.3	Vancomycin[b]	59	0.20	Diaspher-110-Epoxy	300	110	22
2.2.1.4	Vancomycin	120	0.40	LiChrospher-Diol	300	100	70
2.2.1.4	Ristocetin A	—	—	Nucleosil-Diol	35	500	75
2.2.1.4	MDL 63,246	—	—	LiChrospher-Diol	350	100	43
2.2.2.1	Vancomycin	115	0.34	Kromasil	340	100	48
2.2.2.1	Rifamycin B	—	—	—	—	—	47
2.2.2.1	DMP-Vancomycin	—	—	—	—	—	47
2.2.2.1	Teicoplanin A$_2$-2	31	0.10	LiChrosorb	300	100	62
2.2.2.1	Norvancomycin[b]	119	0.35	Kromasil	340	100	17

[a] The authors provided only the carbon content (~17%).
[b] Calculated on the basis of the elemental analyses provided by the authors.

to the refluxing toluene–silica gel slurry. The mixture is allowed to reflux (~110°C) for 2 h and is then cooled, filtered, and washed with methanol, 50% aqueous methanol, and methanol again and then dried. The silanized silica gel can also be slurried in dry DMF. One gram of antibiotic (vancomycin or thiostrepton) is added along with an appropriate carbodiimide dehydrating agent. After 6 h, the CSP is filtered and washed with methanol and then aqueous methanol. Surface coverage data were not provided by the authors [7].

2.2.1.2 Amino-Terminated Organosilanes

Amino-terminated organosilanes were used to immobilize rifamycin B via its active carboxylic acid functionality leading to the formation of stable amide bonds between antibiotics and modified silica [7]. The organosilanes used were (3-aminopropyl)triethoxysilane or (3-aminopropyl)dimethylethoxysilane, and the procedure for preparation was as described in Section 2.2.1.1. Surface coverage data were not provided by the authors [7].

The same chemistry of binding has more recently been used for the immobilization of CDP-1 (see Section 2.1.2.7.5) via the two carboxylic acid groups present in the aglycone portion of the molecule, owing to the hydrolytic loss of ammonia [55].

2.2.1.2.1 Activation of Aminopropylated Silica with
Bifunctional Spacers

A number of glycopeptides containing CSPs have been prepared according to the aforementioned procedure, that is based on the use of 3-aminopropylated silica gel as starting material, but with a preliminary activation step with bifunctional spacers before the surface linking of the macrocycle. In particular, an innovative one-pot synthetic process was developed in 1999 to yield high loaded and stable glycopeptide containing CSPs [61]. Such a process is based on the activation of 3-aminopropylated silica gel with 1,6-diisocyanatohexane to give a monoureido–monoisocyanate intermediate in which the residual, pendant isocyanate groups are used to immobilize the macrocyclic receptors via their free amino groups (Scheme 2.1a). Additional linkage, in which carbamate groups are formed between the glycopeptide alcoholic or phenolic hydroxyls and the surface-linked isocyanate groups, may be present in the final material. In the resulting CSPs, the macrocycles are tethered to the aminopropylated silica via stable ureido bonds and a six-carbon aliphatic spacer. Elemental analysis of the final stationary phases gave typical macrocycle loading of 0.38–0.55 μmol/m^2 (see Table 2.2). This procedure afforded chiral phases with high chemical inertness and effective passivation of the underlying silica by the presence of two ureido functions. Different grafting reagents leading to the formation of one or two stable ureido functions on the final TE CSP structure were also investigated, and the influence of the different spacers used on the chiral performances was evaluated [62]. A parallel

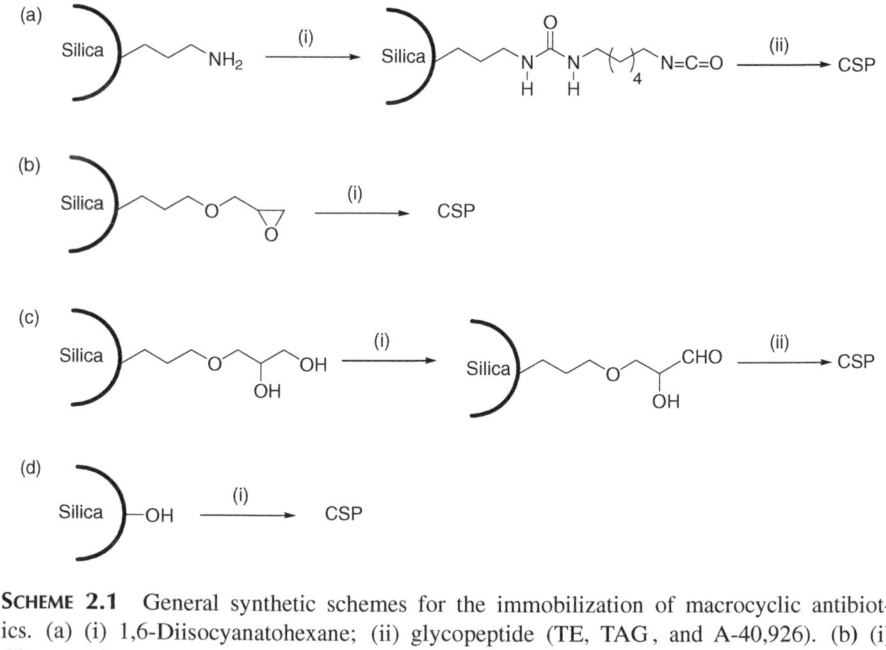

SCHEME 2.1   General synthetic schemes for the immobilization of macrocyclic antibiotics. (a) (i) 1,6-Diisocyanatohexane; (ii) glycopeptide (TE, TAG, and A-40,926). (b) (i) Glycopeptide (eremomycin, ristocetin A, vancomycin). (c) (i) NaIO$_4$; (ii) glycopeptide (vancomycin, ristocetin A, MDL 63,246). (d) (i) 3-Isocyanatopropyl-silyl derivative of macrocyclic antibiotic (TE, rifamycin B, vancomycin, DMP-vancomycin, NVC).

study on the effect of the silica matrix nature was performed, starting from silica gels with different morphological and physicochemical characteristics, such as specific surface areas (ranging from 35 to 450 m²/g), particle sizes (3 and 5 μm), and pore sizes (ranging from 80 to 500 Å). The highest levels of bonded TE were reached with the use of irregular LiChrosorb Si 100, 5 μm silica gel (see Table 2.2). This strategy, first developed to immobilize teicoplanin A_2-2 [61], was then extended to prepare a set of closely related CSPs containing TAG [45] and glycopeptide A-40,926 [40, 41]. It is worthy of note that the last two structures had never been considered as chiral selectors before. This chemistry of binding to the solid matrix has also successfully been scaled for semipreparative applications [40,63], and more recently exploited for the preparation of a new chiral and restricted-access material containing teicoplanin as chiral selector (chiro-RAM-TE), for the direct HPLC injection of biological fluids containing chiral drugs [64]. The external surface was covered with an achiral hydrophilic polymeric network (polyvinyl alcohol 72,000), while the chiral partitioning phase was exclusively confined to the internal region of the porous silica support (Scheme 2.2). The chiro-RAM-TE CSP was applied in the pharmacokinetic monitoring of sotalol in the plasma of healthy volunteers dosed with racemic drug (see Section 2.3.2.5).

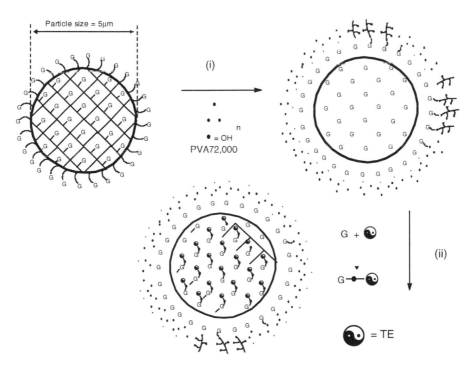

SCHEME 2.2 Graphical representations of the inner and outer phases of the chiro-RAM-TE support. G = grafting group (monoureido–monoisocyanate intermediate), obtained as previously described [61]: (i) LiCl/dry DMSO, 80°C, 3 h. (ii) dry pyridine, 70°C, 12 h.

2.2.1.3 Epoxy-Terminated Organosilanes

Binding to oxirane-activated silica via epoxide ring opening is a highly efficient and generally applicable method for the initial grafting of chiral molecules having nucleophilic centers. Amines [65], amino acids [66], cyclodextrins [67], and proteins [68] have smoothly been grafted to oxirane–silica in both organic- and water-based solvents.

Suitable epoxy-terminated silanes include (3-glycidoxypropyl)trimethoxysilane, (3-glycidoxypropyl)dimethylethoxysilane, and (3-glycidoxypropyl)triethoxysilane (Scheme 2.1b). The attachment of glycopeptides to silica gel via epoxy-terminated organosilanes was first generally envisioned by Armstrong, as a potential synthetic strategy for any macrocycle structure [7]. The full procedure of preparation was as described earlier for cyclodextrins [67]. In the final CSP structure, the glycopeptide is linked via a stable C—N bond to the silica, spaced-out by a dipropyl ether bridge. Very recently, a new CSP was prepared by attachment of eremomycin (A82846A, see Section 2.1.2.2) to epoxy-activated silica gel, but under mild conditions. In a typical reaction, eremomycin sulphate (1 g) was dissolved in water (15 mL) and the pH of the solution was adjusted to 8.6 with 1M potassium hydroxide (KOH). Diaspher-110-Epoxy silica (3 g) was added to the solution and the suspension was stirred at 40°C for 14 h, whereupon modified silica was filtered and washed with water, methanol, and acetone, and dried [22,23]. This chemistry of binding provided a moderate bonding density of 50 μmol/g (0.17 μmol/m^2) (see Table 2.2). For comparative purposes, the same procedure was applied to the immobilization of vancomycin [22]. The eremomycin CSP revealed high enantioselectivity values for the separation of underivatized α-amino acids, in contrast to the vancomycin CSP, which is well known to be not optimal for such substrates. Immobilization onto epoxy-activated silica from water solutions and low temperature significantly improved the enantioselectivity and the resolution of a CSP containing ristocetin A, on purpose prepared under the same conditions (see Table 2.2), in the separation of α-amino acids.

2.2.1.4 Diol-Derivatized Silica Gel

The terminal diol functionality on silica gel can be easily oxidized with sodium periodate to yield a silica surface with aldehyde functions [69]. Chiral molecules having amino groups can be immobilized by reductive amination of the aldehyde-functionalized silica with sodium cyanoborohydride. This chemistry of binding (Scheme 2.1c) was first exploited for the immobilization of vancomycin [70–74], ristocetin A [75] and, more recently, of glycopeptide MDL 63,246 [43,44]. The full procedure is as follows: a slurry of 0.4 g of silica LiChrospher DIOL (Si 100, 5 μm) in 30 mL of a water–methanol (4:1, v/v) mixture containing NaIO$_4$ (60 mM) is sonicated for 1 h. The mixture is centrifuged at 5000 rpm for 5 min and the periodate solution eliminated by washing the silica phase with water (3 × 20 mL). A 3 mM solution of antibiotic (vancomycin or MDL 63,246) is prepared in phosphate buffer (pH = 7), containing NaCNBH$_3$ (10 mM), and added to the modified diol silica. The slurry is sonicated for 1 h, centrifuged, and washed with water (2 mL). Remaining aldehyde moieties are reduced back to hydroxyl groups by adding a solution of

NaCNBH$_3$ (10 mM) in phosphate buffer (pH $=$ 3). The slurry is sonicated for 1 h, centrifuged, and the cyanoborohydride solution eliminated. The modified silica is washed with water and methanol, and then dried. This chemistry of binding provided for vancomycin a substantial bonding density of 120 μmol/g (0.40 μmol/m^2) [70] (see Table 2.2). Surface coverage data for ristocetin A [75] and glycopeptide MDL 63,246 [43,44] were not provided by the authors.

In the case of vancomycin [72], an original study was performed to obtain a well-defined stationary phase structure, since it was reasonably assumed that the antibiotic is randomly linked to the silica by one or both of its amino groups, one belonging to the disaccharide portion (primary), and the other one to the heptapeptide core (secondary). Thus, alternate fluorenylmethyloxycarbonyl (FMOC)-amino-protected derivatives were prepared and immobilized in a packed column, and then vancomycin was recovered by cleavage of the protecting groups. The two defined CSPs obtained, when compared with the CSP produced from native randomly linked vancomycin, showed lower retention and enantioselectivity, also if they still separated the same compounds. Thus, no advantages could be found to choose these phases as an alternative to the native vancomycin CSP.

2.2.2 SYNTHETIC STRATEGY 2

In this synthetic strategy, the macrocyclic antibiotic is covalently bonded to the silica matrix in two steps: (1) chemical modification of the selector via reaction between suitable groups of the antibiotic and proper groups of the spacer, reacting also as a di- or trialkoxysilane; (2) immobilization of the functionalized selector on unmodified silica particles.

2.2.2.1 Isocyanate-Terminated Organosilanes

(3-Isocyanatopropyl)triethoxysilane and (3-isocyanatopropyl)-dimethylchlorosilane are suitable isocyanated-terminated organosilanes, with different functionalities at the opposite terminals: on one end, they offer the highly reactive isocyanate group, and on the other one, they behave as trialkoxy- or dialkyl-monochlorosilanes, respectively [62]. This dual aspect made them very helpful spacers in the preparation of 3,5-dimethylphenyl carbamoylated β-cyclodextrin stationary phases [76]. This chemistry of binding (Scheme 2.1d) was first generally envisioned by Armstrong, as a potential synthetic strategy for any macrocycle structure [7], and then exploited for the immobilization of vancomycin, rifamycin B, and 3,5-dimethylphenyl-derivatized vancomycin [47].

In a typical reaction, the macrocycle is treated with a 2-3 M excess of (3-isocyanatopropyl)triethoxysilane in dry DMF. The derivative is then added to a dry DMF slurry of silica gel (~2 g of functionalized selector to 4 g of silica gel). The solution is stirred, allowed to react for 20 h at 107°C, and then cooled, filtered, and washed with methanol, 50% aqueous methanol, and methanol again, and finally dried. Surface coverage data are reported only in the case of the commercially available vancomycin CSP (see Table 2.2), immobilized by joining on average three linkers per vancomycin molecule, probably via a 3-isocyanatopropyl-silane [48].

A similar reaction was described for the immobilization of TE A$_2$-2 in dry pyridine [62], but the surface coverage accomplished was very low (31 μmol/g) (see Table 2.2), as compared with that obtained by synthetic strategy 1 for the same glycopeptide (165 μmol/g) and starting from the same silica gel. According to this procedure, a novel CSP was recently synthesized containing NVC as chiral selector and providing a substantial bonding density of 119 μmol/g (0.35 μmol/m^2) [16,17].

2.3 ENANTIOSELECTIVE HPLC SEPARATIONS USING THE GLYCOPEPTIDES CONTAINING CSPs

Glycopeptide antibiotics have successfully been used as chiral selectors to resolve the enantiomers of a variety of chiral compounds by means of both chromatographic and electrophoretic techniques. The idea of testing glycopeptide antibiotics as chiral selectors was first introduced by Armstrong and coworkers, at the Pittsburgh Conference in 1994.

As outlined in Section 2.1, these natural compounds contain in their structure multiple stereogenic centers and a variety of functional groups (see Table 2.1). For these reasons, all glycopeptide-containing CSPs are multimodal CSPs, that is, they are capable of operating in three different elution systems: NP, RP, and the new POM [1, 77]. Moreover, as illustrated in Section 2.2, glycopeptide antibiotics are covalently bonded to silica gel through multiple linkages, therefore, there is no detrimental effect when switching from one mobile phase system to another. The NP mode can be applied without any irreversible change in enantioselectivity or denaturation of macrocycles, unlike protein-based CSPs [30]. The RP mode was the first chromatographic system used in the early studies on glycopeptides containing CSPs, owing to the polar nature of the eventual analytes; the new POM refers to the approach when methanol (MeOH) and/or acetonitrile (ACN) are used as the mobile phases with small amounts of acid and/or base as the modifier. It offers the advantages of broad selectivity, high efficiency, low backpressure, short analysis time, high capacity, and excellent prospects for preparative-scale separations [78].

2.3.1 OPTIMIZATION OF CHROMATOGRAPHIC CONDITIONS

The first consideration when investigating HPLC method development protocols is the chemical structure of the analyte, in particular, the presence of functional groups capable of interacting with the stationary phase and containing or in the vicinity of the stereogenic elements [79]. Since the natural target of macrocyclic antibiotics is the N-acyl-D-alanyl-D-alanine terminus (see Section 2.1), the early choice of suitable substrates for this kind of CSPs was that of amino acids [45]. However, it turned out that the macrocyclic CSPs were very successful not only in amino acids enantioresolution, but also in the separation of a wide variety of different structures.

2.3.1.1 Mobile Phase System Choice

If a certain analyte contains more than one polar functional groups capable of interacting with the glycopeptide-containing CSP and at least one of these groups is near the stereogenic element, or even just on it, the new POM system is recommended to be tested first. In this context, suitable functional groups are hydroxyls, halogens, nitrogen in any form (primary, secondary, and tertiary), carbonyl, and carboxyl groups as well as oxidized forms of sulfur (S) and phosphorus (P). In the POM, the ratio of acid to base in the mobile phase affects the enantioselectivity (α), while the concentration of acid and base tunes the retention (k). The optimization of enantiomeric separations in the POM is discussed later in Section 2.3.1.1.1.

If no selectivity is observed at all in this mode, then the RP elution is recommended as the next step in the protocol. In this case, the parameters that affect α and k are (i) the type and percentage of the organic modifier, and (ii) the type, pH, and concentration of the aqueous buffer (see Section 2.3.1.1.2). Fortunately, there is an empirically optimal mobile phase composition for each glycopeptide CSP. Vancomycin CSP has shown its best performance for most compounds when tetrahydrofuran (THF) is used as the organic modifier at a concentration of 10% with 20 mM ammonium nitrate at pH 5.5 (no pH adjustment is needed). MeOH seems to be the preferred organic modifier for teicoplanin and ristocetin A CSPs, and 20% is usually a good starting percentage. Triethylammonium acetate (TEAA) at the concentration of 0.1% is a suitable buffer for both of these latter CSPs, although teicoplanin usually shows better selectivity at pH 4.1 (pH adjustment is needed), while ristocetin A works better at pH 6.8 [79].

When analytes lack the selectivity in the POM or RP mode, then the NP mode can be tested as the last choice. It is obvious that NP systems offer better selectivity with respect to POM and RP, but only for some less polar compounds. In particular, if a carbonyl group is present in the α- or β-position to the stereogenic center, than the substrate will have an excellent chance to be resolved in this mode. The optimization of enantiomeric separations in the NP mode is discussed later in Section 2.3.1.1.3.

For analytes containing only one functional group in the vicinity of the stereogenic centre, typically NP or RP systems are employed, depending on their solubility in such systems.

2.3.1.1.1 Optimization in POM Systems

When a separation must be achieved by POM, it is suggested [78] to start the method development with a medium concentration (0.1%) for both acid and base (1:1 ratio), where some selectivity is typically observed. Then, if retention is too high or too low, the amount of them can be increased up to 1%, or decreased to 0.001%, respectively. Generally, the ratio of acid to base can be manipulated from 4:1 to 1:4 depending on the sample charge. In a few instances, the ratios were higher, but typically, 2:1 is the standard for screening methodology. Usually, glacial acetic acid (AcOH) and triethylamine (TEA) are used as effective acid/base components, but different aliphatic carboxylic acids (formic acid, propionic acid, hexanoic acid) and bases (trimethylamine, diethylamine) were also tested as ionic modifiers for the enantioseparation of a series of alkoxysubstituted esters of phenylcarbamic acid [80]. It must be noted that these additives raise the UV cut-off of the mobile phase, so that they are unsuitable for the analysis of underivatized aliphatic amino acids. In some cases, ACN [81] and ethanol (EtOH) [82] have been used in the place of MeOH or in addition with it, in binary mobile phases mixtures [83]. Smaller amounts of trifluoroacetic acid (TFA) can be used instead of AcOH, whereas ammonia can be used as an alternative base to TEA.

TFA is advantageous in that it enhances the peak shape and efficiency for some polar compounds and its higher volatility is more desirable in liquid chromatography–mass spectrometry (LC-MS) operative conditions. The interfacing of LC methodologies with MS is discussed later in Section 2.3.1.4.

2.3.1.1.2 Optimization in RP Mode Systems

Various organic modifiers have been exploited in the RP separations on glycopeptides containing CSPs. MeOH, EtOH, 2-propanol, ACN, and THF are the most common solvents that give good selectivities for various analytes. At the same percentage, the three alcohols (MeOH, EtOH, and 2-propanol) have shown somewhat different selectivity and resolution in the separation of O-acyl carnitine derivatives [61]. Unlike traditional RP tuning of chromatographic parameters, an increase in the concentration of alcohols results in higher retention and thus better resolution owing to the decreased solubility of the polar amino acids in the MeOH-rich mobile phases. Therefore, most amino acids and small peptides can be easily resolved with an unbuffered mobile phase, at medium percentage (around 50%) of MeOH as organic modifier [84]. For compounds other than amino acids and peptides, the retention factor vs. organic modifier content curves exhibited in most cases is a characteristic U-shape [7]. For example, in the separation of 5-methyl-5-phenylhydantoin enantiomers on a vancomycin CSP, increasing the ACN concentration from 0% to 50% caused a decrease of retention and enantioselectivity; at concentrations ranging from 50% to 80%, there is low retention and the analytes elute near the dead volume of the column; in neat ACN, retention increases and enantioselectivity returns. Therefore, in RP systems, separations are achieved at relatively low percentages of organic modifiers (10–20%).

The effect of the mobile phase modifier was investigated for a series of phenoxypropionic acid (PPA) herbicides on a teicoplanin CSP [85, 86]: an increasing enantioselectivity was found with increasing MeOH content in the mobile phase, attributed to restriction of the solute association with the TE CSP, which led to favorable stereoselective interactions.

A study of the effects of mobile phase composition on retention and selectivity of some carboxylic acids and amino acids was performed on a commercially available teicoplanin CSP, under analytical conditions, on the profile of the adsorption isotherms of the enantiomers and on the overloaded separation [87].

The influence of the mobile phase composition was also studied on the enantioseparation of several cycloaliphatic β-substituted α-quaternary α-amino acids on a teicoplanin CSP [88,89]: the study revealed two distinct enantiomeric and diastereomeric discrimination mechanisms based on different interactions with the stationary phase.

For aromatic amino and hydrazino acids and several other structurally related compounds, the influence of MeOH content in both RP and POM was investigated on a teicoplanin CSP [90]. Using a hydroorganic mobile phase, complete enantioseparations of α-methylamino acids were not attained. However, this type of separation was suitable for the enantiomers of DOPA. Further experiments performed by the same authors in POM allowed the complete enantioseparation of α-methyl-DOPA enantiomers [91].

With regard to mobile phase pH, it must be taken into account that retention and selectivity of molecules possessing ionizable (acidic or basic) or even neutral functional groups can be affected by altering the pH value. Typically, the use of lower pH has the effect of suppressing the unselective retention of chiral analytes on the stationary phase, which in turn enhances the chiral interactions, and leads to higher resolution [78]. The retention and selectivity of a series of dansyl amino acids on a teicoplanin CSP were investigated over a wide range of mobile phase pH (sodium citrate buffer–methanol, 90:10) [92]. The approach, based on the development of various equilibria, was carried out to describe the retention behavior of the solutes in the considered chromatographic system. The equilibrium constants corresponding to the transfer of the anionic and zwitterionic forms of the dansyl amino acids from the mobile to the stationary phase were also determined.

As previously observed, for most free amino acids and small peptides unbuffered hydroorganic mixtures are enough to yield good enantioseparations; however, for some bifunctional amino acids and most other compounds, an aqueous buffer is usually necessary to enhance resolution. TEAA and ammonium nitrate are the most effective buffer systems, while sodium citrate was also effective for the separation of 2-arylpropionic acids (profens) on vancomycin CSPs [78], and ammonium acetate is the most widely used and appropriate in view of LC-MS applications (see Section 2.3.1.4). Small changes in ammonium acetate concentration of MeOH–water (90:10) mobile phases scarcely affected retention and—to a lesser extent—enantioselectivity of carnitine derivatives [61].

2.3.1.1.3 Optimization in NP Mode Systems

Typical NP conditions involve mixtures of n-hexane or n-heptane with alcohols (EtOH and 2-propanol). In many cases, the addition of small amounts (<0.1%) of acid and/or base is necessary to improve peak efficiency and selectivity. Usually, the concentration of alcohols tunes the retention and selectivity: the highest values are reached when the mobile phase consists mainly of the nonpolar component (i.e., n-hexane). Consequently, optimization in NP mode simply consists of finding the ratio n-hexane/alcohol that gives an adequate separation with the shortest possible analysis time [30]. Normally, 20% EtOH gives a reasonable retention factor for most analytes on vancomycin and TE CSPs, while 40% is more appropriate for ristocetin A-based CSPs. Ethanol normally gives the best efficiency and resolution with reasonable backpressures. Other combinations of organic solvents (ACN, dioxane, methyl *tert*-butyl ether) have successfully been used in the separation of chiral sulfoxides on five different glycopeptide CSPs, namely, ristocetin A, teicoplanin, TAG, vancomycin, and VAG CSPs [46].

2.3.1.2 Flow Rate

As a general rule, in the case of CSPs featuring hydrophobic pockets, a decrease of mobile phase flow-rate results in an increase of chromatographic resolution (Rs), as a consequence of better stationary phase mass transfer [78]. This change has significant impact mostly in RP mode [17]. In the NP enantioselective separations of two test solutes (4-hexyl-5-cyano-6-methoxy-3,4-dihydro-2-pyridone and α-methyl-α-phenylsuccinimide) on TE CSP, it has been observed [30] that flow-rate (ranging from 0.5 to 2.0 mL/min) does not affect enantioselectivity, but does affect the separation efficiency, expressed as number of theoretical plates (N). This has a direct implication in the resolution trend, owing to the well-known existing relationship between Rs and N. Decreasing the flow-rate from 2.0 to 1.0 mL/min enhanced the resolution by 20–30%, while further decreases did not produce comparable increases. Some effect has also been observed in the POM for analytes with retention factors lower than 1 [78].

2.3.1.3 Temperature

It is apparent from early observations [93] that there are at least two different effects exerted by temperature on chromatographic separations. One effect is the influence on the viscosity and on the diffusion coefficient of the solute: raising the temperature reduces the viscosity of the mobile phase and also increases the diffusion coefficient of the solute in both the mobile and the stationary phase. This is largely a kinetic effect, which improves the mobile phase mass transfer, and thus the chromatographic efficiency (N). The other completely different temperature effect is the influence on the selectivity factor (α), which usually decreases, as the temperature is increased (thermodynamic effect). This occurs because the partition coefficients and therefore, the Gibbs free energy difference ($\Delta G°$) of the transfer of the analyte between the stationary and the mobile phase vary with temperature.

The effect of temperature on the retention of a series of β-methyl amino acids was investigated on a TE CSP, by using either subambient or elevated temperatures (1.5–50°C) [93]. Linear van't Hoff plots were observed in the studied temperature range, and the apparent changes in enthalpy ($\Delta H°$), entropy ($\Delta S°$), and Gibbs free energy ($\Delta G°$) were calculated. The values of the thermodynamic parameters have been shown to depend on the structures of the investigated compounds.

A parallel study [94] was performed by the same authors on the retention of model compounds (tryptophan, erythro- and threo-β-methyltryptophan, N-carbobenzyloxy-tryptophan, N-(3,5-dinitro-2-pyridyl)-tryptophan, 1-[5-chloro-2-(methylamino) phenyl]-1,2,3,4-tetrahydroisoquinoline, and γ-phenyl-γ-butyrolactone) on a risto-cetin A-based CSP, using the three RP, POM, and NP elution systems. Also, in this case, the natural logarithms of the retention factors ($\ln k$) of the investigated compounds depended linearly on the inverse of temperature ($1/T$).

A reexamination of the chiral discrimination of PPA herbicides was done on a TE CSP using the perturbation method to calculate the solute distribution isotherms [86]. The effect of temperature was well described by the bi-Langmuir model and enabled confirmation of the previous results by the same authors [85].

A comprehensive study on the temperature effect was done in 2004 for 71 chiral compounds on four glycopeptide CSPs: TE, TAG, ristocetin A, and vancomycin phases, using the three RP, POM, and NP elution systems [95]. The separations were studied in the 5–45°C temperature range. Peak efficiencies always increased with temperature, but in only 17% of the separations studied, a small increase of the resolution was observed. In the rest of the cases, the resolution decreased or even vanished when temperature increased. All van't Hoff plots were linear, showing that the selectors did not undergo any conformational changes in the temperature range studied. The calculated enthalpy and entropy variations showed that the interaction of the solute with the stationary phase was always enthalpy driven, with NP and RP mobile phases. It could be enthalpy as well as entropy driven with POM mobile phases, strongly depending on the structure of the solute. The plots of $\Delta(\Delta H)$ *vs.* $\Delta(\Delta S)$ were linear in most cases (enthalpy–entropy compensation).

Comparative studies on thermodynamic parameters were recently performed by the same group of authors toward a set of potential local anesthetic drugs, using POM mobile phases. The investigated chiral selectors were TAG [96], its methylated derivative (Me-TAG) (see Section 2.1.2.7.6) [97], vancomycin, and teicoplanin [98], and the temperature range studied was 0-50°C. In each study, van't Hoff plots were linear and allowed the determination of the thermodynamic parameters of the considered chromatographic system. The gathered information was useful in understanding the thermodynamic driving forces for retention of the investigated compounds, and to give an insight into the chiral recognition mechanism operated by the aforementioned selectors. A detailed discussion on mechanism of enantioseparation is given later in Section 2.4. van't Hoff plots were finally used to determine the thermodynamic data for aryl-substituted β-lactams enantiomers; the thermodynamic data revealed that all the compounds in this study undergo separation via the same enthalpy-driven chiral recognition mechanism [99].

2.3.1.4 Detection Strategies

The analysis of underivatized amino acids has always represented a problem in the choice of the proper detection system, both for chromatographic and electrophoretic applications, for the lack of chromophore groups in the majority of them. They indeed show very weak UV absorption in the same low wavelength UV absorption region (205–220 nm) of commonly used mobile phase additives, with consequent increased background and loss of detection sensitivity [100, 101]. On the other hand, the use of refracting index detection, aside from sensitivity issues, prevents the application of gradient elution for the analysis of complex mixtures containing species with large retention differences [61]. Pre- or postcolumn derivatization procedures with UV absorbing or fluorescent reagents can represent a chance to enhance the detection response of amino acids, but derivatization is always a time-consuming step in the separation process.

The majority of the HPLC chiral separations obtained with glycopeptides-containing CSPs are anyway achieved by UV/visible (see Section 2.3.2) or fluorescence [102] detection.

A simple and efficient alternative to the traditional UV detection of amino acids and related compounds is nowadays represented by the evaporative light scattering (ELS) detector, which allows the direct chromatographic separation, with no need for preliminary derivatization. In the field of glycopeptides-based CSPs, it was applied for the first time in the chromatographic resolution of carnitine and *O*-acylcarnitine enantiomers on a TE CSP [61]. The considered compounds are nonvolatile solids and gave optimal ELS response under a variety of experimental conditions (buffered and unbuffered mobile phases, flow-rates from 0.5 to 1.5 mL/min, different kind and proportion of organic modifiers), with signal-to-noise ratios always much larger than UV detection. ELS detection was also usefully exploited in the enantioseparation of a variety of polar compounds by hydrophilic interaction chromatography (HILIC) on a commercially available TE CSP [103], and in the enantiomeric excess determination of an AMPA receptor antagonist [104].

In addition to ELS, charged aerosol (CA) or corona detector has more recently been introduced as a very promising HPLC detection system [105]: while the sensitivity of the two systems is quite close, CA detector offers the advantage of a nearly linear response factor, particularly crucial for the assessment of enantiomeric purities, whereas ELS provides a nonlinear response at very low or high levels of analytes, resulting from several light scattering mechanisms and particle size distribution.

Mass spectrometric detection has become a widely employed method for the analysis of chiral compounds on glycopeptide-containing CSPs, without the need for preliminary derivatization. Various ionization techniques including electrospray ionization (ESI), atmospheric pressure chemical ionization (APCI), and atmospheric pressure photoionization (APPI) interfaced with chiral liquid chromatographic methods were described in terms of their ionization efficiencies, matrix effects, and limitations [106]. In the case of ESI, care must be taken to avoid situations where nonlinearity effects (such as those due to ion suppression at high concentration) occur; thus, diluted solutions must be prepared to achieve 1:1 areas ratio for a given racemate, especially in the case of high α value. Calibration curves in APCI are frequently linear over five orders of magnitude, whereas four are typical for ESI. Several compounds of different classes have been analyzed by coupling existing HPLC methods with MS detection, or ad hoc developing new LC-MS methods, particularly in the monitoring of biological fluids: sotalol [107], salbutamol [108], propranolol [106, 109], albuterol [110], methylphenidate (MPH) [111–113], 2-hydroxyglutaric acid [114], glyceric acid [115], underivatized amino acids [116], native amino acids and peptides [117,118], fluoxetine [106, 119], terbutaline [106,120], benidipine [121], derivatized and underivatized theanine [122], amphetamine derivatives [123], and pipecolic acid [124].

It must be noted that most existing HPLC methods developed using UV detection cannot be directly used with LC-MS owing to various mobile phase and additive incompatibilities, and simple changes in the mobile phase composition often results in diminished or lost enantioselectivity. For example, NP solvents are highly incompatible when coupled with MS ionization sources such as ESI; however, the compatibility of the RP and POM modes with the MS interface makes them attractive direct approaches for the LC-MS analysis of chiral compounds. Some general rules of thumb when converting HPLC methods to MS amenable methodologies are gathered in a recent study [125], and are as follows: (i) POM mobile phases are the most compatible and easily adaptable to the coupling to ESI-MS, yielding better limits of detection (as low as 100 pg/mL) and sensitivity over RP methods. (ii) NP methods are incompatible with direct coupling to ESI-MS; they can be used if postcolumn dilutions of a large excess of ESI-MS compatible solvents are acceptable in terms of sensitivity and band broadening. (iii) RP methods can be easily switched to APCI-MS with much greater sensitivity, while for ESI interfacing low water contents must be used as it tends to decrease the ionization efficiency. (iv) Ammonium trifluoroacetate enhances ionization for molecules with amine or amide functionalities. (v) Optimized concentrations of additives should be maintained, when compatible with the MS interface.

Very recently, ethoxynonafluorobutane (ENFB) was evaluated as a safe and environmentally friendly NP solvent for the enantioselective separation of 15 compounds on two commercially available glycopeptides CSPs (vancomycin and teicoplanin), with APCI-MS detection [126].

A special mention in the field of enantioselective HPLC separations must be made of chiro-optical detection systems, such as circular dichroism (CD) and optical rotation (OR), which can be also used to circumvent the low UV detectability of chromophore-lacking samples [40, 61]. While sensitivity of chiro-optical detection is not always sufficient to perform enantiomeric trace analysis, the stereochemical information contained in the bisignate spectropolarimetric response is useful in establishing elution order for those compounds not available as single enantiomers of known configuration. An example of application of different online detection systems (UV and CD at 254 nm) in the enantioselective separation of a racemic sulfoxide on a commercially available TAG CSP is reported in Figure 2.12, under NP conditions.

2.3.1.5 Selector Coverage and Binding Chemistry

Only few studies addressed the effect of the silica surface concentration of glycopeptides antibiotics on the chiral performances. The first was performed by Armstrong on a TE CSP toward five racemic test solutes [30]. The study evidenced a modest increase of α and Rs values with increasing the initial selector concentration in the grafting reaction mixture, but no data were reported regarding the effective

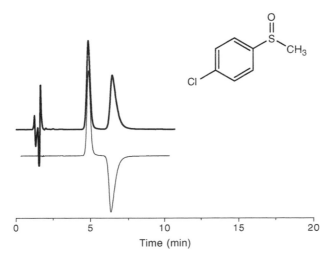

FIGURE 2.12 Traditional (bold line, UV at 254 nm) and chiro-optical (thin line, CD at 254 nm) detection in the enantiomeric resolution of a racemic sulfoxide. Column: Chirobiotic TAG (250×4.6 mm ID); eluent: (EtOH/MeOH 80/20)/n-hexane $= 50/50$; flow-rate: 2 mL/min; $T = 25°C$. $k_1 = 2.19, \alpha = 1.48,$ Rs $= 2.57$.

surface coverage in terms of μmol/g (or μmol/m^2), obtained from the reaction. An increase in the number of theoretical plates (N) was also observed with increasing the concentration of TE. One possible explanation was that a more dense surface coverage of selector could prevent the deep entrapping of the analyte between the silica surface and the bulky chiral selector tethered to it, resulting in a lower resistance to mass transfer.

In 1999, the effect of the selector coverage was studied on ristocetin A by covalently bonding to silica gel three different amounts of macrocyclic antibiotic [127]. Also, in this case, no data were reported regarding the effective surface coverage, in terms of μmol/g, obtained from the grafting reactions. The three CSPs obtained (high, medium, and low loaded) were examined and compared in the RP, POM, and NP modes toward a variety of racemic test compounds. In general, both retention and resolution for most compounds analyzed decreased with decreasing the amounts of ristocetin A. The medium loaded CSP gave good efficiencies (for almost all samples analyzed in the three mobile phase modes) compared to the high loaded CSP. It is important to note that it was the more retained enantiomer of each analyte that appeared to be most affected by changes in ristocetin A loading.

A very recent study [128] deals with the comparison of two commercially available vancomycin-based CSPs with different surface coverage of the chiral selector in the enantioseparation of β-blockers and profens, by RP and POM separation modes. Higher retention and better resolution were obtained on the CSP with higher coverage of vancomycin in both the separation modes. However, in the case of profens, higher retention was not always accompanied by an improvement of the enantioselectivity in the RP mode. An accurate study of the influence of the mobile phase composition was also performed in both the separation modes.

Another factor that remarkably affects the enantioresolution of given enantiomeric pairs has been shown to be the binding chemistry used for the silica immobilization of glycopeptides (see Section 2.2). This was illustrated in the case of ristocetin A, which was covalently bonded to silica microparticles by immobilization onto epoxy-activated silica under mild conditions [22], and compared with the corresponding commercially available CSP, where the macrocycle was immobilized as previously reported for vancomycin, rifamycin B, thiostrepton [7], and teicoplanin [30]. The comparison proved that immobilization of ristocetin A onto epoxy-activated silica could significantly improve the enantioselectivity and the resolution of the corresponding CSP in the separation of α-amino acids under RP conditions [22].

2.3.2 Classes of Compounds Separated

Since the natural target of macrocyclic antibiotics is the N-acyl-D-alanyl-D-alanine terminus (see Section 2.1), the early choice of suitable substrates for this kind of CSPs was that of amino acids [45]. However, it turned out that the macrocyclic CSPs were very successful not only in amino acids enantioresolution, but also in the separation of a wide variety of different structures. The early stages of application of macrocyclic antibiotics have been surveyed in the different fields of chromatography [1, 2]. A summary of the different categories of chiral compounds separated by HPLC on glycopeptides containing CSPs is reported in Table 2.3.

TABLE 2.3
Classes of Compounds Separated by HPLC on Glycopeptides Containing CSPs

Classes	Glycopeptide	References
Amino acids (including underivatized and N-protected α- and β-amino acids, either proteinic and unusual)	Teicoplanin, ristocetin A, vancomycin, avoparcin, TAG, A-40,926, CDP-1, Me-TAG, eremomycin	7, 16, 23, 30, 33, 37, 40, 41, 45, 55–58, 81, 84, 87–90, 92–94, 116, 117, 124, 129–154, 170
Peptides	Teicoplanin, ristocetin A, TAG	30, 33, 84, 118, 155
Cyclic amines, amides and imides	Vancomycin, teicoplanin, ristocetin A	7, 30, 33, 157
α-Hydroxycarboxylic acids	Teicoplanin, A-40,926, MDL 63,246, TAG, Me-TAG	30, 41, 44, 45, 58
β-Adrenoreceptors antagonists	Teicoplanin, vancomycin, TAG, A-40,926, Me-TAG	7, 30, 41, 45, 58, 102, 106–110, 128, 147, 158–164
$β_2$-Adrenoreceptors agonists	Teicoplanin, vancomycin, TAG	82, 83, 120, 165
Carnitine and O-acylcarnitine derivatives	Teicoplanin, A-40,926, TAG	41, 45, 61, 166
Ruthenium(II) polypyridyl octahedral complexes	Teicoplanin, TAG	167
NSAIDs	Vancomycin, teicoplanin, ristocetin A, A-40,926, avoparcin, CDP-1, Me-TAG	7, 30, 33, 37, 55, 58, 63, 128, 168
Sulfur-containing compounds	Ristocetin A, teicoplanin, TAG, vancomycin, VAG	46
β-Lactams	Ristocetin A, teicoplanin, TAG, vancomycin, VAG	99, 169, 170
Amphetamine derivatives	Vancomycin	111–113, 123
Benzodiazepines, phenothiazines, and barbiturates	Vancomycin, DMP vancomycin, teicoplanin, TAG, ristocetin A	7, 33, 157, 171–173
Calcium channel blockers	Vancomycin, teicoplanin, A-40,926, ristocetin A	7, 37, 41, 106, 121, 125, 174

2.3.2.1 Amino Acids

A comprehensive review (with 119 references cited) on the HPLC separation of amino acid enantiomers and small peptides on macrocyclic antibiotics CSPs has only just been published in 2006 [3], including underivatized and N-protected α- and β-amino acids (either proteinic and unusual). A parallel monograph on the separations of α-amino acids on the new glycopeptide eremomycin CSP (see Section 2.1.2.2) completes the literature published till date on this subject [23]. As a general trend, the peak corresponding to the more retained enantiomer of amino acids is characterized by a greater asymmetry factor than that corresponding to the first eluting (and consequently by a lower number of theoretical plates) owing to a slower mass transfer kinetics.

2.3.2.1.1 Underivatized α-Amino Acids

Underivatized α-amino acids have successfully been separated on almost all the gly-copeptides CSPs: teicoplanin [30, 84, 87], ristocetin A [33], TAG [45], A-40,926 [41], CDP-1 [55], and eremomycin [22, 23] in different chromatographic conditions. Vancomycin and avoparcin CSPs are not optimal for such substrates. A commercially available teicoplanin CSP was successfully employed for the determination of the enantiomeric purity of a sample of L-arginine [129]. For the lack of chromophore groups in the majority of them (i.e., aliphatic α-amino acids), UV detection at 205–210 nm usually yielded to loss of detection sensitivity (see Section 2.3.1.4). This problem was circumvented by the recent interfacing to the MS detection [116,117].

An example of separation of underivatized α-amino acids on a TAG CSP is reported in Figure 2.13.

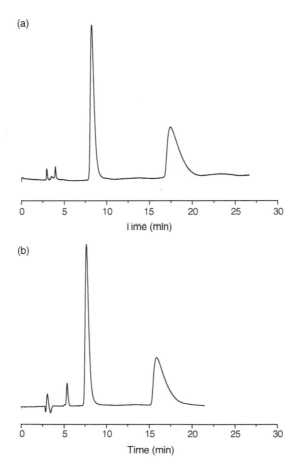

FIGURE 2.13 Separation of underivatized (a) phenylalanine (Phe) and (b) tyrosine (Tyr) . Column: TAG CSP (250 × 4.6 mm ID); eluent: MeOH/H$_2$O (85/15) + ammonium acetate 20 mM; flow-rate: 1 mL/min; $T = 25°C$; UV detection at 254 nm. $k_1 = 1.73$, $\alpha = 2.75$ for Phe; $k_1 = 1.54$, $\alpha = 2.75$ for Tyr.

2.3.2.1.2 N-blocked α-Amino Acids

Derivatization of α-amino acids with N-protecting groups introduces functionalities that could be important for the chiral recognition mechanism and above all improves the UV detectability (UV at 254-280 nm); α-amino acids with the following N-protecting groups have been separated on the glycopeptide-based CSPs: acetyl (Ac) [30, 33, 41, 45, 130], benzoyl (Bz) [7, 30, 33, 37, 58, 131], benzyloxy-carbonyl (CBZ) [7, 30, 33, 37, 131, 132], *tert*-butyloxycarbonyl (*t*-BOC) [30, 33, 37, 57, 58, 133, 134], carbamyl [7, 33, 58], 3,5-dinitrobenzoyl (DNB) [37,81], 2,4-dinitrophenyl (DNP) [30, 33], 3,5-dinitro-2-pyridyl (DNPy) [30, 33, 37], 5-dimethylamino-1-naphthalene-sulfonyl (Dns) [7, 16, 30, 33, 37, 58, 92, 135–138], formyl [30, 33, 58], 9-FMOC [33, 37, 132, 139], methyloxycarbonyl (MOC) [140], phthaloyl (Pht) [7, 33, 37], and phenyl isothiocyanate (PHES) [131]. An example of separation of N-acetyl amino acid is given in Figure 2.14.

2.3.2.1.3 Unusual and Nonproteinic Amino Acids

Unusual amino acids include a class of unnatural α-amino acids such as phenylalanine, tyrosine, alanine, tryptophan, and glycine analogs, and β-amino acid analogs containing 1,2,3,4-tetrahydroisoquinoline, tetraline, 1,2,3,4-tetrahydro-2-carboline, cyclopentane, cyclohexane, cyclohexene, bicyclo[2.2.1]heptane or heptene skeletons. Different selectors were exploited for the separation of unusual amino acids, most of the production being made by Péter and coworkers: teicoplanin [41, 56, 84, 90, 93, 124, 141–144], ristocetin A [33, 94, 145, 146], and TAG [56, 147]. Enantiomeric and diastereomeric separations of cyclic β-substituted α-amino acids were reported by other authors on a teicoplanin CSP [88, 89]. Ester and amide derivatives of tryptophan and phenylalanine were recently analyzed on a Me-TAG CSP [58].

2.3.2.1.4 Unnatural β-Amino Acids

Direct and indirect chromatographic methods were developed and compared in systematic examinations for the enantioseparation of β-amino acids; direct separation of underivatized analytes involved the use of commercially available Crownpak CR(+), teicoplanin, and ristocetin A CSPs [148], while indirect separation was based on precolumn derivatization with 2,3,4,6-tetra-O-acetyl-β-D-glucopyranosyl isothiocyanate (GITC) or N−α-(2,4-dinitro-5-fluorophenyl)-L-alaninamide (FDAA, Marfey's reagent), with subsequent separation on a nonenantioselective column.

Later, a TAG CSP was included in the aforementioned investigation by direct and indirect methods, the chiral derivatizing agent being, in this case, the new reagent (S)-N-(4-nitrophenoxycarbonyl)phenylalanine methoxyethyl ester [149].

Direct and indirect RP chromatographic methods were developed and compared in systematic examinations for the enantioseparation of several unnatural β-amino acids, including some β-3-homo amino acids. Direct separation of underivatized analytes involved the use of commercially available TE and TAG CSPs [150], while indirect separation was based on precolumn derivatization with (1S,2S)-1,3-diacetoxy-1-(4-nitrophenyl)-2-propylisothiocyanate and (S)-N-(4-nitrophenoxycarbonyl)phenylalanine methoxyethyl ester.

A comparative and comprehensive study on the enantioseparation of several unnatural β-amino acids has only just been published in 2006, testing six commercially

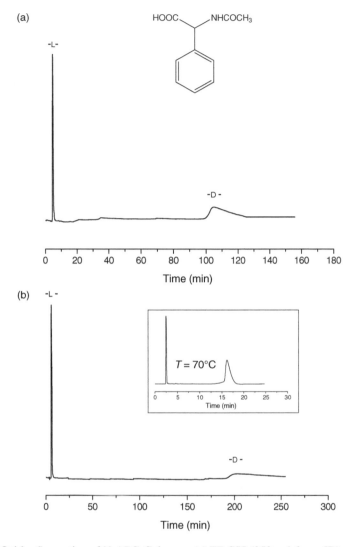

FIGURE 2.14 Separation of N-APG. Columns: (a) TE CSP (250 × 4.6 mm ID) and (b) TAG CSP (250 × 4.6 mm ID); eluent: MeOH/H$_2$O (40/60) + ammonium acetate 20 mM (pH = 5.0); flow-rate: 1 mL/min; $T = 25°C$; UV detection at 220 nm. $k_1 = 0.54$, $k_2 = 34.81$, $\alpha = 64.46$ on TE CSP; $k_1 = 0.80$, $k_2 = 65.33$, $\alpha = 81.66$ on TAG CSP ($\Delta\Delta G = 2.6$ kcal/mol).

available CSPs containing teicoplanin (Chirobiotic T and T2), teicoplanin aglycone (Chirobiotic TAG), vancomycin (Chirobiotic V and V2), and ristocetin A (Chirobiotic R) as chiral selectors [151].

It has finally been described the application of glycopeptide A-40,926 in the analytical and semipreparative separations of linear and cyclic β-amino acids (see Figure 2.15) [40].

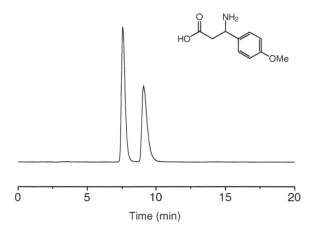

FIGURE 2.15 Separation of a β-amino acid on a Λ-40,926 CSP. (From D'Acquarica, I. et al., *Tetrahedron: Asymmetry*, 11, 2375, 2000. With permission from Elsevier, Copyright (2000).)

2.3.2.1.5 Secondary Amino Acids

Conformationally constrained secondary aromatic amino acids (containing 1,2,3,4-tetrahydroisoquinoline, 1,2,3,4-tetrahydronorharmane-l-carboxylic acid, and 1,2,3,4-tetrahydro-3-carboxy-2-carboline moieties) were separated using three different chromatographic approaches: (i) direct HPLC separation of underivatized analytes on a teicoplanin-containing CSP; (ii) indirect HPLC separation based on precolumn derivatization with GITC or FDAA, with subsequent separation on a noncnantioselective column; (iii) direct GC−MS separation of *N*-trifluoroacetylated isobutyl ester derivatives on a Chirasil-L-Val column [152].

Direct and indirect chromatographic methods were developed and compared in systematic examinations for the enantioseparation of 10 aliphatic secondary α-amino acids; direct separation of underivatized analytes involved the use of commercially available teicoplanin and ristocetin A CSPs [153], while indirect separation was based on precolumn derivatization with (*S*)-*N*-(4-nitrophenoxycarbonyl)phenylalanine methoxyethyl ester, with subsequent separation on a nonenantioselective column.

Later, a commercially available TAG CSP was tested in the enantioseparation of 10 secondary α-amino acids, by using RP mobile mode systems [154]. The chromatographic results, compared with those obtained on a native teicoplanin CSP, were given as the retention, separation, and resolution factors, together with the enantioselective free energy difference corresponding to the separation of the investigated enantiomers.

2.3.2.2 Peptides

2.3.2.2.1 Dipeptides and Tripeptides

The development of chiral separation methods for dipeptides is of relevance for purity controls, for checking racemization processes in peptide syntheses, and for the investigation of peptide and protein hydrolysates. Since their introduction as chiral selectors for CSPs, macrocyclic antibiotics have successfully been applied in the separation of short-chain epimeric peptides. Dipeptides and tripeptides were separated on a teicoplanin CSP by RP mode [30,84], while some N-phthaloyl-protected glycyl-dipeptide epimers were successfully resolved on a ristocetin A CSP [33], by RP and NP mode, together with several unprotected dipeptides and tripeptides. A TAG CSP was also exploited in the separation of some glycyl- and diastereomeric dipeptides and tripeptides by micro-HPLC [155] and CEC [156], and compared with a native TE CSP regarding selectivity, efficiency, and separation time. The two phases were obtained starting from 3.5-μm silica microparticles.

2.3.2.2.2 Small Peptides

Forty-two peptides (with up to 13 amino acids in length) with single amino acid polymorphisms, belonging to 11 peptide families were separated on three commercially available CSPs (teicoplanin, TAG, and ristocetin A) in the RP mode using ESI-MS-compatible mobile phases [118]. The peptides families studied were angiotensins, bradykinins, α-bag cell peptides, β,γ-bag cell factors, β-casomorphins, dynorphins, enkephalins, leucokinins, neurotensins, substance P, vasopressins, and lutinizing hormone releasing factor. High selectivity was observed for single amino acid substitutions (achiral and chiral), regardless of their position in the peptide sequence.

2.3.2.3 Cyclic Amines, Amides, and Imides

Cyclic amines (including local anesthetic drugs) and amides were among the first classes of chiral compounds investigated in the early stages of the application of macrocyclic antibiotics as chiral selectors; therefore, they were screened on vancomycin [7], teicoplanin [30], and ristocetin A [33] CSPs, under RP mode systems. Cyclic imides (including barbiturates, piperidine-2,6-diones, and mephenytoin) have been separated on a vancomycin CSP [157], under NP and RP mobile phase conditions.

2.3.2.4 α-Hydroxycarboxylic Acids

α-Hydroxycarboxylic acids [mandelic acid, 4-hydroxymandelic acid, 4-chloromandelic acid, 3-hydroxy-4-methoxymandelic acid, 3-(4-hydroxyphenyl) lactic acid] have first been investigated on a TE CSP [30]; later, they were tested on A-40,926 [41], TAG [45], and MDL 63,246 CSPs [44]. Mandelic acid derivatives were recently analyzed on a Me-TAG CSP [58].

2.3.2.5 β-Adrenoreceptors Antagonists

The β-adrenoreceptors antagonists (also called β-blockers) comprise a group of chiral drugs that are mostly used in the treatment of cardiovascular disorders such as hypertension, cardiac arrhythmia, or ischemic heart disease. Teicoplanin is the chiral selector most exploited for the enantioseparation of this class of compounds, followed by vancomycin. Several β-blockers have been analyzed, particularly in the monitoring of biological fluids: oxprenolol [7, 30, 58, 107, 128], alprenolol [7, 58, 107, 128], salbutamol [108, 158], sotalol [107], albuterol [110, 159], atenolol [41, 45, 58, 107, 128, 160], metoprolol [30, 58, 102, 106, 107], α-hydroxymetoprolol [102], propranolol [7, 58, 106, 107, 109, 128, 161], arotinol [162], bambuterol [163], pindolol [41, 45, 58, 106, 128], acebutolol [58, 128], and carvedilol [164].

An example of LC-MS separation of four racemic β-blockers achieved on a TE CSP is reported in Figure 2.16. The chromatographic assay developed was applied in the direct HPLC injection of human plasma containing sotalol on a chiro-RAM-TE support [64] (Figure 2.17).

2.3.2.6 β_2-Adrenoreceptors Agonists

The β_2-adrenoreceptors agonists such as clenbuterol, cimaterol, and mabuterol are a recent group of chiral drugs, originally used for the treatment of pulmonary diseases. Clenbuterol was first quantitatively determined in human plasma on a TE CSP [82, 165]; afterward, the investigation was extended to commercially available TAG and vancomycin CSPs, and included also cimaterol and mabuterol [83]. It has been shown by a comparison between the three CSPs that enantioselectivity for this class of compounds was in the order TAG > vancomycin > teicoplanin.

Online-dual-column extraction coupled to enantioselective LC-MS was used for the determination of terbutaline enantiomers in human plasma on a commercially available TE CSP [120].

2.3.2.7 Carnitine and *O*-Acylcarnitine Derivatives

Carnitine is a vitamin-like quaternary ammonium salt, playing an important role in the human energy metabolism by facilitating the transport of long-chained fatty acids across the mitochondrial membranes. An easy, fast, and convenient procedure for the separation of the enantiomers of carnitine and *O*-acylcarnitines has been reported on a lab-made teicoplanin-containing CSP [61]. The enantioresolution of carnitine and acetyl carnitine was enhanced when tested on a TAG CSP, prepared in an identical way [45]. Higher α values were reached also in the case of A-40,926 CSP [41].

A new chiral–achiral tandem-columns arrangement based on a commercially available TAG CSP connected in series with a Spherisorb S5 SCX column was developed for the enantio- and chemoselective dosage of propionyl L-carnitine and relative impurities in pharmaceutical batches [166]. Some chromatograms relative to the separation of carnitine and carnitine derivatives on a teicoplanin CSP are collected in Figure 2.18.

2.3.2.8 Ruthenium(II) Polypyridyl Octahedral Complexes

While the enantiorecognition ability of glycopeptides CSPs toward chiral species carrying free carboxylic moieties was not surprising, considering the established propensity of the glycopeptides to bind the natural target at the carboxy-terminus, the large selectivity for the enantiomers of octahedral ruthenium complexes was totally unexpected [167]. Ruthenium(II) polypyridyl complexes with ancillary

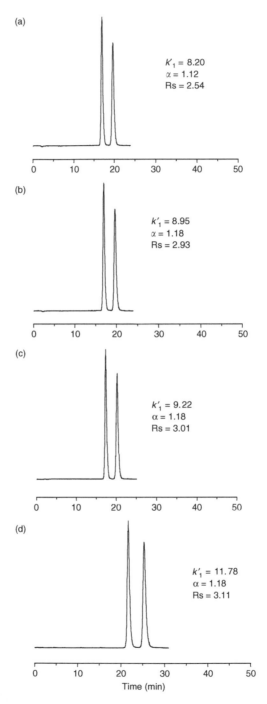

FIGURE 2.16 Direct chromatographic resolution of four racemic β-adrenergic blockers: (a) oxprenolol, (b) alprenolol, (c) metoprolol, and (d) propranolol. (From Badaloni, E. et al., *J. Chromatogr. B*, 796, 45, 2003. With permission from Elsevier, Copyright (2003).)

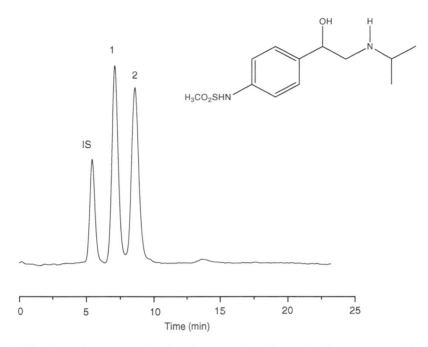

FIGURE 2.17 Direct HPLC injection of human plasma from a healthy volunteer at 4 h after 80 mg single oral administration of racemic sotalol. IS = internal standard (L-norephedrine, 0.50 μg/ml); **1** = (+)-(*S*)-sotalol; **2** = (−)-(*R*)-sotalol. Column: chiro-RAM-TE (250 × 4.5 mm ID); eluent: MeOH + ammonium formate 2.5 mM; flow-rate: 1.5 mL/min; *T* = 30°C; detection: MS (ESI +), source block at 100°C, nitrogen at a flow-rate of 600 L/h, desolvation temperature at 150°C; cone and capillary voltages at 40 V and 3.6 kV, respectively. SIR mode at 134, 255, and 273 *m/z*.

ligands like 2,2'-bipyridine ($[Ru(bpy)_3]^{2+}$), 1,10-phenanthroline ($[Ru(phen)_3]^{2+}$), and 4,7-diphenyl-1,10-phenanthroline ($[Ru(dpphen)_3]^{2+}$) showed enantioselective associations with immobilized teicoplanin and TAG (Figure 2.19), the extent of discrimination between the Δ and Λ enantiomers increasing with the size of the flat portion of the aromatic ligands and decreasing with the steric bulk at the distal positions.

2.3.2.9 Nonsteroidal Antiinflammatory Drugs

Non steroidal antiinflammatory drugs were among the first classes of chiral compounds investigated in the early stages of the application of macrocyclic antibiotics as chiral selectors; therefore, they were screened on vancomycin [7], teicoplanin [30], ristocetin A [33] CSPs under RP mode systems, and on avoparcin CSP under NP mode systems [37]. The enantioresolution of a variety of profens was later reported on commercially available vancomycin CSPs [128, 168], and recently on a ME-TAG CSP [58]. Ibuprofen enantiomers were also separated on a CDP-1-containing CSP [55]. Glycopeptide A-40,926 CSP was successfully employed in the analytical and semipreparative separation of 2-arylpropionic acids [63].

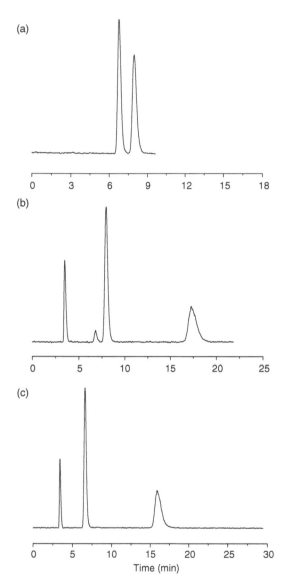

FIGURE 2.18 Separation of (a) carnitine, (b) acetyl carnitine, and (c) propionyl carnitine. Column: TE CSP (250×4.6 mm ID); eluent: MeOH/H_2O (85/15) + ammonium acetate 20 mM; flow-rate: 1 mL/min; $T = 25°C$; ELSD detection. $k_1 = 1.10$, $\alpha = 1.33$ for (a); $k_1 = 1.46$, $\alpha = 3.04$ for (b), $k_1 = 1.25$, $\alpha = 3.55$ for (c). Peaks at 4 min in (b) and (c) correspond to NH$_4$Cl. Elution order: (R) before (S).

2.3.2.10 Sulfur-Containing Compounds

A set of sulfoxides, tosylated sulfilimines, and sulfinate esters were separated using five different commercially available glycopeptides CSPs, namely ristocetin A, teicoplanin, TAG, vancomycin, and VAG, and seven mobile phases (three NP, two RP,

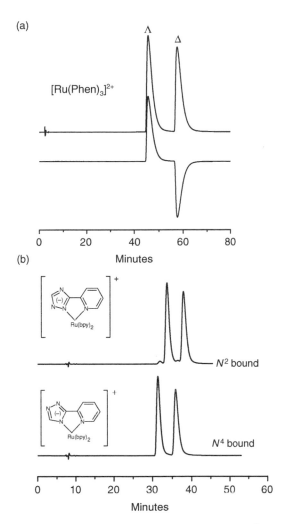

FIGURE 2.19 Enantioselective chromatography of (a) $[Ru(phen)_3]^{2+}$ and (b) $[Ru(bpy)_2]^+$ complexes on a TE CSP. (From Gasparrini, F. et al., *Tetrahedron: Asymmetry*, 11, 3535, 2000. With permission from Elsevier, Copyright (2000).)

and two POM) [46]. Notably, this study was the first example of application of the VAG CSP in enantioselective HPLC separations.

A series of unusual cysteine derivatives were easily resolved on a teicoplanin CSP under RP conditions (Figure 2.20). As it can be observed, the TE CSP shows a good chemo- and enantioselectivity toward a mixture of chemically different pairs.

2.3.2.11 β-Lactams

This class of compounds has been investigating on glycopeptides-containing CSPs for the past 2 years. Twelve underivatized β-lactams were separated on commercially

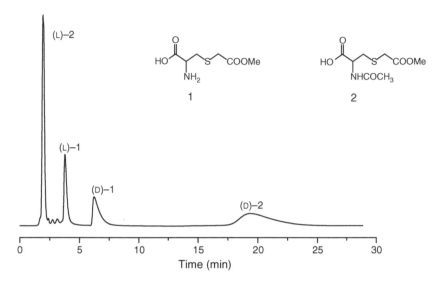

FIGURE 2.20 Direct chromatographic resolution of a mixture of 3-[(methoxycarbonyl)-methylthio]-2-amino-propanoic acid (**1**) and 3-[(methoxycarbonyl)methylthio]-2-acetamido-propanoic acid (**2**). Column: TE CSP (250 × 4.5 mm ID); eluent: MeOH/ammonium acetate 20 mM 85/15 (v/v); flow-rate: 1 mL/min; $T = 25°C$; UV detection at 220 nm.

available CSPs containing teicoplanin and TAG as selectors, in comparison with commercially available CSPs containing cellulose-tris-3,5-dimethylphenyl carbamate [169]. It was clearly established that, within the teicoplanin-based CSPs, TAG CSP was the most often responsible for the enantioseparation of the investigated compounds.

Another study focused on aryl-substituted β-lactams, using the same set of teicoplanin-based CSPs and variable-temperature conditions [99]. Tricyclic β-lactams were investigated by the same group of authors, together with some bicyclic β-amino acids, on five different commercially available glycopeptides CSPs, namely ristocetin A, TE, TAG, vancomycin, and VAG, and on a new dimethylphenyl carbamate-derivatized β-cyclodextrin-based CSP. The chromatographic results, achieved with different methods, were compared in systematic examinations [170].

2.3.2.12 Amphetamine Derivatives

Enantioseparation of nine amphetamine derivatives, methorphan, and propoxyphene was studied by comparing two different CSP typologies, a macrocyclic antibiotic CSP (vancomycin) and a native β-cyclodextrin CSP [123]. The suitability of the eluent systems to ESI interfacing was discussed, and a tandem mass spectrometric (MS/MS) detection method was developed.

MPH is an amphetamine-like prescription stimulant commonly used to treat Attention Deficit Hyperactivity Disorder (ADHD) and narcolepsy in children and adults. LC/APCI-MS enantiomeric separations of racemic MPH (Ritalin®) were reported using a commercially available vancomycin CSP [111–113].

2.3.2.13 Benzodiazepines, Phenothiazines, and Barbiturates

Temazepam was the first benzodiazepine (BZDP) analyzed on macrocyclic antibiotics CSPs [7], while thioridazine was the first phenothiazine derivative tested [*ibidem*], followed by promethazine [33]. Enantioselective separation of some BZDPs and phenothiazine derivatives was later studied on six different CSPs, based either on cyclodextrins (underivatized β-cyclodextrin and hydroxypropyl ether β-cyclodextrin), or on glycopeptide antibiotics (vancomycin, TE, TAG, and ristocetin A) in a RP separation mode [171]. BZDPs could be enantioresolved with almost all the CSPs used, except for the vancomycin CSP, which was more effective for the enantioseparation of promethazine, as previously demonstrated [172]. Enantioseparation of phenothiazine derivatives was more difficult to achieve, but it was partly successful, with both the typologies of CSPs used (except for levomepromazine). On-column racemization observed for oxazepam and lorazepam at room temperature during the chromatographic time-scale was suppressed in some instances at reduced temperature. Interconversion of the individual enantiomers of oxazepam and lorazepam was further investigated [173], centering the attention only on glycopeptides CSPs. The interconversion phenomenon occurred for the enantioresolution of lorazepam at room temperature on a commercially available teicoplanin CSP is shown in Figure 2.21.

Barbiturates were among the first classes of chiral compounds investigated in the early stages of the application of macrocyclic antibiotics as chiral selectors; they were screened on a vancomycin CSP under RP [7] and NP mode systems [157], and on 3,5-dimethylphenyl-derivatized vancomycin CSP [7].

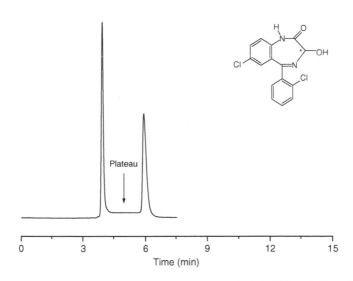

Plateau

0   3   6   9   12   15
Time (min)

FIGURE 2.21 Separation of lorazepam. Column: Chirobiotic T (250 × 4.6 mm ID); eluent: MeOH/AcOH/TEA (100/0.1/0.1); flow-rate: 1 mL/min, $T = 25°C$; UV detection at 254 nm. $k_1 = 0.25$, $\alpha = 3.53$.

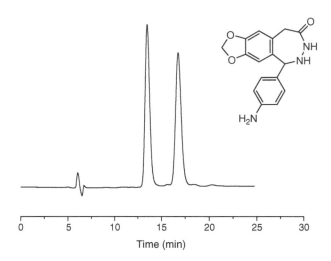

FIGURE 2.22 Enantioseparation of a new AMPA receptor antagonist. Column: A-40,926 CSP (250 × 4.5 mm ID); eluent: THF/water (70/30) + ammonium formate 20 mM; flow-rate: 0.5 mL/min; $T = 25°C$; UV detection at 254 nm. $k_1 = 1.29$, $\alpha = 1.15$, Rs $= 1.80$.

Several other cyclic imidic compounds containing a piperidine-2,6-dione skeleton (glutethimide, pyrido- and amino-glutethimide, thalidomide) and having similar hypnotic-sedative activity as barbiturates were resolved on a vancomycin CSP [7,157].

A-40,926 CSP proved to be suitable for the enantiomeric separation of a selected set of putative AMPA receptor antagonists, having a 2,3-benzodiazepin-4-one skeleton and an anticonvulsant activity (Figure 2.22).

2.3.2.14 Calcium Channel Blockers

Calcium channel blockers (also called calcium antagonists) are drugs used to treat high blood pressure, angina, and some arrhythmias. Among these, verapamil [7, 37, 41], norverapamil [7], nicardipine [106, 125], and benidipine [121] were enantioresolved on glycopeptides-containing CSPs. Separation of the enantiomers of 4-aryldihydropyrimidine-5-carboxylates, which are aza-analogs of nifedipine-type dihydropyridine calcium channel blockers, was also reported on commercially available vancomycin and TE CSPs [174].

2.3.2.15 Miscellaneous

In this section, HPLC applications of glycopeptides-containing CSPs purposely dedicated to single compounds of pharmaceutical and biological interest are briefly collected and cited in chronological order of appearance in the literature. Most of these applications are gathered in a comprehensive review on the evaluation of generic chiral HPLC screens for pharmaceutical analysis [163]. Substituted pyridones [30], substituted 2-methoxy-6-oxo-1,4,5,6-tetrahydropyridine-3-carbanitriles [175], 1,4-piperazine derivatives of aryloxy-aminopropanol [176], semisynthetic ergot alkaloids

[177], chiral thiol compounds [178], citalopram [179], non-nucleoside reverse transcriptase inhibitors [180], captopril [181], substituted 2-(4-pyridylalkyl)-1-tetralone derivatives [182], plant growth regulators [183], diperodon [184], cromakalim [185], substituted dihydrofurocoumarins [186], vesamicol [187], butaclamol [188], methotrexate [189], pterocarpans [190], 1- and 3-methyl-substituted tetrahydroisoquinoline analogs [191].

2.4 RECOGNITION MECHANISM

Vancomycin (Figure 2.2) is the representative member of the family of glycopeptide antibiotics that has thoroughly been investigated in terms of structure, stereochemistry, ligand recognition, and propensity to autoaggregation and dimerization. TE (Figure 2.4), ristocetin A (Figure 2.5), avoparcin (Figure 2.6), and some other related glycopeptides have also received attention and have been studied for their ability to bind small molecules that mimic the natural ligand. As a common feature, glycopeptide antibiotics enclose a heptapeptide core whose side chains are covalently joined to form macrocycles. The structures also include biaryl (AB) and bisaryl ether (CDE, and FG in teicoplanin-type antibiotics) systems that further rigidify the aglycone fragments. The bisaryl ether CDE forms both the rear wall and the ceiling of the ligand-binding cavity, while the biaryl AB unit and the additional bisaryl ether in TE-type antibiotics delimit the cavity at the lateral sites of the heptapeptide chain. The macrocycles are decorated with sugar substituents at various positions, with the sugars playing a crucial role in controlling water solubility, aggregation, dimerization, and surface recognition.

2.4.1 RECOGNITION OF NATURAL LIGAND AND OF ITS MODELS

The free solution structure of glycopeptide antibiotic vancomycin has been determined by combining extensive nuclear magnetic resonance (NMR) studies with X-ray diffraction data and computer modeling. With the structure of vancomycin and related glycopeptides in hand, the molecular basis for the binding of the antibiotics to mucopeptide precursors terminating in Lys-D-Ala-D-Ala was later established again taking advantage of the converging evidences coming from NMR and crystallographic studies.

2.4.1.1 Energetics

Association constants for the complexation of glycopeptides and acetylated amino acids, di- and tripeptides have been determined by standard UV and NMR titrations, and CE. In each case, large free energy changes are recorded for ligands presenting the minimum structural motif of acetyl-D-Ala, with values approaching 8.4 kcal/mol for the elongated tripeptide Ac_2-L-Lys-D-Ala-D-Ala [192]. Free energy changes for ligand binding to vancomycin span from 2.0 kcal/mol for acetate anion [193] to 8.4 kcal/mol for Ac_2-L-Lys-D-Ala-D-Ala [192] in water buffered to pH 6 and 5.1, respectively. Truncated ligand Ac-D-Ala-D-Ala binds to vancomycin with a free energy change of 6.3 kcal/mol [194]. Several related ligands bind weakly to vancomycin with free energy changes close to the value of acetate: D-lactate (2.6 kcal/mol) [194], 2-acetoxy-D-propanoic acid (2.2 kcal/mol) [195], and N-acetylglycine (2.6 kcal/mol) [194] bind loosely to vancomycin, while N-acetyl-D-Ala is more strongly bound (3.4 kcal/mol) [194], as revealed by ^{13}C NMR titrations in aqueous solution buffered to pH 6. The 0.8 kcal/mol increase in binding energy on passing from N-acetylglycine to N-acetyl-D-Ala gives an estimate of the magnitude for the hydrophobic effect of the methyl group, which is located inside the binding cavity and faces one of the aryl rings of residue 4.

Free energy, enthalpy, and entropy changes for the binding of vancomycin to Ac_2-L-Lys-D-Ala-D-Ala in water buffered to pH 4.7 are $-\Delta G° = 7.7$ kcal/mol, $-\Delta H° = 9.6$ kcal/mol, and $-T\Delta S25°C = 1.8$ kcal/mol, respectively, as determined by isothermal titration calorimetry. Similar values have been recorded for more complex ligands carrying one or two sugar residues also, indicating that the saccharide units of the synthetic ligands do not contribute significantly to complexation [196]. Moreover, the sugar residues of vancomycin have no direct effect on complexation of Ac_2-L-Lys-D-Ala-D-Ala that binds to vancomycin (see Figure 2.23) and its aglycone VAG with free energy changes of 7.7 and 7.8 kcal/mol, respectively, in water buffered to pH 5.1 at 25°C [197]. N-acetylvancomycin, a derivative of the antibiotic in which the leucine N-terminus is acetylated, forms a complex with Ac_2-L-Lys-D-Ala-D-Ala

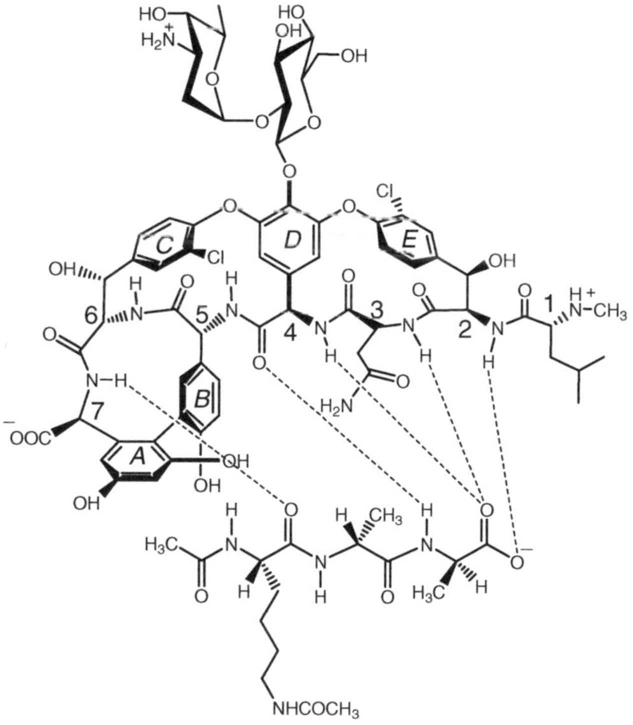

FIGURE 2.23 Exploded view of the complex formed between Ac_2-L-Lys-D-Ala-D-Ala and vancomycin. Dotted lines indicate H-bonding interactions.

with a decrease in free energy of binding, compared to vancomycin, of 1.8 kcal/mol, as determined by UV titration in water buffered to pH 5.1 [192], and of 1.4 kcal/mol, as determined by CE [198]. Thus, electrostatic interactions between the positive charge on the protonated N-methyl group of vancomycin and the negative charge on the carboxylate anion of the ligand contribute only for a small fraction to the overall binding energy. This is consistent with simple calculations of the interaction enthalpy assuming a 5 Å charge separation and a dielectric constant $\varepsilon \sim 50$. Accordingly, a large fraction of the binding energy is retained in the acylated antibiotic, a result that has direct implications in the study of the chiral recognition mechanism operated by the immobilized glycopeptides and their derivatives (see Section 2.4.2).

Structural variations on the Ac$_2$-L-Lys-D-Ala-D-Ala ligand also have a sizeable impact on the energetics of binding with vancomycin: thus, replacing the D-Ala terminal with D-lactate [197] or D-Ser [198] reduces the binding affinity by 4.1 and 1.8 kcal/mol, respectively, while changing the terminal amide NH with a methylene (CH$_2$) costs 1.5 kcal/mol [197].

Complexation studies carried out on teicoplanin and short peptides that mimic the natural ligand yielded thermodynamic results that are closely related to those observed for vancomycin. Binding free energies of teicoplanin to Ac$_2$-L-Lys-D-Ala-D-Ala [198], Ac$_2$-L-Lys-D-Ala-D-Ser [198], Ac-D-Ala-D-Ala [194], and Ac-D-Ala-D-Ser [194] are 8.5, 6.5, 7.2, and 6.6 kcal/mol, respectively. Similar to vancomycin, modifications at the side arms of the glycopeptide (e.g., acylation of the N-terminus amine or esterification of the carboxylate group at residue 7) account for a maximum reduction in free energy of binding of 2.7 kcal/mol for tripeptide ligands [199].

2.4.1.2 Binding Models Based on NMR and X-Ray Studies

Extensive NMR studies based on the analysis of chemical shift changes on ligand binding and of intermolecular NOEs effects have shown the ligands are aligned antiparallel to the antibiotic backbone in the concave portion of the molecular surface, with the terminal carboxylate fragment of the ligands involved in a network of three H-bonding interactions with amide NH groups of residues 2, 3, and 4 of the heptapeptide. The N-terminal portion of the ligands forms two additional amide-amide H-bonding interactions [200–202]. This model also places the methyl groups of the D-Ala-D-Ala fragments close to the aromatic rings of residue 4 of the antibiotics, adding a hydrophobic contribution to the overall binding. The oppositely charged carboxylate group of the ligand and the protonated amino terminus on residue 1 of the antibiotic are located at a distance that results in a modest attractive interaction. Solid-state structure of the complexes between vancomycin and several peptide ligands (see Figure 2.24a) confirm these findings [195, 203, 204].

Partitioning of the different contributions (H-bonding, charge–charge interaction, hydrophobic effect, adverse entropy changes) to the overall binding is hampered by the complex nature of cooperativity between weak interactions. Nevertheless, thermodynamic data for several systems clearly indicate that the main driving force for the association is provided by the interaction of the ligand carboxylate with the three aligned amide NHs inside the antibiotic binding pocket, contributing 5–7 kcal/mol to the overall binding.

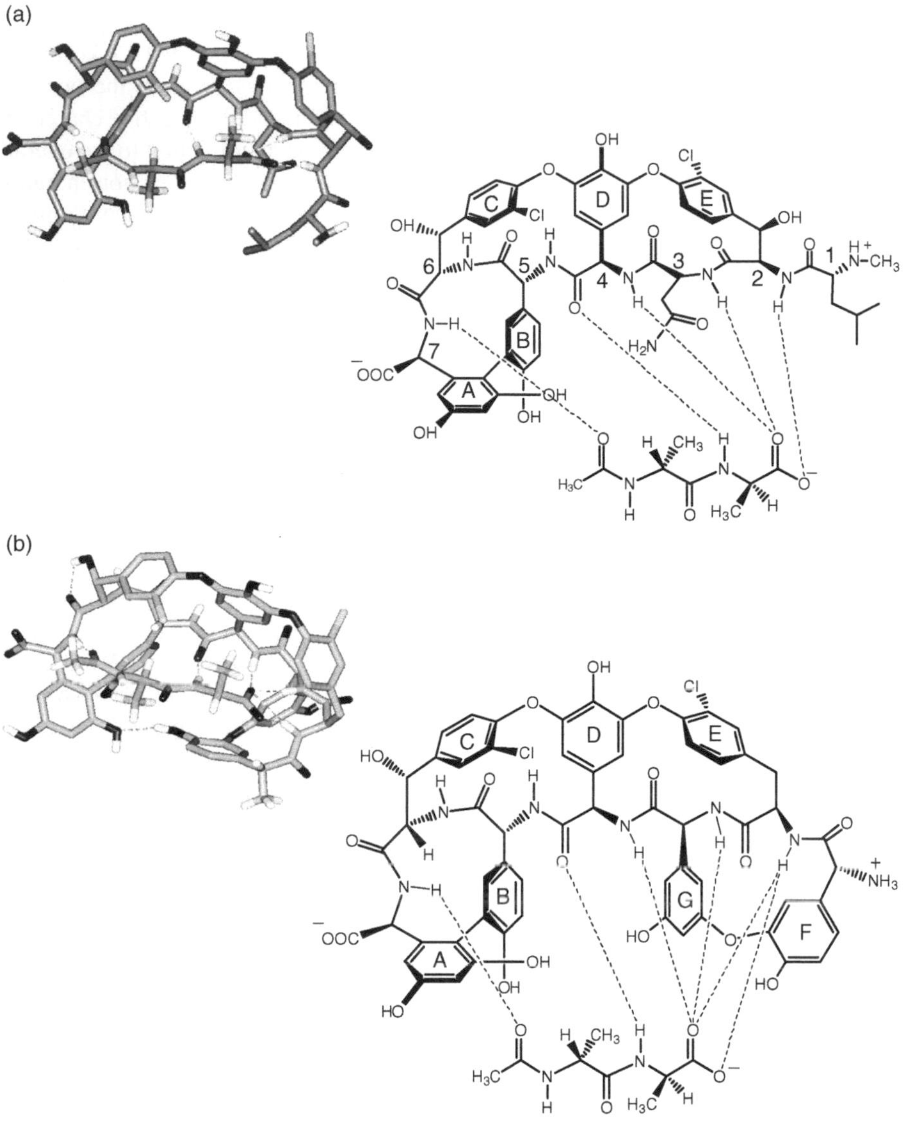

FIGURE 2.24 (a) Polytube model and exploded view of the complex between VAG and
N-acetyl-D-Ala-D-Ala as obtained by computer editing of the X-ray PDB structure 1FVM
.The experimental structure was edited by removing the sugar residues from vancomycin
and the Ac₂-L-Lys fragment from the Ac₂-L-Lys-D-Ala-D-Ala ligand. Dotted lines indicate
H-bonding interactions (some hydrogens are omitted for clarity). (b) Polytube model and
exploded view of the complex between TAG and *N*-acetyl-D-Ala-D-Ala as obtained by molecular modeling. Dotted lines indicate H-bonding interactions (some hydrogens are omitted for
clarity).

2.4.2 ENANTIORECOGNITION STUDIES BY SILICA-BOUND GLYCOPEPTIDES AND AGLYCONES

2.4.2.1 Models Based on Chromatographic Data

A detailed picture at molecular level of the events that govern the enantioselectivity of HPLC systems based on silica-bound glycopeptides is not yet available. This is mainly due to the complications arising from the underlying achiral matrix to which antibiotics are covalently bound (see Chapter 2). In addition, the solid matrix may aspecifically contribute to the overall binding, thereby "diluting" the enantioselectivity of the grafted antibiotics, or it can act by limiting their conformational mobility and forcing them to adopt unfavorable orientations for analyte binding. While classical spectroscopic techniques have fruitfully been used to gain insights into the mechanism of free solution complexation, the same techniques cannot be used in the case of surface-linked glycopeptides. However, in a recent study, vancomycin was covalently linked to amino-functionalized resins through the antibiotic carboxyl terminus on residue 7 [205], and high-resolution magic angle spinning (HR MAS) NMR studies showed the solid phase-bound vancomycin adopting a conformation close to that of free vancomycin in solution. Very similar broadening and shifts of proton signals were observed in resin bound and free vancomycin on addition of the Ac_2-L-Lys-D-Ala-D-Ala ligand, thus demonstrating a substantial equivalence between free and matrix-bound vancomycin, in terms of molecular interactions with its target ligand.

Yet, indirect information on enantioselective complexation by silica-bound antibiotics in HPLC can be extracted from the analysis of retention data of several ligands whose structure is systematically varied to explore chemical diversity in terms of functional groups, stereogenic elements, molecular complexity, and rigidity–flexibility.

Early studies on mechanistic details of silica-bound antibiotics enantioselectivity have focused on the HPLC resolution of free amino acids and of their N-acylated derivatives that resembled truncated mimics of the D-Ala terminating natural ligand. Accordingly, the L- before D-elution sequence is to be expected in the separations of chiral analytes that are structurally similar to the natural ligand. Indeed, several studies reported larger retentions for the D-enantiomers of natural amino acids, their N-acetyl derivatives, and in general, a greater affinity for the CSP in the case of peptides featuring D-configurated residues at the carboxyl terminus [33, 37, 156]. On the contrary, elution order of unnatural amino acids, cyclic amino acids, and β-amino acids was found to be reversed (i.e., D- before L-) on TE, TAG, and ristocetin A stationary phases [141, 149].

The recognition of amino acids enantiomers by vancomycin was investigated using a CE system, where the antibiotic was used as chiral additive in the background electrolyte (BGE) [206]. The enantioselectivity of such CE systems was compared to that obtained when copper(II)-bound vancomycin was used as chiral selector in the BGE, yielding a recognition model in which the secondary amine of the N-methyl leucine of vancomycin appeared as a necessary interaction site for chiral recognition (the primary amino group on the disaccharide residue is not involved in the copper(II) complexation). Later, the model was extended to teicoplanin-based CSPs for HPLC, where it was assumed that the primary predominant step in the chiral recognition is the strong charge–charge interaction between either the carboxyl group of the amino acid and the ammonium group of teicoplanin or the ammonium group of the amino acid and the teicoplanin carboxylate [84]. However, in view of the aforementioned energetic studies on antibiotic–ligand complexation, the role of charge–charge interaction in this model appears overestimated. In addition, it cannot be invoked when the antibiotic is tethered to silica via amide (see Section 2.2.1.1) or ureido (see Sections 2.2.1.2.1 and 2.2.2.1) bonds involving the amino group on residue 1 of the antibiotic. Yet, the preparation and testing of a CSP containing an amide derivative of teicoplanin (mideplanin or MDL 62,873) clearly demonstrated that the carboxylate group is not necessary for the chiral recognition; in fact, similar retention and selectivity values were obtained on the mideplanin-based CSP and on the native teicoplanin CSP, toward a selected set of α- and β-amino acids, and carnitines [207].

In a recent study carried out on ureido-linked teicoplanin CSP and free amino acids or N-blocked derivatives as chiral guests [208], it was shown that the deprotonated carboxylic group of the guest is pivotal in the interaction process between analyte enantiomers and the CSP. Compounds able to fit the aglycone-binding cavity establish several H-bonds between their carboxylate fragment and the glycopeptide amide NHs. Once in the cavity, the onset of weak hydrophobic interactions stabilizes, at different extents, the different complexes. For N-acetylated amino acids, it was found that the L-enantiomers were excluded by the CSP (i.e., eluted before the void volume), while the corresponding D-enantiomers were strongly retained. This was explained assuming the presence of repulsive ionic forces arising between negatively charged analytes and the silica-bound antibiotic, which is also negatively charged: L-enantiomers cannot enter the aglycone basket and only experience repulsive interactions, whereas D-enantiomers are able to penetrate the cavity, where they are strongly stabilized through the formation of an array of H bonds. The effect of the ionic strength of the medium on the behavior of the different classes of compounds investigated was explained in light of the simple Donnan model. Steric exclusion, as opposed to ion-exclusion, was observed for N-BOC protected amino acids, whose enantiomers were scarcely retained under a variety of experimental conditions.

A different approach to the study of recognition mechanism is based on the determination of the adsorption isotherms of the two enantiomers on a CSP. The retention behavior of the enantiomers of phenylglycine was studied on a commercially available teicoplanin CSP [209]. Analysis of the band profiles for the L- (first eluting, symmetrical shape) and D- (second eluting, tailed profile) enantiomers using the stochastic theory of chromatography suggested the presence of at least two additional adsorption sites in the stationary phase contributing to the retention of the D-enantiomer in addition to the adsorption site of the less retained L-enantiomer. In a series of related studies, the retention of D,L-tryptophan enantiomers was studied on CSPs containing silica-bound TE or silica-bound TAG. From the experimental data, a bi-Langmuir model was found to adequately describe D- and L-enantiomer retention on both the CSPs. The sugar units had a detrimental effect on enantioselectivity, mainly due to the inhibition of some enantioselective contacts with low-affinity binding regions of the aglycone and to a reduction of the intrinsic enantioselective properties of the aglycone high-affinity binding region [210, 211].

2.4.2.2 Models Based on Molecular Modeling

Glycopeptide antibiotics have been the subject of molecular modeling investigations focusing on both structure and binding to models of the natural ligands [212–215]. Nevertheless, no direct theoretical studies have till date addressed the explanation of the impressive capabilities showed by these natural antibiotics to perform enantio-discriminations. Recently, we started an accurate investigation aimed to clarify some aspects of the enantiodiscrimination mechanism operated by glycopeptide antibiotics. Our study is based on the use of a systematic and multiconformational molecular docking procedure [216, 217], integrated with an automated complete structure optimization of all the complexes obtained. A final statistical elaboration of the calculated energetic data yields the thermodynamic quantities $\Delta G°$, $\Delta H°$, and $\Delta S°$ for the complexation process of each enantiomer, and thus the $\Delta\Delta G°$, $\Delta\Delta H°$, and $\Delta\Delta S°$ quantities related to the enantiodiscrimination process. Application of this protocol to the association between N-acetyl-D-Ala-D-Ala and TAG gave a low energy structure for the 1:1 complex in which the dipeptide is held inside the binding pocket of TAG by a network of H-bonding interactions. In the computed structure (Figure 2.24b), the conformation of the guest and its orientation relative to the host match exactly those experimentally observed by X-ray crystallography of the related complex between the same guest and vancomycin. Encouraged by these results, we extended the computational approach to study the chiral discrimination performed by TAG toward N-acetylphenylglycine (N-APG), whose enantiomers were nicely separated by silica-bound TAG (see Figure 2.14) with an enantioselectivity value $\alpha = 82$ that translates into a free energy difference of complexation $\Delta\Delta G = 2.6$ kcal/mol at $T = 25°C$ in favor of the D-enantiomer. In the docking simulations, 10 different conformations of both enantiomers of N-APG were successively moved toward six different conformations of TAG. The docking simulation indicated a stability difference between the two diastereomeric complexes of 2.8 kcal/mol in favor of the D-enantiomer, in qualitative agreement with the HPLC results. From the geometric point of view, the two complexes closely resemble the solid-state structure of a related complex formed by N-acetylglycine and vancomycin [195].

Irrespective of the guest configuration, both enantiomers of N-APG form very strong interactions with TAG by establishing a wide pattern of H-bonds between their carboxylate group and three amide NHs lined inside the binding pocket of TAG (see Figure 2.25, top). These interactions may be considered as a nonselective contribution to the enantiodiscrimination, but are very important to correctly orient the ligand along the cleft of the TAG recognition site.

On the contrary, a clear selective contribution to the enantiodiscrimination arises by the establishment of an effective H-bonding interaction between the amide NH of N-APG and the frontal carbonyl oxygen of the amide group of residue 4 of TAG. While this H-bonding interaction can be easily formed by the D-enantiomer (NH-O distance of 1.9 Å), in the computed complex with L-N-APG, the amide NH diverges from the TAG carbonyl oxygen (see Figure 2.25, bottom). Moreover, the phenyl ring

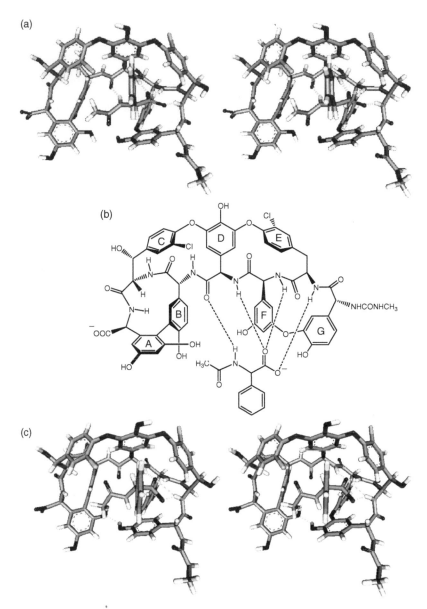

FIGURE 2.25 Stereoviews of the complexes between (a) TAG and *N*-acetyl-D-phenylglycine or (c) *N*-acetyl-L-phenylglycine, generated by computer modeling. (b) Exploded view of the complex between TAG and *N*-acetyl-phenylglycine (unspecified stereochemistry at the ligand). The primary amino group of TAG is converted to a ureido function to model the TAG CSP.

of the ligand is found, in both complexes, in an optimal disposition to equally establish effective dispersive interactions with the surrounding aromatic rings of TAG. These findings suggest that, in principle, the aforementioned TAG enantiodiscrimination mechanism could also operate in the general case of racemic amino acid derivatives (e.g., di- and polypeptides) having a terminal free carboxylate function and their amino groups derivatized as amide functions.

In conclusion, molecular recognition by glycopeptide antibiotics is based on a range of cooperative interactions, accompanied by shape and size complementarities. Spectroscopic and computational analysis proved to be functional to uncover geometric and energetic details about the molecular mechanisms that regulate the complexation with small peptide molecules that mimic the natural ligand. It is expected that additional studies on a broader set of ligands will reveal new aspects and insights into the mode of actions of these naturally occurring chiral receptors, and will be useful to make predictions about the potential enantioselectivity of eventual separation systems that incorporate glycopeptide antibiotics or their derivatives.

ACKNOWLEDGMENTS

We acknowledge financial supports from Università "La Sapienza," Roma, Italy (Funds for selected research topics 2003-2005), Ministero dell'Istruzione, dell'Università e della Ricerca, MIUR, Italy (PRIN 2005, contract n. 2005037725_001), and the "Istituto Pasteur—Fondazione Cenci Bolognetti", Università "La Sapienza", Roma, Italy.

REFERENCES

1. Aboul-Enein, H.Y. and Ali, I., Macrocyclic antibiotics as effective chiral selectors for enantiomeric resolution by liquid chromatography and capillary electrophoresis, *Chromatographia*, 52, 679, 2000.
2. Ward, T.J. and Farris III, A.B., Chiral separations using the macrocyclic antibiotics: a review, *J. Chromatogr. A*, 906, 73, 2001.
3. Ilisz, I., Berkecz, R., and Péter, A., HPLC separation of amino acid enantiomers and small peptides on macrocyclic antibiotic-based chiral stationary phases: a review, *J. Sep. Sci.*, 29, 1305, 2006.
4. Ward, T.J. and Oswald, T.M., Enantioselectivity in capillary electrophoresis using the macrocyclic antibiotics, *J. Chromatogr. A*, 792, 309, 1997.
5. Ward, T.J., Dann III, C., and Blaylock, A., Enantiomeric resolution using the macrocyclic antibiotics rifamycin B and rifamycin SV as chiral selectors for capillary electrophoresis, *J. Chromatogr. A*, 715, 337, 1995.
6. Ward, T.J., Macrocyclic antibiotics—the newest class of chiral selectors, *LC-GC*, 14, 886, 1996.
7. Armstrong, D.W. et al., Macrocyclic antibiotics as a new class of chiral selectors for liquid chromatography, *Anal. Chem.*, 66, 1473, 1994.
8. Gasparrini, F. et al., Natural and totally synthetic receptors in the innovative design of HPLC chiral stationary phases, *Pure Appl. Chem.*, 75, 407, 2003.
9. Nicolaou, K.C. et al., Chemistry, biology, and medicine of the glycopeptide antibiotics, *Angew. Chem. Int. Ed.*, 38, 2096, 1999.
10. Marshall, F.J., Structure studies on vancomycin, *J. Med. Chem.*, 8, 18, 1965.
11. Aboul-Enein, H.Y. and Ali, I. Macrocyclic glycopeptide antibiotics-based chiral stationary phase, in *Chiral Separation by Liquid Chromatography and Related Technologies*, Marcel Dekker; New York, 2003, chap. 2.

12. McCormick, M.H. et al., Vancomycin, a new antibiotic. I. Chemical and biologic properties, *Antibiot. Annu. 1955–1956*, 606.
13. Best, G.K., Best, N.H., and Durham, N.N., Chromatographic separation of the vancomycin complex, *Antimicrob. Agents Chemother.*, 8, 115, 1968.
14. Nieto, M. and Perkins, H.R., Physicochemical properties of vancomycin and iodovancomycin and their complexes with diacetyl-L-lysyl-D-alanyl-D-alanine, *Biochem. J.*, 123, 773, 1971.
15. Gao, W. et al., Ototoxicity of a new glycopeptide, norvancomycin with multiple intravenous administrations in guinea pigs, *J. Antibiot. (Tokyo)*, 57, 45, 2004.
16. Ding, G.S. et al., Chiral separation of enantiomers of amino acid derivatives by high-performance liquid chromatography on a norvancomycin-bonded chiral stationary phase, *Talanta*, 62, 997, 2004.
17. Ding, G.S. et al., Chiral separation of racemates of drugs and amino acid derivatives by high-performance liquid chromatography on a norvancomycin-bonded chiral stationary phase, *Chromatographia*, 59, 443, 2004.
18. Guo, Z., Wang, H., and Zhang, Y., Chiral separation of ketoprofen on an achiral C8 column by HPLC using norvancomycin as chiral mobile phase additives, *J. Pharm. Biomed. Anal.*, 41, 310, 2006.
19. Nicas, T.I. et al. Semisynthetic glycopeptide antibiotics derived from LY264826 active against vancomycin-resistant enterococci, *Antimicrob. Agents Chemother.*, 40, 2194, 1996.
20. Debono, M. and Molloy, R.M., Isolation and structure of the novel branched-chain amino sugar derived from antibiotic A35512B, *J. Org. Chem.*, 45, 4685, 1980.
21. Debono, M. et al., A35512, a complex of new antibacterial antibiotics produced by Streptomyces candidus. II. Chemical studies on A35512B, *J. Antibiot. (Tokyo)* 33, 1407, 1980.
22. Staroverov, S.M. et al., New chiral stationary phase with macrocyclic glycopeptide antibiotic eremomycin chemically bonded to silica, *J. Chromatogr. A*, 1108, 263, 2006.
23. Petrusevska, K. et al., Chromatographic enantioseparation of amino acids using a new chiral stationary phase based on a macrocyclic glycopeptide antibiotic, *J. Sep. Sci.*, 29, 1447, 2006.
24. Strege, M.A., Huff, B.E., and Risley, D.S., Evaluation of macrocyclic antibiotic A82846B as a chiral selector for capillary electrophoresis separations, *LC-GC*, 14, 144, 1996.
25. Sharp, V.S. et al., Evaluation of a new macrocyclic antibiotic as a chiral selector for use in capillary electrophoresis, *J. Liq. Chromatogr. Rel. Technol.*, 20, 887, 1997.
26. Sharp, V.S. and Risley, D.S., Evaluation of the macrocyclic antibiotic LY333328 as a chiral selector when used as a mobile phase additive in narrow bore HPLC, *Chirality*, 11, 75, 1999.
27. Sharp, V.S. et al., Enantiomeric separation of dansyl amino acids using macrocyclic antibiotics as chiral mobile phase additives by narrow-bore high-performance liquid chromatography, *Chirality*, 16, 153, 2004.
28. Parenti, F. et al., Teichomycin, new antibiotic from Actinoplanes teichomyceticus NOV. SP., *J. Antibiotics*, 31, 276, 1978.
29. Lancini, G. and Cavalleri, B., in *Biochemistry of Peptide Antibiotics*, Kleinhauf, H. and Dohren, H.V., Eds., W. De Gruyter, Berlin, 1990, 159.

30. Armstrong, D.W., Liu, Y., and Ekborg-Ott, K.H., A covalently bonded teicoplanin chiral stationary phase for HPLC enantioseparations, *Chirality*, 7, 474, 1995.
31. Jordan, D.C., Ristocetin, in *Antibiotics*, Gottlieb, D. and Shaw, P., Eds., Springer, New York, 1967, 84.
32. Sztaricskai, F. et al., Structural studies of ristocetin A (ristomycin A): carbohydrate–aglycone linkages, *J. Am. Chem. Soc.*, 102, 7093, 1980.
33. Ekborg-Ott, K.H., Liu, Y., and Armstrong, D.W., Highly enantioselective HPLC separations using the covalently bonded macrocyclic antibiotic, ristocetin A, chiral stationary phase, *Chirality*, 10, 434, 1998.
34. McGahren, W.J. et al., Avoparcin, *J. Am. Chem. Soc.*, 101, 2237, 1979.
35. McGahren, W.J. et al., Structure of avoparcin components, *J. Am. Chem. Soc.*, 102, 1671, 1980.
36. Ellestad, G.A. et al., Avoparcin and epiavoparcin, *J. Am. Chem. Soc.*, 103, 6522, 1981.
37. Ekborg-Ott, K.H., et al., Evaluation of the macrocyclic antibiotic avoparcin as a new chiral selector for HPLC, *Chirality*, 10, 627, 1998.
38. Waltho, J.P. et al., Structure elucidation of the glycopeptide antibiotic complex A-40,926, *J. Chem. Soc., Perkin Trans. 1*, 2103, 1987.
39. Goldstein, B.P. et al., A40926, a new glycopeptide antibiotic with anti-Neisseria activity, *Antimicrob. Agents Chemother.*, 31, 1961, 1987.
40. D'Acquarica, I. et al., Application of a new chiral stationary phase containing the glycopeptide antibiotic A-40,926 in the direct chromatographic resolution of β-amino acids, *Tetrahedron: Asymmetry*, 11, 2375, 2000.
41. Berthod, A. et al., Evaluation of the macrocyclic glycopeptide A-40,926 as a high-performance liquid chromatographic chiral selector and comparison with teicoplanin chiral stationary phase, *J. Chromatogr. A*, 897, 113, 2000.
42. Goldstein, B.P. et al., Antimicrobial activity of MDL 63,246, a new semisynthetic glycopeptide, *Antimicrob. Agents Chemother.*, 39, 1580, 1995.
43. Fanali, S., Catarcini, P., and Presutti, C., Enantiomeric separation of acidic compounds of pharmaceutical interest by capillary electrochromatography employing glycopeptide antibiotic stationary phases, *J. Chromatogr. A*, 994, 227, 2003.
44. Fanali, S. et al., Use of short-end injection capillary packed with a glycopeptide antibiotic stationary phase in electrochromatography and capillary liquid chromatography for the enantiomeric separation of hydroxy acids, *J. Chromatogr. A*, 990, 143, 2003.
45. Berthod, A. et al., Role of the carbohydrate moieties in chiral recognition on teicoplanin-based LC stationary phases, *Anal. Chem.*, 72, 1767, 2000.
46. Berthod, A. et al., Separation of chiral sulfoxides by liquid chromatography using macrocyclic glycopeptide chiral stationary phases, *J. Chromatogr. A*, 955, 53, 2002.
47. Armstrong, D.W., US Patent No. US 7,008,533 B2, 2006.
48. Berthod, A. et al., Derivatized vancomycin stationary phases for LC chiral separations, *Talanta*, 43, 1767, 1996.
49. Berthod, A., Chang, S.-C., and Armstrong, D.W., Empirical procedure that uses molecular structure to predict enantioselectivity of chiral stationary phases, *Anal. Chem.*, 64, 395, 1992.
50. Sheldrick, G.M. et al., Structure of vancomycin and its complex with acetyl-D-alanyl-D-alanine, *Nature*, 271, 223, 1978.

Advances in Chromatography, Volume 46

51. Harris, C.M. and Harris, T.M., Structure of the glycopeptide antibiotic vancomycin. Evidence for an asparagine residue in the peptide, *J. Am. Chem. Soc.*, 104, 4293, 1982.

52. Harris, C.M., Kopecka, H., and Harris, T.M., Vancomycin: structure and transformation to CDP-I, *J. Am. Chem. Soc.*, 105, 6915, 1983.

53. Ghassempour, A., Darbandi, M.K., and Asghari, F.S., Comparison of pyrolysis-mass spectrometry with high performance liquid chromatography for the analysis of vancomycin in serum, *Talanta*, 55, 573, 2001.

54. Backes, D.W., Aboleneen, H.I., and Simpson, J.I., Quantitation of vancomycin and its crystalline degradation product (CDP-1) in human serum by high performance liquid chromatography, *J. Pharm. Biomed. Anal.*, 16, 1281, 1998.

55. Ghassempour, A. et al., Crystalline degradation products of vancomycin as a new chiral stationary phase for liquid chromatography, *Chromatographia*, 61, 151, 2005.

56. Péter, A. et al., Comparison of the separation efficiencies of CHIROBIOTIC T and TAG columns in the separation of unusual amino acids, *J. Chromatogr. A*, 1031, 159, 2004.

57. Lokajová, J., Tesařová, E., and Armstrong, D.W., Comparative study of three teicoplanin-based chiral stationary phases using the linear free energy relationship model, *J. Chromatogr. A*, 1088, 57, 2005.

58. Xiao, T.L. et al., Evaluation and comparison of a methylated teicoplanin aglycone to teicoplanin aglycone and natural teicoplanin chiral stationary phases, *J. Sep. Sci.*, 29, 429, 2006.

59. Donovick, R. et al., Thiostrepton, a new antibiotic. I. In vitro studies, *Antibiot. Annu. 1955-1956*, 554.

60. Nicolaou, K.C. et al., Total synthesis of thiostrepton, Part 1: construction of the dehydropiperidine/thiazoline-containing macrocycle, *Angew. Chem. Int. Ed.*, 43, 5087, 2004.

61. D'Acquarica, I. et al., Direct chromatographic resolution of carnitine and O-acylcarnitine enantiomers on a teicoplanin-bonded chiral stationary phase, *J. Chromatogr. A*, 875, 145, 1999.

62. D'Acquarica, I., New synthetic strategies for the preparation of novel chiral stationary phases for high-performance liquid chromatography containing natural pool selectors, *J. Pharm. Biomed. Anal.*, 23, 3, 2000.

63. Alcaro, S. et al., Enantioselective semi-preparative HPLC of two 2-arylpropionic acids on glycopeptides containing chiral stationary phases, *Tetrahedron: Asymmetry*, 13, 69, 2002.

64. Ciogli, A. et al., A new chiral and restricted access material containing teicoplanin as selector (chiro-RAM-TE) for the HPLC determination of chiral drugs in biological matrices, presented at 16th Int. Symp. on Chirality, New York, July 11–14, 2004, 62.

65. Gasparrini, F., Misiti, D., and Villani, C., Chromatographic optical resolution on *trans*-1,2-diaminocyclohexane derivatives: theory and applications, *Chirality*, 4, 447, 1992.

66. Galli, B. et al., Enantiomeric separation of DNS-amino acids and DBS-amino acids by ligand exchange chromatography with (*S*)- and (*R*)-phenylalaninamide modified silica gel, *J. Chromatogr. A*, 666, 77, 1994.

67. Armstrong, D.W. and DeMond, W., Cyclodextrin bonded phases for the liquid chromatographic separation of optical, geometrical, and structural isomers, *J. Chromatogr. Sci.*, 22, 411, 1984.

68. Stratilová, E., Èapka, M., and Rexová-Benková, L., Endopolygalacturonase immobilized on epoxide-containing supports, *Biotechnol. Lett.*, 9, 1987.
69. Ernst-Cabrera, K. and Wilchek, M., Silica containing primary hydroxyl groups for high-performance affinity chromatography, *Anal. Biochem.*, 159, 267, 1986.
70. Svensson, L.A. et al., Immobilized vancomycin as chiral stationary phase in packed capillary liquid chromatography, *Chirality*, 10, 273, 1998.
71. Dönnecke, J. et al., Evaluation of a vancomycin chiral stationary phase in packed capillary supercritical fluid chromatography, *J. Microcol. Sep.*, 11, 521, 1999.
72. Svensson, L.A. et al., Vancomycin-based chiral stationary phases for micro-column liquid chromatography, *Chirality*, 11, 121, 1999.
73. Wikstrom, H. et al., Immobilisation and evaluation of a vancomycin chiral stationary phase for capillary electrochromatography, *J. Chromatogr. A*, 869, 395, 2000.
74. Desiderio, C., Aturki, Z., and Fanali, S., Use of vancomycin silica stationary phase in packed capillary electrochromatography I. Enantiomer separation of basic compounds, *Electrophoresis*, 22, 535, 2001.
75. Svensson, L.A. and Owens, P.K., Enantioselective supercritical fluid chromatography using ristocetin A chiral stationary phases, *Analyst*, 125, 1037, 2000.
76. Hargitai, T., Kaida, Y., and Okamoto, Y., Preparation and chromatographic evaluation of 3,5-dimethylphenyl carbamoylated beta-cyclodextrin stationary phases for normal-phase high-performance liquid-chromatographic separation of enantiomers, *J. Chromatogr.*, 628, 11, 1993.
77. Aboul-Enein, H.Y. and Ali, I., Optimization strategies for HPLC enantioseparation of racemic drugs using polysaccharides and macrocyclic antibiotic chiral stationary phases, *Il Farmaco*, 57, 513, 2002.
78. Beesley, T.E., Lee, J.T., and Wang, A.X., Method development and optimization of enantiomeric separations using macrocyclic glycopeptide chiral stationary phases, in *Chiral Separation Techniques*, Second completely revised and updated edition, Subramanian, G., Ed., Wiley-VCH; Weinheim, 2001, 25.
79. Scott, R.P.W. and Beesley, T.E., Optimum operating conditions for chiral separations in liquid chromatography, *Analyst*, 124, 713, 1999.
80. Lehotay, J. et al., Modification of the chiral bonding properties of teicoplanin chiral stationary phase by organic additives. HPLC separation of enantiomers of alkoxysubstituted esters of phenylcarbamic acid, *J. Liq. Chrom. Rel. Technol.*, 24, 609, 2001.
81. Chen, S. and Ward, T., Comparison of the chiral separation of amino-acid derivatives by a teicoplanin and RN-β-cyclodextrin CSPs using waterless mobile phases: factors that enhance resolution, *Chirality*, 16, 318, 2004.
82. Aboul-Enein, H.Y. and Serignese, V., Optimized enantioselective separation of clenbuterol on macrocyclic antibiotic teicoplanin chiral stationary phase, *J. Liq. Chrom. Rel. Technol.*, 22, 2177, 1999.
83. Aboul-Enein, H.Y. and Imran, A., Chiral resolution of clenbuterol, cimaterol, and mabuterol on CHIROBIOTIC V, T and TAG columns, *J. Sep. Sci.*, 25, 851, 2002.
84. Berthod, A. et al., Facile RPLC enantioresolution of native amino-acids and peptides using a teicoplanin chiral stationary phase, *J. Chromatogr. A*, 731, 123, 1996.
85. Guillaume, Y.C. et al., Chiral discrimination of phenoxypropionic acid herbicides on teicoplanin phase: effect of mobile phase modifier, *Chromatographia*, 55, 143, 2002.

86. André, C. and Guillaume, Y.C., Reanalysis of chiral discrimination of phenoxypropionic acid herbicides on a teicoplanin phase using a bi-Langmuir approach, *Chromatographia*, 58, 201, 2003.

87. Jandera, P. et al., Effect of the mobile phase on the retention behaviour of optical isomers of carboxylic acids and amino acids in liquid chromatography on bonded teicoplanin columns, *J. Chromatogr. A*, 917, 123, 2001.

88. Schlauch, M. and Frahm, A.W., Enantiomeric and diastereomeric high-performance liquid chromatographic separation of cyclic β-substituted α-amino acids on a teicoplanin chiral stationary phase, *J. Chromatogr. A*, 868, 197, 2000.

89. Schlauch, M., Kos, O., and Frahm, A.W., Comparison of three chiral stationary phases with respect to their enantio- and diastereoselectivity for cyclic β-substituted α-amino acids, *J. Pharm. Biomed. Anal.*, 27, 409, 2002.

90. Doležalová, M. and Tkaczyková, M., HPLC enantioselective separation of aromatic amino and hydrazino acids on a teicoplanin stationary phase and the enantiomeric purity determination of L-isomers used as drugs, *Chirality*, 11, 394, 1999.

91. Doležalová, M. and Tkaczyková, M., Direct high-performance liquid chromatographic determination of the enantiomeric purity of levodopa and methyldopa: comparison with pharmacopoeial polarimetric methods, *J. Pharm. Biomed. Anal.*, 19, 555, 1999.

92. Peyrin, E. et al., Dansyl amino acid enantiomer separation on a teicoplanin chiral stationary phase: effect of eluent pH, *J. Chromatogr. A*, 923, 37, 2001.

93. Péter, A. et al., Effect of temperature on retention of enantiomers of β-methyl amino acids on a teicoplanin chiral stationary phase, *J. Chromatogr. A*, 828, 177, 1998.

94. Péter, A., Vékes, E. and Armstrong, D.W., Effects of temperature on retention of chiral compounds on a ristocetin A chiral stationary phase, *J. Chromatogr. A*, 958, 89, 2002.

95. Berthod, A., He, B.L., and Beesley, T.E., Temperature and enantioseparation by macrocyclic glycopeptide chiral stationary phases, *J. Chromatogr. A*, 1060, 205, 2004.

96. Rojkoviéová, T. et al., Study of the mechanism of enantioseparation. VII. Effect of temperature on retention of some enantiomers of phenylcarbamic acid derivates on a teicoplanin aglycone chiral stationary phase, *J. Liq. Chrom. Rel. Technol.*, 27, 1653, 2004.

97. Rojkovièová, T. et al., Study of the mechanism of enantioseparation. IX. Effect of temperature on retention of chiral compounds on a methylated teicoplanin chiral stationary phase, *J. Liq. Chrom. Rel. Technol.*, 27, 2477, 2004.

98. Rojkovièová, T. et al., Study of the mechanism of enantioseparation. X. Comparison study of thermodynamic parameters on separation of phenylcarbamic acid derivatives using vancomycin and teicoplanin CSPs, *J. Liq. Chrom. Rel. Technol.*, 27, 3213, 2004.

99. Berkecz, R. et al., LC enantioseparation of aryl-substituted β-lactams using variable-temperature conditions, *Chromatographia*, 63, S29, 2006.

100. Husek, P. and Simek, P., Advances in amino acid analysis, *LC-GC*, 19, 986, 2001.

101. Dash, A.K. and Sawhney, A., A simple LC method with UV detection for the analysis of creatine and creatinine and its application to several creatine formulations, *J. Pharm. Biomed. Anal.*, 29, 939, 2002.

102. Mistry, B., Leslie, J.L., and Eddington, N.D., Enantiomeric separation of metoprolol and α-hydroxymetoprolol by liquid chromatography and fluorescence detection using a chiral stationary phase, *J. Chromatogr. B*, 758, 153, 2001.

103. Risley, D.S. and Strege, M.A., Chiral separations of polar compounds by hydrophilic interaction chromatography with evaporative light scattering detection, *Anal. Chem.*, 72, 1736, 2000.
104. Guisbert, A.L. et al., Enantiomeric separation of an AMPA antagonist using a CHIROBIOTIC T column with HPLC and evaporative light-scattering detection, *J. Liq. Chrom. Rel. Technol.*, 23, 1019, 2000.
105. Wipf, P. et al., HPLC determinations of enantiomeric ratios, *Chirality*, 19, 5, 2007.
106. Bakhtiar, R. and Tse, F.L.S., High-throughput chiral liquid chromatography/tandem mass spectrometry, *Rapid Commun. Mass Spectrom.*, 14, 1128, 2000.
107. Badaloni, E. et al., Enantioselective liquid chromatographic-electrospray mass spectrometric assay of β-adrenergic blockers: application to a pharmacokinetic study of sotalol in human plasma, *J. Chromatogr. B*, 796, 45, 2003.
108. Joyce, K.B. et al., Determination of the enantiomers of salbutamol and its 4-*O*-sulphate metabolites in biological matrices by chiral liquid chromatography tandem mass spectrometry, *Rapid Commun. Mass Spectrom.*, 12, 1899, 1998.
109. Xia, Y.-Q., Bakhtiar, R., and Franklin, R.B., Automated online dual-column extraction coupled with teicoplanin stationary phase for simultaneous determination of (*R*)- and (*S*)-propranolol in rat plasma using liquid chromatography-tandem mass spectrometry, *J. Chromatogr. B*, 788, 317, 2003.
110. Jacobson, G.A., Chong, F.V., and Davies, N.W., LC-MS method for the determination of albuterol enantiomers in human plasma using manual solid-phase extraction and a non-deuterated internal standard, *J. Pharm. Biomed. Anal.*, 31, 1237, 2003.
111. Ramos, L. et al., Liquid chromatographic/atmospheric pressure chemical ionization tandem mass spectrometry enantiomeric separation of DL-*threo*-methylphenidate (Ritalin®) using a macrocyclic antibiotic as the chiral selector, *Rapid Commun. Mass Spectrom.*, 13, 2054, 1999.
112. Bakhtiar, R., Ramos, L., and Tse, F.L.S., Quantification of methylphenidate (Ritalin®) in rabbit fetal tissue using a chiral liquid chromatography/tandem mass spectrometry assay, *Rapid. Commun. Mass Spectrom.*, 16, 81, 2002.
113. Bakhtiar, R., Ramos, L., and Tse, F.L.S., Quantification of methylphenidate in rat, rabbit and dog plasma using a chiral liquid-chromatography/tandem mass spectrometry method. Application to toxicokinetic studies, *Anal. Chim. Acta*, 469, 261, 2002.
114. Rashed, M.S., AlAmoudi, M., and Aboul-Enein, H.Y., Chiral liquid chromatography tandem mass spectrometry in the determination of the configuration of 2-hydroxyglutaric acid in urine, *Biomed. Chromatogr.*, 14, 317, 2000.
115. Rashed, M.S. et al., Chiral liquid chromatography tandem mass spectrometry in the determination of the configurations of glyceric acid in urine of patients with D-glyceric and L-glyceric acidurias, *Biomed. Chromatogr.*, 16, 191, 2002.
116. Petritis, K. et al., Simultaneous analysis of underivatized chiral amino acids by liquid chromatography—ionspray tandem mass spectrometry using a teicoplanin chiral stationary phase, *J. Chromatogr. A*, 913, 331, 2001.
117. Desai, M.J. and Armstrong, D.W., Analysis of native amino acid and peptide enantiomers by high-performance liquid chromatography/atmospheric pressure chemical ionization mass spectrometry, *J. Mass Spectrom.*, 39, 177, 2004.
118. Zhang, B., Soukup, R., and Armstrong, D.W., Selective separations of peptides with sequence deletions, single amino acid polymorphisms, and/or epimeric centers using macrocyclic glycopeptide liquid chromatography stationary phases, *J. Chromatogr. A*, 1053, 89, 2004.

119. Shen, Z., Wang, S., and Bakhtiar, R., Enantiomeric separation and quantification of fluoxetine (Prozac®) in human plasma by liquid chromatography/tandem mass spectrometry using liquid-liquid extraction in 96-well plate format, *Rapid Commun. Mass Spectrom.*, 16, 332, 2002.

120. Xia, Y.-Q., Liu, D.Q., and Bakhtiar, R., Use of online-dual-column extraction in conjunction with chiral liquid chromatography tandem mass spectrometry for determination of terbutaline enantiomers in human plasma, *Chirality*, 14, 742, 2002.

121. Kang, W. et al., Analysis of benidipine enantiomers in human plasma by liquid chromatography—mass spectrometry using a macrocyclic antibiotic (vancomycin) chiral stationary phase column, *J. Chromatogr. B*, 814, 75, 2005.

122. Desai, M.J. and Armstrong, D.W., Analysis of derivatized and underivatized theanine enantiomers by high-performance liquid chromatography/atmospheric pressure ionization-mass spectrometry, *Rapid Commun. Mass Spectrom.*, 18, 251, 2004.

123. Pihlainen, K. and Kostiainen, R., Effect of the eluent on enantiomer separation of controlled drugs by liquid chromatography-ultraviolet absorbance detection-electrospray ionisation tandem mass spectrometry using vancomycin and native β-cyclodextrin chiral stationary phases, *J. Chromatogr. A*, 1033, 91, 2004.

124. Rashed, M.S. et al., Determination of L-pipecolic acid in plasma using chiral liquid chromatography-electrospray tandem mass spectrometry, *Clin. Chem.*, 47, 2124, 2001.

125. Desai, M.J. and Armstrong, D.W., Transforming chiral liquid chromatography methodologies into more sensitive liquid chromatography—electrospray ionization mass spectrometry without losing enantioselectivity, *J. Chromatogr. A*, 1035, 203, 2004.

126. Ding, J., Desai, M., and Armstrong, D.W., Evaluation of ethoxynonafluorobutane as a safe and environmentally friendly solvent for chiral normal-phase LC-atmospheric pressure chemical ionization/electrospray ionization-mass spectrometry, *J. Chromatogr. A*, 1076, 34, 2005.

127. Ekborg-Ott, K.H., Wang, X., and Armstrong, D.W., Effect of selector coverage and mobile phase composition on enantiomeric separations with ristocetin A chiral stationary phases, *Microchem. J.*, 62, 26, 1999.

128. Bosáková, Z., Cuřinová, E., and Tesařová, E., Comparison of vancomycin-based stationary phases with different chiral selector coverage for enantioselective separation of selected drugs in high-performance liquid chromatography, *J. Chromatogr. A*, 1088, 94, 2005.

129. Aboul-Enein, H.Y., Hefnawy, M.M., and Hoenen, H., LC determination of the enantiomeric purity of L-arginine using a teicoplanin chiral stationary phase, *J. Liq. Chrom. Rel. Technol.*, 27, 1681, 2004.

130. Yu, Y-P. and Wu, W-H., Simultaneous analysis of enantiomeric composition of amino acids and *N*-acetyl-amino acids by enantioselective chromatography, *Chirality*, 13, 231, 2001.

131. Chen, S., HPLC enantiomeric resolution of phenyl isothiocyanated amino acids on teicoplainin-bonded phase using an acetonitrile-based mobile phase: a structural consideration, *J. Liq. Chrom. Rel. Technol.*, 26, 3475, 2003.

132. Piccinini, A.-M., Chiral separation of natural and unnatural amino acid derivatives by micro-HPLC on a ristocetin A stationary phase, *J. Biochem. Biophys. Methods*, 61, 11, 2004.

133. Tesařová, E., Bosáková, Z., and Pacáková, V., Comparison of enantioselective separation of N-tert-butyloxycarbonyl amino acids and their non-blocked analogues on teicoplanin-based chiral stationary phase, *J. Chromatogr. A*, 838, 121, 1999.
134. Tesařová, E., Bosáková, Z., and Zusková, I., Enantioseparation of selected N-tert-butyloxycarbonyl amino acids in high-performance liquid chromatography and capillary electrophoresis with a teicoplanin chiral selector. *J. Chromatogr. A*, 879, 147, 2000.
135. Lehotay, J. et al., Chiral separation of enantiomers of amino acid derivatives by HPLC on vancomycin and teicoplanin chiral stationary phases, *Pharmazie*, 53, 863, 1998.
136. Xiao, T.L., Reversal of enantiomeric elution order on macrocyclic glycopeptide chiral stationary phases, *J. Liq. Chrom. Rel. Technol.*, 24, 2673, 2001.
137. Steffeck, R.J. and Zelechonok, Y., Enantioselective ion-exclusion chromatography on teicoplanin aglycone and (+)-(18-crown-6)-2,3,11,12-tetracarboxylic acid stationary phases, *J. Chromatogr. A*, 983, 91, 2003.
138. Peyrin, E. et al., Interactions between D,L dansyl amino acids and immobilized teicoplanin: study of the dual effect of sodium citrate on chiral recognition, *Chromatographia*, 53, 645, 2001.
139. Sun, Q. and Olesik, S.V., Chiral separation by simultaneous use of vancomycin as stationary phase chiral selector and chiral mobile phase additive, *J. Chromatogr. B*, 745, 159, 2000.
140. Boesten, J.M.M. et al., Enantioselective high-performance liquid chromatographic separation of N-methyloxycarbonyl unsaturated amino acids on macrocyclic glycopeptide stationary phases, *J. Chromatogr. A*, 1108, 26, 2006.
141. Péter, A., Török, G., and Armstrong, D.W., High-performance liquid chromatographic separation of enantiomers of unusual amino acids on a teicoplanin chiral stationary phase, *J. Chromatogr. A*, 793, 283, 1998.
142. Török, G. et al., Enantiomeric high-performance liquid chromatographic separation of β-substituted tryptophan analogues, *Chromatographia*, 51, S165, 2000.
143. Péter, A. et al., High-performance liquid chromatographic separation of the enantiomers of unusual α-amino acid analogues, *J. Chromatogr. A*, 871, 105, 2000.
144. Péter, A. et al., A comparison of the direct and indirect LC methods for separating enantiomers of unusual glycine and alanine amino acid analogues, *Chromatographia*, 56, S79, 2002.
145. Török, G. et al., Direct chiral separation of unnatural amino acids by high performance liquid chromatography on a ristocetin A-bonded stationary phase, *Chirality*, 13, 648, 2001.
146. Peter, A., et al., High-performance liquid chromatographic separation of enantiomers of synthetic amino acids on a ristocetin A chiral stationary phase, *J. Chromatogr. A*, 904, 1, 2000.
147. Grobuschek, N. et al., Enantioseparation of amino acids and drugs by CEC, pressure supported CEC, and micro-HPLC using a teicoplanin aglycone stationary phase, *J. Sep. Sci.*, 25, 1297, 2002.
148. Péter, A. et al., High-performance liquid chromatographic enantioseparation of β-amino acids, *J. Chromatogr. A*, 926, 229, 2001.
149. Péter, A. et al., Direct and indirect high performance liquid chromatographic enantioseparations of β-amino acids, *J. Chromatogr. A*, 1031, 171, 2004.
150. Árki, A. et al., High-performance liquid chromatographic separation of stereoisomers of β-amino acids and a comparison of separation efficiencies on CHIROBIOTIC T and TAG columns, *Chromatographia*, 60, S43, 2004.

151. Sztojkov-Ivanov, A. et al., Comparison of separation efficiency of macrocyclic glycopeptide-based chiral stationary phases for the LC enantioseparation of β-amino acids, *Chromatographia*, 64, 89, 2006.
152. Péter, A. et al., Enantiomeric separation of unusual secondary aromatic amino acids, *Chromatographia*, 48, 53, 1998.
153. Péter, A. et al., Enantioseparation by HPLC of imino acids on macrocyclic glycopeptide stationary phases and as their (*S*)-*N*-(4-nitrophenoxycarbonyl)-phenylalanine methoxyethyl ester derivatives, *Chromatographia*, 56, S41, 2002.
154. Péter, A., Török, R., and Armstrong, D.W., Direct high-performance liquid chromatographic separation of unusual secondary amino acids and a comparison of the performances of CHIROBIOTIC T and TAG columns, *J. Chromatogr. A*, 1057, 229, 2004.
155. Schmid, M.G. et al., Enantioseparation of dipeptides and tripeptides by micro-HPLC comparing teicoplanin and teicoplanin aglycone as chiral selectors, *J. Biochem. Biophys. Methods*, 61, 1, 2004.
156. Schmid, M.G., et al., Enantioseparation of dipeptides by capillary electrochromatography on a teicoplanin aglycone chiral stationary phase, *J. Chromatogr. A.*, 990, 83, 2003.
157. Aboul-Enein, H.Y. and Serignese, V., Enantiomeric separation of several cyclic imides on a macrocyclic antibiotic (vancomycin) chiral stationary phase under normal and reversed phase conditions, *Chirality*, 10, 358, 1998.
158. Halabi, A. et al., Validation of a chiral HPLC assay for (*R*)-salbutamol sulfate, *J. Pharm. Biomed. Anal.*, 34, 45, 2004.
159. Fried, K.M., Koch, P., and Wainer, I.W., Determination of the enantiomers of albuterol in human and canine plasma by enantioselective high-performance liquid chromatography on a teicoplanin-based chiral stationary phase, *Chirality*, 10, 484, 1998.
160. Lamprecht, G. et al., Enantioselective analysis of (*R*)- and (*S*)-atenolol in urine samples by a high-performance liquid chromatography column-switching setup, *J. Chromatogr. B*, 740, 219, 2000.
161. Misl'anova, C. and Stefancova, A., Comparison of two different approaches of sample pretreatment for stereoselective determination of (*R,S*)-propranolol in human plasma, *J. Trace Microprobe Techn.*, 19, 173, 2001.
162. Aboul-Enein, H.Y. and Hefnawy, M.M., Enantioselective determination of arotinolol in human plasma by HPLC using teicoplanin chiral stationary phase, *Biomed. Chromatogr.*, 17, 453, 2003.
163. Andersson, M.E. et al., Evaluation of generic chiral liquid chromatography screens for pharmaceutical analysis, *J. Chromatogr. A*, 1005, 83, 2003.
164. Lamprecht, G. et al., Enantioselective analysis of (*R*)- and (*S*)-carvedilol in human plasma by high-performance liquid chromatography, *Chromatographia*, 56, S25, 2002.
165. Aboul-Enein, H.Y. and Serignese, V., Quantitative determination of clenbuterol enantiomers in human plasma by high-performance liquid chromatography using the macrocyclic antibiotic chiral stationary phase teicoplanin, *Biomed. Chromatogr.*, 13, 520, 1999.
166. D'Acquarica, I. et al., Enantio- and chemo-selective HPLC separations by chiral–achiral tandem-columns approach: the combination of CHIROBIOTIC TAG™ and SCX columns for the analysis of propionyl carnitine and related impurities, *J. Chromatogr. A*, 1061, 167, 2004.

167. Gasparrini, F. et al., Efficient enantiorecognition of ruthenium(II) complexes by silica-bound teicoplanin, *Tetrahedron: Asymmetry*, 11, 3535, 2000.
168. Péhourcq, F., Jarry, C., and Bannwarth, B., Chiral resolution of flurbiprofen and ketoprofen enantiomers by HPLC on a glycopeptide-type column chiral stationary phase, *Biomed. Chromatogr.*, 15, 217, 2001.
169. Péter, A. et al., Direct high-performance liquid chromatographic enantioseparation of β-lactams stereoisomers, *Chirality*, 17, 193, 2005.
170. Berkecz, R. et al., LC enantioseparation of β-lactam and β-amino acid stereoisomers and a comparison of macrocyclic glycopeptide and β-cyclodextrin-based columns, *Chromatographia*, 63, S37, 2006.
171. Tesařová, E. and Bosáková, Z., Comparison of enantioseparation of selected benzodiazepine and phenothiazine derivatives on chiral stationary phases based on β-cyclodextrin and macrocyclic antibiotics, *J. Sep. Sci.*, 26, 661, 2003.
172. Bosáková, Z., Kloučková, I., and Tesařová, E., Study of the stability of promethazine enantiomers by liquid chromatography using a vancomycin-bonded chiral stationary phase, *J. Chromatogr. B*, 770, 63, 2002.
173. Tesařová, E. and Bosáková, Z., The factors affecting the enantiomeric resolution and racemization of oxazepam, lorazepam and promethazine on macrocyclic antibiotics-bonded chiral stationary phases, *Chemia Analityczna*, 48, 439, 2003.
174. Kleidernigg, O.P. and Kappe, C.O., Separation of enantiomers of 4-aryldihydropyrimidines by direct enantioselective HPLC. A critical comparison of chiral stationary phases, *Tetrahedron: Asymmetry*, 8, 2057, 1997.
175. Chen, S. et al., Enantioresolution of substituted 2-methoxy-6-oxo-1,4,5,6-tetrahydropyridine-3-carbanitriles on macrocyclic antibiotic and cyclodextrin stationary phases, *J. Liq. Chromatogr.*, 18, 1495, 1995.
176. Lehotay, J. et al., Separation of enantiomers of some 1,4-piperazine derivatives of aryloxy-aminopropanols on a vancomycin chiral stationary phase, *Pharmazie*, 54, 743, 1999.
177. Tesařová, E., Zaruba, K., and Flieger, M., Enantioseparation of semisynthetic ergot alkaloids on vancomycin and teicoplanin stationary phases, *J. Chromatogr. A*, 844, 137, 1999.
178. Kullman, J.P. et al., Resolution of chiral thiol compounds derivatized with *N*-(1-pyrenyl)-maleimide and Thioglo™3, *J. Liq. Chrom. Rel. Technol.*, 23, 1941, 2000.
179. Zheng, Z., Jamour, M., and Klotz, U., Stereoselective HPLC-assay for citalopram and its metabolites, *Ther. Drug Monitor.*, 22, 219, 2000.
180. Aubry, A.-F. et al., Column selection and method development for the determination of the enantiomeric purity of investigational non-nucleoside reverse transcriptase inhibitors, *Chirality*, 13, 193, 2001.
181. Owens, P.K., Svensson, L.A., and Vessman, J., Direct separation of captopril diastereoisomers including their rotational isomers by RP-LC using a teicoplanin column, *J. Pharm. Biomed. Anal.*, 25, 453, 2001.
182. Aboul-Enein, H.Y. and Ali, I., A comparative study of the enantiomeric resolution of several tetralone derivatives on macrocyclic antibiotic chiral stationary phases using HPLC under normal phase mode, *Pharmazie*, 334, 258, 2001.
183. Hui, F., Ekborg-Ott, K.H., and Armstrong, D.W., High-performance liquid chromatographic and capillary electrophoretic enantioseparation of plant growth regulators and related indole compounds using macrocyclic antibiotics as chiral selectors, *J. Chromatogr. A*, 906, 91, 2001.

184. Horoboňová, K. et al., In vitro study of enzymatic hydrolysis of diperodon enantiomers in blood serum by two-dimensional LC, *J. Pharm. Biomed. Anal.*, 30, 875, 2002.

185. Aboul-Enein, H.Y. and Ali, I., Chiral resolution of cromakalim by HPLC on teicoplanin aglycone chiral stationary phases, *J. Liq. Chrom. Rel. Technol.*, 25, 2337, 2002.

186. Xiao, T.L. et al., Separation of enantiomers of substituted dihydrofurocoumarins by HPLC using macrocyclic glycopeptide chiral stationary phases, *Anal. Bioanal. Chem.*, 377, 639, 2003.

187. Hefnawy, M.M. and Aboul-Enein, H.Y., A validated LC method for the determination of vesamicol enantiomers in human plasma using vancomycin chiral stationary phase and solid phase extraction, *J. Pharm. Biomed. Anal.*, 35, 535, 2004.

188. Aboul-Enein, H.Y. and Hefnawy, M.M., Chiral analysis of butaclamol enantiomers in human plasma by HPLC using a macrocyclic antibiotic (vancomycin) chiral stationary phase and solid phase extraction, *Chirality*, 16, 147, 2004.

189. Abd El-Hady, D. et al., Methotrexate determination in pharmaceuticals by enantioselective HPLC, *J. Pharm. Biomed. Anal.*, 37, 919, 2005.

190. Warnke, M.M. et al., Use of native and derivatized cyclodextrin based and macrocyclic glycopeptide based chiral stationary phases for the enantioseparation of pterocarpans by HPLC, *J. Liq. Chrom. Rel. Technol.*, 28, 823, 2005.

191. Péter, A. et al., Comparison of column performances in direct high performance liquid chromatographic enantioseparation of 1– or 3-methyl-substituted tetrahydroisoquinoline analogs. Application of direct and indirect methods, *Biomed. Chromatogr.*, 19, 459, 2005.

192. Kannan, R. et al., Function of the aminosugar and *N*-terminal amino acid of the antibiotic vancomycin in its complexation with cell wall peptides, *J. Am. Chem. Soc.*, 110, 2946, 1988.

193. Pearce, C.M., Gerhard, U., and Williams, D.H., Ligands, which bind weakly to vancomycin: studies by ^{13}C NMR spectroscopy, *J. Chem. Soc., Perkin Trans. 2*, 159, 1995.

194. Billot-Klein, D. et al., Association constants for the binding of vancomycin and teicoplanin to *N*-acetyl-D-alanyl-D-alanine and *N*-acetyl-D-alanyl-D-serine, *Biochem. J.*, 304, 1021, 1994.

195. Loll, P.J. et al., Vancomycin binding to low-affinity ligands: delineating a minimum set of interactions necessary for high-affinity binding, *J. Med. Chem.*, 42, 4714, 1999.

196. Rekharsky, M. et al., Thermodynamics of interactions of vancomycin and synthetic surrogates of bacterial cell wall, *J. Am. Chem. Soc.*, 128, 7736, 2006.

197. McComas, C.C., Crowley, B.M., and Boger, D.L., Partitioning the loss in vancomycin binding affinity for D-Ala-D-Lac into lost H-bond and repulsive lone pair contributions, *J. Am. Chem. Soc.*, 125, 9314, 2003.

198. van Wageningen, A.M.A., Staroske, T., and Williams D.H., Binding of D-serine-terminating cell-wall analogues to glycopeptide antibiotics, *Chem. Commun.*, 1171, 1998.

199. Scrimin, P. et al., Kinetics and thermodynamics of binding of a model tripeptide to teicoplanin and analogous semisynthetic antibiotics, *J. Org. Chem.*, 61, 6268, 1996.

200. Fesik, S.W. et al., Determining the structure of a glycopeptide-Ac$_2$-Lys-D-Ala-D-Ala complex using NMR parameters and molecular modeling, *J. Am. Chem. Soc.*, 108, 3165, 1986.

201. Williams, D.H. and Bardsley, B., The vancomycin group of antibiotics and the fight against resistant bacteria, *Angew. Chem. Int. Ed.*, 38, 1172, 1999.
202. Molinari, H., Structure of vancomycin and a vancomycin/D-Ala-D-Ala complex in solution, *Biochemistry*, 29, 2271, 1990.
203. Loll, P.J. et al., Simultaneous recognition of a carboxylate-containing ligand and an intramolecular surrogate ligand in the crystal structure of an asymmetric vancomycin dimer, *J. Am. Chem. Soc.*, 119, 1516, 1997.
204. Loll, P.J. et al., A ligand-mediated dimerization mode for vancomycin, *Chem. Biol.*, 5, 293, 1998.
205. Yao, N-H. et al., Conformational studies of resin-bound vancomycin and the complex of vancomycin and Ac$_2$-L-lys-D-Ala-D-Ala, *J. Comb. Chem.*, 7, 123, 2005.
206. Nair, U.B. et al., Elucidation of vancomycin's enantioselective binding site using its copper complex, *Chirality*, 8, 590, 1996.
207. D'Acquarica, I., New synthetic strategies for the preparation of novel chiral stationary phases for HPLC containing natural pool selectors, presented at 8th Int. Meeting on Recent Developments in Pharmaceutical Analysis, Roma, June 29–July 3, 1999, 37.
208. Cavazzini, A. et al., Study of mechanisms of chiral discrimination of amino acids and their derivatives on a teicoplanin-based chiral stationary phase, *J. Chromatogr. A*, 1031, 143, 2004.
209. Jandera, P., Bačkovská, V., and Felinger A., Analysis of the band profiles of the enantiomers of phenylglycine in liquid chromatography on bonded teicoplanin columns using the stochastic theory of chromatography, *J. Chromatogr. A*, 919, 67, 2001.
210. Loukili, B. et al., Study of tryptophan enantiomer binding to a teicoplanin-based stationary phase using the perturbation technique. Investigation of the role of sodium perchlorate in solute retention and enantioselectivity, *J. Chromatogr. A*, 986, 45, 2003.
211. Haroun, M. et al., Thermodynamic origin of the chiral recognition of tryptophan on teicoplanin and teicoplanin aglycone stationary phases, *J. Sep. Sci.*, 28, 409, 2005.
212. Li, D. et al., Simulated dipeptide recognition by vancomycin, *J. Mol. Recognit.*, 10, 73, 1997.
213. Jusuf, S., Loll, P.J., and Axelsen, P.H., The role of configurational entropy in biochemical cooperativity, *J. Am. Chem. Soc.*, 124, 3490, 2002.
214. Lee, J.-G., Sagui, C., and Roland, C., First principles investigation of vancomycin and teicoplanin binding to bacterial cell wall termini, *J. Am. Chem. Soc.*, 126, 8384, 2004.
215. Lee, J.-G., Sagui, C., and Roland, C., Quantum simulations of the structure and binding of glycopeptide antibiotic aglycones to cell wall analogues, *J. Phys. Chem. B*, 109, 20588, 2005.
216. Alcaro, S et al., Glycopeptide-containing CSP: new synthesis, applications and insight into mechanism of chiral recognition, presented at 22nd Int. Symp. on Chromatography, Roma, September 13–18, 1998, 39.
217. Alcaro, S. et al., A quasi-flexible automatic docking processing for studying stereoselective recognition mechanisms. Part I. Protocol validation, *J. Comp. Chem.*, 21, 515, 2000.

3 Screening Approaches for Chiral Separations

Debby Mangelings and Yvan Vander Heyden

CONTENTS

ABSTRACT

The publications of the past 10 years in which generic screening approaches or separation strategies were developed in the field of chiral separations are reviewed. The idea is to give an overview of the different techniques that were used along with the possibilities of performing chiral separations applying these techniques. Research papers in which generic conditions are defined, that is, applicable on large sets of structurally diverse substances, either as a screening experiment or as a strategy including some optimization possibilities after the screening step, were considered. This should allow giving the reader an idea about the techniques used in the chiral separation field that enable generic analysis and about the conditions that can be considered as a first step in chiral method development.

3.1 INTRODUCTION

Chiral method development is often referred to as one of the most difficult fields in terms of development time. Interaction with a chiral selector is required to achieve separation but the enantioselectivity of a given selector for a given chiral molecule is a priori unknown. For some compounds, it can take several days to find suitable separation conditions when using sequential approaches. Therefore, industry most often defines generic separation strategies, which are often kept internally or are described in the literature. This review will focus on the latter, as it is our opinion that not only industry but also starting researchers can benefit from the use of these strategies for their separation problems. Therefore, the most suitable approaches from the past 10 years are summarized in this paper. The generic applicability of the strategies implies that they were and can be used on a broad range of compounds, thus the described conditions give a good chance for a successful separation.

The review is divided into several sections, depending on the technique. The considered techniques are capillary electrophoresis, high-performance liquid chromatography (HPLC), super- and subcritical fluid chromatography (SFC), capillary electrochromatography (CEC), preparative liquid chromatography, and some miscellaneous approaches. Techniques other than those mentioned earlier, such as gas chromatography (GC) [1], can be used to obtain enantioseparations, but we preferred to focus on published generic screening conditions or separation strategies. For gas chromatography, no generic screening approaches were found.

In this way, we aim to give an overview of what can be used as a separation technique and which conditions will most likely give an (beginning of) enantiomer separation after a first screening. Chiral method development starter kits are also available and evaluated in some papers [2], but we will not focus on this kind of applications.

3.2 CAPILLARY ELECTROPHORESIS

In [3] an ultra-fast screening of two basic drugs was performed based on earlier results that were published as a flow chart [4]. This chart indicated that for the substances investigated, metaprotenerol and isoprotenerol, dimethyl-beta-cyclodextrin (DM-β-CD) should be used as chiral selector in a 25 mM phosphate buffer of pH 2.5. Further, an applied voltage of 30 kV and a temperature of 15°C, in combination with a neutral coated capillary with 20 cm effective length, are used. Then, the cyclodextrin (CD) concentration can be optimized. Only this concentration is varied and no other conditions are changed. Once the optimal CD concentration is found, the effective length of the capillary is reduced to further decrease analysis time. Using this approach, complete separation of metaprotenerol and isoprotenerol enantiomers was achieved within 45 and 40 s, respectively, with around 1 million plates/m.

Guttman [5] presented a separation scheme for the capillary electrophoretic separation of basic, amphoteric, and acidic solutes. Using two model compounds of each class of molecules, the separation was executed according to the Vigh theory [5–10], that is, using ionoselective/duoselective or desionoselective separations. In the first situation, the pH of the electrolyte is chosen at least two units from the pK_a value of the solute on making sure that it is charged. In the second case, it is as close as possible to the pK_a of the analyte. Using a neutral coated capillary with 20 cm effective length, four types of CDs (β-CD, γ-CD, HP-β-CD and DM-β-CD) were used at low and high concentrations, applying normal polarity for basic solutes and reversed polarity for acidic ones. It was seen that for the basic test compounds, a low pH allowed the separation, while for the amphoteric molecules, an acidic pH provided good results. However, for the acidic test compounds, a pH = pK_a buffer was more efficient, thus desionoselective interactions were necessary. When the appropriate pH and CD type are known, the authors suggest further optimization of the separation by fine-tuning the chiral selector concentration and the applied electrical field. It must be noted that the used buffers at pH 2.4 and 8 were 25 mM phosphate buffers, whereas the buffer with pH 4.66 was prepared either from 25 mm acetic acid or 200 mM 2-(N-morpholino)ethanosulfonic acid.

A paper by Fillet et al. [11] describes strategies in CE developed for basic, acidic, or neutral drugs, which are displayed in Figure 3.1. For all compounds, 100 mM phosphoric acid adjusted to pH 3 with triethanolamine (TEOA) was used as electrolyte. For basic compounds, experiments were conducted in normal polarity mode. Five CD derivatives, carboxymethyl-beta-cyclodextrin (CM-β-CD), sulfobutyl-beta-cyclodextrin (SB-β-CD), DM-β-CD, trimethyl-beta-cyclodextrin (TM-β-CD), and hydroxypropyl-beta-cyclodextrin (HP-β-CD), were tested at a given concentration and the obtained separation was evaluated. The concentration of the CD giving the highest resolution was then optimized. When CM-β-CD was selected as the best chiral selector, an increase of the pH to 5 was possible to enhance the resolution. Another possibility for basic compounds was the addition of an organic modifier, such as methanol or cyclohexanol, to the background electrolyte. For acidic and neutral drugs, anionic CM-β-CD and SB-β-CD were tested first in reversed polarity mode. To increase the resolution, dual systems of these anionic CD derivatives with a neutral CD, such as TM-β-CD and DM-β-CD, can be tested. Again, pH 5 can be considered to increase the resolution when CM-β-CD is used in the electrolyte, especially for weak acids and neutrals. The defined screening and optimization steps proved to be very useful, as 48 of 50 drugs were fully separated with analysis times usually below 10 min.

In a series of 11 papers [12–22] by the group of Koppenhoefer, Zhu et al., large test sets of drugs were screened using general analysis conditions on polymer coated capillaries combined with different selectors in the background electrolyte, of which some are very rarely used, for example, hydroxypropyl-α-cyclodextrin (HP-α-CD) and hexakis-(2,3,6-tri-O-methyl)-alpha-cyclodextrin. However, the success rates for some selectors were rather poor (Table 3.1). No optimization steps were included in these papers; sometimes a possibility for further optimization was indicated. Some interaction mechanisms of the substances with the selectors were revealed and explained, but we will not go further into detail on this topic. A short overview of the obtained enantioselective results is given in Table 3.1. Heptakis (2,3,6-tri-O-methyl)-beta-cyclodextrin performed overall best as chiral selector under the tested analysis conditions, as it showed enantioselectivity for more than 50% of the test set substances. All other selectors gave inferior results.

Liu and Nussbaum [23] proposed a screening approach for small, amine-containing enantiomers possessing chromophores and primarily one or two chiral centers. Using an operating temperature of 20°C and detection at 214 nm, the following five screening experiments are proposed: (1) 30-mM phosphate of pH 2.5, adjusted with tetrabutyl ammonium (TBA) hydroxide to determine the migration of the uncomplexed substance; (2) 15-mM DM-β-CD, (3) 30-mM HP-β-CD; (4) 30-mM HP-α-CD; and (5) 30-mM hydroxypropyl-gamma-cyclodextrin (HP-γ-CD). For steps 2–5, the CD dissolved in the electrolyte of step 1 is used. All steps are to be

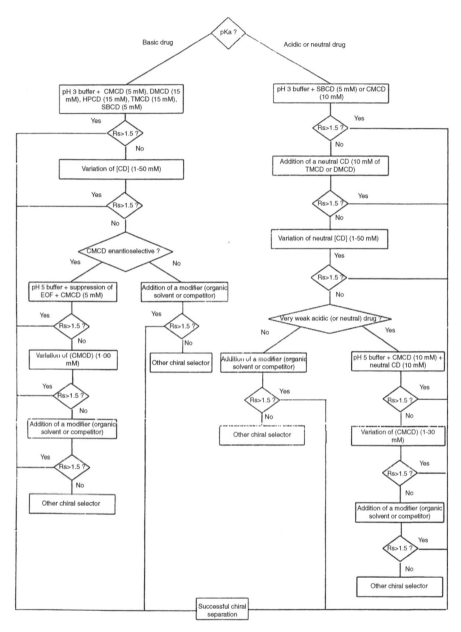

FIGURE 3.1 Method development scheme followed in Fillet et al. [11]. (Reprinted from Fillet et al. *Electrophoresis* 1998, 19, 2834–2840. With permission from Wiley VCH.)

tested for each compound. Further, 30 kV normal polarity is used as applied voltage, combined with a 50 μm × 57 cm bare fused silica capillary. In a sixth step, the electrolyte consists of 25-mM phosphoric acid adjusted to pH 2.5 by triethylamine (TEA), containing 32-mM sulphated-beta-cyclodextrin (S-β-CD). Further conditions

TABLE 3.1

Enantioselective Results Obtained by Koppenhoefer, Zhu et al. [12–22]

Chiral selector	Concentration (mM)	Success Rate[a] (Percentage)	Conditions	Reference
γ-Cyclodextrin	15	7/57 (12)	1	12
α-Cyclodextrin	15	6/59 (10)	1	13
β-Cyclodextrin	15	7/34 (20)	1	14
HP-γ-CD	45	30/86 (35)	2	15
HP-α-CD	45	34/86 (39)	2	16
HP-β-CD	45	42/86 (49)	2	17
γ-Cyclodextrin	45	18/86 (21)	2	18
β-Cyclodextrin	15	21/86 (24)	2	19
Hexakis-(2,3,6-tri-O-methyl)-α-cyclodextrin	45	23/86 (27)	2	20
Heptakis (2,3,6-tri-O-methyl)-β-cyclodextrin	45	47/86 (55)	3	21
Octakis (2,3,6-tri-O-methyl)-γ-cyclodextrin	45	15/86 (17)	3	22

[a] Number of partially or fully separated compounds over the total number of tested compounds.

General analysis conditions: (1) 0.1 M phosphate buffer pH 2.5; injection: 8 kV, 6 s; applied voltage: 14 kV, temperature 30°C. *Capillary:* 44.5 cm × 50 μm, polymer coated. (2) 0.1 M phosphate buffer pH 2.5; injection: 15 kV, 3 s; applied voltage: 15 kV, temperature 25°C. *Capillary:* 29, 30, 32, or 36 cm × 50 μm, polymer coated, (3) 0.1 M phosphate buffer pH 2.5; injection: 15 kV, 3 s; applied voltage: 15 kV, temperature 25°C. *Capillary:* 30 cm × 50 μm, polymer coated.

were an applied voltage of 12 kV, reverse polarity, and a bare silica capillary of 25 μm × 27 cm. This screening approach provided very satisfying results because a partial or baseline separation was achieved for 96% of the 49 tested compounds, of which 55% had a resolution ≥1.40.

An experimental design approach was used in Reference 24 to investigate the effects of different factors on the separation of several β-blocking agents. Briefly, experimental designs are tools to explore an experimental domain in a predefined and relatively low number of experiments. They cover usually the experimental domain as good as possible and allow estimating the effect of a factor on a given response. More information about experimental designs and their interpretation can be found in Reference 25. In the study, a screening design for four factors at three levels (nine experiments) was used to select a good combination of the factors CD type, CD concentration, buffer pH, and percentage methanol in the running buffer. Next, a fractional factorial design for three factors at two levels (four experiments) was used to improve the separation. The experimental setup of these designs is displayed in Table 3.2. It was shown that the first design allows making a decision about the CD type to be used; the second design can be used for further optimization of the other factors. It was observed that anionic derivatives provided the best results for the 11 investigated compounds: for 8 of them, it was CM-β-CD gave the best results,

TABLE 3.2
Setup of the Experimental Designs Used in Reference 24

(a) Factorial Design for Four Factors at Three Levels

	Factors			
Experiment	CD Type	Concentration CD	pH	MeOH (%)
1	−1	0	0	1
2	−1	1	−1	0
3	−1	−1	1	−1
4	0	1	0	−1
5	0	0	1	0
6	0	−1	−1	1
7	1	0	−1	−1
8	1	−1	0	0
9	1	1	1	1
1	CM-β-CD	15 mM	4	30
2	CM-β-CD	30 mM	2.5	15
3	CM-β-CD	5 mM	5.5	0
4	DM-β-CD	30 mM	4	0
5	DM-β-CD	15 mM	5.5	15
6	DM-β-CD	5 mM	2.5	30
7	HP-β-CD	15 mM	2.5	0
8	HP-β-CD	5 mM	4	15
9	HP-β-CD	30 mM	5.5	30

(b) Factorial Design for Three Factors at Two Levels

	Factors		
Experiment	Concentration CD	pH	MeOH (%)
1	1	1	1
2	−1	1	−1
3	1	−1	−1
4	−1	−1	1
1	30 mM	5.5	30
2	5 mM	5.5	0
3	30 mM	2.5	0
4	5 mM	2.5	30

Note: Top of a design: theoretical levels; bottom: substituted factor levels.

Electrolyte: 0.1 M phosphate buffer brought to the desired pH with TEOA; injection 0.8 psi, 10 s; *Applied voltage:* 25 kV; Temperature: 15°C; Fused-silica capillary 43.3 cm × 50 μm.

and for 3 of the investigated compounds, sulfobutyl ether-beta-cyclodextrin (SBE-β-CD). However, not all substances were brought to a baseline separation at the best CE conditions predicted by the design.

An experimental design approach was also used in Reference 26 for the chiral analysis of amino acid derivatives. The screening and optimization schedule followed

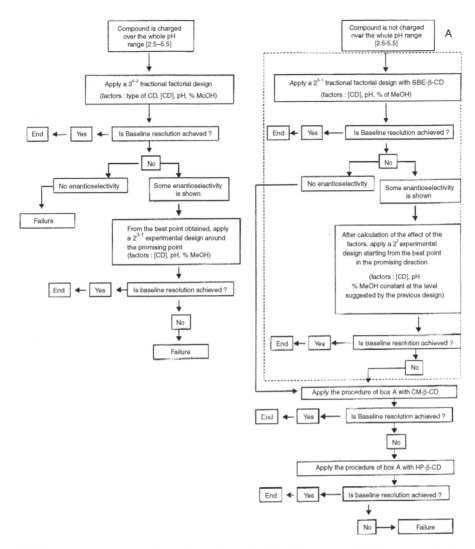

FIGURE 3.2 Schedule followed in Perrin et al. [26]. Electrolyte: 0.1 M phosphate buffer brought to the desired pH with TEOA; injection 0.8 psi, 4 s; Applied voltage: ±613 V/cm (amino alkyl derivatives) or −490 V/cm (N-CBZ or N-t-Boc derivatives); Temperature: 20°C; Fused-silica capillary 40.8 cm × 50 μm. (Reprinted from Perrin et al., *J. Chromatogr. A* 2000, 883, 249–265. Copyright (2000). With permission from Elsevier.)

is given in Figure 3.2. A distinction is made between compounds that are charged or uncharged in the considered pH range (2.5–5.5). The types of CD indicated in the flow chart are HP-β-CD, CM-β-CD, and SBE-β-CD. Of the 14 compounds studied, 12 compounds could be separated in less than 9 experiments. However, no general optimal conditions were defined for the investigated amino acids; best conditions were determined for each compound individually.

Bergholdt et al. [27] used a short-end injection method combined with S-β-CD as chiral selector to obtain fast enantioseparation of ormeloxifene and 15 of its analogs.

The electrolyte consisted of 2% (w/v) of the selector in 10 mM phosphate buffer pH 3, and the effective length of the CE setup was only 8.5 cm. Because the mobilities of the anionic selector and the complex towards the anode are greater than the mobility of the analytes towards the cathode, the polarity remained in normal mode when applying the short end of the capillary as separation channel and it was set on 25 kV. Further, a bare silica capillary of 28.5 cm × 50 μm was used. The 16 investigated analytes were all separated using the proposed conditions, with resolutions ranging from 1.3 to 26.5 and analysis times between 46 and 73 s. A comparison was made between the long- and short-end methods and it was concluded that no differences in interaction mechanisms occur. Due to the fast analysis of the proposed short-end method, these conditions seem suitable for a fast screening experiment. To obtain more resolution, the authors suggest trying the long-end method using the same conditions, but with reversed polarity. The latter can be considered as an optimization experiment.

The short-end injection was also used in a paper by Perrin et al. [28]. They saw a very high chiral recognition capability of highly sulfated cyclodextrins (HS-CD). Using a test set of 27 amino acid derivatives, the application of HS-α-CD, HS-β-CD, and HS-γ-CD in a 5% w/v concentration allowed the separation of 26 compounds, of which 22 had a Rs > 2. From their experiments, a screening and optimization scheme was derived (Figure 3.3), and based on this scheme, a separation strategy was defined

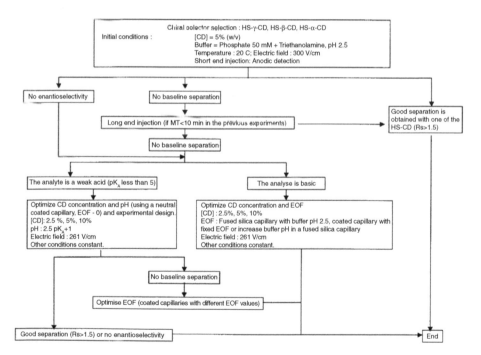

FIGURE 3.3 Screening and optimization scheme developed in Perrin et al. [28]. Other conditions: Injection 0.5 psi, 3.5 s; Fused-silica capillary or polyacrylamide-coated capillary of 30 cm × 50 μm. (Reprinted from Perrin et al. *Electrophoresis* 2001, 22, 3203–3215. With permission from Wiley VCH.)

(a)

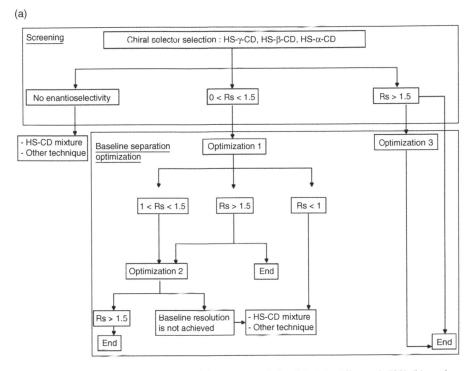

FIGURE 3.4 (a) General structure of the strategy defined in Matthijs et al. [29] (b) optimization 1 in case of a partial separations, (c) optimization 2 for peak shape and efficiency optimization in case of Rs ≥ 1, and (d) optimization 3 for migration time optimization and global optimization in case of a baseline separation. (Reprinted from Matthijs, N., et al., *J. Pharm. Biomed. An.* 2002, 27, 515–529. Copyright (2002). With permission from Elsevier.)

by Matthijs et al. [29]. The resulting strategy and the different optimization steps are displayed in Figure 3.4. For the screening step of the strategy, the same conditions as in Reference 28 were used, but an uncoated fused silica capillary was applied for all experiments. The strategy in Figure 3.4 also refers to the use of dual selector systems (CD mixture) when no successful method development is achieved. This part of the strategy was developed in a later paper [30] and is given in more detail in Figure 3.5. The dual CD related parts are indicated in bold. When dual systems are used, the authors recommend using a long-end injection in the screening step, because efficient separations not always were obtained using the short end of the capillary. The use of short-end injections is indicated when an Rs > 1.5 is obtained after the screening. When a partial separation is obtained, the strategy indicates that the concentration of neutral CD should be altered, which can result as well in an increased as in a decreased separation, because the outcome seems compound dependent and unpredictable. For a test set of 25 compounds, 89% of the basic compounds were better separated using a dual CD system.

Another separation strategy in CE using HS-CD was presented by Chapman and Chen [31]. For the screening, they recommend three experiments with a starting

(b)

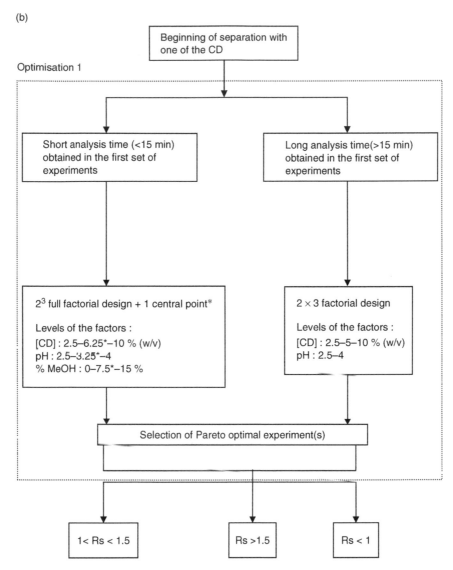

FIGURE 3.4 Continued.

concentration of 5% of HS-α-CD, HS-β-CD, and HS-γ-CD in a 25-mM phosphate buffer at pH 2.5. They also specify the use of a 50-μm bare fused-silica capillary with a relatively short effective length (i.e., 10 or 20 cm to the detector). Other analyzing conditions were a temperature of 22°C, field strength of 500 V/cm, detection at 200 nm, and simultaneously recording a PDA spectrum, and an injection at 0.3 psi during 4 s. Using this screening procedure, 156 out of 160 drugs showed Rs > 1.0, indicating the generic applicability of the proposed screening. Optimization steps are also described: In case of a partial resolution, the CD content can be varied between 2% and 10%. The capillary temperature and length, and the field strength can be used

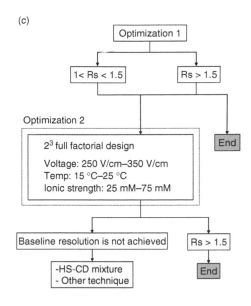

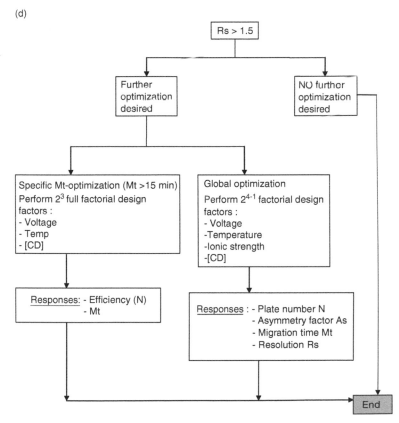

FIGURE 3.4   Continued.

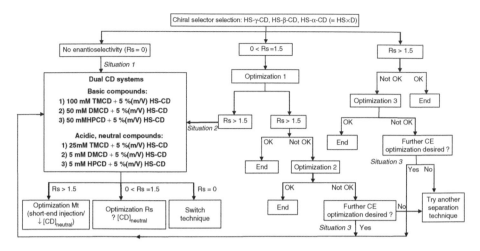

FIGURE 3.5   CE strategy proposed in Matthijs et al. [30] including dual CD systems. Steps to be performed in optimizations 1, 2, and 3, see Figure 3.4b–d. (Reprinted from Matthijs, N. et al. *An. Chim. Acta* 2004, 525, 247–263. Copyright (2004). With permission from Elsevier.)

to further fine tune the separation, although the field strength is preferably varied in the last instance. The authors also indicate that when no resolution is seen and broad peaks are observed, it is most likely that the CD concentration is too high, thus lower concentrations should be tested.

An extended separation strategy was developed by Jimidar et al. [32]. The general structure of the strategy is given in Figure 3.6. Different routes are followed depending on the compound's acidic, basic, or neutral character. Before application of this strategy, a pH screening in CE can optionally be performed when the pK_a values are unavailable. The buffer type depends on the pH. For pH 2.5, 3.0, 7.0, and 8.0, phosphate buffers are used, prepared by adjusting the pH of phosphoric acid by means of TEOA. For pH 4.0 and 5.0, an acetate buffer is used, prepared by the titration of acetic acid with TEOA. For pH 9.0, a borate buffer is used, prepared by adjusting the pH of boric acid with TEOA. The steps A–G (Figure 3.6) all use a Taguchi orthogonal array mixed level $(18\ (6^1,\ 3^3))$ design for four factors requiring 18 experiments, in which the type of selector, the concentration of selector, the concentration of the buffer, and the amount of organic modifier are varied. Because the designs are presented in detail in [32], we refer to this manuscript for more details. Regarding experimental conditions, a fused silica capillary of $32/40\ cm \times 50\ \mu m$ was used, and the applied voltage was set on 25 kV, either in positive or negative polarity. Further, the temperature was set on 20°C and an injection of 34.5 mbar during 5 s was used, followed by an electrolyte plug at the same pressure during 1 s. The authors claim that by using this strategy, satisfactory results are usually achieved. This was also demonstrated by the fact that the 10 reported test components were baseline separated after application of the strategy. When no satisfying separation is obtained, one is recommended to use another technique. The optimization step at the bottom of Figure 3.6 is explained more in detail in Reference 33, where for a

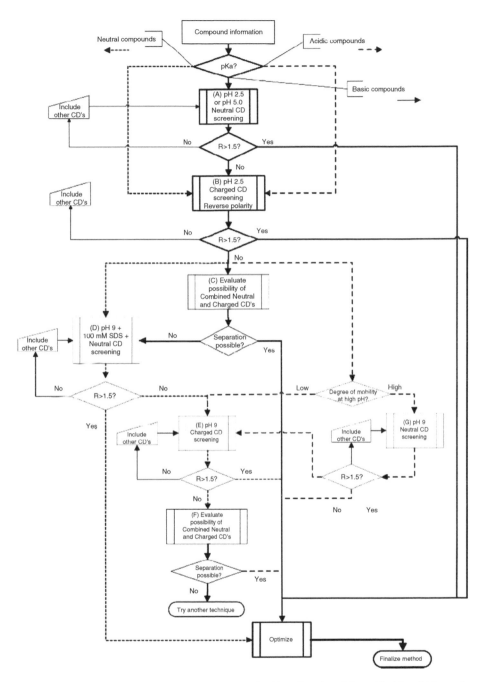

FIGURE 3.6 Strategy proposed by Jimidar et al. [32]. (Reprinted from Jimidar, M.I. et al. *Electrophoresis* 2004, 25, 2772–2785. With permission from Wiley VCH.)

particular compound the separation conditions were optimized. This was done using a Box-Behnken design for four factors, including three center point replicates, which gives a total of 27 experiments to be performed. The varied factors were the concentrations of two CDs, the buffer concentration, and the applied voltage. The optimum was found using statistical software (Minitab Statistical Software, Minitab Inc.). The final method was then subjected to a quality control performance test and validated.

Rocheleau [34] compared the separation performance of HS-β-CD from Beckman, S-β-CD from Aldrich, and heptakis (6-sulfato)-β-CD from Regis Technologies for weak bases. The first two were found superior over the CD from Regis Technologies, as they were able to induce a separation of the two test compounds, which is a very limited set to draw generalized conclusions. However, for screening purposes, S-β-CD from Aldrich was preferred because its between-batch variability was smaller when three batches were evaluated. The proposed screening conditions include a fused-silica capillary with 50 cm effective length and 50 μm ID, an applied voltage of 15 kV at reversed polarity, 10% w/v S-β-CD in a 50 mM phosphate-TEOA buffer pH 2.5, an operating temperature of 25°C, and an injection of 0.5 psi during 10 s. Because these conditions were only applied on five compounds, their generic character was not really demonstrated, although baseline resolution was obtained for four compounds.

A strategy for the enantioseparation of basic compounds was described by Sokolieβ and Köller [35] and is displayed in Figure 3.7. All steps from method development till validation are included in the flow chart. Perhaps a disadvantage in this approach is that sequential screening and optimization steps are used (i.e., every factor is optimized individually). The use of the developed scheme was demonstrated for one compound, for which the method was developed, optimized, and validated. The generic applicability of this approach was not considered and is unknown.

Zhou et al. [36] developed a strategic approach for pharmaceutical basic compounds using sulfated CDs as chiral selectors. First, various kinds of selectors—mostly derivatives of β-CD—from different suppliers were screened for their separation performances. The used conditions in this screening step were a 25 mM phosphate buffer at pH 2.5, 2.5 w/v% of selector, a capillary of 41.5/48.5 cm × 75 μm, a temperature of 20°C, an injection of 50 mbar, 3 s (plus a buffer plug during 4 s), and an applied voltage of −10.5 kV. Using these conditions, baseline Rs was achieved for the five compounds that were tested with at least one type of CD. For more information considering the used CD, we refer to Reference 36. Regarding optimization, the authors recommend to optimize sequentially CD concentration, pH, organic modifier content (acetonitrile and methanol), buffer concentration and-type, capillary temperature and-length, and finally the applied voltage. The approach was tested on 17 compounds and baseline resolution was seen for all of them, thus the generic applicability is demonstrated to a certain extent.

Some special approaches defined in the CE enantioseparation field can be found in References 37 and 38. In Reference 37, the authors tested four approaches to decrease the analysis time of a previously developed chiral separation method for amphetamine and its related compounds. The considered possibilities were (i) the short-end injection technique or (ii) increased electrical field combined with a capillary length reduction,

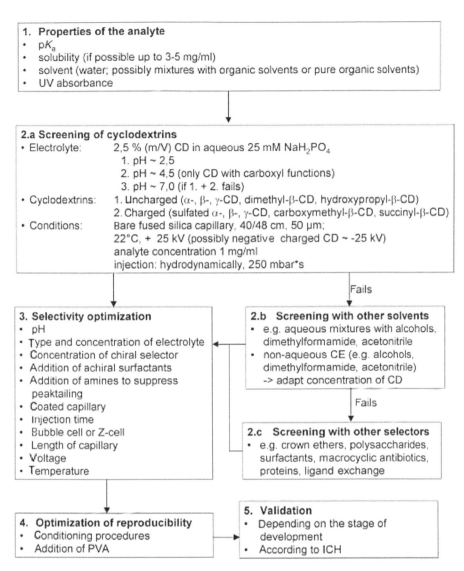

1. **Properties of the analyte**
   - pK_a
   - solubility (if possible up to 3-5 mg/ml)
   - solvent (water; possibly mixtures with organic solvents or pure organic solvents)
   - UV absorbance

2.a Screening of cyclodextrins
- Electrolyte: 2,5 % (m/V) CD in aqueous 25 mM NaH$_2$PO$_4$
                   1. pH ~ 2,5
                   2. pH ~ 4,5 (only CD with carboxyl functions)
                   3. pH ~ 7,0 (if 1. + 2. fails)
- Cyclodextrins: 1. Uncharged (α-, β-, γ-CD, dimethyl-β-CD, hydroxypropyl-β-CD)
                   2. Charged (sulfated α-, β-, γ-CD, carboxymethyl-β-CD, succinyl-β-CD)
- Conditions: Bare fused silica capillary, 40/48 cm, 50 µm;
                   22°C, + 25 kV (possibly negative charged CD ~ -25 kV)
                   analyte concentration 1 mg/ml
                   injection: hydrodynamically, 250 mbar·s

Fails

3. **Selectivity optimization**
- pH
- Type and concentration of electrolyte
- Concentration of chiral selector
- Addition of achiral surfactants
- Addition of amines to suppress peaktailing
- Coated capillary
- Injection time
- Bubble cell or Z-cell
- Length of capillary
- Voltage
- Temperature

2.b Screening with other solvents
- e.g. aqueous mixtures with alcohols, dimethylformamide, acetonitrile
- non-aqueous CE (e.g. alcohols, dimethylformamide, acetonitrile) -> adapt concentration of CD

Fails

2.c Screening with other selectors
- e.g. crown ethers, polysaccharides, surfactants, macrocyclic antibiotics, proteins, ligand exchange

4. **Optimization of reproducibility**
- Conditioning procedures
- Addition of PVA

5. **Validation**
- Depending on the stage of development
- According to ICH

FIGURE 3.7   The strategy proposed in Sokolieβ et al. [35] for basic active pharmaceutical ingredients. PVA: Polyvinylalcohol; Injection in step 2a: 50 mbar, 5 s. (Reprinted from Sokolieβ, T., Köller, G. *Electrophoresis* 2005, 26, 2330–2341. With permission from Wiley VCH.)

which are often described in the literature. Also, (iii) an external pressure application or (iv) dynamic coating of the capillary to generate an increased EOF were tested, which is seldom described. The latter approach gave the overall best results (i.e., low retention times, and sufficient resolutions and efficiency).

In Reference 38, a revolutionary approach was developed, keeping in mind a high-throughput context. Here, a cePRO9600 system was used to screen several selectors for their separation performance in one run using a 96-capillary array, a 96-well sample tray plate containing eight samples, and a 96-well buffer tray containing the electrolytes with various selectors to be tested. The special feature of this approach is the fact that 96 analyses are performed simultaneously within the time of one run on any ordinary CE instrument. It also allows screening a higher number of selector types without spending too much time, because a maximum of 12 selectors can be screened simultaneously if only one run per selector is considered. A schematic diagram of the setup of the sample and buffer well plates is given in Figure 3.8. For the optimization procedures, this approach can also be useful, as long as factors such as applied voltage and temperature do not have to be varied individually. The drawbacks are the necessity for such a specific instrument and the fact that all 96 analyses have to be performed at the same voltage and temperature. The latter parameters can thus only be optimized in a sequential manner. Finally, we can also mention a drawback for the capillary

Position	1	2	3	4	5	6	7	8	9	10	11	12
A	Atenolol											
B	Alprenolol											
C	p-Chloroamphetamine											
D	Isoprotenerol											
E	Metaproterenol											
F	Terbutaline											
G	Nelopam											
H	Warfarin											

Position	1	2	3	4	5	6	7	8	9	10	11	12
A												
B												
C												
D	5 % HS-α-CD		5 % HS-β-CD		5 % HS-γ-CD		5 % S-β-CD					
E												
F												
G												
H												

FIGURE 3.8 Possible set-up of the well plates of samples and buffer for a screening of the types of selector [38].

length: if one wants to change, all 96 capillaries have to be replaced by longer or shorter ones.

In summary, it can be stated that for screening approaches in CE, only CD derivatives have been used. Other chiral selectors occasionally applied in CE, such as crown ethers, macrocyclic antibiotics, and chiral surfactants [39], were not found to be involved in (generic) screening approaches for chiral separations.

3.3 HIGH-PERFORMANCE LIQUID CHROMATOGRAPHY

Most screening and optimization approaches in HPLC were defined using polysaccharide chiral stationary phases (CSP), thanks to their broad chiral recognition ability toward a large number of compounds.

De la Puente et al. [40] have developed a gradient screening procedure in normal-phase mode combined with an experiment in high polar mode for drug discovery compounds. Four CSP were included in their screening procedure: Chiralpak AD and AS, and Chiralcel OJ and OD using an amylose tris(3,5-dimethylphenylcarbamate)-, an amylose tris[(S)-α-methylbenzylcarbamate]-, a cellulose tris(4-methylbenzoate)-, and a cellulose tris(3,5-dimethylphenylcarbamate) selector, respectively. Two gradient systems were used, that is, hexane modified with either isopropanol or ethanol. As mobile phase additive, 0.05% trifluoroacetic acid (TFA) was added. The gradient ranged from 20% to 70% modifier over 20 min. The highest concentration of organic modifier was kept constant for 2 min then returned to starting conditions and after 5 min of isocratic equilibration at 20%, the next analysis was started. The flow rate was kept constant at 0.75 mL/min. Using an eight-port column-switching valve, this procedure could be performed routinely. However, Chiralpak AD was excluded from the gradient screening with ethanol due to pressure restrictions. For this column an alternative screening with the same chromatographic parameters was used but with a high-polar mobile phase ranging between 70% and 100% ethanol. When the acidic conditions affect the stability of the compound, the separation is performed under neutral or basic (using 0.2% dimethylethylamine (DMEA) as mobile phase additive) conditions. This screening procedure was successfully applied for over 85% of 800 synthesized compounds, all differing in structural properties and polarities, showing the generic character of the proposed conditions. When no enantioselectivity is seen, isocratic optimization is proposed, but no further details concerning the followed procedure were included. In Reference 41, a procedure slightly modified from that in Reference 40 was proposed, as the gradient range only went up to 60% modifier in 15 min for the normal-phase conditions; for the high-polar mode, only the gradient time changed from 20 to 15 min. For the latter mode, 0.05% TFA is added to obtain acidic conditions and 0.2% DMEA for basic conditions. When methanol is used in the polar mode, only neutral or basic elution conditions are tested. The LC system was coupled with an Atmospheric Pressure Chemical Ionisation (APCI) mass spectrometer. Using these adapted conditions, the chiral resolutions obtained in the standard gradient mode could be reproduced adequately.

Andersson et al. [42] have developed a screening procedure that includes both polysaccharide and macrocylic antibiotic CSP. The first screening procedure is executed using the four classical polysaccharide CSP (AD, OD, OJ, and AS) and five mobile phases. Acidic compounds are analyzed using two mobile phases consisting of organic modifier/isohexane/TFA (15/85/0.1, v/v/v) with either ethanol or 2-propanol as organic modifier. Basic and neutral compounds are analyzed with three mobile phases: two composed of organic modifier/isohexane/diethylamine (15/85/0.1, v/v/v) with either ethanol or 2-propanol as organic modifier, and one composed of methanol/ethanol (50/50, v/v). A mixture of 2-propanol/isohexane (10/90, v/v) is used as washing solvent to rinse the columns. All experiments of the first screening step use a flow rate of 1 mL/min. The second screening uses Chirobiotic R (ristocetin A as selector), Chirobiotic V (vancomycin selector), and Chirobiotic T (teicoplanin selector), in that order, combined with two mobile phases. Basics are analyzed in polar organic mode using methanol/acetic acid/TEA (100/0.02/0.01, v/v/v), neutrals in reversed-phase mode with methanol/triethylammonium acetate (0.1%, pH 6) (25/75, v/v), and acids with both mobile phases. The flow rates here are 2 mL/min in polar organic mode and 1 mL/min in reversed-phase mode. Using these proposed screenings on 53 compounds, of which two had two pairs of isomers (55 enantiomeric pairs in total), resulted in enantioselectivity for 87% of the set in the first screening and for 65% in the second. Both screenings resulted in the partial or baseline separation of 96% of the test set, which shows the generic applicability of the strategy.

A special kind of screening procedure was applied in Reference 43. Five columns were analyzed simultaneously using a column-switching device. All five columns were coupled to one pump and analyzed in parallel using a particular mobile phase. The separations were monitored with five ultraviolet (UV) detectors. Behind the UV detectors, all channels were again combined into one and then entered in a circular dichroism detector. This detector allowed observing which enantiomer elutes first. This approach seems interesting, but will only work when the retention times on the five columns differ enough, otherwise it will result in an uninterpretable result. The considered stationary phases were Chiralpak AD and AS, Chiralcel OD and OJ, and Whelk-O1. Five mobile phases were considered in the screening procedure, but no further details were given. It is only stated that both normal-phase and polar-organic solvent modes were used.

Different screening approaches and separation strategies were also developed by our group [44–48]. In a first paper [44], a screening step was developed for the chiral separation of pharmaceuticals in normal-phase liquid chromatography (NPLC). The considered stationary phases were Chiralcel OD-H, Chiralpak AD, and Chiralcel OJ. Two mobile phases are tested using a flow rate of 1 mL/min at ambient temperature: *n*-hexane/2-propanol (90/10) and *n*-hexane/ethanol (90/10). For the separation of basic compounds, 0.1% (v/v) diethylamine (DEA) is added to the mobile phases, and it is recommended to test Chiralcel OD and Chiralpak AD first (four experiments), Chiralcel OJ (two experiments) is only tested when no separation occurred on the first two phases. For acidic compounds, 0.1% TFA is added to the mobile phases and the three phases are screened (i.e., six experiments are conducted). Neutral and bifunctional compounds are treated as acidic. The generic applicability of this screening step was demonstrated by applying the proposed conditions on a set of 36 compounds. Thirty-two of them (89%) showed enantioselectivity, of which 24 were baseline separated. Relatively short analysis times were obtained i.e., usually below 20 min.

Secondly, a screening step was developed in reversed-phase liquid chromatography (RPLC) [45]. Compounds are to be screened, in a first instance, with an acidic mobile phase, a 50 mM phosphate buffer pH 2 containing 100 mM, potassium hexafluorophosphate (KPF$_6$), mixed with 40% (v/v) acetonitrile, on Chiralcel OD-RH, Chiralpak AD-RH, and Chiralcel OJ-RH columns. Potassium hexafluorophosphate, KPF$_6$, is a chaotropic salt which forms ion pairs with the positively charged basics, hence enabling interaction of the apparently neutral complex with the CSP. When no or only a beginning of separation is seen for the compound, a second mobile phase is tested, that is, 20 mM borate buffer pH 9/ACN (60/40, v/v). The flow rate is set at 0.5 mL/min and the experiments are conducted at ambient temperature. The applicability of this screening step was tested by means of a test set of 37 compounds. It was shown that 33 were enantioresolved using the proposed screening step, with analysis times usually below 30 min. Both screening steps were implemented in separation strategies by Matthijs et al. [46]. In these strategies, optimization steps were also included. The strategies for NPLC and RPLC are displayed in Figures 3.9 and 3.10, respectively. The screening step defined in NPLC [44] was translated into experimental designs to be executed (Figure 3.9a), but the essential part remains the same; that is, in a 2^2 full factorial design, four experiments are conducted, in which each combination of the proposed factor levels is tested. For the 3×2 experimental

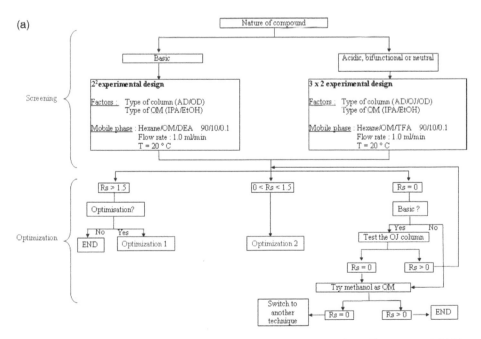

FIGURE 3.9 Strategy in NPLC. (Reprinted from Matthijs, N. et al. *J. Chromatogr. A* 2004, 1041, 119–133. Copyright (2004). With permission from Elsevier.)

(b)

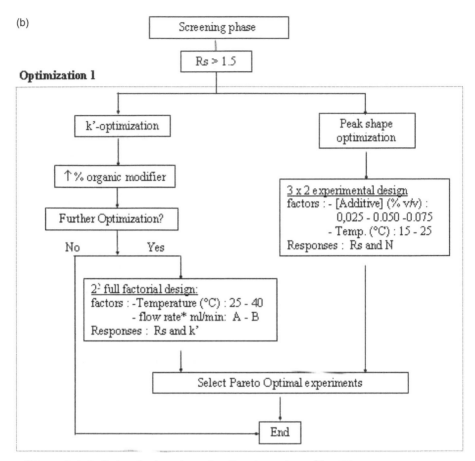

*Flow rate is limited and values depend on the chromatographic system.
Values above 0.5 mL/min are used.

FIGURE 3.9 Continued.

design, six experiments are performed. The same designs are also applied in optimizations 1 and 2. Optimization 1 (Figure 3.9b) includes a retention factor and peak-shape optimization step for compounds that are baseline separated after the screening. For the indicated Pareto optimality to be selected, we refer to Reference 25 for more details. Briefly, it can be stated that the best experiments will be those that exhibit the best compromise between acceptable resolution and analysis time. In optimization 2 (Figure 3.9c), the aim is to bring compounds partially separated in the screening step to a baseline separation. Here, we can mention that the change of the organic modifier can result either in an increase or a decrease in resolution.

For RPLC, the general strategy (Figure 3.10a) is less complex because no distinction between acidic and basic compounds is made. The optimization stage is also less complicated. In case of a baseline separation, the retention factor can be optimized based on the fact that a linear relationship is assumed between $\log k'$ and the fraction

(c)

Optimization 2

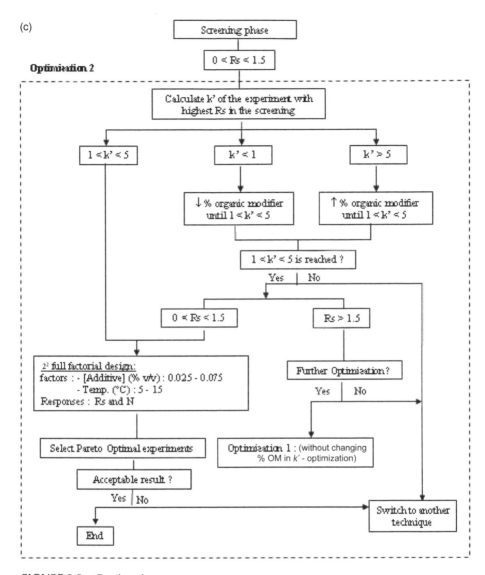

FIGURE 3.9 Continued.

of organic modifier in binary reversed-phase systems. After defining the desired k' value, the analyst has to perform an experiment with a changed fraction of organic modifier, the fraction depends on whether a lower or higher k' is wanted. From this and the screening experiment, the coefficients of the linear relationship are derived and the organic modifier content for the desired k' value predicted.

For a partial separation situation after screening, the organic modifier content and temperature are decreased according to a 2^2 full factorial design. When baseline separation is obtained, the retention factor can be further optimized by changing the

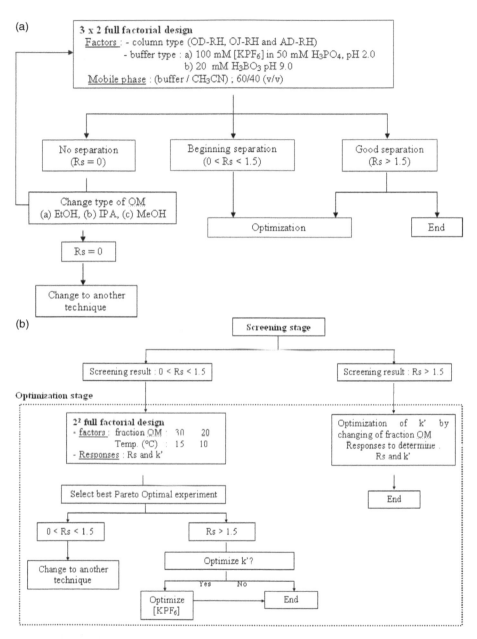

FIGURE 3.10 RPLC strategy. (Reprinted from Matthijs, N. et al. *J. Chromatogr. A* 2004, 1041, 119–133. Copyright (2004), With permission from Elsevier)

concentration of the chaotropic salt in the buffer, however, with the risk of a rapid decrease of the selectivity. Both strategies were implemented in a Chiral Knowledge-based System, a software that has the aim to guide the analyst through the method development of a chiral separation. Application of the strategies on nine components in NPLC and eight substances in RPLC resulted in baseline separation for eight and seven of them, respectively, i.e., an overall success rate of 88%.

Thirdly, a screening approach (Figure 3.11a) was developed in another mode of liquid chromatography, i.e., polar organic solvent chromatography (POSC) [47]. The screening is executed sequentially on four polysaccharide columns, in the following order: Chiralpak AD-RH, Chiralcel OD-RH, Chiralpak AS-RH, and finally, Chiralcel

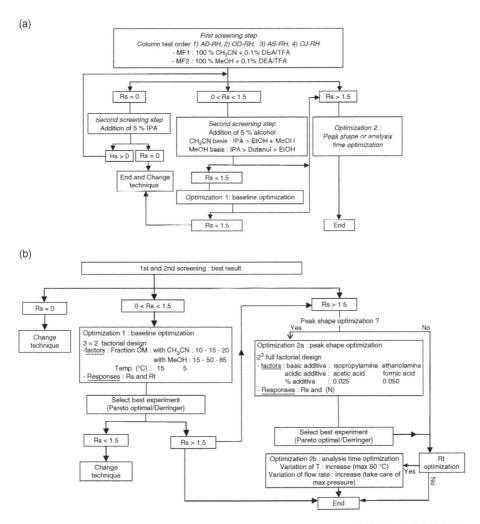

FIGURE 3.11 Strategy in POSC. (Reproduced from Matthijs, N., et al. *J. Sep. Sci.* 2006, 29, 1353–1362. With permission from Wiley VCH.)

OJ-RH. Two mobile phases are used in the first screening step: 100% methanol or 100% acetonitrile, both containing 0.1% (v/v) DEA and 0.1% (v/v) TFA. The flow rate is set at 0.5 mL/min. When a resolution below 1.5 is obtained, a second screening is performed, where 5% (v/v) of an organic modifier is added to the mobile phase. For an acetonitrile-based mobile phase, 2-propanol, ethanol, and methanol are tested, in that order, for a methanol-based mobile phase, 2-propanol, 1-butanol, and ethanol are sequentially considered. The defined screening conditions were applied on a set of 25 components. Only for three components, no enantioselectivity was seen using the proposed screening steps. Thus, 22 compounds (88%) showed enantioselectivity, of which 20 were already baseline separated without any further optimization. The complete separation strategy in POSC, including optimization steps, was defined in Reference 48. Its structure is given in Figure 3.11. The first screening step is the one defined in Reference 47 (Figure 3.11a). The second screening step of Reference 47 is used for compounds with a partial separation (see earlier text), or without observed enantioselectivity. For the latter, the addition of 5% (v/v) 2-propanol is recommended. When a partial separation is obtained after the second screening step, a baseline separation is aimed at in optimization 1 using a 3×2 experimental design (six experiments) where the concentration organic modifier is increased and the temperature decreased (Figure 3.11b). For compounds with a Rs > 1.5, a peak-shape optimization and an analysis time optimization might be performed. For the first, a 2^3 full factorial design is proposed, meaning that eight experiments are performed. The factors varied are the types and total percentage of additive. For the analysis time optimization, temperature and flow rate are increased individually.

3.4 SUPER- AND SUBCRITICAL FLUID CHROMATOGRAPHY

The development of a screening strategy for supercritical fluid chromatography (SFC) coupled to mass spectrometry is reported in Reference 49. The experimental conditions used in this screening reveal that CSP based on polysaccharide selectors can also be used in SFC mode. The columns considered in the screening [49] are Chiralpak AD, Chiralpak AS, Chiralcel OJ, and Chiralcel OD, in that order. Temperature, outlet pressure, and flow rate are set on 35°C, 110 bar, and 2.5 mL/min, respectively. In total, 32 conditions are tested in the screening step (four CSP with eight organic modifier concentrations). The mobile phases consisted of CO_2 combined with MeOH as organic modifier. Depending on the acidic or basic character of the test substance, TFA or isopropylamine (IPA) were used as additives to improve the peak shape. Screening was performed with eight organic modifier concentrations, starting with one of the following percentages of methanol: 40%, 35%, 30%, 25%, 20%, 15%, 10%, or 5%. The starting conditions were held isocratically for 20 min using a flow rate of 2.5 mL/min. Then the organic modifier concentration and the flow rate were ramped up to 45% and 4.5 mL/min, respectively, within 0.25 min. These conditions remained for 4.5 min to elute all compounds, and then returned to a percentage and flow rate of 40% and 2.5 mL/min, respectively, within 0.25 min. For the fine-tuning process, the replacement of methanol by isopropanol was described, and to improve the peak shape, the flow rate was increased to 4 mL/min, the temperature to 40°C, and the outlet pressure to 120 bars. This screening approach was used on six compounds, which were all brought to baseline separation. However, given the small number of compounds, a conclusion concerning the generic applicability of the screening approach cannot be drawn from this paper. Nevertheless, lower detection limits than in LC were found due to the use of MS detection, and shorter analysis times were observed due to the use of SFC.

Another application of rapid chiral method development in SFC was presented by Villeneuve and Anderegg [50]. The same columns as in Reference 49 are used, combined with four organic modifiers, including methanol, methanol with 0.1% TEA, ethanol, and isopropanol. Using a six-way column switcher, the four columns can remain constantly inside the device. For some separations, 0.1% DEA or TFA was added. The separations were generated at a flow rate of 2 mL/min, 205 atm of pressure, and a temperature of 40°C. This approach seems applicable, based on the baseline separation that was obtained for the four analyzed compounds, but lacks detailed sequential steps. Therefore, no real strategy can be derived.

A detailed screening and optimization procedure in SFC was described by Maftouh et al. [51]. Polysaccharide CSP is again selected. For the screening step, four columns were used in the sequence AD, OD, OJ, and AS. The basis of the mobile phase was CO_2. Two organic modifiers were used in the screening step: 10% (v/v) methanol and 20% isopropanol. For basic, bifunctional and neutral compounds, 0.5% IPA was used as additive to the mobile phase; for compounds with an acidic functionality, 0.5% TFA. The change from one modifier to another and from one column to another was performed using a switching valve. Other operating parameters consisted of a temperature of 30°C, a pressure of 200 bar, and a flow rate of 3 mL/min. Eight experiments are performed in the screening step (4 columns × 2 mobile phases). If no baseline separation is obtained at one of the screening conditions, an optimization step to improve the resolution is included in the separation strategy. First, the concentration of organic modifier is changed, with percentages ranging between 2% and 20%, and 5% and 30% for methanol and isopropanol, respectively. If this step appears unsuccessful, 10% (v/v) ethanol is tested as organic modifier. In addition, the pressure can be reduced to 100 bars and the flow rate to 1.5 mL/min in case of an unacceptable separation. When no separation is seen after application of the proposed optimization steps, it is recommended to switch to another technique. It is also stated that replacement of the 10-μm particles (AD, OD, OJ, and AS phases) used in the screening by 5-μm particle columns results in a more efficient separation (i.e., higher resolutions are obtained). The generic applicability of the separation strategy was demonstrated on 40 marketed drugs. Baseline separation was achieved for 28 compounds at the screening step, and this number could be increased to 39 after application of the optimization steps. Thus, a success rate of 98% was seen, showing the generic character of the proposed approach.

A screening approach was described in subcritical fluid chromatography for basic compounds by Stringham [52]. The analyzing conditions of the screening step are the following: a flow rate of 2 mL/min, 180 bars of backpressure, and room temperature. The column for screening was Chiralpak AD-H, i.e. 5-μm material was used. The mobile phase contained CO_2, 20% of ethanol, and 0.1% ethanesulfonic acid (ESA) as additive. ESA is an acidic additive that forms ion pairs with the basic analytes, which then can be separated on the neutral CSP. Applying this screening step on a set of 45 basic compounds resulted in a separation of 36 substances (80%), of which 30 were baseline resolved. These results already show the generic character of the screening step. Decreasing the ethanol content can be used as optimization step in case of a partial separation. Increasing the organic modifier concentration resulted in faster analyses. Substitution of ethanol by 20% methanol or 2-propanol allows obtaining different selectivities. The replacement of ESA by methanesulfonic acid is another possibility to change the selectivity, but the authors warn not to expect spectacular effects. Chiralcel OD-H was also used to screen 19 compounds maintaining the mobile phase and operating conditions as for Chiralpak AD-H. At least a partial separation was obtained for 13 compounds (68%).

3.5 CAPILLARY ELECTROCHROMATOGRAPHY

Our group has published several papers [53–56] concerning the definition of screening and optimization strategies in CEC on polysaccharide CSP. To our knowledge, these are the only attempts in the literature regarding generic separation approaches using CEC. In Reference 53, basic and bifunctional compounds were focused on. From a study of the most influential factors on analysis time and resolution, a screening mobile phase was derived. In the following two papers [54,55], separation strategies, including a screening and some optimization steps for no-, partial-, or baseline separation situations were defined for acidic [54] and nonacidic [55] compounds, respectively. The two strategies were combined into one in Reference 56. The general structure can be seen in Figure 3.12a. Depending on the nature of the compound, the screening experiment will either use an ammoniumformate buffer at pH 2.9 for acidic compounds, or use a phosphate buffer at pH 11.5 for the other compounds. The amount of acetonitrile in the screening mobile phases is 65% for acidic and 70% for nonacidic species. Other operating conditions are equal for all compounds, that is, an applied voltage of 15 kV over a column with 31.2 cm total length (packed section 20 cm), and a temperature of 25°C. The column testing order is different for acidic and nonacidic species. For the former, Chiralcel OJ-RH and Chiralpak AD-RH are both screened in a first instance, then Chiralpak AS-RH is considered, and finally Chiralcel OD-RH. For the latter compounds, the column testing order is AD-RH and OD-RH, followed by OJ-RH and AS-RH. The proposed screening step resulted in a partial– or baseline resolution of 42 out of 63 tested compounds, which represents more than 65% of the test set. Optimization 1 is applied in case of a partial separation. It differs for acidic and nonacidic compounds and the schedules followed are shown in Figure 3.12b,c. Optimization step 2, possibly applied when Rs > 1.5 at screening conditions, is displayed in Figure 3.12d and is the same for all types of compounds. After application of the optimization steps, the number of at least partially separated substances was increased up to 52, i.e. around 82%. These results show the generic applicability of the proposed strategy. A drawback of the proposed approach was the success rate of baseline-separated compounds of the chosen test set, which was only 49%. Some electrochromatograms of separations at screening conditions and

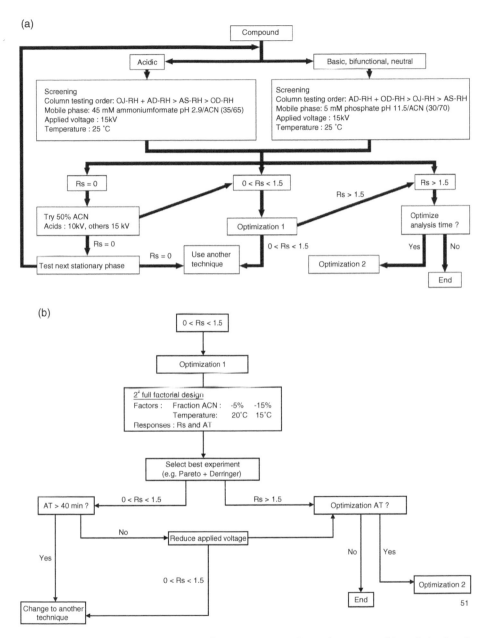

FIGURE 3.12   Separation strategy defined in CEC. (a) General structure, (b) optimization 1 for acidic compounds, (c) optimization 1 for nonacidic compounds, (d) optimization 2. (Mangelings, D., et al. *Electrophoresis* 2005, 26, 818–832; Mangelings, D., et al. *Electrophoresis* 2005, 26, 3930–3941; Mangelings, D., et al. *LC–GC Europe*, 2006, 19, 40–47. Figure 3.12a was printed with permission from *LC–GC Europe*, 19, 40–47 (2006). Figures 3.12b–d were reprinted with permission from Wiley VCH.)

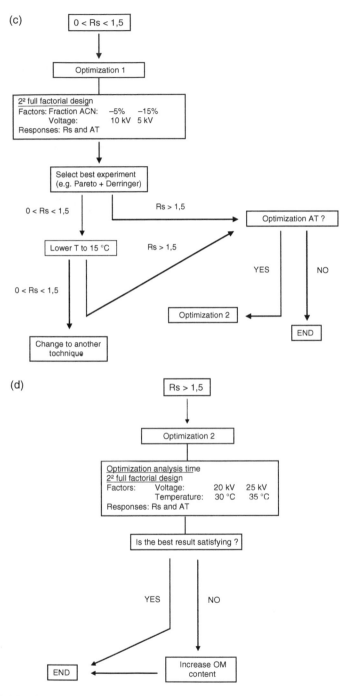

FIGURE 3.12 Continued.

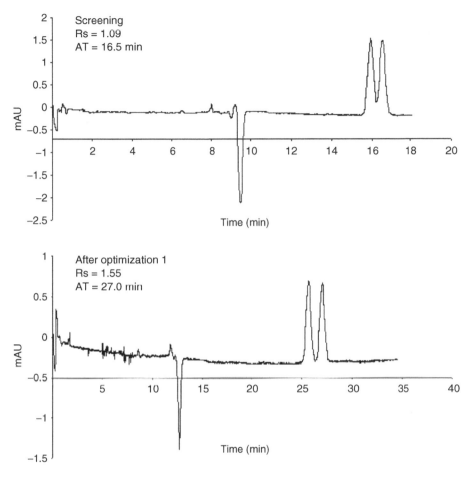

FIGURE 3.13 Electrochromatograms of fenoprofen on Chiralcel OJ-RH at screening conditions and after application of optimization 1 for acidic compounds when a partial resolution is achieved at screening conditions. Rs: Resolution, AT: Analysis time.

after application of the proposed optimization steps are given in Figures 3.13 and 3.14. For the detailed analysis conditions of these substances, we refer to the strategy presented in Figure 3.12, and to the related papers [54,55].

3.6 PREPARATIVE LIQUID CHROMATOGRAPHY

Kennedy et al. [57] have described the use of an intelligent chiral resolution system using a rapid screening of conditions on polysaccharide CSP, aiming to transfer the separation to preparative LC afterward. For the analytical part of the strategy, 10-μm particles were used; for the preparative section, the particles had a 20-μm diameter. The screening performs 11 experiments on analytical columns, which are displayed in Table 3.3. Chiralpak AD, Chiralcel OD, Chiralcel OJ, and Chiralpak AS are the considered stationary phases, and they are analyzed either in POSC or NPLC mode. The

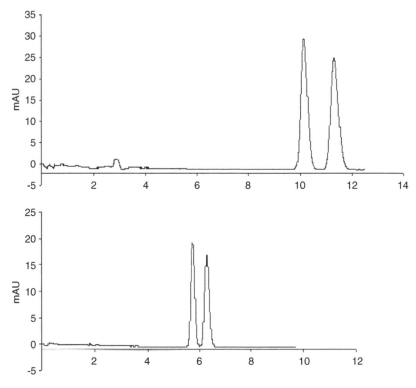

FIGURE 3.14   Electrochromatograms of mebeverine on Chiralcel OD-RH at screening conditions and after application of the proposed optimization 2 when a baseline resolution is achieved at screening conditions (Rs: Resolution, AT: Analysis Time).

TABLE 3.3
Screening Performed in Analytical Mode [57]

Experiment	Column	Eluent[a]
1	Chiralpak AD	Ethanol 100%
2	Chiralpak AD	Methanol 100%
3	Chiralpak AD	Acetonitrile 100%
4	Chiralcel OD	Isopropanol/heptane (10/90)
5	Chiralpak AD	Isopropanol/heptane (10/90)
6	Chiralcel OJ	Isopropanol/heptane (20/80)
7	Chiralpak AS	Isopropanol/heptane (20/80)
8	Chiralcel OD	Ethanol/heptane (10/90)
9	Chiralpak AD	Ethanol/heptane (10/90)
10	Chiralcel OJ	Ethanol/heptane (20/80)
11	Chiralpak AS	Ethanol/heptane (20/80)

[a] Contains 0.1% TFA for acids or 0.1% DMEA for basics.

authors also stress on the importance to know the compound's solubility before starting method development, as it is useless to develop, for instance, a method in NPLC
if the substance is only partially soluble in solvents such as heptane. The experiments
in POSC have a run time of 10 min, with an equilibration time of 5 min between
injections. For NPLC, this is 15 and 5 min, respectively. Running all 11 experiments
from the proposed screening on 84 compounds resulted in the complete or partial separation of 95% of the tested compounds, indicating its generic applicability. When no
separation is seen after the screening, OJ and AS are evaluated in the POSC mode.
The proposed screening will usually indicate the best CSP. However, it is possible
that some extra adaptations to the mobile phase are necessary to obtain a baseline
separation. An example is described for an unspecified pharmaceutical intermediate,
for which the starting conditions were defined in NPLC by means of the screening
procedure, but the extra addition of 10% methanol to a heptane/isopropanol system
allowed baseline separation of the enantiomers.

3.7 MISCELLANEOUS

Although no screening approach is presented in the following papers, the concepts
applied seem quite interesting in a screening context.

In Reference 58, a simulated moving-column technique was developed to achieve
baseline resolution on columns that do not show large enantioselectivity. This is done
using two or three short columns of 5 cm connected in series by a switching valve
and forcing the enantiomers to recycle through the columns until an acceptable separation is obtained. The enantiomers elute from column A to B to C to again A, B,
C, and so forth. Actually, this approach can be considered as virtually extending the
column length without experiencing limiting backpressure problems. Thus, a compound that is only partially separated on a column of regular length can be completely
separated using this technique. Another advantage of this approach can be found in
the possible use of shorter columns, which allow using more viscous solvents. Finally,
the enantiomers do not recycle through pump and detector systems, excluding peak
dispersion or extra-column band broadening. The used columns in this manuscript
were polysaccharide based, Chiralcel OD and OJ, Chiralpak AD and AS, but also two
brush type phases, Chirex 3005 and Chirex 3022, were used. The applications were
developed in either POSC or NPLC mode, but no systematic approach was described.
The approach was tested on five compounds, which were better separated using the
simulated moving column technique compared with the use of one column only.

Chip-based enantioseparations using an electrophoresis principle were presented by Gao et al. [59]. They used mono-, two-, and four-channel chips to develop
chiral separations of fluorescein isothiocyanate (FITC)-labeled basic compounds. To
obtain the chiral separations, seven neutral CD were screened (i.e., α-CD, β-CD,
γ-CD, HP-α-CD, HP-γ-CD, and DM-β-CD). Using the monochannel chip, the seven
selectors were screened sequentially. Using the two-channel chip, between-channel
repeatability could be demonstrated using the same separation conditions. Using
two different selectors in the channels, the analysis time for the screening of the
seven CD can be reduced to half, compared to the time needed in the monochannel

chip. The four-channel chip offered two possibilities: either one sample can be ana-lyzed using four selectors at a time, or four samples can be analyzed using the same selector. Although this approach is still far from mature, it offers future possibilities for high-throughput screening when some practical issues are resolved.

3.8 CONCLUSIONS

Most screening approaches and separation strategies for chiral separations were developed in CE. The CDs and their derivatives are then undoubtedly the most used and preferred chiral selectors to enable chiral separations. HS-CDs seem very popu-lar in this field, and exhibit a very broad enantioselectivity, making them suitable for generic screening approaches.

The next most used technique for screening approaches is liquid chromatography. Although the number of papers published using this technique is smaller than with CE, in our opinion, industry still prefers chromatographic techniques to perform chiral separations, because an occasional later transfer to preparative scale applica-tions is easier. However, strategies developed and used in an industrial environment are often kept confidential or are only very vaguely discussed in the literature. For liquid chromatographic techniques, including RPLC, NPLC, POSC, SFC, preparative LC, and also for CEC polysaccharide-based columns exhibit undoubtedly the largest enantioselectivity. They are used in all screening approaches or separation strategies that were previously defined. The four considered columns for these approaches are Chiralpak AD, Chiralcel OD, Chiralcel OJ, and Chiralpak AS, either in normal- or reversed-phase version, and usually evaluated in that sequence. Other CSP are only rarely used in screening approaches or generic separation strategies.

Another remarkable fact is that the number of published papers largely depends on the maturity of the technique and its frequency of use in the industry. For example, CE and HPLC are widely accepted techniques and take the majority of papers for their account, while CEC, a rather immature technique, and SFC, a less frequently applied one, are much less studied. Therefore, CE and HPLC can be recommended as the two techniques to be tested in a first instance, using either CD or polysaccharide CSP, respectively.

ABBREVIATIONS

APCI:	Atmospheric pressure chemical ionisation
AT:	Analysis time
atm:	Atmosphere
CD:	Cyclodextrin(s)
CEC:	Capillary electrochromatography
CM-β-CD:	Carboxymethyl-beta-cyclodextrin
CSP:	Chiral stationary phase(s)
DEA:	Diethylamine
DM-β-CD:	Dimethyl-beta-cyclodextrin
DMEA:	Dimethylethylamine
ESA:	Ethanesulfonic acid

FITC:	Fluorescein isothiocyanate
HP-α-CD:	Hydroxypropyl-alpha-cyclodextrin
HP-β-CD:	Hydroxypropyl-beta-cyclodextrin
HP-γ-CD:	Hydroxypropyl-gamma-cyclodextrin
HS-CD:	Highly sulfated cyclodextrins
HS-α-CD:	Highly sulfated alpha-cyclodextrin
HS-β-CD:	Highly sulfated beta-cyclodextrin
HS-γ-CD:	Highly sulfated gamma-cyclodextrin
IPA:	Isopropylamine
LC:	Liquid chromatography
N-CBZ:	N-carboxybenzyl
N-t-Boc:	N-tert-butoxycarbonyl
NPLC:	Normal-phase liquid chromatography
POSC:	Polar organic solvent chromatography
RPLC:	Reversed-phase liquid chromatography
SB-β-CD:	Sulfobutyl-beta-cyclodextrin
SBE-β-CD:	Sulfobutyl ether-beta-cyclodextrin
S-β-CD:	Sulfated-beta-cyclodextrin
SFC:	Supercritical fluid chromatography
T:	Temperature
TEA:	Triethylamine
TEOA:	Triethanolamine
TFA:	Trifluoroacetic acid
TM-β-CD:	Trimethyl-beta-cyclodextrin

REFERENCES

1. He, L., Beesley, T.E. Applications of enantiomeric gas chromatography: A review. *J. Liq. Chromatogr. Rel. Techn.* 2005, 28, 1075–1114.
2. Roos, N., Ganzler, K., Szemàn, J., Fanali, S. Systematic approach to cost- and time-effective method development with a starter kit for chiral separations by capillary electrophoresis. *J. Chromatogr. A* 1997, 782, 257–269.
3. Aumatell, A., Guttman, A. Ultra-fast chiral separation of basic drugs by capillary electrophoresis. *J. Chromatogr. A* 1995, 717, 229–234.
4. Guttmann, A., Jurado, C., Brunet, S., Cooke, N. Presented at the 7th International Symposium on High Performance Capillary Electrophoresis, Würzburg, Germany, January 1995, paper 204.
5. Guttman, A. Novel separation scheme for capillary electrophoresis of enantiomers. *Electrophoresis* 1995, 16, 1900–1905.
6. Rawjee, Y.Y., Stark, D.U., Vigh, G. Capillary electrophoretic chiral separations with cyclodextrin additives: I. acids: Chiral selectivity as a function of pH and the concentration of β-cyclodextrin for fenoprofen and ibuprofen. *J. Chromatogr.* 1993, 635, 291–306.
7. Rawjee, Y.Y., Williams, R.L., Vigh, G. Capillary electrophoretic chiral separations using β-cyclodextrin as resolving agent: II. Bases: Chiral selectivity as a function of pH and the concentration of β-cyclodextrin. *J. Chromatogr.* 1993, 652, 233–245.

8. Rawjee, Y.Y., Williams, R.L., Vigh, G. Capillary electrophoretic chiral separations using cyclodextrin additives: III. Peak resolution surfaces for ibuprofen and homatropine as a function of the pH and the concentration of β-cyclodextrin. *J. Chromatogr.* 1994, 680, 599–611.

9. Rawjee, Y.Y., Williams, R.L., Buckingham, L.A., Vigh, G. Effects of pH and hydroxypropyl β-cyclodextrin concentration on peak resolution in the capillary electrophoretic separation of the enantiomers of weak bases. *J. Chromatogr.* 1994, 688, 273–282.

10. Rawjee, Y.Y., Vigh, G. A peak resolution model for the capillary electrophoretic separation of the enantiomers of weak acids with hydroxypropyl-β-cyclodextrin containing background electrolytes. *Anal. Chem.* 1994, 66, 619–627.

11. Fillet, M., Hubert, P., Crommen, J. Method development strategies for the enantioseparation of drugs by capillary electrophoresis using cyclodextrins as chiral additives. *Electrophoresis* 1998, 19, 2834–2840.

12. Koppenhoeffer, B., Epperlein, U., Christian, B., Yibing, J., Yujing, C., Bingcheng, L. Separation of enantiomers of drugs by capillary electrophoresis. I. γ-Cyclodextrin as chiral solvating agent. *J. Chromatogr. A* 1995, 717, 181–190.

13. Bingcheng, L., Yibing, J., Yujing, C., Epperlein, U., Koppenhoeffer, B. Separation of drug enantiomers by capillary electrophoresis: α-Cyclodextrin as chiral solvating agent. *Chromatographia* 1996, 42, 106–110.

14. Koppenhoeffer, B., Epperlein, U., Christian, B., Lin, B., Ji, Y., Chen, Y. Separation of enantiomers of drugs by capillary electrophoresis. III. β-Cyclodextrin as chiral solvating agent. *J. Chromatogr. A* 1996, 735, 333–343.

15. Koppenhoeffer, B., Epperlein, U., Zhu, X., Lin, B. Separation of enantiomers of drugs by capillary electrophoresis. IV. Hydroxypropyl-γ-cyclodextrin as chiral solvating agent. *Electrophoresis* 1997, 18, 924–930.

16. Koppenhoeffer, B., Epperlein, U., Schlunk, R., Zhu, X., Lin, B. Separation of enantiomers of drugs by capillary electrophoresis. V. Hydroxypropyl-α-cyclodextrin as chiral solvating agent. *J. Chromatogr. A* 1998, 793, 153–164.

17. Lin, B., Zhu, X., Epperlein, U., Schwierskott, M., Schlunk, R., Koppenhoeffer, B. Separation of enantiomers of drugs by capillary electrophoresis, part 6. Hydroxypropyl-β-cyclodextrin as chiral solvating agent. *J. High Resol. Chromatogr.* 1998, 21, 215–224.

18. Koppenhoeffer, B., Epperlein, U., Jakob, A., Wuerthner, S., Xiaofeng, Z., Bingcheng, L. Separation of enantiomers of drugs by capillary electrophoresis, part 7. Gamma-cyclodextrin as chiral solvating agent. *Chirality* 1998, 10, 548–554.

19. Lin, B., Zhu, X., Wuerthner, S., Epperlein, U., Koppenhoeffer, B. Separation of enantiomers of drugs by capillary electrophoresis, part 8. β-Cyclodextrin as chiral solvating agent. *Talanta* 1998, 46, 743–749.

20. Zhu, X., Lin, B., Jakob, A., Wuerthner, S., Koppenhoefer, B. Separation of drugs by capillary electrophoresis, part 10. Permethyl-alpha-cyclodextrin as chiral solvating agent. *Electrophoresis* 1999, 20, 1878–1889.

21. Koppenhoefer, B., Zhu, X., Jakob, A., Wuerthner, S., Lin, B. Separation of drug enantiomers by capillary electrophoresis in the presence of neutral cyclodextrins. *J. Chromatogr. A* 2000, 875, 135–161.

22. Zhu, X., Bingcheng, L. Separation of enantiomers of drugs by capillary electrophoresis with permethyl-gamma-cyclodextrin as chiral solvating agent. *J. High Resol. Chromatogr.* 2000, 23, 413–429.

23. Liu, L., Nussbaum, M.A. Systematic screening approach for chiral separations of basic compounds by capillary electrophoresis with modified cyclodextrins. *J. Pharm. Biomed. An.* 1999, 19, 679–694.

24. Vargas, M.G., Vander Heyden, Y., Maftouh, M., Massart, D.L. Rapid development of the enantiomeric separation of β-blockers using an experimental design approach. *J. Chromatogr. A* 1999, 855, 681–693.

25. Massart, D.L., Vandeginste, B.G.M., Buydens, L.M.C., De Jong, S., Lewi, P.J., Smeyers-Verbeke, J. *Handbook of Chemometrics and Qualimetrics: Part A*; Elsevier, Netherlands, 1997; 643 pp.

26. Perrin, C., Vargas, M.G., Vander Heyden, Y., Maftouh, M., Massart, D.L. Fast development of separation methods for the chiral analysis of amino acid derivatives using capillary electrophoresis and experimental designs. *J. Chromatogr. A* 2000, 883, 249–265.

27. Bergholdt, A.B., Jørgensen, K.W., Wendel, L., Lehmann, S.V. Fast chiral separations using sulfated β-cyclodextrin and short-end injection in capillary electrophoresis. *J. Chromatogr. A* 2000, 875, 403–410.

28. Perrin, C., Vander Heyden, Y., Maftouh, M., Massart, D.L. Rapid screening for chiral separations by short-end injection capillary electrophoresis using highly sulfated cyclodextrins as chiral selectors. *Electrophoresis* 2001, 22, 3203–3215.

29. Matthijs, N., Perrin, C., Maftouh, M., Massart, D.L., Vander Heyden, Y. Knowledge-based system for method development of chiral separations with capillary electrophoresis using highly sulphated cyclodextrins. *J. Pharm. Biomed. An.* 2002, 27, 515–529.

30. Matthijs, N., Van Hemelryck, S., Maftouh, M., Massart, D.L., Vander Heyden, Y. Electrophoretic separation strategy for chiral pharmaceuticals using highly-sulfated and neutral cyclodextrins based dual selector systems. *An. Chim. Acta* 2004, 525, 247–263.

31. Chapman, J., Chen, F.A. Implementing a generic methods development strategy for enantiomer analysis. *LC–GC Europe*, 2001, January, 2–6.

32. Jimidar, M.I., Van Ael, W., Van Nyen, P., Peeters, M., Redlich, D., De Smet, M. A screening strategy for the development of enantiomeric separation methods in capillary electrophoresis. *Electrophoresis* 2004, 25, 2772–2785.

33. Jimidar, M.I., Vennekens, T., Van Ael, W., Redlich, D., De Smet, M. Optimization and validation of an enantioselective method for a chiral drug with eight stereo-isomers in capillary electrophoresis. *Electrophoresis* 2004, 25, 2876–2884.

34. Rocheleau, M.-J. Generic capillary electrophoresis conditions for chiral assay in early pharmaceutical development. *Electrophoresis* 2005, 26, 2320–2329.

35. Sokoließ, T., Köller, G. Approach to method development and validation in capillary electrophoresis for enantiomeric purity testing of active basic pharmaceutical ingredients. *Electrophoresis* 2005, 26, 2330–2341.

36. Zhou, L., Thompson, R., Song, S., Ellison, D., Wyvratt, J.M. A strategic approach to the development of capillary electrophoresis methods for pharmaceutical basic compounds using sulfated cyclodextrins. *J. Pharm. Biomed. An.* 2002, 27, 541–553.

37. Souverain, S., Geiser, L., Rudaz, S., Veuthey, J.-L. Strategies for rapid chiral analysis by capillary electrophoresis. *J. Pharm. Biomed. An.* 2006, 40, 235–241.

38. Kenseth, J., Bastin, A. High throughput chiral separations using the cePRO9600 system. Application note 207400, 2004. http://www.combisep.com/applicationNotes.htm

39. Vander Heyden, Y., Mangelings, D., Matthijs, N.,; Perrin, C. Chiral separations, in Ahuja, S., Dong, M. (eds.), *Handbook of Pharmaceutical Analysis by HPLC*, Elsevier, Netherlands, 2005; pp. 447–498.
40. de la Puente, M.L., White, C.T., Rivera-Sagredo, A., Reilly, J., Burton, K., Harvey, G. Impact of normal-phase gradient elution in chiral chromatography: A novel, robust, efficient and rapid chiral screening procedure. *J. Chromatogr. A* 2003, 983, 101–114.
41. de la Puente, M.L. Highly sensitive and rapid normal-phase chiral screen using high-performance liquid chromatography-atmospheric pressure ionization tandem mass spectrometry (HPLC/MS). *J. Chromatogr. A* 2004, 1055, 55–62.
42. Andersson, M.E., Aslan, D., Clarke, A., Roeraade, J., Hagman, G. Evaluation of generic chiral liquid chromatography screens for pharmaceutical analysis. *J. Chromatogr. A* 2003, 1005, 83–101.
43. Zhang, Y., Watts, W., Nogle, L., McConnell, O. Rapid development for chiral separation in drug discovery using multi-column parallel screening and circular dichroism signal pooling. *J. Chromatogr. A* 2004, 1049, 75–84.
44. Perrin, C., Vu, V.A., Maftouh, M., Massart, D.L., Vander Heyden, Y. Screening approach for chiral separation of pharmaceuticals Part I. Normal-phase liquid chromatography. *J. Chromatogr. A* 2002, 947, 69–83.
45. Perrin, C., Matthijs, N., Mangelings, D., Granier-Loyaux, C., Maftouh, M., Massart, D.L., Vander Heyden, Y. Screening approach for chiral separation of pharmaceuticals Part II. Reversed-phase liquid chromatography. *J. Chromatogr. A* 2002, 966, 119–134.
46. Matthijs, N., Perrin, C., Maftouh, M., Massart, D.L., Vander Heyden, Y. Definition and system implementation of strategies for method development of chiral separations in normal- or reversed-phase liquid chromatography using polysaccharide-based stationary phases. *J. Chromatogr. A* 2004, 1041, 119–133.
47. Matthijs, N., Maftouh, M., Vander Heyden, Y. Screening approach for chiral separation of pharmaceuticals Part IV. Polar organic solvent chromatography. *J. Chromatogr. A* 2006, 1111, 48–61.
48. Matthijs, N., Maftouh, M., Vander Heyden, Y. Chiral separation strategy in polar organic solvent chromatography and performance comparison with normal-phase liquid and supercritical-fluid chromatography. *J. Sep. Sci.* 2006, 29, 1353–1362.
49. Zhao, Y., Woo, G., Thomas, S., Semin, D., Sandra, P. Rapid method development for chiral separation in drug discovery using sample pooling and supercritical fluid chromatography-mass spectrometry. *J. Chromatogr. A* 2003, 1003, 157–166.
50. Villeneuve, M.S., Anderegg, R.J. Analytical supercritical fluid chromatography using fully automated column and modifier selection valves for the rapid development of chiral separations. *J. Chromatogr. A* 1998, 826, 217–225.
51. Maftouh, M., Granier-Loyaux, C., Chavana, E., Marini, J., Pradines, A., Vander Heyden, Y., Picard, C. Screening approach for chiral separation of pharmaceuticals Part III. Supercritical fluid chromatography for analysis and purification in drug discovery. *J. Chromatogr. A* 2005, 1088, 67–81.
52. Stringham, R.W. Chiral separation of amines in subcritical fluid chromatography using polysaccharide stationary phases and acidic additives. *J. Chromatogr. A* 2005, 1070, 163–170.
53. Mangelings, D., Hardies, N., Maftouh, M., Suteu, C., Massart, D.L., Vander Heyden, Y. Enantioseparations of basic and bifunctional compounds by capillary electrochromatography using polysaccharide stationary phases. *Electrophoresis* 2003, 24, 2567–2576.

54. Mangelings, D., Tanret, I., Matthijs, N., Maftouh, M., Massart, D.L., Vander Heyden, Y. Separation strategy for acidic chiral pharmaceuticals with capillary electrochromatography on polysaccharide stationary phases. *Electrophoresis* 2005, 26, 818–832.

55. Mangelings, D., Discry, J., Maftouh, M., Massart, D.L., Vander Heyden, Y. Strategy for the chiral separation of nonacidic pharmaceuticals using capillary electrochromatography. *Electrophoresis* 2005, 26, 3930–3941.

56. Mangelings, D., Maftouh, M., Massart, D.L., Vander Heyden, Y. Generic capillary electrochromatographic screening and optimization strategies for chiral method development. *LC–GC Europe*, 2006, 19, 40–47.

57. Kennedy, J.H., Sharp, V.S., Williams, J.D. Development of preparative chiral separations using an intelligent chiral resolution system. *J. Liq. Chrom. Rel. Techn.* 2003, 26, 529–543.

58. Zhang, Y., McConnell, O. Simulated moving columns technique for chiral liquid chromatography. *J. Chromatogr. A* 2004, 1028, 227–238.

59. Gao, Y., Shen, Z., Wang, H., Dai, Z., Lin, B. Chiral separations on multichannel microfluidic chips. *Electrophoresis* 2005, 26, 4774–4779.

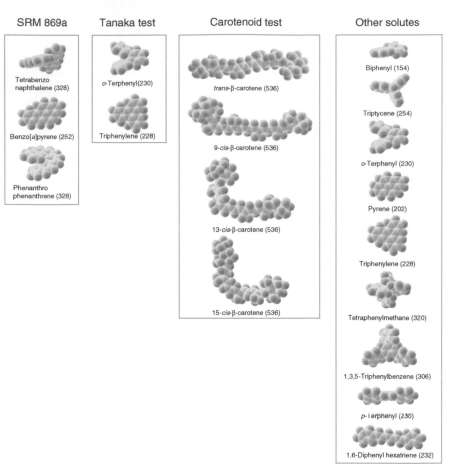

COLOR FIGURE 5.3 Solute test mixtures used for shape-selectivity characterization.

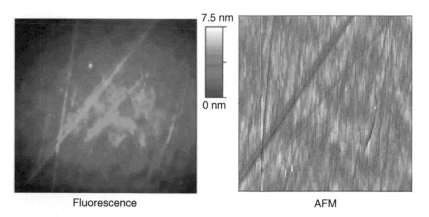

COLOR FIGURE 5.19 Fluorescence image for cationic dye molecule (DiI) on a C$_{18}$ surface at high concentration measured using a charge-coupled device (CCD) camera (left). AFM image of the same region (20 μm × 20 μm) (right). (Reproduced from Wirth, M.J., et al., *J. Phys. Chem. B*, 107, 6258, 2003. With permission.)

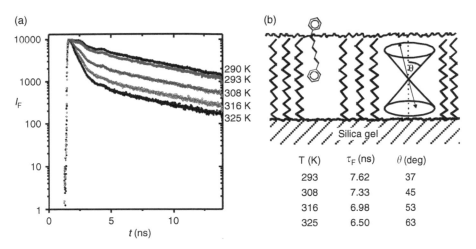

T (K)	τ_F (ns)	θ (deg)
293	7.62	37
308	7.33	45
316	6.98	53
325	6.50	63

COLOR FIGURE 5.20 Temperature influence for fluorescence anisotropy measurements on C_{22}-modified silica. (a) Decay curves for DPH at different temperatures, (b) schematic of wobble motion of DPH and resulting fluorescence lifetimes (τ_F) and half-cone angles (θ). (Reproduced from Pursch, M., et al., *J. Am. Chem. Soc.*, 121, 3201, 1999. With permission.)

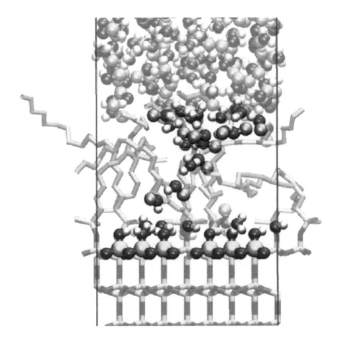

COLOR FIGURE 5.21 Snapshot of a methanol bridge cluster for a MC simulated monomeric C_{18} chromatographic surfaces. Solutes involved in the cluster are highlighted in darker, opaque coloring. (Reproduced from Zhang, L., et al., *J. Chromatogr. A*, 1126, 219, 2006. With permission.)

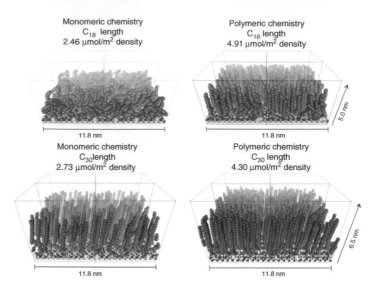

COLOR FIGURE 5.22 Perspective four-cell view snapshots of various stationary phase models, which illustrate the contrasting surface topography with bonding density and alkyl chain length. (Adapted from Lippa, K.A., et al., *Anal. Chem.*, 77, 7862, 2005.)

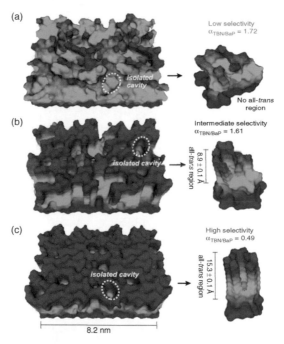

COLOR FIGURE 5.23 Molecular surfaces generated with a 2.0 Å radius probe for MD simulation snapshots of (a) C_{18} monomeric model at 3.28 μmol/m^2 surface coverage representing an actual RPLC phase with low shape selectivity ($\alpha_{TBN/BaP} = 1.72$) and two C_{18} polymeric models at (b) 3.89 μmol/m^2 and (c) 5.94 μmol/m^2 surface coverages also representing actual phases with intermediate to high shape selectivity ($\alpha_{TBN/BaP} = 1.61$ and 0.49, respectively). Surfaces are color coded according to the height along the z-axis. Included are the side-view surface areas generated for the isolated cavities with an average diameter of $\approx$10 Å. (Adapted from Lippa, K.A., et al., *J. Chromatogr. A*, 1128, 79, 2006. With permission.)

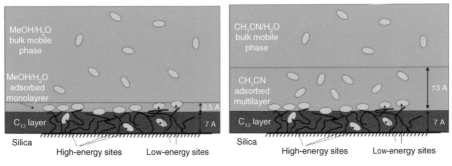

(a) Methanol/water

MeOH/H$_2$O
bulk mobile
phase

MeOH/H$_2$O
adsorbed
monolayrer

C$_{13}$ layer   2.5 A

7 A

Silica

High-energy sites   Low-energy sites

(b) Acetonitrile/water

CH$_3$CN/H$_2$O
bulk mobile
phase

CH$_3$CN
adsorbed
multilayer

13 A

C$_{13}$ layer   7 A

Silica

High-energy sites   Low-energy sites

COLOR FIGURE 5.24 Schematic comparison of the adsorption mechanisms of a solute from aqueous solutions of (a) methanol and (b) acetonitrile onto a RPLC material. Three different "phases" (bulk mobile phase, the adsorbed mono- or multilayer of organic modifier molecules, and the C$_{18}$ phase) are involved in the chromatographic system. The solute is represented by small ovals. (Reproduced from Gritti, F. and Guiochon, G., *Anal. Chem.*, 77, 4257, 2005. With permission.)

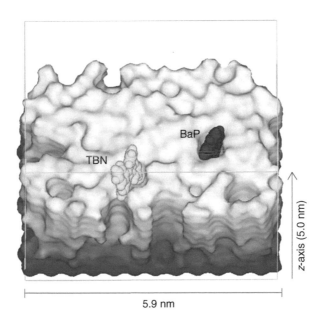

COLOR FIGURE 5.26 A representation of the "slot model" illustrating potential constrained-shape solute (BaP) interactions with the conformational ordered cavities of a polymeric C$_{18}$ stationary phase simulation model. Also included on the chromatographic surface is an identical-scale molecular structure of 1,2:3,4:5,6:7,8-tetrabenzonaphthalene (TBN).

4 Increasing Speed of Enantiomeric Separations Using Supercritical Fluid Chromatography

Naijun Wu

CONTENTS

ABSTRACT

Enantioselective separation by supercritical fluid chromatography (SFC) has been a field of great progress since the first demonstration of a chiral separation by SFC in the 1980s. The unique properties of supercritical fluids make packed column SFC the most favorable choice for fast enantiomeric separation among all of the separation techniques. In this chapter, the effect of chiral stationary phases, modifiers, and additives on enantioseparation are discussed in terms of speed and resolution in SFC. Fundamental considerations and thermodynamic aspects are also presented.

4.1 INTRODUCTION

There is an increasing demand for simple and fast analytical methods for enantiomers. Today, the pharmaceutical industry is facing the major challenges of shortening drug discovery and development cycle time to push new drug candidates into the market [1–4]. Various new technologies have been utilized to increase the speed and improve the efficiency of drug discovery and development [2, 4]. Accordingly, the most urgent goal for analytical scientists is to minimize assay time and maximize analytical information using newly developed technologies [5–8]. Nearly 60–70 % of the most frequently prescribed drugs and the drug candidates that entered the development process in the United States are single enantiomers [9, 10]. To ensure the efficacy and safety of currently used and newly developed drugs, it is important to isolate the enantiomers and to examine each one separately [11]. Furthermore, it is necessary to check the chiral purity of drug raw materials and to monitor the stereochemical composition of drugs during development, manufacturing, and storage. Conventional synthese and modern combinatorial libraries used in drug discovery and development require high-throughput screening methods to handle the large number of samples [13–15].

Separation of enantiomers has been performed by various techniques [16], including gas chromatography (GC) [17], high-performance liquid chromatography (HPLC) [18, 19], supercritical fluid chromatography (SFC) [20, 21], capillary electrophoresis (CE) [22, 23], capillary electrochromatography (CEC) [24], and thin layer chromatography (TLC) [25]. Among these approaches, packed column SFC possesses important advantages for rapid separation of enantiomers [5, 26–29]. Because enantiomeric resolution comes primarily from a close interaction between the analyte and the stationary phase, the normal phase-like properties of supercritical fluids such as carbon dioxide facilitate high interaction and high enantioselectivity. Separations in SFC can generally be performed at temperatures well below the temperatures required for GC analysis. Supercritical or subcritical fluids have solvating power and, thus, speed the elution of solutes even at low temperatures [27, 28]. Low temperature is beneficial because chiral selectivity usually increases with decreasing temperature [30, 31]. Furthermore, analysis at low temperatures reduces the risk of racemization and thermal decomposition of the chiral stationary phase and analytes [32]. Compared to HPLC, SFC often provides higher resolution per unit time [33, 34], because mobile phases have low viscosities and high diffusivity. These two characteristics allow for higher optimum linear velocities and high efficiency at high flow rates [26, 35]. Finally, SFC has proven to be a "green" technique for chiral separation by reducing organic solvent consumption, which becomes more significant at preparative scales [5, 8, 36].

For a single channel chromatographic separation, the most straightforward way to reduce the analysis time is to use a short column and/or high flow rates with a highly selective chiral stationary phase. Enantiomeric selectivity primarily depends on diastereomeric interactions between the enantiomers and the chiral stationary phases. To find a highly selective chiral column is a key step in method development of chiral separation. In addition, modifiers or additives usually play a significant role in enantiomeric recognition of SFC, since they can significantly improve peak shapes and often selectivity [37]. Finally, column temperatures and pressures can also have a significant effect on enantioseparation in SFC [38–40].

Chiral SFC can be performed in open tubular [41, 42], and packed column [43, 44] modes. Packed column SFC can be further categorized into analytical, semipreparative, and preparative SFC [7, 8]. Packed column SFC is more suitable for fast separations than open tubular column SFC, since a packed column generally provides low mass transfer resistance and high selectivity [45, 46]. Packed column SFC also provides high sample loading capacity [27, 47], which can increase sensitivity. Only packed column SFC is suitable for preparative-scale enantioseparation. This chapter will focus on chiral separation using packed column SFC in the analytical scale.

In this chapter, approaches to fast chiral separations using SFC, including fundamental considerations, influences of chiral stationary phases, modifiers, and additives are discussed. The thermodynamic aspects of SFC are also presented.

4.2 FUNDAMENTAL CONSIDERATIONS

4.2.1 EFFECT OF EFFICIENCY, SELECTIVITY, AND RETENTION ON SEPARATION TIME

From fundamental considerations, the resolution of a column is given by [48]:

$$R = \frac{\sqrt{N}}{4} \left(\frac{\alpha - 1}{\alpha} \right) \left(\frac{k}{k + 1} \right), \tag{4.1}$$

where N is the number of theoretical plates, α is the selectivity factor, and k is the retention factor. This relationship can be illustrated in Figure 4.1 by using a reference point at which the resolution of two peaks with a selectivity factor of 1.05 on a column with a plate number of 20,000 is 1.4 [48]. It can be seen that the retention factor has a significant effect on enhancing resolution at the k value less than 2. The resolution gain levels off at greater k values. Similarly, the resolution increase is significant at the N value less than 10,000 and less significant at higher efficiencies. It can be seen that the resolution increases significantly with an increasing selectivity factor over a wide range of selectivity factor values. Thus, the selectivity factor is the most significant parameter to improve the resolution. Resolution nearly doubles as the selectivity factor increases from 1.05 to 1.11, holding all other factors constant.

From Equation 4.1, the required number of theoretical plates (N_{req}) to provide a given resolution can be calculated:

$$N_{req} = 16R^2 \left(\frac{\alpha}{\alpha - 1} \right)^2 \left(\frac{k + 1}{k} \right)^2. \tag{4.2}$$

Taking $k = 2$ and $R = 1.5$, the required theoretical plate numbers for $\alpha = 1.05$ and 1.25 are 35,721 and 2,025, respectively. Therefore, a column with high selectivity can produce sufficient resolution, even with low theoretical plate numbers (short length).

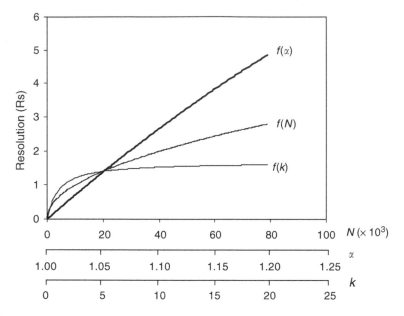

FIGURE 4.1 Effect of the plate number (N), the separation factor (α), and the retention factor (k) on resolution (Rs). (Adapted from Sandra, P.J. 1989. *High Resolut. Chromatogr.* 12: 82–86. With permission.)

Similarly, a high retention factor also favors high resolution. However, a high retention factor results in increased retention time. The separation time is given as

$$t_R = 16R^2 \left(\frac{\alpha}{\alpha - 1} \right)^2 \left[\frac{(1 + k_2)^3}{k_2^2} \frac{H_2}{u} \right], \tag{4.3}$$

where H_2 is the plate height of the second component and u is the average mobile phase linear velocity. The term in Equation 4.3 has a minimum of 6.75 for $k_2 = 2$. Equation 4.3 suggests that a column with high selectivity and a k value of approximately 2 can provide fast separations of enantiomers.

4.2.2 SFC vs. LC

Supercritical fluids possess favorable physical properties that result in good behavior for mass transfer of solutes in a column. Some important physical properties of liquids, gases, and supercritical fluids are compared in Table 4.1 [49]. It can be seen that solute diffusion coefficients are greater in a supercritical fluid than in a liquid phase. When compared to HPLC, higher analyte diffusivity leads to lower mass transfer resistance, which results in sharper peaks. Higher diffusivity also results in higher optimum linear velocities, since the optimum linear velocity for a packed column is proportional to the diffusion coefficient of the mobile phase for liquid-like fluids [50, 51].

TABLE 4.1

Properties of Supercritical Fluids vs. Gases and Liquids

	Density (g/cm^3)	Diffusion (cm^2/s)	Viscosity (g/cm · s)
Gas	10^{-3}	0.1	10^{-4}
Supercritical fluid	0.1–1	10^{-3}–10^{-4}	10^{-3}–10^{-4}
Liquid	1	$<10^{-5}$	10^{-2}

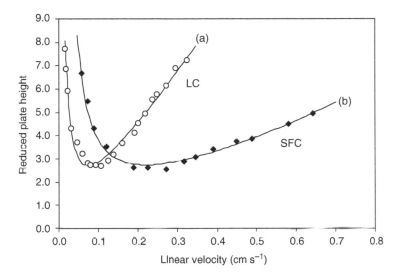

FIGURE 4.2 The van Deemter curves for HPLC and SFC. Conditions: (a) LC, 55 cm × 250 μm ID fused-silica capillary columns packed with 5 μm porous particles, nitromethane test solute, 25°C, UV detector (214 nm); (b) SFC, 75 cm × 250 μm ID fused-silica capillary columns packed with 5-μm porous particles, 45°C, 230 atm, carbon dioxide mobile phase, methane test solute, FID. (Adapted from Wu, N. et al. *Anal. Chem.* 71: 5084–5092. With permission.)

Gere [52] compared the performance of SFC and HPLC using the same column temperature and the same stationary phase. The mobile phases for the HPLC and SFC were acetonitrile/water and carbon dioxide, respectively. It was observed that rapid diffusivity and high average linear velocity combined to produce flat van Deemter curves in SFC. Wu et al. [53] compared the van Deemter curves for HPLC and SFC using typical SFC and HPLC conditions, respectively, as shown in Figure 4.2. The optimum linear velocity for SFC is approximately three times as high as for HPLC. The curve for SFC is also much flatter than at higher velocities when compared to HPLC. These two characteristics can be mainly attributed to the higher diffusion coefficient of supercritical fluids. For a gradient separation, the faster diffusion of mobile phase means faster re-equilibration and, thus, shorter cycle time.

Table 4.1 also shows that supercritical fluids have viscosities that are similar to those of gas phases but much lower than those of liquid phases. The lower viscosity

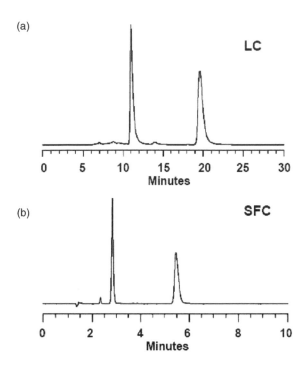

FIGURE 4.3   Separation of metoprolol enantiomers by LC and SFC on a Chiralcel OD stationary phase. Conditions: (a) The LC separation was performed with 20% 2-propanol in hexane with 0.1% (v/v) diethylamine at 0.5 mL/min, selectivity (α) = 2.67, and resolution (Rs) = 4.8. (b) The SFC separation was performed with 20% methanol, which contained 0.5% isopropylamine, in carbon dioxide at 2.0 mL/min, 15 MPa, and 30°C; α = 2.77, and Rs = 12.7. (Adapted from Phinney, K.V. 2000. *Anal. Chem.* 72: 204A–211A. With permission.)

of the mobile phase leads to lower pressure drops and, hence, higher flow rates or longer columns can be used in SFC. Overall, SFC can allow for a three- to fivefold improvement in separation speed with maintained or improved resolution as compared to HPLC [54–57]. Figure 4.3 shows the separations of metoprolol enantionmers using HPLC and SFC on a polysaccharide-based chiral stationary phase [58]. It can be seen that SFC not only provided three times faster separation but also resulted in narrower and more symmetric peaks. The better performance in SFC was ascribed to the low viscosity and high diffusivity of the carbon dioxide supercritical fluid.

Pirkle and coworkers [59] compared retention and selectivity factors between HPLC and SFC using Poly Whelk-O chiral stationary phases and α-naphthyl-1-ethylamine carbamates. The results indicate that both retention and selectivity factors in SFC were higher than those in HPLC. This can be mainly attributed to the weaker solvating power of the carbon dioxide supercritical fluid as compared to a liquid such as methanol or hexane.

Hoke et al. [60] demonstrated fast analysis of ultratrace drugs in biomatrixes using packed column SFC tandem mass spectrometry (SFC/MS/MS). Compared to an LC-MS/MS method, the SFC separation provided a roughly threefold reduction in

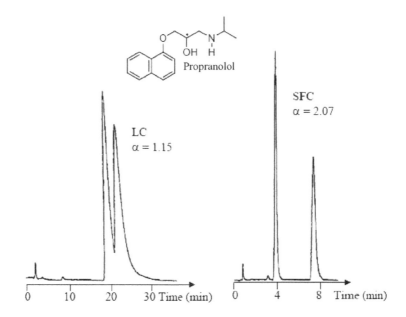

FIGURE 4.4 Comparison of enantiomeric separation of propranolol by LC and sub-SFC. Conditions: A 150 × 4.6 mm ID column with ChyRoSine-A stationary phase. For LC: 95:5 v/v hexane/ethanol containing 1% v/v of n-propylamine; flow rate: 1 mL/min; UV detection at 224 nm; room temperature. For SFC, 90:10 v/v CO_2/methanol containing 1% v/v of n-propylamine; flow rate: 4 mL/min; UV detection at 224 nm; 0°C; average column pressure: 200 bar. (Reprinted from Siret, L. Bargmann, N. et al. 1992. *Chirality* 4: 252. With permission.)

analytical time. Both a 2.3-min SFC separation and a 6.5-min LC separation provided near-baseline-resolved peaks for the (R)- and (S)-enantiomers of ketoprofen, a potent anti-inflammatory drug, in human plasma. After preparation using automated solid-phase extraction in the 96-well format, ketoprofen enantiomers were resolved on a Chirex 3005 column using isocratic conditions. The validation data showed that specificity, linearity, sensitivity, accuracy, precision, and ruggedness, for both of these methods were comparable with the exception of significant time saving achieved for analysis of a large number of pharmacokinetic samples by using SFC.

Siret et al. [61] separated underivatized enantiomers of 1,2-amino alcohol (β-blockers) on a π-acidic chiral stationary phase using SFC and normal-phase LC conditions, as shown in Figure 4.4. Obviously, SFC provided much faster separation with better peak shape than normal-phase LC. The difference in performance between two systems is unexpectedly significant. The authors believed that normal-phase LC and SFC are no longer the interchangeable systems in this case. NMR studies suggest that carbon dioxide plays an important role in the enantiomeric separation process of β-blockers. It was proposed that carbon dioxide could form a complex with the amino and the hydroxyl groups of the solute, which may lead to a conformation change and favor the enantiomeric discrimination. Furthermore, the formation of carbon dioxide–solute complex reduced its polarity and resulted in symmetric peak shapes.

They achieved fast separation of β-blockers in less than 2 min on a short column using supercritical carbon dioxide modified with 20% ethanol and small amount of n-propylamine as the mobile phase.

Note that not all enantioseparations in SFC are better than in HPLC [34]. Bernal et al. [62] described the enantiomeric separation of several pharmaceutical-related compounds on a polysaccharide-based column using HPLC and SFC. They showed that most of the separations obtained by SFC are better, in terms of resolution and analysis time, than the separations obtained by HPLC. However, one compound could not be resolved using SFC, but LC provided baseline resolution.

4.2.3 SFC vs. GC

The solubility of an analyte is usually related to mobile phase density at constant temperature. The densities of supercritical fluids are 10^2–10^3 times as high as the densities of liquids, as shown in Table 4.1. This means that supercritical fluids possess very high solubility or strong solvating power while gaseous mobile phases have little interactions with analytes, since GC is usually operated at high temperatures. The unique solvating power of supercritical fluids makes it possible to use low temperatures for sample introduction and chromatographic separation in SFC. More important, low temperatures are often favorable for high enantioselectivity and low retention factors, which can allow fast and high-resolution enantioseparations.

Extensive comparisons between GC and SFC have been reported in chiral separation [63–66]. Zoltan investigated the performance of SFC and GC using the same chiral capillary columns coated with cyclodextrin-based stationary phases. It was observed that chiral selectivity was higher in GC than in SFC using the same open tubular column at the identical temperature (e.g., >100°C). However, the selectivity in SFC was significantly increased at low temperatures, especially for polar compounds [67].

Petersson et al. [68] compared the performance of GC and SFC using copolymeric stationary phases and observed that SFC could produce higher resolution because of its lower operating temperature, which facilitates solute-stationary phase interactions. Furthermore, packed columns are most commonly used in SFC, which provided high selectivity and sample capacity. Figure 4.5 shows two fast chiral separations performed by open tubular column GC and packed column SFC, respectively [44]. For similar resolution, α-phenylethanol was separated in 36 s in SFC as compared to 210 s in open tubular column GC. In other words, the resolution per unit time for SFC is approximately four times higher than that for GC. Note that the same chiral stationary phase and both separations were performed at the near optimum retention factor (~2). It is obvious that the packed SFC provided much higher selectivity, although it produced 18 times lower column efficiency. The high selectivity in packed column SFC can be ascribed to the lower operation temperature and the high surface area of the packed column. In contrast, open tubular column GC required higher efficiency to obtain the same resolution and thus a long column was used and the separation time was longer. Therefore, highly selective short columns can provide fast separation of enantiomers in SFC.

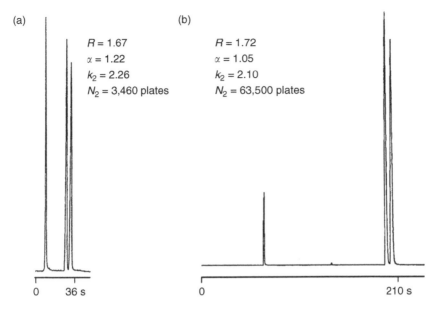

FIGURE 4.5 Chromatograms of α-phenylethanol enantiomers using (a) SFC and (b) open tubular column GC. Conditions: (a) 12 cm×250 μm ID capillary packed with 5-μm porous (300 Å) silica particles encapsulated with β-CD polymethylsiloxane (10% w/w) and end-capped with HMDS, 30°C, 140 atm, CO_2, FID, 10 cm×12 μm ID restrictor. (b) 25 m×250 μm ID cyano-deactivated capillary cross-linked with β-CD polymethylsiloxane (0.25 μm d$_f$); 130°C; He; FID. (Reprinted from Wu, N. et al. 2000. *J. Microcol. Sep.* 12: 454–461. With permission.)

It should be pointed out that capillary GC is more suitable for the separation of some volatile enantiomers in a complex matrix, such as essential oils, where high peak capacity and hence high column efficiency is required [69, 70]. Approximately 50 components including enantionmers in the natural peppermint oil were separated using a GC method [70]. Eight pairs of volatile lactone emantiomers were resolved by GC with a γ-cyclodextrin-based column [71].

4.3 EFFECT OF CHIRAL STATIONARY PHASES

Chiral separations result from the formation of transient diastereomeric complexes between stationary phases, analytes, and mobile phases. Therefore, a column is the "heart" of chiral chromatography as in other forms of chromatography. Most chiral stationary phases designed for normal phase HPLC are also suitable for packed column SFC with the exception of protein-based chiral stationary phases. It was estimated that over 200 chiral stationary phases are commercially available [72]. Typical chiral stationary phases used in SFC include Pirkle-type, polysaccharide-based, inclusion-type, and cross-linked polymer-based phases.

Finding the most selective separation conditions for a chiral compound can be time consuming [73, 74]. Enantioselectivity is compound-specific and a slight change in one functional moiety on a molecule may require totally different chiral stationary phases and separation conditions. Chiral SFC screening using automated column and mobile phase selection devices has been widely used over a decade [2, 36, 75–78]. Terfloth [78] and Phinney [79] summarized the applications of different chiral stationary phases to specific compounds in SFC. Thompson [80] reviewed some practical guides to enantioseparations for pharmaceutical compounds. Pirkle and Welch [81, 82] reviewed the evolution and applications of Pirkle-type chiral stationary phases in enantioseparation. Empirical rules for successful chiral recognition candidates using various chiral stationary phases have evolved on the basis of extensive chromatographic data. For example, the presence in the guest molecule of at least one aromatic ring favors chiral resolution with β-cyclodextrins [83]. ChirBase, a database specializing in enantioseparation including SFC, can often aid in reducing the developing effort [84]. Recently, Grinberg [5] provided an extensive review on the various interactions between enantiomeric solutes and stationary phases. The understanding of these interactions can lead to a rationale for development of selective, efficient, and fast methods. For a highly selective and efficient column, short column lengths and/or high flow rates can be used to increase separation speed.

Once the most selective conditions are identified by automated or manual screening, they can usually be optimized for fast separation, which includes increasing the flow rate, shortening the column length, and adjusting the concentration of modifiers. A good example for this application is shown in Figure 4.6 [36]. The result of screening 12 chiral columns reveals that the outstanding separation was observed with the Chiralcel OF column, while essentially no separation with other columns. The separation time was reduced to less than 5 min (Figure 4.6b) by the initial conversion to an isocratic elution from a gradient condition. The increased flow rate (3.5 from 1.5 mL/min) leads to a fast separation within 2 min. The wide expanse between the two peaks suggests that further decrease in separation time may be possible. When a 50-mm Chiralcel OF column was used with a high flow rate, the time for a baseline separation was reduced to 18 s (Figure 4.6c).

Lynam and Stringham [85] showed fast SFC separations of *trans*-stillbene oxide enantiomers using an increased flow rate and a higher methanol concentration with a 10-cm column, as illustrated in Figure 4.7. The separation was seven times faster with high resolution when the flow rate was increased by fivefold and the methanol concentration was doubled. Terfloth [86] showed that the separation of the two enantiomers was achieved in 0.6 min using a 50 × 3 mm Chiralpak AD column. Gyllenhaal [87] reported applications of 5-cm columns to fast separations. Ten injections and baseline separations of a racemic dihydropyridine-substituted acid were achieved in less than 6 min on a 5-cm Chiralpak AD column. Armstrong and coworkers [88] reported fast enantiomeric separations of 24 chiral dihydrofurocoumarin derivatives with macrocyclic glycopeptide stationary phases. Peculiar chromatographic behavior of β-blockers on tyrosine-derived chiral stationary phases under supercritical conditions was evidenced by studying the van Deemter curves [89]. Most separations were completed in less than 10 min when a flow rate of 3 mL/min was used. Separations for β-blockers were completed within 2 min using a short column packed with a ChyRoSine-A chiral stationary phase.

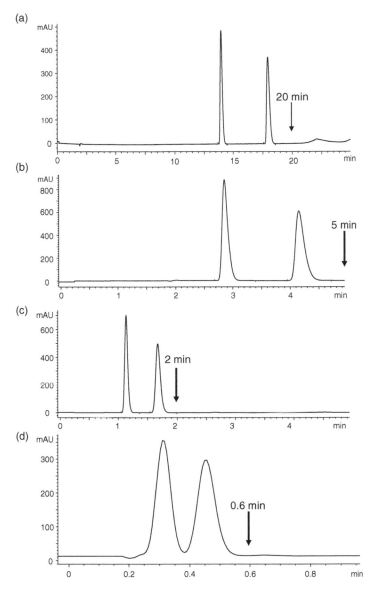

FIGURE 4.6 Stages in conversion of a screening result for separation of enantiomers of 2-alkynyl-5-pyrimidyl alcohol. Conditions: (a) standard gradient method used in chiral SFC screening: Chiralcel OF column (250 × 4.6 mm); 4% MeOH/CO$_2$ for 4 min, then ramp to 40% MeOH/CO$_2$ at 2%/min; 1.5 mL/min; UV 215 nm; 200 bar outlet pressure, 35°C. (b) Isocratic method using polar eluent: 40% MeOH/CO$_2$ (c) Isocratic method using polar eluent and faster flow rate: 40% MeOH/CO$_2$; 3.5 mL/min. (d) Optimized method using short column: Chiralcel OF column (50 × 4.6 mm); 30% MeOH/CO$_2$ at 4.0 mL/min; all examples using 215 nm UV detection, 200 bar outlet pressure and 35°C column temperature. (Reprinted from Welch, C.J. et al. *Chirality*, 2007, 19, 34–43. With permission.)

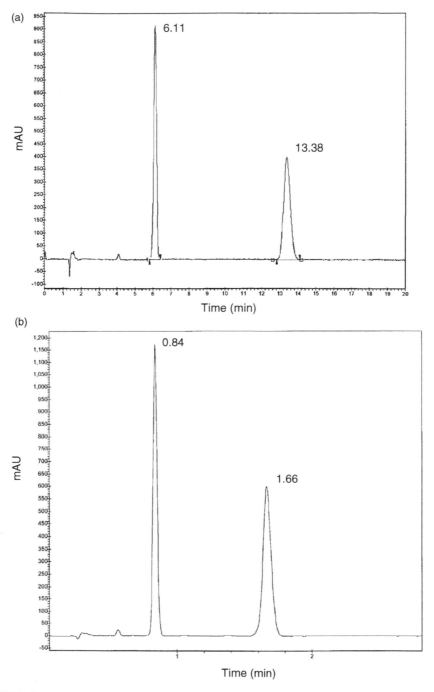

FIGURE 4.7 Fast SFC separation using a 10 cm column. Conditions: 100 × 4.6 mm (5 μm particles) with a Chiralpak AD-H phase; *trans*-stillbene oxide as analyte. (a) 10% Methanol; 1 mL/min flow rate. (b) 20% Methanol; 5 mL/min flow rate. (Adapted from Kenneth G. Lynam and Rodger W. Stringham, *HPLC2006*, June 17–23, 2006, San Francisco, CA, USA. With permission.)

It has been shown, theoretically and experimentally, that fast and efficient chromatographic separations can be achieved by using short columns packed with small packing particles [90, 91]. Sub-2-μm particles become increasingly acceptable for achiral separations as high-pressure LC instruments are commercially available. However, the development of small packing materials for chiral separation has not been appreciated as much as for achiral separation. Nevertheless, the evolution in using smaller packings in chiral chromatography can still be seen. Some commercial chiral columns packed with 10–20 μm particles have been gradually replaced by 5 μm particles, such as Chiralcel series (10 μm) by Chiracel-H series (5 μm). Several 3-μm packings for chiral separation such as Sapapak phases are currently commercially available. It can be expected that short columns packed with smaller packing particles would further improve the speed of enantioseparation in SFC.

4.4 EFFECT OF MODIFIERS AND ADDITIVES

In packed column SFC, polar solutes such as amines and carboxylic acids often have too much retention or elute with poor peak shapes when neat carbon dioxide is used as a mobile phase [28, 92]. This is mainly due to the weak solvent strength of neat carbon dioxide compared to a liquid solvent. The use of modifiers is often necessary to enhance the solvating power of the mobile phase in SFC. Various alcohols such as methanol and isopropanol are commonly used modifiers in SFC, but other solvents such as acetonitrile was also utilized [92]. The concentrations of modifiers are usually less than 50%. The technique in which the concentrations of modifiers are greater than 50% is often called enhanced-fluidity liquid chromatography [93].

The alcohol modifiers not only affect the retention of enantiomers but also enantioselectivity, both of which can significantly improve resolution per unit time for a chiral separation. Figure 4.8 presents some examples of how the chiral selectivity changes with amount and type of alcohols in the mobile phase at constant pressure and temperature for the separation of an aromatic amide on a cellulose-based stationary phase [94]. A maximum value of selectivity was observed for all types of alcohols at low modifier content. In this case, branched alcohols such as isopropanol provided better selectivity than linear alcohols such as 1-propanol. Toribio et al. [95] investigated the effect of modifier on elution order of the omeprazole enantiomers in SFC, as illustrated in Figure 4.9. The retention time and selectivity of the enantiomers were decreased and the elution order remained the same when the modifier was changed to ethanol from methanol. The elution order was reversed and the retention and selectivity were changed, when the methanol was replaced by isopropanol. The separation time was reduced from 25 to less than 7 min. Gyllenhaal [96] also observed the reversal of the elution order of several 2-propionic acids in SFC by varying organic modifier.

The use of additives in the mobile phase can further extend the polarity range of analytes in SFC. An additive is usually more polar than a modifier and is often immiscible with carbon dioxide by itself, but mixed with a modifier at 0.1–1% levels [92]. Possible roles for additives include modifying the bonded stationary

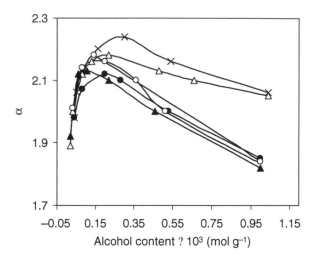

FIGURE 4.8 Chiral selectivity as a function of amount and type of alcohol modifiers in the mobile phase in SFC. Conditions: N-(2-heptyl)-p-tolylamide enantiomers as the analyte; 250 mm × 4.6 mm ID column, 10 μm silica particles coated with celhtlose tribenxoate; carbon dioxide and various types and amounts of modifiers: ● = methanol, ○ = ethanol, ▲ = 1-butanol, × = 2-propanol, and = 2-butanol; 25°C; flow rate 4.5 mL/min at 0°C; UV detection at 229 nm; average column pressure 140 bar. (Reprinted from Macaudiére, P. et al. 1989. *J. Chromatogr. Sci.* 27: 383–394. With permission.)

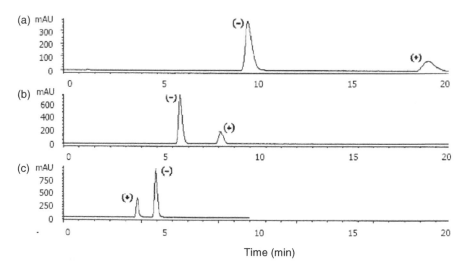

FIGURE 4.9 The effect of modifier on elution order of the enantiomers of omeprazole in SFC. Conditions: 20 MPa, 35°C, 2 mL/min, 30% of modifier, 70% (−)-omeprazole and 30% (+)-omeprazole. (a) Methanol, (b) ethanol, and (c) 2-propanol. (Reprinted from Toribio, L. et al. 2006. *J. Sep. Sci.* 29: 1363–1372. With permission.)

phase, suppressing ionization of acidic or basic analytes, deactivating the active sites in the stationary phase, and ion-pairing with ionic analytes [97]. Most effective additives usually belong to the same family as the solutes of interest [98]. For example, stronger bases improve the peak shapes of weaker bases and, similarly, stronger acids improve the peak shapes of weaker acids. Stringham [99], however, recently reported the use of a strong acid as additive for the separation of amine enantiomers. Screening with ethanol containing 0.1% ethanesulfonic acid (ESA) on a CHIRALPAK AD-H phase gave separation of 36 of 45 basic compounds previously not separated in SFC. The mechanism appears to involve the formation of an intact salt pair between the basic compounds and the acidic additive.

Figure 4.10 shows the effect of additive concentration on the separation of clenbuterol enantiomers on a polysaccharide-based chiral stationary phase [79]. The peak shapes were dramatically improved by adding an amine additive and the separation time was also reduced from 14 to 7 min when 1.0% amine was added to the mobile phase. Phinney and Sander [100] investigated the effect of amine additives using chiral stationary phases having either a macrocyclic glycopeptide or a

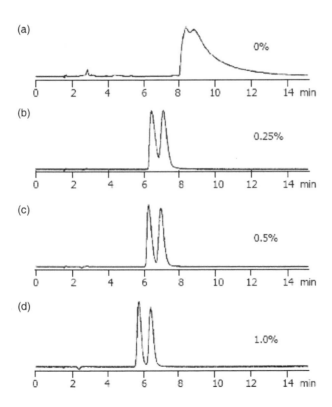

FIGURE 4.10 Separation of clenbuterol enantiomers at various isopropylamine additive concentration. Conditions: a Chiralpak AD chiral stationary phase (a) without any additive, (b) with 0.25%, (c) 0.5%, and (d) 1.0% isopropylamine. (Reprinted from Phinney, K. 2005. *Anal. Bioanal. Chem.* 382: 639–645. With permission.)

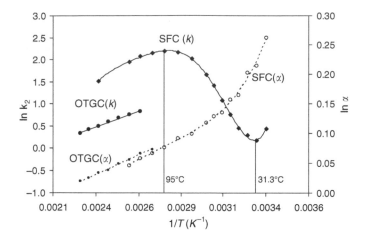

FIGURE 4.11 The effect of temperature on retention and selectivity in packed column SFC and open tubular GC. Conditions: 15 cm × 250 μm ID capillary packed with 5 μm porous (300 Å) silica particles encapsulated with β-CD polymethylsiloxane (10% w/w) and end capped with HMDS, 160 atm, CO_2, FID, 10 cm × 12 μm ID restrictor. (B) 25 m × 250 μm ID cyano-deactivated capillary cross-linked with β-CD polymethylsiloxane (0.25 μm d_f); He; FID. (Adapted from Wu, N. et al. 2000. *J. Microcol. Sep.* 12: 454–461. With permission.)

derivatized polysaccharide as the chiral selector. Two basic additives, isopropylamine and triethylamine, were incorporated into the methanol modifier at various concentrations and the effects on retention, selectivity, and resolution were monitored. Many of the analytes failed to elute from the macrocyclic glycopeptide stationary phase in the absence of an additive and the most noticeable effect of increasing additive concentration was a significant decrease in retention and thus in separation time. On the derivatized polysaccharide stationary phase the additives had little effect on retention, but they did provide significant improvements in peak shape and resolution.

4.5 THERMODYNAMIC ASPECTS

Thermodynamic behaviors and retention mechanisms are unique in SFC [101, 102]. At constant pressure, solute retention in SFC depends on temperature in a very characteristic manner [44, 103], as illustrated in Figure 4.11 [44]. The curve of temperature ($1/T$) vs. retention factor, k, for SFC is significantly deviated from the linear relationship predicted by the van't Hoff equation for typical GC and HPLC. At moderate pressures (e.g., 150 atm), increasing the temperature from ambient to critical (T_c, 31.3°C), k values decrease and reach a minimum (the right part of the SFC curves in Figure 4.11), which results from an increase in solute solubility with temperature in liquid carbon dioxide. This trend was also observed in a lower temperature range (−40°C to T_c) [59, 104]. Thermodynamically, an enthalpical-driven is dominant for the solute retention in this temperature range. After passing T_c, k values increase significantly, as shown in Figure 4.11. The density and thus solvating power of CO_2 decreases as the temperature increases, which results in an increase in k. The vapor pressure of the analyte also increases at higher temperatures, leading to an increase in solubility and thus a decrease in k. The second effect overcompensates for the first around 95°C, resulting in a maximum in retention factor. At temperatures higher than 95°C, the second effect becomes more pronounced and k values decrease with increasing temperature. In other words, an entropical-driven is dominant for the solute retention, which is very similar to that in open tubular GC in the range of high temperatures. At high pressure or high levels of organic modifier in the mobile phase, the decrease in density with increasing temperature is not as important as the effect of temperature on solvating power. Therefore, plots of ln k vs. $1/T$ are usually more flat.

As shown in Figure 4.11, the effect of temperature on enantiomeric selectivity for SFC is very similar to that for open tubular column GC, although the van't Hoff curve for selectivity is not linear in SFC. The selectivity increases significantly with decreasing temperature in both SFC and GC, and to a lesser degree in GC. These results indicate that low temperature is favored for fast chiral separation in SFC, at which, both low retention factor and high selectivity can be obtained. Figure 4.12 shows the comparison of resolution per unit time between SFC and GC at various temperatures [44]. The maximum resolution per minute is 1.5 for SFC at 30°C, compared to 0.5 for GC at 140°C. In this case, SFC is approximately three times faster than GC under respective optimum conditions, although GC provides much higher theoretical plates. Low temperatures can also reduce the risk of racemization and thermal decomposition of the chiral stationary phase and analytes. In open tubular column GC, higher temperatures are necessary because the solutes must be vaporized for solute elution and sample introduction. For example, it is difficult to elute α-phenylethanol at temperatures lower than 70°C in GC, since the mobile phase such as helium has no solvating power.

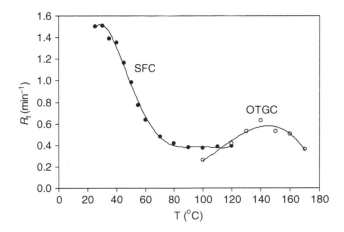

FIGURE 4.12 Comparison of separation speed between packed column SFC and OTGC. Conditions are the same as in Figure 4.11. (Adapted from Wu, N. et al. 2000. *J. Microcol. Sep.* 12: 454–461. With permission.)

Stringham and Blackwell [105] reported the reversal of elution order in SFC when the column temperature is higher than the isoelution temperature. At the isoelution temperature, T_{iso}, the enthalpic and entropic contributions to the enatioseparation are equal. Enantiomeric separations at $T < T_{iso}$ are called "enthalpically driven" and chiral separations at $T > T_{iso}$ entropically driven. In most cases, the isoelution temperature is above the working temperature. Recently, Toribio et al. [95] observed that the isoelution temperatures for some compounds are below the working temperature. The isoelution temperature for a given compound can be tuned by changing modifiers. The resolution of two pairs of enantiomers was markedly improved by performing at temperatures above the isoelution temperature. The elution order change could be useful in method development, as it is often favorable to elute the minor component first.

Column pressure usually has little effect on enantioselectivity in SFC. However, pressure affects the density of the mobile phase and thus retention factor [44]. Therefore, similar to a modifier gradient, pressure or density programming can be used in fast separation of complex samples [106]. Later et al. [51] used density/temperature programming in capillary SFC. Berger and Deye [107] demonstrated that, in packed column SFC, the effect of modifier on retention was more significant than that of pressure. They also showed that the enhanced solvent strength of polar solvent-modified fluid was not due to an increase in density, caused by the addition of the liquid phase modifier, but mainly due to the change in composition.

4.6 CONCLUSIONS

Excellent characteristics of supercritical fluids make SFC the most favorable choice for fast chiral separation among all of the separation techniques. The solvating power of critical fluids allows for low operation temperatures, which lead to high selectivity and low retention factors for enantiomers in SFC. Thus, packed column SFC is more suitable for fast separations of enantiomers than open tubular column GC. Supercritical fluids have higher diffusivity and lower viscosity than liquids. Therefore, packed column SFC frequently provides a higher resolution of enantiomers in shorter times than for the corresponding HPLC separations. Chiral stationary phases, modifiers, and additives play a key role in the fast SFC separation of enantiomers. Once the most selective stationary phase and mobile phase are identified, the conditions can be further optimized for fast separation, which includes increasing the flow rate, shortening the column length, and adjusting the concentration of modifiers or additives.

Thermodynamic behaviors and retention mechanisms for SFC are unique. Low temperatures and high pressures or high densities usually favor fast separation of enantiomers in SFC. In the case that the isoelution temperature is below the working temperature, the selectivity increases as temperature increases and higher temperatures are favorable for chiral separation. Future development in SFC will likely include new chiral column technologies and instrumentation refinement. A greater variety of chiral columns packed with smaller particles will open up more areas of application for fast chiral separations. In addition, improvement in signal-to-noise ratio of detection is highly desirable to trace minor analysis in SFC. Finally, understanding of interactions between chiral stationary phases, enantiomers, and mobile phase will facilitate method development of chiral separation.

ACKNOWLEDGMENTS

The author would like to thank Professor Milton Lee of Brigham Yound University for his guidance and thank Dr. Andrew Clausen, Dr. Christopher Welch, Dr. Lili Zhou, and Dr. Thomas Dowling for invaluable conversations and technical expertise.

REFERENCES

1. Pauwels, R. 2006. *Antiviral Res.* 71: 77–89.
2. Workman, P. 2003. *Cur. Pharm. Design* 9: 891–902.
3. Ghosh, S., Nie, A., An, J., Huang, Z. 2006. *Cur. Opin. Chem. Biol.* 10: 194–202.
4. Hausheer, F.H., Kochat, H., Parker, A.R., Ding, D., Yao, S., Hamilton, S.E., Petluru, P.N., Leverett, B.D., Bain, S.H., Saxe, J.D. 2003. *Cancer Chemother. Pharmacol.* 52 (Suppl. 1): S3–S15.
5. Grinberg, N. 2006. *Am. Pharm. Rev.* 9: 65–71.
6. Wu, N., Clausen, A. 2007. *J. Sep. Sci.* 30: 1167–1182.
7. Zhang, Y., Wu, D.R., Wang-Iverson, D.B., Tymiak, A.A. 2005. *Drug Discovery Today*, 10: 571–577.
8. Welch, C., Leonard, W. R., DaSilva, J.O., Biba, M., Albaneze-Walker, J., Henderson, D.W., Laing, B., Mathre, D.J. 2005. *LC•GC*, May: 264–272.
9. Stevenson, D., Williams, G.A. In *Chiral Separations*, Stevenson, D., Wilson, I.D., Eds., Plenum Press, New York, 1988.
10. Caner, H., Groner, E., Levy, L., Agranat, I. 2004. *Drug Discovery Today*, 9: 105–110.
11. Branch, S.K. In *Chiral Separation Techniques: A Practical Approach*, Subramanian, Ed., Wiley-VCH, 2001, p. 319
12. Alexander, A.J., Staab, A. 2006. *Anal. Chem.* 78: 3835–3838.
13. White, C., Burnett, J. 2005. *J. Chromatogr. A* 1074: 175–185.
14. Borman, S. 1998. *Chem. Eng. News* 76 (April): 47–68.
15. Czarnik, A.W. 1998. *Anal. Chem.* 70: 378–368A
16. Ward, T.J. 2006. *Anal. Chem.* 78: 3947–3956.
17. He, L., Beesley, T.E. 2005. *J. Liq. Chromatogr. Rel. Technol.* 28: 1075–1114.
18. Roussel, C., Del Rio, A., Pierrot-Sanders, J., Piras, P., Vanthuyne, N. 2004. *J. Chromatogr. A* 1037: 311–328.
19. Xiao, T.L. Armstrong, D.W. 2004. *Methods Mol. Biol.* 243: 113–171.
20. Lee, M.L., Markides, K.E. 1987. *Science* 235: 1342–1347.
21. Maftouh, M., Granier-Loyaux, C., Chavana, E., Marini, J., Pradines, A., Vander Heyden, Y., Picard, C. 2005. *J. Chromatogr. A* 1088: 67–81.
22. Patel, B.K., Hanna-Brown, M., Hadley, M.R., Hutt, A.2004. *Electrophoresis* 25: 2625–2656.
23. Shamsi, S.A., Miller, B.E. 2004. *Electrophoresis* 25: 3927–3961.
24. Wistuba, D., Kang, J., Schurig, V. 2004. *Methods Mol. Biol.* 243: 401–409.
25. Siouffi, A.-M., Piras, P., Roussel, C. 2005. *J. Planar Chromatogr. Mod. TLC* 18: 5–12.
26. Petersson, P., Markides, K.E. 1994. *J. Chromatogr. A* 666: 381–394.

27. Wolf, C., Pirkle, W.H. 1997. *LC?GC* 15: 352–363.
28. Berger T.A. In *Supercritical Fluid Chromatography with Packed columns, Techniques and Applications*, Anton, K., Berger, C., Eds., Marcel Dekker, New York, 1998, Chapter 2.
29. Smith, R.M. In *Supercritical Fluid Chromatography with Packed Columns, Techniques and Applications*, Anton, K., Berger, C., Eds., Marcel Dekker, New York, 1998, Chapter 8.
30. Koppenhoefer, B., Bayer, E. 1985. *Chromatographia* 19: 123–136.
31. Mourier, P., Sassiant, P., Caude, M., Rosset, R. 1986. *J. Chromatogr*. 353: 61–75.
32. White, C.M., Houck R.K. 1986. *J. High Resolut. Chromatogr.* 9: 4–17.
33. Gere, D.R., Board, R., McManigill, D. 1982. *Anal. Chem.* 54: 736–740.
34. Anton, K., Eppinger, J., Frederiksen, L., Francotte, E., Berger, T.A., Wilson, W.H. 1994. *J. Chromatogr. A* 666: 395–401.
35. Kot, A., Sandra, P., Venema, A. 1992. *J. Chromatogr. Sci.* 32: 439–448.
36. Welch, C.J., Biba, M., Sajonz, P. 2007. *Chirality* 19: 34–43.
37. Lesellier, E., Tchapla, A. In *Supercritical Fluid Chromatography with Packed columns, Techniques and Applications*, Anton, K., Berger, C., Eds., Marcel Dekker, New York, 1998, p. 195.
38. Sandra, P., Medvedovici, A., Kot, A., David, F. In *Supercritical Fluid Chromatography with Packed Columns, Techniques and Applications*, Anton, K., Berger, C., Eds., Marcel Dekker, New York, 1998, p. 161.
39. Lee, M.L., Markides, K.E. *Analytical Supercritical Fluid Chromatography and Extraction*; Chromatography Conferences; Provo, UT, 1990.
40. Roth, M. 2004. *J. Chromatogr. A* 1037: 369–391.
41. Petersson, P., Markides, K.E. 1994. *J. Chromatogr. A* 684: 297–309.
42. Juvancz, Z., Grolimund, K., Schurig, V. 1993. *J. Microcol. Sep.* 5: 459–468.
43. Berger, T.A., Wilson, W.H. 1993. *Anal. Chem.* 65: 1451–1455.
44. Wu, N., Chen, Z., Medina, J.C., Bradshaw, J.S., Lee, M.L. 2000. *J. Microcol. Sep.* 12: 454–461.
45. Schoenmakers, P.J. 1988. *J. High Resol. Chromatogr.* 11: 278–282.
46. Wu, N., Yee, R., Lee, M.L. 2001. *Chromatographia* 53: 197–200.
47. Shen, Y., Lee, M.L. In *Supercritical Fluid Chromatography with Packed Columns, Techniques and Applications*, Anton, K., Berger, C., Eds., Marcel Dekker, New York, 1998, Chapter 5.
48. Sandra, P.J. 1989. *High Resolut. Chromatogr.* 12: 82–86.
49. Kroon, R., Baggen, M., Lagendijk, A. 1989. *J. Chem. Phys.* 91: 74–78.
50. Giddings, J.C. *Unified Separation Science.* John Wiley & Sons, Inc., New York, 1991, p. 276.
51. Later, D.W., Campbell, E.R., Richter, B.E. 1988. *J. High Resolut. Chromatogr.* 11: 65–69.
52. Gere, D.R. 1983. *Science* 222: 253.
53. Wu, N., Tang, Q., Shen, Y., Lee, M.L. 1999. *Anal. Chem.* 71: 5084–5092.
54. Yaku, K., Morishita, F. 2000. *J. Biochem. Biophys. Methods* 43: 59–76.
55. Nishikawa, Y. 1993. *Anal. Sci.* 9: 33–37.
56. Williams, K.L., Sander, L.C., Wise, S.A. 1996. *J. Chromatogr. A* 746: 91–101.
57. Yaku, K., Aoe, K., Nishimura, N., Sato, T., Morishita, F. 1997. *J. Chromatogr. A* 785: 185–193.
58. Phinney, K.V. 2000. *Anal. Chem.* 72: 204A–211A.
59. Pirkle, W.H., Brice, J., Terfloth, G.J. 1996. *J. Chromatogr. A* 753: 109–119.

60. Hoke, S.H., Pinkston, J.D., Bailey, R.E., Tanguay, S.L., Eichhold, T.H. 2000. *Anal. Chem.* 72: 4235–4241.
61. Siret, L., Bargmann, N., Tambute, A., Caude, M. 1992. *Chirality* 4: 252.
62. Bernal, J.L., Toribio, L., del Nozal, M.J., Nieto, E.M., Montequi, M.I. 2002. *J. Biochem. Biophys. Methods* 54: 245–254.
63. Juvancz, Z., Szejtli, J. 2002. *Trends Anal. Chem.* 21: 379–388.
64. Juvancz, Z., Markides, K.E., Rouse, C.A., Jones, K., Tarbet, B.J., Bradshaw, J.S., Lee, M.L. 1998. *Enantiomer* 3: 89–94.
65. Majors, R.E. 1997. *LC?GC* 15: 412–422.
66. Jung, M., Schurig, V.J. 1993. *High Resolut. Chromatogr.* 16: 215–223.
67. Zoltan, J. 2002. *Olaj Szappan Kozmetika* 51: 23–29.
68. Petersson, P., Markides, K.E., Johnson, D.F., Rossiter, B.E., Bradshaw, J.S., Lee, M.L. 1992. *J. Microcol. Sep.* 4: 155–162.
69. Schurig, V. 1994. *J. Chromatogr. A* 666: 111–129.
70. Faber, B., Dietrich, A., Mosandl, A. 1994. *J. Chromatogr. A* 666: 161–165.
71. Konig, W.A., Fricke, C., Saritas, Y., Momeni, B., Hohenfeld, G. 1997. *J. High Resolut. Chromatogr.* 20: 55–61.
72. Francotte, E.R. *Enantioselective Chromatography for the Preparation of Drug Enantiomers*, ISCD-15, Shizuoka, Japan, October 2003.
73. Perrin, C., Vu, V.A., Matthijs, N., Maftouh, M., Massart, D.L., Vander Heyden, Y. 2002. *J. Chromatogr. A* 947: 69–83.
74. Villeneuve, M.S., Anderegg, R.J. 1998. *J. Chromatogr. A* 826: 217–215.
75. Mourier, P.A. 1985. *Anal. Chem.* 57: 2819.
76. Zhao, Y., Woo, G., Thomas, S., Semin, D., Sandra, P. 2003. *J. Chromatogr. A* 1003: 157–166.
77. Zhang, Y., Watts, W., Nogle, L., McConnell, O. 2004. *J. Chromatogr. A* 1049: 75–84.
78. Terfloth, D. 2001. *J. Chromatogr. A* 906: 301–307.
79. Phinney, K. 2005. *Anal. Bioanal. Chem.* 382: 639–645.
80. Thompson, R. 2005. *J. Liq. Chromatogr. Rel. Technol.* 28: 1215–1231.
81. Pirkle, W.H., Pochapsky, T.C. 1989. *Chem. Rev.* 89: 347–362.
82. Welch, C.J. 1994. *J. Chromatogr. A* 666: 3–26.
83. Armstrong, D.W., Ward, T.J., Armstrong, R.D., Beesley, T.E. 1986. *Science* 232: 1132–1135.
84. http://www.acdlabs.com/products/chrom_lab/chirbase/
85. Kenneth G. Lynam, Rodger W. Stringham, *HPLC2006*, June 17–23, 2006, San Francisco, CA, USA.
86. Terfloth, G. 1990. *LC•GC* 17: 400–405.
87. Gyllenhaal, O. 2001. *Fresen. J. Anal. Chem.* 369: 54–56.
88. Liu, Y., Rozhkov, R.V., Larock, R.C., Xiao, T.L., Armstrong, D.W. 2003. *Chromatographia* 58: 775–779.
89. Bargmann-Leyder, N., Thiebaut, D., Vergne, F., Begos, A., Tambute, A., Caude, M. 1994. *Chromatographia* 39: 673–681.
90. MacNair, J.E., Patel, K.D., Jorgenson, J.W. 1999. *Anal. Chem.* 71: 700–708.
91. Wu, N., Lippert, A.J., Lee, M.L. 2001. *J. Chromatogr. A* 911: 1–12.
92. Berger, T.A. 1997. *J. Chromatogr. A* 785: 3–33.
93. Sun, Q., Olesik, S.V. 1999. *Anal. Chem.* 71: 2139–2145.
94. Macaudiére, P., Caude, M., Rosset, R., Tambute, A. 1989. *J. Chromatogr. Sci.* 27: 383–394.

95. Toribio, L., Alonso, C., del Nozal, M.J., Bernal, J.L., Jiménez, J.J. 2006. *J. Sep. Sci.* 29: 1363–1372.
96. Gyllenhaal, O. 2005. *Chirality* 17: 257–265.
97. Zheng, J., Taylor, L.T., Pinkston J. David, Mangels, M.L. 2005. *J. Chromatogr. A* 1082: 220–229.
98. Berger, T.A., Deye, J.F. 1991. *J. Chromatogr. Sci.* 29: 310.
99. Stringham, R.W. 2005. *J. Chromatogr. A* 1070: 163–170.
100. Phinney, K.W., Sander, L.C. 2003. *Chirality* 15: 287–294.
101. Roth, M. 2004. *J. Chromatogr. A* 1037: 369–391.
102. Wu, Y. 2004. *J. Liq. Chromatogr. Rel. Tech.* 27: 1203–1236.
103. Leyendecker, D., Schmitz, F.P., Klesper, E. 1986. *J. High Resolut. Chromatogr.* 9: 566–571.
104. Loughlin, T., Thompson, R., Bicker, G., Grinberg, N. 1996. *Chirality* 8: 157–162.
105. Stringham, R.W. 1996. *Anal. Chem.* 68: 2179–2185.
106. Smith, R.D., Chapman, E.G., Wright, B.W. 1985. *Anal. Chem.* 57: 2829–2836.
107. Berger, T.A., Deye, J.F. 1990. *Anal. Chem.* 62: 1181–1185.

5 Shape Selectivity in Reversed-Phase Liquid Chromatography*

Katrice A. Lippa, Catherine A. Rimmer, and Lane C. Sander

CONTENTS

---

* Certain commercial equipment, instruments, or materials are identified in this paper to specify adequately the experimental procedure. Such identification does not imply recommendation or endorsement by the National Institute of Standards and Technology, nor does it imply that the materials or equipment identified are necessarily the best available for the purpose.

5.1 INTRODUCTION

Reversed-phase liquid chromatography (RPLC) separations provide the foundation in a wide range of chemical analytical techniques for the determination of individual components in complex mixtures and matrices. Separations of nonvolatile organic compounds, polar pharmaceutical compounds, and other larger-size biomolecules are routine, and selectivity can be controlled by the choice of stationary and mobile phases as well as by modification of operational parameters such as temperature and the application of gradients. The discriminating power of RPLC is exemplified by the superior separation of polycyclic aromatic hydrocarbon (PAH) isomers on bonded oct-adecylsilane (C_{18}) stationary phases, for which subtle differences in molecular shape controls their separation. Over 25 years ago, Ogan and Katz illustrated the improved separation of certain PAH isomers on higher-density C_{18} stationary phases [1–3]. During the same time period, the U.S. Environmental Protection Agency method 610 [4] required a specific commercial C_{18} column for the improved RPLC separation of 16 priority pollutant PAHs. It was unclear, however, which properties of the stationary-phase were driving the enhanced selectivity for the PAH isomers as many details regarding the stationary-phase and its preparation were not available. During early investigations of the mechanism of retention for these chromatographic materials [1,3,5,6], it became apparent that the specific chemical modifications of the silica surface in addition to the shape of the PAH isomer were important parameters that influenced selectivity.

The majority of RPLC stationary phases are prepared by alkylsilanization of microparticulate silica through reaction of a monofunctional alkylsilane (e.g., chlo-rooctadecylsilane) with the silanol groups on the silica surface. Monofunctional or "monomeric" C_{18} phase materials have been successfully applied to a wide range of analytical separations but are limited in their capacity to separate solutes with similar polarity and fixed conformational structures, such as PAHs and polychlorin-ated biphenyl (PCB) congeners or for molecules with constrained geometric shapes (carotenoids, steroids). For such solutes, in which molecular shape influences selectiv-ity, the use of a solution-polymerized (originating from trichlorooctadecylsilane starting material) or "polymeric" C_{18} stationary-phase with its typically greater sur-face coverage can yield significantly improved separations. These polymer-based modified surfaces are designated "shape-selective" chromatographic sorbents and have been developed [7] and extensively utilized for the enhanced separation of shape-constrained solutes. Typical chromatograms for the separation of PAH isomers of molecular mass 302 in Figure 5.1 illustrate the differences between low and high shape-selective chromatographic materials [8], which often correspond to monomeric and polymeric stationary phases, respectively.

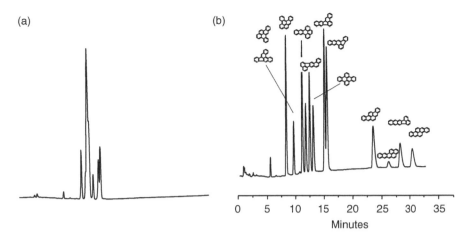

FIGURE 5.1 Separation of PAH isomers of relative molecular mass 302 on (a) monomeric C_{18} column and (b) polymeric C_{18} column. Conditions: 90:10 acetonitrile/water (volume fraction) to acetonitrile over 10 min. (Adapted from Wise, S.A., et al., *Anal. Chem.*, 60, 630, 1988. With permission.)

The separation of shape-constrained solutes on alkyl-based columns is generally influenced by subtle differences in molecular shape together with stationary-phase and operational parameters that promote stationary-phase conformational order [9]. Mechanistic models that describe solute retention by the difference in various solute hydrophobic (or solvophobic) and electronic interactions as they partition between the stationary and mobile phases or alternatively as selective solute sorption onto stationary-phase surface sites can be sufficient to characterize the separation of compounds with differing polarity. Numerous reviews describing the details within and the differences between these models are provided elsewhere [10–14]. These models, however, are limited to account for the separation differences between mixtures of solutes with similar polarity and constrained molecular structures, such as the separation of the geometric isomers of PAHs and PCB congeners. It is important to recognize that chromatographic separations, including highly selective shape-recognition processes, are ultimately governed by free energy differences between a solute in the mobile phase and the solute incorporated into the stationary-phase. It is also important to note that a baseline separation of two components ($R_S > 1.5$) can be the result of very small changes in free energy (on the order of 0.3 kJ/mol) [15] comprising enthalpic and entropic contributions that may be challenging to quantify independently. As highlighted in a recent review [16], the retentive process may be a combination of adsorption and partitioning on and into the stationary-phase. Thus, the challenge remains to account for any individual contributions from simultaneous interactions (e.g., hydrophobic, steric) of the shape-constrained solute in both the mobile and stationary phases to fully understand the driving forces involved in RPLC shape-recognition separations.

This chapter will examine the wealth of chromatographic and spectroscopic investigations that elucidate the dominant factors that control shape-selective processes in RPLC. The review will center on alkyl chain-based chromatographic materials that have been utilized for the reversed-phase mode separation of solutes that differ only in molecular shape. Other shape-recognition chromatographic materials such as molecular imprinted polymers (MIPs) and chiral stationary phases that may involve additional interactions (H-bonding, ion exchange) in the retentive process will not be reviewed. The various stationary-phase properties and experimental conditions that influence chromatographic shape recognition will be discussed. Experimental evidence of stationary-phase conformational order and disorder via spectroscopy investigations as well as complementary evidence provided through molecular simulations will be examined and related to the shape discrimination capability of "shape-selective" chromatographic sorbents. Lastly, the information provided herein will be used to propose plausible mechanisms of retention that may advance our current understanding of the shape-selective process in RPLC.

5.2 DEFINITION OF SHAPE SELECTIVITY

The term "shape selectivity" is used to describe a chromatographic condition for a particular stationary-phase that results in the enhanced separation of geometric isomers or structurally related compounds. The primary interest in shape-selective separations—and hence an evaluation of a column's ability to resolve solutes based on shape—is largely driven by the various biological activities that are associated with each individual geometric isomer. For instance, the U.S. Department of Health and Human Services has recently listed four of the molecular mass 302 PAH isomers, dibenzo[a,e]pyrene, dibenzo[a,h]pyrene, dibenzo[a,i]pyrene and dibenzo[a,l]pyrene, and one methylated PAH (5-methylchrysene) as potential human carcinogens [17]. Differences in molecular shape have also been related to the toxicity of PAH and MPAH isomers [18]. The various isomeric forms of vitamins result in differing biological activity; for example, the conversion of carotenes to vitamin A is less efficient for the *cis* isomer than for the all-*trans* isomers [19]. The improved characterization of food, environmental, or biological matrices through enhanced analytical separations of individual isomeric species will thus facilitate the health benefit and toxicological assessments of these materials.

From the perspective of the solute, isomeric separation is based solely on molecular structure, rather than on other chemical and physical properties, such as polarity and H-bonding capability. Shape selectivity has been demonstrated to be important in the separation of several categories of shape-constrained solutes including PAHs [20–24] and methyl-substituted polycyclic aromatic hydrocarbons (MPAHs) [24,25], carotenoids [26–30], steroids [31], and polychlorinated biphenyl congeners [32]. These compounds exhibit rigid conformation structures imposed by fused ring structures (PAHs, steroids), conjugated bonds (carotenoids), or other bond rotation restrictions (PCBs). In contrast, little or no effect of shape selectivity is observed for the separation of solutes with more flexible conformation structures (e.g., alkylbenzene isomers).

5.2.1 MOLECULAR DESCRIPTORS

Earlier observations of shape recognition on liquid crystalline phases in gas chromatography [33–36] in which rod-like PAHs were more retained than the square-shaped solutes were used to initially explain shape-selectivity behavior on polymeric C_{18} liquid chromatography (LC) materials. Radecki et al. [36] developed a solute molecular descriptor (L/B) defined by the ratio of the length to the breadth of a box drawn to enclose the solute. This descriptor was used to correlate with PAH retention in gas chromatography. Wise et al. [6] further refined the L/B descriptor by orienting the box to provide a maximum value (see Figure 5.2a), which provided improved correlations with PAH retention on polymeric C_{18} LC columns. Solutes with larger L/B ratios (long, narrow) are retained longer than are solutes with smaller L/B ratios (squared or bulkier-shaped) on shape-selective materials. A listing of calculated L/B values and related data for over 600 PAHs has been previously published [37].

Other molecular descriptors have been utilized to describe the molecular shape of constrained solutes (Figure 5.2b,c). Yan and Martire [38,39] have described a

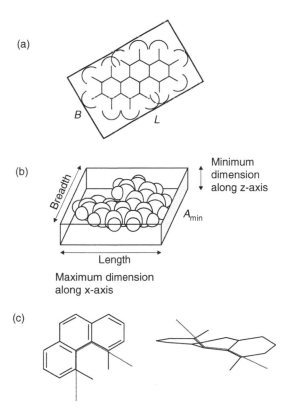

FIGURE 5.2 Molecular descriptors: (a) length-to-breadth (L/B) ratio; (b) minimum area, A_{min}; (c) dihedral angle of distortion. (Adapted from Sander, L.C. and Wise, S.A., in Smith, R.M. (Ed.), *Retention and Selectivity Studies in HPLC*, Elsevier, Amsterdam, 1995, p. 337. With permission.)

solute cross-sectional area descriptor, A_{min}, which is calculated from the side with the smallest area of a box drawn to enclose the molecular shape of the solute. Smaller values of A_{min} are typical of rod-like solutes, whereas more squared-shaped solutes have larger values of A_{min}. A linear relationship between the retention factor, $\ln k'$, and A_{min} has been reported [24]. For planar molecules, A_{min} values are identical to L/B ratios as described by Wise et al. [6]. A more detailed comparison between the molecular descriptors and their evaluation in correlations of LC retention are provided by Wise and Sander [24]. The "dihedral angle of distortion" as identified by Garrigues et al. [40] provides an alternative description of solute nonplanarity, in which the deviation from planarity is calculated from four contiguous atoms within the solute. A dihedral angle of $0°$ is indicative of planarity, whereas nonzero values indicate a nonplanar or a twisted solute shape.

5.2.2 TEST SOLUTE MIXTURES

Several empirical chromatographic test mixtures exist for the evaluation of a column's ability to resolve solutes based on molecular shape [29,41–43]. These include a Standard Reference Material (SRM) available from the National Institute of Standards and Technology (NIST) SRM 869a Column Selectivity Test Mixture for Liquid Chromatography, the Tanaka test, and the Carotenoid (β-carotene) isomer test; each test mixture contains two or more structurally constrained compounds with dissimilar molecular shapes. The solutes in each test mixture are illustrated in Figure 5.3. Other mixtures of shape-constrained solutes (e.g., 1,6-diphenylhexatriene) can be used to assess shape selectivity; however, the use of such solutes for chromatographic evaluations has not been extensively applied to produce a standard reference scale of selectivity results. In general, planar molecules are retained longer than bulky solutes and linear solutes are preferentially retained over nonlinear solutes on highly shape selective stationary phases.

SRM 869a Column Selectivity Test Mixture for Liquid Chromatography [44] is composed of three shape-constrained PAHs (phenanthro[3,4-c]phenanthrene, PhPh; 1,2:3,4:5,6:7,8-tetrabenzonaphthalene, TBN; and benzo[a]pyrene, BaP) and is routinely employed to evaluate the shape selectivity of stationary phases. The retention differences between the nonplanar TBN and planar BaP solutes (expressed as a selectivity factor $\alpha_{TBN/BaP} = k'_{TBN}/k'_{BaP}$) provide a numerical assessment of the phase's shape recognition ability. For phases with low shape selectivity, BaP elutes before the nonplanar PAHs (PhPh and TBN) and has a value of $\alpha_{TBN/BaP} > 1.7$, whereas for phases with high shape selectivity, the planar BaP elutes last and values of $\alpha_{TBN/BaP}$ are less than 1.0 (typically polymeric phases). Phases with intermediate shape selectivity typically have $\alpha_{TBN/BaP}$ values between 1.0 and 1.7. Overall, polymeric or "shape-selective" chromatographic sorbents are better able to discriminate between long planar BaP solute (and with increased retention) from the more square-shaped and nonplanar solutes.

Another informative test mixture is that described by Tanaka [43], in which selectivity between triphenylene (TRI) and o-terphenyl (o-TER) is used to characterize the shape recognition capability of LC stationary phases. The primary difference between these two solutes is their planarity; TRI is a planar PAH and o-TER possesses

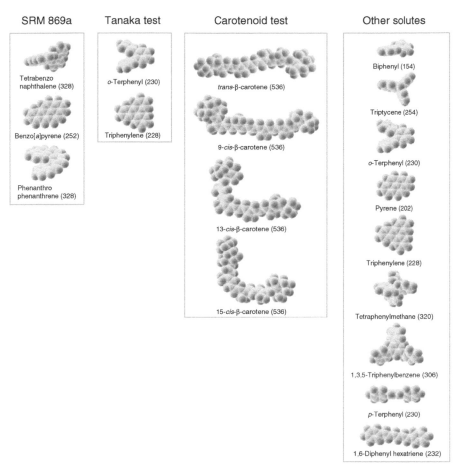

FIGURE 5.3 **(See color insert following page 280.)** Solute test mixtures used for shape-selectivity characterization.

a highly twisted shape (Figure 5.3). Ratios of TRI retention to o-TER retention ($\alpha_{\text{TRI}/o\text{-TER}} = k'_{\text{TRI}}/k'_{o\text{-TER}}$) are used as indications of planar shape recognition; higher values ($\alpha_{\text{TRI}/o\text{-TER}} > 3$) are indicative of high shape selectivity, whereas lower values ($\alpha_{\text{TRI}/o\text{-TER}} < 2$) correspond to low shape selectivity.

Engelhardt et al. [45] have evaluated the performance of both the SRM 869a and Tanaka chromatographic tests for the characterization of column shape selectivity. From a comparison of ten various columns, ranging in chain length, bonding chemistry, and phase type (i.e., bonded cholestane), the two test procedures produced differing results (Figure 5.4). Within data obtained from n-alkane-modified surfaces, however, the data are more consistent. Four monomeric-type columns that have chain lengths ranging from 4 to 18 carbons result in an average $\alpha_{\text{TBN}/\text{BaP}} \approx 1.8$ and an average $\alpha_{\text{TRI}/o\text{-TER}} \approx 2.2$. Five polymeric-type columns prepared from both C_{18} and C_{30} silanes result in relatively low and consistent $\alpha_{\text{TBN}/\text{BaP}}$ selectivity values (average $\alpha_{\text{TBN}/\text{BaP}} \approx 0.7$). The Tanaka test applied to these materials, however, results in an

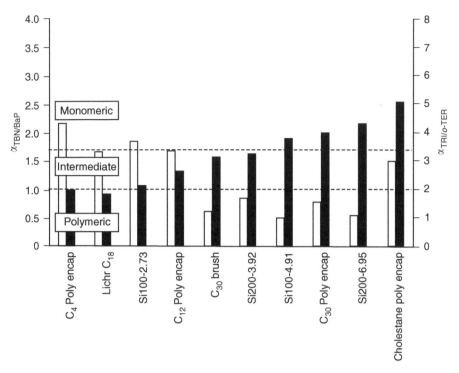

FIGURE 5.4 Comparison of shape selectivity and planar recognition test procedures: □ SRM 869a test (solutes PhPh, TBN, BaP; eluent acetonitrile/water 85/15 (v/v), flow 2 ml/min, detection UV 254 nm) and ■ Tanaka test (solutes o-TER, TRI; eluent methanol/water 80/20 (v/v), flow 1 ml/min, detection UV 254 nm. (Reproduced from Engelhardt, H., et al., *Chromatographia*, 48, 183, 1998. With permission.)

average $\alpha_{TRI/o\text{-}TER} \approx 3.7$ with a considerably larger dispersion of values (ranging from $\approx$3.2 to 5.1).

It is apparent from Figure 5.4 that $\alpha_{TRI/o\text{-}TER}$ values from the Tanaka test illustrate more of a continuum in selectivity among these columns examined, whereas the $\alpha_{TBN/BaP}$ values obtained from the SRM 869a test mixture provide a clearer delineation between these two groups of columns. This differentiation may be attributable to the phase's ability to selectively retain the relatively long and narrow BaP solute (molecular dimensions of 13.8 Å × 9.2 Å × 3.9 Å [46]) used as a planarity measure in the SRM 869a test rather than the more triangular-shaped TRI solute (molecular dimensions of 11.6 Å × 10.4 Å × 4.1 Å [46]) used in the Tanaka test. It has been suggested that polymeric-type alkyl-modified surfaces contain rigid and narrow "slots" into which solutes of a particular size and shape can penetrate [47]. If such features exist within the polymeric-type stationary phases, then it is plausible that narrow solutes (in this case, BaP) would be selectively retained over less narrow solutes (such as TRI). The differences between these two test solutes' molecular shape may also be responsible for the selectivity differences observed for the column containing bonded cholestane groups. The enhanced retention of TRI ($\alpha_{TRI/o\text{-}TER}$ of 5.1) suggests that the embedded cholestane moieties within the stationary-phase provide the appropriate contact surface size and shape that contributes to shape recognition. However, the molecular-level interactions between the stationary-phase and the retained solute that control such shape-recognition process are not yet identified.

The shape-recognition ability of an LC column toward larger-sized, constrained solutes can also be evaluated through the separation of carotenoid isomers [29]. Carotenoids are a class of polyolefins that occur naturally in fruits and vegetables and exist in various isomeric forms [19]. Given that the length of the all-*trans* forms of carotenoid molecules generally exceeds the length of C_{18} phases ($\approx$30 vs. 21 Å, respectively), specific stationary phases of extended length (C_{30}) have been designed for the determination of carotenoid isomers in complex natural matrices [29]. A common shape-recognition test for columns is the separation of β-carotene isomers (Figure 5.3). In general, the more linear 9-*cis* β-carotene and *trans* β-carotene isomers exhibit longer retention on polymeric C_{30} columns as compared with the 13- and 15-*cis* β-carotene isomers, the latter of which are curvilinear near the midpoint of the olefin structure.

5.2.3 Steric versus Shape Selectivity

Shape selectivity has also been described recently [48] as a restricted retention process for which rigid solute molecules are retained within a relatively rigid stationary-phase, which stems from a linear free-energy relationship (LFER) investigation of column selectivity for a wide range of solutes on both monomeric and polymeric C_{18} materials [49]. This clarification has been proposed to distinguish "shape selectivity" from a related restricted retention mechanism term "steric selectivity" [48], in which a steric interaction term ($\sigma'S$) characterizes the hindrance to insertion of a more flexible solute into a less-rigid stationary-phase (i.e., lower-density monomeric-type phase). Solute molecules of longer molecular length and significant molecular "thickness" invoke a greater steric resistance as they associate with the stationary-phase and result in larger values of the steric solute parameter (σ'). Stationary phases that are less penetrable to these larger solutes will have lower values of the column steric parameter (S). It is important to note that values of σ' and S are relative to an "average" solute and column value, respectively, as determined by the LFER characterization. Therefore, a measure of "steric selectivity" in terms of $\sigma'S$ represents a relative increase in entropic cost associated with both the constraint imposed on the flexible solute's conformational changes together with restriction of the normally flexible alkyl ligands of the stationary-phase on retaining the larger-sized solute.

When comparing "steric selectivity" to "shape selectivity," the relevant solutes for the latter mechanism already exist in a rigid, shape-constrained state. Therefore, any entropic cost associated with solute conformational changes during the retentive process are minimized. In addition, highly shape-selective alkyl-modified chromatography surfaces typically contain considerable alkyl chain conformation order; restricted alkyl motion is also correlated within this increase in conformational order [9]. These relatively rigid stationary phases likely possess well-defined adsorption sites that may not change as significantly as more flexible stationary phases on solute retention. To borrow from the protein literature, the mechanism of retention for the shape recognition on polymeric-type phases may be represented as more of a rigid-body docking process. This contrasts with a representation of "steric selectivity" as more of a flexible docking process to describe the interaction of more flexible solutes with less-rigid monomeric-type stationary phases.

5.2.4 Additional Molecular Interactions

Reversed-phase liquid chromatography shape-recognition processes are distinctly limited to describe the enhanced separation of geometric isomers or structurally related compounds that result primarily from the differences between molecular shapes rather than from additional interactions within the stationary-phase and/or silica support. For example, residual silanol activity of the base silica on nonendcapped polymeric C_{18} phases was found to enhance the separation of the polar carotenoids lutein and zeaxanthin [29]. In contrast, the separations of both the nonpolar carotenoid probes (α- and β-carotene and lycopene) and the SRM 869 column test mixture on endcapped and nonendcapped polymeric C_{18} phases exhibited no appreciable difference in retention. The nonpolar probes are subject to shape-selective interactions with the alkyl component of the stationary-phase (irrespective of endcapping), whereas the polar carotenoids containing hydroxyl moieties are subject to an additional level of retentive interactions via H-bonding with the surface silanols. Therefore, a direct comparison between the retention behavior of nonpolar and polar carotenoid solutes of similar shape and size that vary by the addition of polar substituents (e.g., all-*trans* β-carotene vs. all-*trans* β-cryptoxanthin) may not always be appropriate in the context of shape selectivity.

Several chromatographic materials have been recently employed that provide significant improvements in selectivity for a range of solutes, including some shape-constrained solutes. The ligands of these materials, however, contain embedded functional groups that may interact with the shape-constrained solutes through additional mechanisms (e.g., induced dipole interactions). Nogueira et al. [50] have synthesized a mixed-mode reversed-phase/weak anion exchange (RP/WAX) phase *in situ* on a silica monolith material (Chromolith). Shape selectivity was evaluated with the Tanaka test mixture and compared for three RP/WAX columns and a C_{18} monolith. An increase in shape selectivity ($\alpha_{TRI/o\text{-}TER}$) was observed for the RP/WAX columns that the authors attribute to electronic interactions between amide and thiol groups of the stationary-phase ligand and the π-electrons of the planar TRI solute. The origin of the shape selectivity is not clear as the surface coverage of ligands for these phases were not reported. In a recent study by Kayillo et al. [51], the selectivity factors for a range of PAHs on various reversed-phase chromatographic materials were described. Unfortunately, the isomer solutes examined (BaP, benzo[*e*]pyrene, and perylene) did not differ greatly in shape; therefore, a direct shape-selectivity comparison between these materials could not be assessed. However, the phenyl-type phases did provide an improved selectivity for the PAH homolog series compared with the monomeric C_{18} phases investigated and suggests a separation mechanism based on π–π interactions. Other branched polymer phases comprising embedded functional groups (i.e., ester) with longer-chain alkyl ligands (C_{18}) have been shown to exhibit shape selectivity [52–55]. It is unclear whether the selectivity of these materials originates from carbonyl–π interactions with the polar group or from the conformational order of the alkyl chains; details regarding these phases are presented in a later Section 5.3.4.

5.3 STATIONARY-PHASE CHEMISTRY

The synthesis of nonpolar stationary phases for use in RPLC has been the subject of numerous chromatographic and analytical science investigations over the past few decades. The majority of reversed-phase stationary phases are prepared through alkyl- modification of surface silanols of microparticulate silica, although alternative materials (e.g., titania, zirconia) have been successfully employed as viable substrates [56]. Chromatographic materials containing other types of surface coverings (e.g., aromatic, ionic ligands) can be prepared in a similar fashion. However, the focus within this review will be on the alkyl chain modification of silica-based support materials for use in shape-selective separations.

The molecular order of alkyl-modified stationary phases can be described as intermediate to that of a liquid or crystalline solid [9]. The motion of the stationary-phase is reduced for tethered alkyl chains compared with the relatively free translation of an analogous alkane in the liquid state. The covalent attachment of the alkyl chains to surface silanols imposes physical constraints on the relative position of the chains to each other as well as to the surface. In addition, the irregularity of the microparticulate silica surface together with the extent and type (i.e., free, geminal, vicinal) of silanols present will influence the distribution of alkyl chains across the surface and thus the relative spacing between the tethered alkyl chains of the stationary-phase. The influence of the substrate properties including composition, pore size, particle size, and surface area, together with the effects of the bonded-phase chemistry, surface coverage, and alkyl phase length on the formation of stationary phases is the subject of numerous book chapters [22,24,57] and earlier review articles [9,58].

5.3.1 MONOMERIC PHASES

Monomeric phases are prepared from the reaction of monofunctional reactive chloro- and alkoxysilanes (typically a dimethylchloroalkylsilane) with available silanols on the silica surface (Figure 5.5a). A commonly reported silanol concentration of amorphous silica material is $8.0 \pm 1.0 \, \mu mol/m^2$ and is present in a mixture of vicinal, isolated, and geminal forms [59]. A single siloxane bond is created from the silicon atom of the reactive silane and the oxygen of the silanol, followed by elimination of HCl or alcohol. Monomeric synthesis reactions tend to be limited by the steric bulk imposed by dimethyl substituents and result in a maximum covalent modification of approximately half of the available surface silanols ($\approx 4 \, \mu mol/m^2$). It is also anticipated that the steric bulk of the monomeric ligand's dimethyl substituents will result in a relatively even distribution of alkyl ligands across the silica surface.

Bonding density (or surface coverage) for a stationary-phase is calculated from the percent carbon loading, substrate surface area, and estimated relative molecular

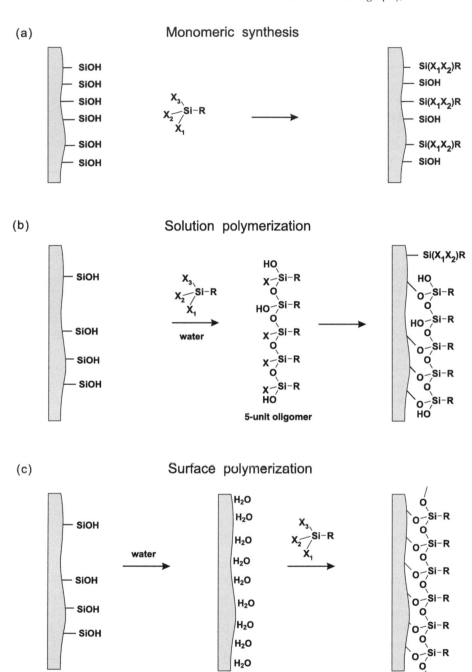

FIGURE 5.5   Synthesis schemes for chromatographic stationary phases: (a) monomeric synthesis where X represents reactive (e.g., chloro or alkoxy) or nonreactive (methyl) substituents, (b) solution polymerization, in which water is added to the slurry and (c) surface polymerization, in which water is added to the silica surface.

mass of the bonded species [60]. Bonding densities are reported in units of μmol/m^2 and groups/nm^2. The majority of commercial C_{18} RPLC columns are composed of monomeric-type stationary phases and typically have surface coverages ranging from 2.5 to 3.3 μmol/m^2. It is important to note that columns are commonly evaluated on the basis of percent carbon; however, the ligand bonding density on the silica surface is the predominate factor for determining the shape-selective capacity of a particular RPLC column.

Di- and trifunctional silanes are also used for the preparation of alkyl-modified silicas to be employed as chromatographic sorbents. Under anhydrous reaction conditions, sorbents prepared from these silanes have similar surface coverages and chromatographic behavior to monomeric stationary phases prepared with monofunctional silanes.

5.3.2 SOLUTION POLYMERIZATION

Polymeric stationary phases are assembled by covalent linkage of solution-polymerized alkylsilanes to a silica surface. In the presence of water, silane polymerization occurs within a reaction solution of di- or trifunctional silane in organic solvent (Figure 5.5b). Following polymerization, the remaining and accessible Cl-Si reactive bonds within the silane polymer are then anticipated to react with surface silanols of dispersed silica in the reaction solution to form the stable stationary-phase. This synthesis approach is described as "solution polymerization" and the stationary-phase is generally referred to as a polymeric phase [7]. The extent to which polymerization occurs is controlled by the amount of water added to the reaction mixture along with the addition of the reagents and application of reflux conditions [61].

A limited number of investigations have focused on the extent of silane polymerization that occurs during the solution polymerization process. Andrianov and Izmaylov [62] isolated the polycondensation products of alkyltrichlorosilanes in the presence of water and have observed the formation of cyclic oligomers six to eight units in size. An octyltrichlorosilane was found to form six-unit oligomers in the presence of water. Reaction of a trichlorosilane solution with the addition of water readily forms a white solid at room temperature, which dissolves under reflux conditions in toluene or xylene [61]. A C_{18} silane polymer formed from an n-octadecyltrichlorosilane solution in the presence of water exhibited a similar elution volume to a 5000 g/mol polystyrene standard by size exclusion chromatography [63]. Polymeric stationary phases have been characterized using ^{29}Si nuclear magnetic resonance (NMR) and, on average, are composed of five-unit oligomers that are bound to the surface through three to four siloxane bonds [64]. The oligomers are primarily linear with limited branching; however, circular oligomer structures are anticipated to yield comparable ^{29}Si NMR results. These initial investigations suggest that the silane polymers formed under typical solution polymerization conditions exhibit a low degree of polymerization with minimal three-dimensional (3D) branching. The oligomers immobilized on the silica surface are also anticipated to have alkyl chains present with relatively close interchain spacing. Ulman [65] has described the average distance between alkyl chains of crystalline alkylpolysiloxanes to be similar to the distance imposed by the Si—O—Si bond length.

In general, polymeric C_{18} stationary phases prepared via solution polymerization synthetic routes have surface coverages of approximately $5\,\mu mol/m^2$, which is equivalent to approximately three alkyl chains/nm^2. The surfaces of solution-polymerized stationary phases are anticipated to be heterogeneous overall, with regions of oligomer units with an average distance between alkyl chains similar to the bond lengths imposed by the polysiloxane network together with areas with individually bonded alkyl chains. Several preparative conditions can influence the range of bonding densities (typically 4.0–$5.5\,\mu mol/m^2$) that results for these materials. These include the amount of water in the reaction solution or adsorbed onto the silica surface, the stoichiometric quantity of silane available for reaction, and the relative size of substrate pores (i.e., smaller pores may preferentially exclude larger-sized polymers). Alternatively, Orendorff and Pemberton [66] have demonstrated the preparation of C_{18} polymeric phases with a range of surface coverages (≈ 3.5–$6\,\mu mol/m^2$) with the use of a long-chain alcohol as a "displaceable surface template." The surface coverage of octadecyltrimethoxysilane on a silica substrate was controlled through the addition of varying amounts of octanol; in contrast, the addition of hexanol was determined to decrease the bonding density of the phase. This recent study demonstrates a viable synthetic route for polymeric alkylsilane stationary phases that also provides an additional advantage of surface coverage control during the synthesis of the stationary-phase.

5.3.3 SURFACE POLYMERIZATION

The controlled reaction of a trifunctional silane on a humidified silica surface containing a thin film of absorbed water (Figure 5.5c) [67] can result in the formation of a dense alkyl monolayer or self-assembled monolayer (SAM) [64,68–72]. Parikh et al. [73] have described the mechanism of octadecylsilane SAM formation on a SiO_2 surface through spectroscopic, diffraction, and thermal gravimetric analyses. Polymerized octadecylsilane is first formed through hydrolysis of n-octadecyltrichlorosilane at the water interface to form alkyl silanols ($C_{18}Si(OH)_3$), which then rapidly organize via a Langmuir film formation. This two-dimensional (2D) assembly then undergoes Si—O—Si condensation and is immobilized on the surface through cross-linked siloxane bonds. As noted by Parikh et al. [73], the crucial feature of this mechanism is the rapid lateral diffusion of the surface-adsorbed species ($RSi(OH)_3$) on the thin water film present on the surface and is driven by the formation of an organized 2D structure to maximize chain packing. The authors also suggest that a random distribution of silica silanols may introduce irregularity between the cross-linking of the polymerized monolayer and the silica surface. The application and additional preparation methodologies for these silane-based monolayers within the field of surface science have been extensively reviewed [74–76].

The high surface coverage and near-crystalline structure of SAMs has limited their application as stationary phases in RPLC. Wirth et al. [70,71] have described the preparation of SAM-based chromatographic materials with the use of mixed length alkyltrifunctional silanes to increase the spacing between the alkyl chains while maintaining the 2D siloxane bond network. n-Octadecyltrichlorosilane was mixed with a shorter-chain alkylsilane (C_3) to generate SAMs on both flat silica and amorphous silica particle surfaces. Both materials exhibited an increase in acid and base hydrolytic stability. In addition, the chromatographically relevant material [71] yielded comparable separation performance to a conventional monomeric C_{18} stationary-phase.

For this mixed monolayer phase, the total surface coverage ($C_{18}+C_3$) was determined to be $8.0 \pm 0.5 \, \mu mol/m^2$, with $\approx 40\%$ attributable to the C_{18} chains and the remaining 60% attributable to the C_3 spacers. This bonding density coverage is consistent with the surface coverage limits imposed by SAMs ($\approx 8 \, \mu mol/m^2$). Sander and Wise [61] have also demonstrated that the bonding densities of "surface polymerized" stationary phases of homogenous alkyl length can be controlled by reaction of less than stoichiometric quantities of trichlorosilane reagents with humidified silica. Additional materials with lower surface coverages ($\approx 5 \, \mu mol/m^2$) have been prepared with dichlorosilane reagents [61]. The nature and extent of cross-linking of these monolayers with the silica surface may be significantly different when compared with phases prepared from trichlorosilanes.

5.3.4 BRANCHED POLYMER PHASES

Fujimoto et al. [77] synthesized a novel phase by coupling a dodecylamino-substituted β-cyclodextrin (β-CD) to 3-glycidoxypropyl-derivatized silica gel. The surface coverage of this phase was reported as $0.37 \, \mu mol/m^2$, which amounts to a surface coverage of $2.6 \, \mu mol/m^2$ for C_{12} chains (seven chains per β-CD). An increase in shape selectivity was observed when compared with a conventional C_{18} monomeric phase as determined by selectivity differences between p/m-terphenyl, p/o-terphenyl, and coronene/phenanthro[3,4-c]phenanthrene solute pairs and was attributed to the localized high ligand density as constrained by the β-CD platform structure.

Chowdhury et al. [52–55] developed a comb-shaped polymer stationary-phase intended for application toward shape-recognition separations. Comb-shaped polymer structures were prepared by telomerization of poly(octadecyl) acrylate (ODA_n) via reaction with 3-mercaptopropyltrimethoxysilane. The reactive terminal group of the polymer was then immobilized onto porous silica gel to generate the stationary-phase (Figure 5.6a). Chromatographic separations on this material were demonstrated to be highly temperature dependent and were attributed to a phase transition behavior [54]. For example, sharp changes in phase selectivity ($\alpha_{TBN/BaP}$) were observed over a modest range of column operating temperatures (5–60°C). In support of chromatographic observations, a clear crystalline-to-isotropic phase transition was observed with differential scanning calorimetry (DSC) for these phases (Figure 5.6b). Further studies have examined the effect of solvent on the degree of polymerization and together with the influence of the transition temperature on the shape-recognition characteristics of the phase [78]. In general, comb-shaped polymers with higher degrees of polymerization exhibited better shape recognition. Similar phases were then prepared via multianchoring points to the silica surface (co-ODA_n) [79]. A sharp transition

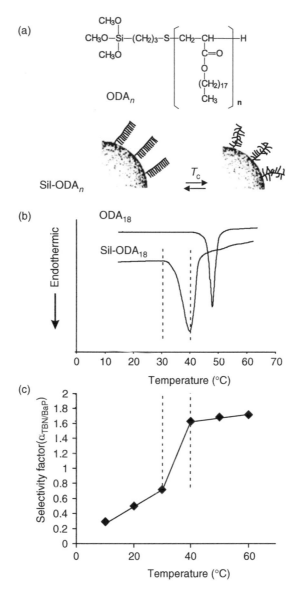

FIGURE 5.6 (a) Schematic illustration and (b) DSC thermograms of silica-supported, comb-shaped polymer (Si-ODA$_{18}$). DSC was measured in methanol–water (80:20) dispersion in the heating process. (Reproduced from Chowdhury, M.A.J., et al., *Chromatographia*, 52, 45, 2000. With permission.). (c) Selectivity factor $\alpha_{TBN/BaP}$ plotted as a function of temperature for the Si-ODA$_{18}$ stationary-phase. (Adapted from Sander, L.C. and Wise, S.A., *J. Sep. Sci.*, 24, 910, 2001. With permission.). Changes in selectivity are greatest for the temperature interval of phase transition.

was observed at the critical temperature (T_c) of the polymer; chromatographic shape selectivity was only observed at temperatures below the T_c [79].

Rahman et al. [80,81] have created novel stationary phases composed of polymerized disteryl glutamide (DSG) and dibutyl glutamide (DBG) oligomers that are tethered to a silica substrate via an aminopropyl silane linkage. These chromatographic materials are described as lipid membrane analogs and are composed of a two-unit branching structure, in comparison to the extensive branching of the ODA_n phases as illustrated in Figure 5.6. The chromatographic behavior of these two materials (DSG, DBG) were compared with both the related ODA_n stationary phases and the more conventional ODS stationary phases based on the Tanaka test and SRM 869a. The DSG phases illustrated improved shape selectivity when compared with ODS, ODA_n, and the shorter chain length (C_4) DBG phase, but did not demonstrate any clear phase transition behavior typical of the ODA_n material. It should be noted that all of the phases in these studies have been carefully synthesized and characterized through NMR spectroscopy and chromatographic methods. The phases demonstrated similar chromatographic trends as traditional alkyl phases in that increased phase length, regional density, and alkyl chain ordering lead to increased shape recognition.

Rahman et al. [80,81] offer a possible secondary mechanism for the origin of shape selectivity in these phases. NMR measurements have indicated that the ODA_n and co-ODA_n alkyl ligands are highly ordered below the transition temperature (T_c) [82]; the authors hypothesize that the C_{18} chains promote a molecular alignment of the ester moieties at the base of the ligands that may provide a site for carbonyl–π interactions between the phase and planar aromatic hydrocarbon solutes. It has been suggested that the carbonyl groups of the DSG phases remain ordered even under conditions that result in significant alkyl chain disorder (i.e., higher temperature); a clear retention mechanism under such conditions is not evident. It is also difficult to rationalize the origin of carbonyl–π interactions within co-ODA_n and DSG phases in which solute access to the electron donor–acceptor sites may be precluded by the bulk of the C_{18} alkyl chains. However, these phases do appear to have some common features with electron acceptor phases that have been specifically designed for the separation of aromatic hydrocarbon solutes in both normal-phase [32] and reversed-phase [83] chromatography. Sander et al. [32] demonstrated the preferential retention of planar PAHs over nonplanar PAHs on phases composed of pyrene, tetrachlorophthalimidopropyl, or tetranitrofluoreniminopropyl ligands; the enhanced selectivity was attributed to the accessibility of the electron donor/acceptor capability of the phase's conjugated π system under normal-phase conditions (i.e., dichloromethane solvent). An analogous charge-transfer phase composed of naphthalimide ligands operated under reversed-phase conditions also illustrated preferential retention of planar PAHs [83]. The authors did note a difference in reduction in retention when changing the organic modifier from methanol to acetonitrile, the latter of which may reduce π–π interactions. These charge-transfer phases represent alternative chromatographic materials that result in shape-selective separations but likely involve stronger molecular interactions (i.e., carbonyl–π or π–π that are dependent not only on the shape of solute but also on the solvent environment).

5.3.5 Other Alkyl-Modified Surfaces

Alternative methods in the preparation of alkyl stationary phases through the use of hydride-modified silica surfaces have been reported. Pesek and coworker [84] have prepared a silica intermediate containing silicon hydride surface species via reaction with thionyl chloride followed by reduction with $LiAlH_4$. The stationary-phase is then prepared by a hydrosilation reaction, in which a catalytic addition of a SiH group to an alkyl ligand possessing a terminal α-olefin [85–87] or alkyne [88] moiety produces a direct Si—C bond between the silica surface and the alkyl ligand. The resultant chromatographic materials have comparable selectivity to conventionally prepared silanized phases, with an additional advantage of increased hydrolytic and thermal stability. Sunseri et al. [89,90] have also used a similar approach for the preparation of alkyl-modified surfaces with near-quantitative methylation of surface silanol groups with little degradation of the silica surface. The authors state that a future goal is to increase the phase length on these materials for use in chromatographic separations.

Kobayashi et al. [91] recently synthesized a phase that is comprised of thiol-modified gold-coated polystyrene particles. An increase in the selectivity of the anthracene–phenanthrene pair was observed on the C_{18}-Au particle when compared with a traditional monomeric C_{18} phase with a surface coverage of $\approx 3.0\,\mu mol/m^2$. This isomer pair is not the ideal choice for the determination of shape selectivity; however, this synthetic technique should in general lead to dense, ordered phases that are anticipated to yield relatively highly shape-selective chromatographic separations.

Liu et al. [92] synthesized a low density ($2.4\,\mu mol/m^2$) polar embedded stationary-phase that consists of a distal C_{16} alkyl segment bound to a sulfonamide-ether moiety tethered via a silane linkage to the silica surface. The phase was evaluated with both the Tanaka and SRM 869a column selectivity tests and demonstrated an intermediate level of shape selectivity ($\alpha_{TRI/o\text{-}TER} = 2.3$ and $\alpha_{TBN/BaP} = 1.6$). Overall, an increase in shape recognition was observed when compared with several other embedded polar phases and two conventionally synthesized monomeric C_{18} phases with relatively low shape selectivity ($\alpha_{TRI/o\text{-}TER} < 2.1$). The authors attribute the increase in shape selectivity to the increased length of the ligand (approximately the length of a C_{25} phase); this also suggests that the longer (C_{16}) distal segment may be participating in the shape-recognition process.

The synthesis of long-alkyl chain stationary phases has been accomplished via the immobilization of poly(ethylene coacrylic acid) copolymers onto glycidoxypropyl- or aminopropyl-modified silica surfaces [93] (Figure 5.7). Several copolymers were used and the 5% acrylic acid was found to produce a more efficient, shape-selective phase than the 10% or 15% acrylic acid poly(ethylene coacrylic acid). Enhanced separation of carotenoid and PAH isomers has been demonstrated [28] for these chromatographic sorbents, comparable to that observed with extended-length (C_{30}) alkyl polymeric phases. On the basis of relative molecular mass and composition of the 5% acrylic acid copolymer reagent, Wegmann et al. [93] estimated the chain length of such materials to be on the order of a C_{100} phase with approximately one acid group per polymer molecule.

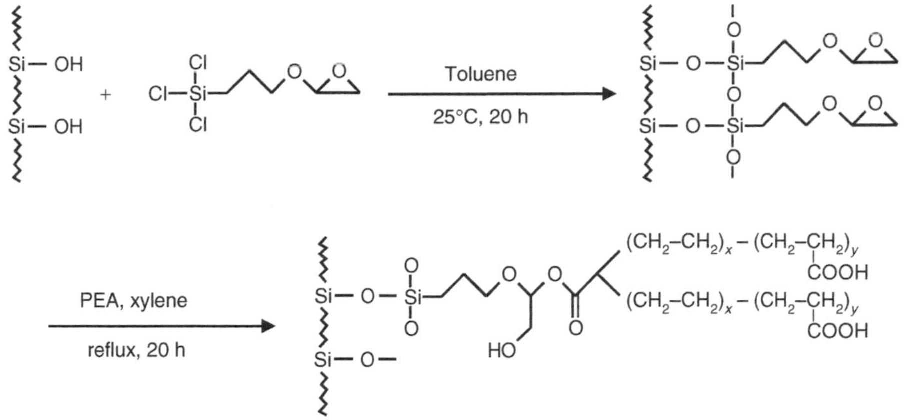

FIGURE 5.7 Reaction schemes for the synthesis of a poly(ethylene *co*-acrylic acid)-glycidoxypropyl-silica (PEAGlyP) stationary-phase. (Reproduced from Wegmann, J., et al., *Anal. Chem.*, 73, 1814, 2001. With permission.)

5.4 FACTORS INFLUENCING SHAPE SELECTIVITY

The selectivity differences between monomeric phases and those prepared via polymeric (solution or surface) chemistries are readily apparent in the RPLC separation of PAH isomers (Figures 5.1 and 5.8). This enhancement in selectivity provided by polymeric chromatographic materials is well known to analysts involved in the measurement of PAHs and related shape-constrained solutes. Several stationary-phase and substrate characteristics (alkyl chain length, surface coverage, pore size) together with experimental conditions (temperature, mobile-phase composition) can influence the shape-selectivity performance of these chromatographic materials. In general, both experimental conditions and phase compositional factors that promote alkyl phase order tend to promote shape selectivity [9]. For applications involving the separation of solutes possessing more flexible molecular structures, the differences in retentive behavior between monomeric and polymeric phases tend to be less dramatic. Likewise, the selectivity of these separations is not as sensitive to factors such as bonding density and alkyl chain length that are deemed important in shape-selective separations.

5.4.1 BONDING DENSITY

Shape-selective separations of geometric isomers are strongly correlated with the bonding density (or surface coverage) of the alkyl ligands of the stationary-phase [47,61,94,95]. A plot of $\alpha_{TBN/BaP}$ as a function of surface coverage for both monomeric and polymeric C_{18} phases clearly illustrates this trend for the PAH isomers (Figure 5.9). Initially, the selectivity difference was thought to originate from differences in phase types. However, the extensive relationship between selectivity and surface coverage (irrespective of preparation chemistry) in Figure 5.9 suggests that

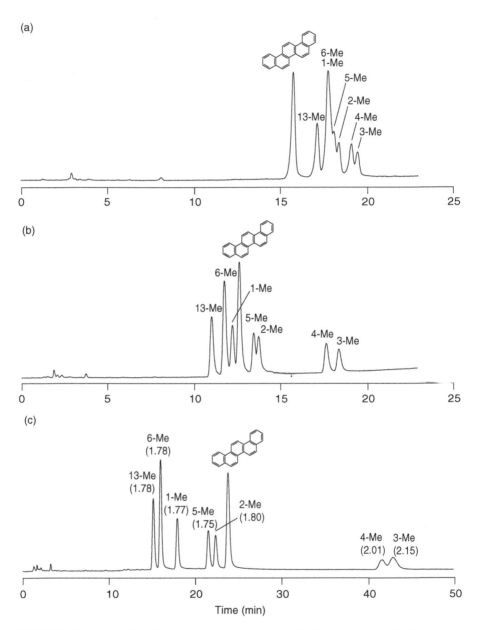

FIGURE 5.8 Separation of picene and methyl-substituted picene isomers with (a) a mono-meric C_{18} column, (b) an "intermediate" C_{18} column, and (c) a polymeric C_{18} column; L/B values given in parentheses. (Adapted from Wise, S.A. and Sander, L.C., in Jinno, K. (Ed.), *Chromatographic Separations Based on Molecular Recognition*, Wiley-VCH, Inc., New York, 1997, p. 1. With permission.)

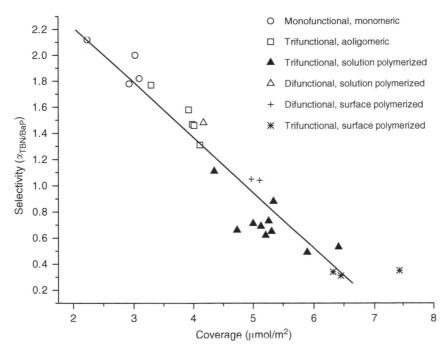

FIGURE 5.9   Selectivity ($\alpha_{TBN/BaP}$) plotted as a function of bonding density for various monomeric and polymeric C_{18} columns. (Adapted from Sander, L.C., et al., *Anal. Chem.*, 71, 4821, 1999. With permission.)

shape recognition is dependent on alkyl chain spacing that may stem from the ligand bonding density. Polymeric syntheses routinely result in phases with higher bonding densities, which are typically more shape selective. Bonding densities for SAMs can approach the theoretical limit for close-packed alkyl chains for surface polymerization reactions ($\approx 8\ \mu mol/m^2$). Phases with such high surface coverages can exhibit very high shape selectivity but also demonstrate poor efficiency and asymmetric peak shape. SAM-based chromatographic materials with the use of mixed-length alkyltrifunctional silanes have demonstrated shape selectivity comparable to monomeric C_{18} phases but with superior hydrolytic stability [70,71,96]. The limited shape-recognition capacity of these materials is attributable to the low "effective" bonding density of the C_{18} chains ($\approx 3\ \mu mol/m^2$) within the complete alkyl monolayer coverage on the silica surface.

The influence of bonding density on the partitioning and retention of nonpolar solutes (e.g., naphthalene) on monomeric-type C_{18} phases was first determined by Sentell and Dorsey [97]; a maximum thermodynamic partition coefficient (K) was observed at a surface coverage of $3.1\ \mu mol/m^2$, with a decrease in retention at higher bonding densities. However, a slight increase in shape selectivity was observed for a high-density ($4.07\ \mu mol/m^2$) monomeric phase when tested with SRM 869a [94]. A comparison between the absolute retention of various shape-constrained solutes over a range of bonding densities (from both monomeric and polymeric C_{18} stationary

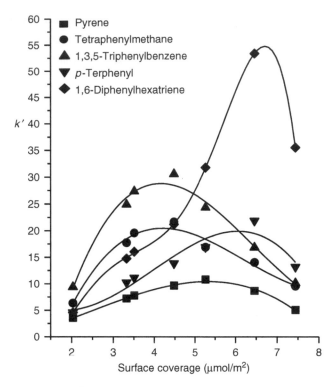

FIGURE 5.10   Retention k' plotted as a function of surface coverage for various C_{18} columns and selectivity probes. (Adapted from Wise, S.A. and Sander, L.C., in Jinno, K. (Ed.), *Chromatographic Separations Based on Molecular Recognition*, Wiley-VCH, Inc., New York, 1997, p. 1. With permission.)

phases) can provide an alternative view into shape-selective retention mechanisms. A plot of retention k' for several solutes with respect to bonding densities is illustrated in Figure 5.10. The maximum retention occurs at different surface coverages for different solutes. A maximum retention is observed at lower bonding densities (near $4\,\mu mol/m^2$) for the bulkier solutes, such as *o*-TER, triphenylbenzene, and tetraphenylmethane. In contrast, the more narrow solutes, such as *p*-terphenyl and 1,6-diphenylhexatriene, exhibit a maximum retention at higher surface coverages ($\approx 7\,\mu mol/m^2$). These results suggest that the available interchain distances within the alkyl phase become significantly reduced for high-bonding density materials and thus limit the retention for the bulky solutes. Conversely, the more linear solutes are able to fit and thus be retained within the narrower interchain spaces that exist on phases with relatively high surface coverages.

5.4.2   CHAIN LENGTH

The influence of alkyl chain length on shape selectivity has been the topic of numerous investigations [98–105]. Through the use of SRM 869a test mixture, the shape

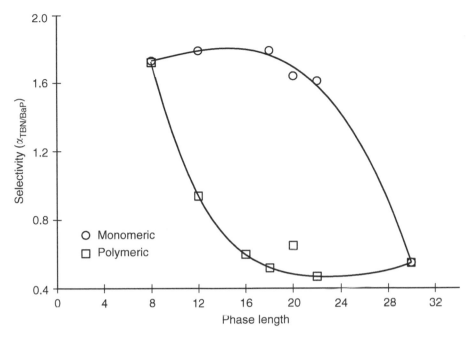

FIGURE 5.11 Variations in column shape selectivity ($\alpha_{TBN/BaP}$) as a function of alkyl-phase length for monomeric and polymeric phases. (Adapted from Sander, L.C. and Wise, S.A., *Anal. Chem.*, 59, 2309, 1987. With permission.)

selectivity of phases with alkyl ligands ranging from C_8 to C_{30} have been documented [104]. It was observed for monomeric phases that PAH isomer shape recognition did not change appreciably among phases of the C_8, C_{12}, and C_{18} length but selectivity ($\alpha_{TBN/BaP}$) increased dramatically when the alkyl chain was increased to 30 carbons in length (Figure 5.11). For the polymeric phases, the influence of chain length is not as significant over the carbon numbers of 16–30, but values of $\alpha_{TBN/BaP}$ increase sharply for shorter alkyl chain length (C_8). Overall, C_8 phases of both monomeric and polymeric types exhibit little shape discrimination, whereas C_{30} phases of both types are highly shape selective. For the intermediate alkyl chain lengths (e.g., C_{18}), the polymeric phase synthesis with its generally high surface coverage exhibits greater shape recognition compared with the lower-density monomeric phase.

Phases of extended length (C_{30}) have been utilized for the separation of larger-size constrained solutes, such as carotenoids and steroids [27–29,93,106,107]. A practical limit of alkyl chain length of C_{34} to C_{36} is imposed by the commonly employed silanization chemistry techniques [106]. Immobilization of longer alkyl stationary phases has been achieved through the use of poly(ethylene-*co*-acrylic acid) materials for use in carotenoid separations [27,28,93]. Rimmer et al. [28] have recently compared the selectivity of both alkyl and poly(ethylene-*co*-acrylic acid) stationary phases on the basis of separations of carotenoids in food matrices (Figure 5.12), in addition to mixtures of tocopherols and PAHs.

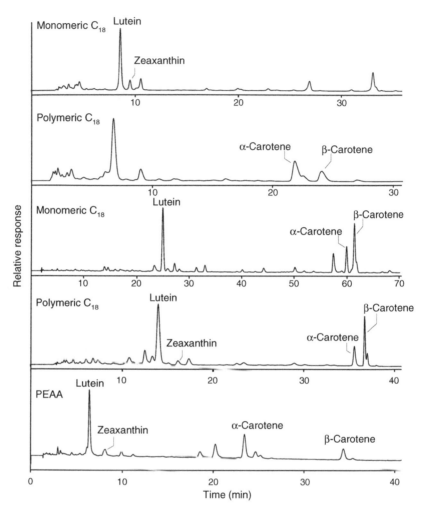

FIGURE 5.12 Gradient separation of carotenoids in SRM 2385 Slurried Spinach. (Adapted from Rimmer, C.A., et al., *Anal. Bioanal. Chem.*, 382, 698, 2005.) Solvent and gradient conditions described in Reference 28.

5.4.3 PORE SIZE

Sander et al. [63] investigated the effect of microparticulate silica pore size on the properties of solution-polymerized C_{18} stationary phases and observed both an increase in bonding density and shape recognition for wider pore (>120 Å) silica. A size-exclusion mechanism was proposed, in which the reaction of the silane polymer on the surface is enhanced for wide pores and reduced for narrow pores. Polymeric C_{18} phases prepared on substrates with narrow pores exhibited monomeric-like chromatographic properties. This effect may be the result of an increase in competitive surface linkage with the less sterically hindered monomers that coexist with the bulkier oligomers that have polymerized in the reaction solution (Figure 5.13).

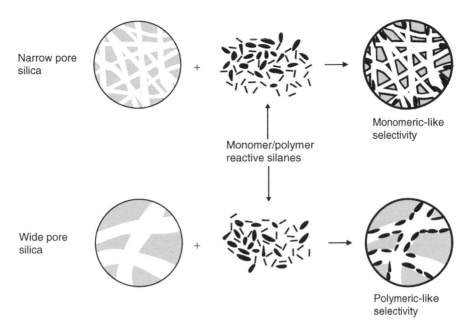

Narrow pore silica

Wide pore silica

+

Monomer/polymer reactive silanes

Monomeric-like selectivity

Polymeric-like selectivity

FIGURE 5.13 Proposed size exclusion mechanism for the reduced reaction of the silane polymer on narrow pores and an enhanced reaction for wide pores.

5.4.4 TEMPERATURE

Column temperature has been demonstrated to have a strong effect on shape-selective separations [15,108]. Increased shape recognition as measured by the selectivity factor ($\alpha_{TBN/BaP}$) is observed at reduced temperature for both monomeric and polymeric C_{18} phases (Figure 5.14). At any given temperature, however, polymeric phases exhibit greater shape selectivity ($\alpha_{TBN/BaP}$) than that exhibited by monomeric phases. The control of column temperature provides an opportunity to tune selectivity and thus attain comparable separations on quite different stationary phases. It is also important to highlight that the change in shape selectivity with respect to temperature is relatively smooth (Figure 5.14) and therefore suggests that a phase transition does not readily occur for these materials. This is not the case for the comb-shaped polymer stationary phases prepared by Chowdhury et al. [53] for which sharp changes in phase selectivity ($\alpha_{TBN/BaP}$) were observed over a temperature range of 5–60°C. In addition, a clear crystalline-to-isotropic phase transition was observed with DSC for these stationary-phase materials (Figure 5.6b).

5.4.5 MOBILE-PHASE COMPOSITION

A very minor effect on shape selectivity has been observed for changes in mobile phase composition [109]. Shape selectivity increased slightly with an increase in percent organic modifier for water–organic mobile phase systems and increased in order of methanol ≈ acetonitrile < ethanol. Changes in the shape selectivity factor

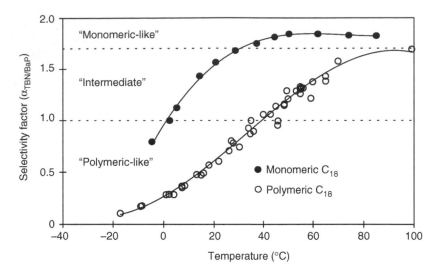

FIGURE 5.14 Column shape selectivity ($\alpha_{TBN/BaP}$) plotted as a function of temperature for monomeric and polymeric C_{18} columns. (Adapted from Sander, L.C. and Wise, S.A., *Anal. Chem.*, 61, 1749, 1989. With permission.)

($\alpha_{TBN/BaP}$) were relatively small in the context of other variables and therefore the effect is considered negligible with respect to method development. Larger differences in shape selectivity resulted from the use of stronger solvents (e.g., acetone and ethyl-*t*-butyl ether) [20]. The use of strong solvents has been recently demonstrated by Saito et al. [110] to benefit the separation of highly retained shape-constrained species, such as fullerene and high relative molecular mass PAH mixtures.

5.5 PHASE CHARACTERIZATION

The most common approach to chromatographic stationary-phase characterization is in terms of bulk-phase properties, such as percent carbon loading onto the silica substrate. This property together with the surface area of the substrate and the molecular characteristics of the bonded silane can be used to calculate the bonding density (N) of the chromatographic sorbent [60]:

$$N(\mu mol/m^2) = \frac{10^6 P_c}{1200 n_c - P_c(M-1)} \frac{1}{S},\qquad(5.1)$$

where M is the molecular mass, n_c is the number of carbon atoms in the bonded species, P_c is the percent carbon of the stationary-phase, and S is the specific surface area of the silica substrate. At best, this information provides an assessment of only the bulk stationary-phase. A more direct description of the stationary-phase can be obtained by the application of spectroscopic techniques for the molecular-level analyses of inter- and intrachain conformations and their relative motions. Several techniques such as Fourier tansform infrarerd (FTIR) and Raman spectroscopy,

TABLE 5.1
An Overview of Spectroscopy (and Other Experimental) Techniques Used to Determine Stationary-Phase Conformational Order

Technique	Phase Characterization Result
FTIR	Chain conformation (*gauche* vs. *trans*)
SFG	Degree of conformational order
Raman	Chain conformation (*gauche* vs. *trans*); interchain coupling
AFM	Surface topography
^{1}H NMR	Chain dynamics
^{13}C NMR	Chain dynamics; regional domain classification; chain conformation (*gauche* vs. *trans*)
^{29}Si NMR	Silane bonding chemistry
Fluorescence	Polarity of chain environment; chain conformation (*indirect*)
EPR	Chain conformation (*indirect*)
SANS;	
X-ray reflectivity; ellipsometry	Thickness of chains
DSC	Phase transition
AED	Surface adsorption sites

nuclear magnetic resonance (NMR) spectroscopy, fluorescence spectroscopy, differential scanning calorimetry (DSC), ellipsometry, and X-ray and neutron scattering have been applied to the characterization of alkyl-based stationary phases. An overview of commonly applied techniques with a summary of alkyl conformational order indicators is provided in Table 5.1.

A complete review of spectroscopic methods applied to the analysis of alkyl-modified surfaces with a comprehensive list of spectroscopic indicators of alkyl chain conformational order is provided elsewhere [9]; this review will focus on the application of spectroscopic and other relevant experimental techniques for the characterization of shape-selective chromatographic materials. On the whole, it has been observed experimentally that any increase in alkyl stationary-phase conformational order promotes an increase in selectivity for shape-constrained solutes in RPLC separations [9]. As a complement to the wealth of spectroscopic and chromatographic data, the use of molecular simulation techniques to visualize and characterize alkyl-modified surfaces may also provide new insights into molecular-level features that control shape selectivity. A review of progress in the field of chromatographic material simulations will also be discussed.

5.5.1 FOURIER TRANSFORM INFRARED SPECTROSCOPY

One of the initial spectroscopic methods applied to stationary-phase characterization was Fourier transform infrared spectroscopy (FTIR). This originated from several important studies of phase conformational order in crystalline *n*-alkanes conducted in the late 1960s and early 1980s by Snyder, Maroncelli, and coworkers [111–114]. In this work, assignments of C—H bond wagging modes were associated with chain

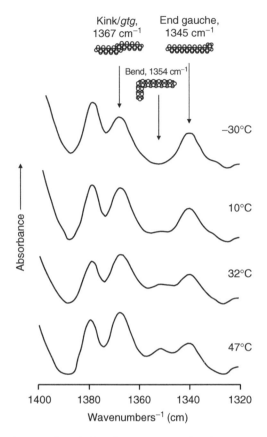

FIGURE 5.15 Fourier transform infrared spectroscopy absorbance spectra at different temperatures for a C_{22} monomeric stationary-phase. (Reproduced from Sander, L.C., et al., *Anal. Chem.*, 55, 1068, 1983. With permission.)

conformational defects. Sander et al. [115] describe the first application of FTIR to study a series of C_1 to C_{22} monomeric alkyl stationary phases at various temperatures. A reduction in signal intensity at 1367 and 1354 cm^{-1} (corresponding to chain-kink and double-*gauche* defects, respectively) with a decrease in temperature was observed for a C_{22} phase (Figure 5.15). No evidence of a melting or freezing phase transition was observed. Jinno et al. [116] describe a similar reduction in these conformation defects in the FTIR spectra for a large set of monomeric and polymeric C_{18} phases. Monomeric phases exhibit an increase in double-*gauche* effects, in contrast to the more ordered phases prepared via polymeric functionalities.

More recently, Singh et al. [117] has applied FTIR spectroscopy to determine the influence of bonding density, alkyl chain length, and temperature on overall stationary-phase order. In a broad study that included a range of phases (C_8 to C_{30}), the kink/*gauche-trans-gauche (gtg)* and double-*gauche* conformers were reduced with increasing chain length and the effect was especially distinctive for the phases above 18 carbons in length. An increase in conformational order with a reduction in temperature was observed for various stationary phases; spectral changes as a function of temperature were consistent with the original investigations of Sander et al. [115]. Interestingly, the end-*gauche* conformers did not change appreciably with either chain length or temperature. Singh et al. [117] identified the methylene symmetric (2850 cm^{-1}) and antisymmetric (2917 cm^{-1}) stretching model signals of *n*-alkyl chains tethered to silica gel in the dry state as indicators of chain conformational structure. The FTIR signals were compared with assignments previously determined for *trans* conformational states in crystalline alkanes. The frequency shift of the band maxima in the stretching regions was correlated with changes in conformational order as a function of temperature; increased alkyl order was observed for an increase in surface coverage and chain length, and for a decrease in temperature. For a C$_{18}$-modified silica at room temperature, it was estimated that ≈ 3.6 *gauche* defects per chain were present, in contrast to ≈ 2.5 *gauche* defects for a C$_{30}$ chromatographic material.

In addition, Singh et al. [117] prepared selectively deuterated C$_{18}$-modified silicas at surface coverages of ≈ 4 µmol/m^2 to determine the percentage of *gauche* conformers for positions C-4, C-6, and C-12 along the alkyl chain. An analysis of the CD$_2$ rocking vibrational bands in the interval $600–680$ cm^{-1} is used to assign *trans–trans* and *trans–gauche* conformations of a three-carbon segment of the chain. At room temperature, the percentage of *gauche* conformers was 40%, 20%, and 25%, respectively, indicating significantly higher disorder for the carbons proximal to the silica surface. This difference was more prominent at reduced temperature (approximately 30%, 9%, and 8% for C-4, C-6, and C-12 at $-18°C$). For comparison, a shorter-chain C$_9$-modified silica (≈ 5 µmol/m^2) with deuteration at position C-4 resulted in $>65\%$ *gauche* conformers, and the disorder did not increase significantly with an increase in temperature (up to $70°C$).

FTIR has also been widely applied for the conformational analysis of SAMs. The influence of the bonding substrate on the alkyl conformational order for C$_{30}$ SAMs prepared on silica, titania, and zirconia was investigated by Srinivasan et al. [56]. The highest degree of order was observed for C$_{30}$ phases prepared on titania, followed by zirconia and two phases prepared on silica supports. The CH$_2$ wagging bands were used to determine relative amounts of kink/*gtg*, double- and end-*gauche* conformations along the chain. Herein, the observed temperature influence on alkyl chain disorder stemmed primarily from changes in the number of kink/*gtg* chain conformations. In a further study, Srinivasan et al. [118] demonstrated a comparable trend for C$_{18}$ phases prepared on zirconia: a higher bonding density C$_{18}$ phase (4.0 µmol/m^2) illustrated more order than a 2.5 µmol/m^2 C$_{18}$ phase and both of these zirconia phases exhibited more alkyl chain order than a C$_{18}$ phase on silica at a higher surface coverage (4.2 µmol/m^2). It is important to note that the FTIR analysis of these materials was performed on dry pellets as the spectral quality was found to be poor in solvated pellets.

An FTIR comparison between C$_{18}$ and C$_{30}$ sorbents prepared by both surface- and solution-polymerized synthetic routes over a range of temperatures was recently described [119]. Little difference in conformational order was observed for C$_{30}$ sorbents prepared by both surface and solution-polymerized synthetic routes, as indicated by kink/*gtg*, double- and end-*gauche* conformations. Differences in conformational order in C_{18} sorbents were attributed to surface coverage differences rather than synthetic routes, with higher order observed for an increase in bonding density. Moreover, a consistent two-fold increase in the percentage of *gauche* conformers was observed for the extended-length C_{30} phase in contrast to the C_{18} phase with a comparable surface coverage over the temperature range investigated. In a similar study, Srinivasan et al. [120] investigated a series of C_{18} stationary phases with a range of bonding densities (2.0–8.2 μmol/m^2). As anticipated, the alkyl chain order increased with decreasing temperature, primarily from kink/*gtg* conformational changes. An interesting exception was the increase in order observed for a low-density (2.0 μmol/m^2) C_{18} phase prepared with a stoichometrically limited amount of silane in the presence of water compared with the more disordered, yet more densely bonded (2.7 μmol/m^2), C_{18} phase synthesized under anhydrous conditions. The authors suggest that the differences in order may be attributed to the differing synthetic routes, in which the lower density phase may be polymerized in island formations, leading to more interaction between the alkyl ligands, whereas the 2.7 μmol/m^2 may be more homogenously distributed across the surface.

Henry et al. [121] investigated potential solvent-induced conformational changes in monomeric, polymeric, and SAM stationary phases with the application of sum-frequency generation (SFG) spectroscopy. This technique is related to infrared spectroscopy in that it utilizes polarized infrared radiation over the interval 2600–3200 cm^{-1} in combination with polarized radiation at 532 nm to elucidate molecular orientations at interfaces. The ratio of the CH_3 symmetric stretch peak at $\approx$2872 cm^{-1} to the CH_2 symmetric stretch peak at $\approx$2840 cm^{-1} (I_{CH3}/I_{CH2}) is an indication of alkyl conformation order, with larger values indicative of increased order. A related measure of the difference of CH_3 symmetric stretching between *s*- and *p*-polarized infrared (I_{ssp}/I_{sps}) indicates the orientation of CH_3 groups at the surface. The conformation of alkyl chains of mixed C_{18} and C_1 monolayers did not change appreciably under different solvent conditions, including 100% D_2O. It was therefore concluded that phase collapse under aqueous conditions did not occur. The influence of various solvents on the relative order of the C_{18} ligands over a range of temperatures has also been recently investigated by FTIR spectroscopy [122]. Interestingly, a comparable increase in alkyl chain conformational disorder was observed for acetone, cyclohexane, and dichloromethane relative to the dry C_{18} phase with an increase in temperature. This suggests that solvent polarity alone cannot explain the increase in the alkyl chain disorder upon solvation. A significant increase in alkyl chain disorder was also observed for a series of perdeuterated solvents (including *n*-hexane-d$_{14}$). This is attributed to a deuterium isotope effect for which solvents with a C—D bond do not interact as strongly with the alkyl stationary-phase as do solvents with C—H bonds. In contrast, *n*-hexane generally promotes order with the alkyl chains (likely via hydrophobic interactions) and is less sensitive to temperature effects.

5.5.2 RAMAN SPECTROSCOPY

Raman spectroscopy is considered a complementary technique to FTIR for the conformational analysis of chromatographic materials. Raman spectroscopy relies on scattering of radiation by the sample, rather than an absorption process, as does FTIR. Many of the interferences common in FTIR studies (e.g., background water absorption) are not present in Raman spectroscopy. On the other hand, Raman can be subjected to fluorescence interferences from impurities in silica and can have a weaker response for alkane materials. Spectral indicators of alkyl chain conformation disorder, such as the intensity ratio of the antisymmetric to symmetric methylene bands $I[\nu_a(CH_2)]/I[\nu_s(CH_2)]$, have been identified through investigations of lipids [123] and bulk hydrocarbons [124]. A more complete description of intensity ratios and additional spectral bands in the $\nu(C-H)$, $\nu(C-C)$, and $\delta(C-H)$ spectral regions used for the quantitative determination of alkyl chain conformational order are summarized by Sander et al. [9].

The studies of Pemberton, Dorsey and coworkers [124–137] represent the majority of Raman spectroscopy applications for the characterization of alkyl stationary phases. Doyle and coworkers [130–132] have described an instrumental configuration for the on-column characterization of alkyl chromatographic materials under typical flow-rate and pressure conditions. The influence of surface coverage, temperature, and solvent conditions on alkyl chain conformational order for a range of chromatographic sorbents was measured through the intensity ratio $I[\nu_s(CH_3)]/I[\nu_s(CH_2)]$. Values obtained from chromatographic sorbents indicated an intermediate level of order, and little difference in conformational order was observed under different solvent environments for relatively low-density monomeric phases (2.34 and 3.52 $\mu mol/m^2$) [132]. Larger differences were observed for changes in column temperature, with an increase in order at reduced temperatures [131]. Similarly, Ho et al. [129,133,134] describe the conformational order of relatively low-density (1.6–1.9 $\mu mol/m^2$) monomeric and polymeric columns over changes in temperature and solvent conditions through an analysis of several Raman spectral indicators. An increase in alkyl chain order resulted at lower temperatures, and a residual number of *gauche* conformers were still detected upon cooling with liquid N_2 [133]. For both the monomeric and polymeric columns, the presence of polar solvents typically used in RPLC had little influence on alkyl chain conformation order in contrast to air [129].

An investigation of higher surface coverage (3.09–6.45 $\mu mol/m^2$) polymeric C_{18} stationary phases by Raman spectroscopy was conducted by Ducey et al. [135,136]. The influence of surface coverage and temperature on alkyl chain conformational order was assessed using the intensity ratio of the antisymmetric and symmetric $\nu(CH_2)$ modes; lower values of $I[\nu_a(CH_2)]/I[\nu_s(CH_2)]$ are indicative of greater disorder (Figure 5.16). Overall, alkyl chain disorder increased with decreasing bonding density and with increasing temperature. A minimum value of $I[\nu_a(CH_2)]/I[\nu_s(CH_2)]$ was observed for the entire range of surface coverages and temperatures that was greater than what is typically observed for bulk liquid alkanes. This suggests that stationary phases exist in a more ordered molecular state than the analogous liquids. A slight decrease in conformational order was observed in the presence of polar solvents (methanol, acetone, acetonitrile, and water) [136] when compared with spectral data obtained from dry stationary-phase samples. In addition, the authors concluded that alkyl chain order is largely dependent on surface coverage rather than stationary-phase preparation (surface vs. solution polymerization).

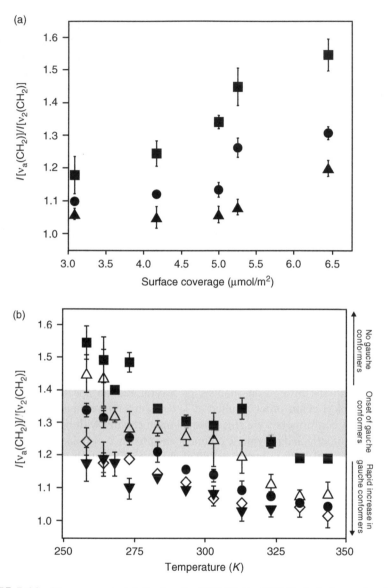

FIGURE 5.16 Raman spectral indicator $I[\nu_a(CH_2)]/ I[\nu_s(CH_2)]$ plotted as a function of temperature and surface coverage for various monomeric and polymeric C_{18} columns. (a) Surface coverage (■) 258 K, (●) 293 K, and (▲) 333 K; (b) Temperature (■) 6.4 μmol/m^2, (△) 5.2 μmol/m^2, (●) 5.0 μmol/m^2, (◇) 4.2 μmol/m^2 and (▼) 3.1 μmol/m^2. Error bars represent one standard deviation. (Reproduced from Ducey, M.W., Jr., et al., *Anal. Chem.*, 5576, 2002. With permission.)

Liao et al. [125] utilized this technique to evaluate the differences in four C_{22} phases. It was found that 4.89 μmol/m^2 solution-polymerized phase (prepared with a trichlorosilane) was more ordered than the 4.65 μmol/m^2 surface-polymerized phase (prepared with a methyldichlorosilane); the increased degree of ordering was attributed to the synthetic route, as the overall ligand densities are quite similar. It has been suggested that this increase in order may be attributable to "island" clustering of the ligands on the surface, whereas the methyldichlorosilane ligands may be more uniformly distributed across the surface. Additional spectroscopic studies (e.g., ^{29}Si NMR) to determine the degree of cross-linking of the ligands would be necessary to understand such effects originating from the silane reagent type.

Raman spectroscopy has also been used for the investigation of more minor conformational ordering effects in stationary phases; both monomeric and polymeric C_{18} ligands exhibited a slight increase in ordering when salt (NH_4Cl and $NaCl$) was added to the mobile phase [126]. In the same study, pH effects were probed; however, relatively insignificant changes in the stationary-phase ordering were observed. Finally, Raman spectroscopy was used to show that there is a 2–7% decrease in alkyl chain ordering when the stationary-phase is exposed to solvent and pressure ($\approx$14,000 kPa for 12 h) [137]. Unfortunately, the spectroscopic measurements were conducted at atmospheric pressure; therefore, it is difficult to extrapolate these data to typical chromatographic conditions.

5.5.3 ATOMIC FORCE MICROSCOPY

Atomic force microscopy (AFM) is a commonly employed imaging technique for the characterization of the topography of material surfaces. In contrast to other microscopy techniques (e.g., scanning electron microscopy), AFM provides additional quantitative surface depth information and therefore yields a 3D profile of the material surface. AFM is routinely applied for the nanoscale surface characterization of materials and has been previously applied to determine surface heterogeneity of alkylsilane thin films prepared on planar surfaces [74,75,138].

Pursch et al. [139] report AFM images of chromatographically relevant C_{18} and C_{30} trichlorosilane monolayers prepared on silicon wafers under restricted volume conditions to be compared with their chromatographically relevant monolayers; unfortunately, the monolayer bonding densities used in the AFM studies are not available as the exact surface area of the material was unknown. The C_{18} monolayer is described as relatively uniform and smooth (with roughness <0.09 nm), whereas the C_{30} monolayer is more heterogeneous with a measured roughness of 0.77 nm. The increase in roughness for the latter is attributed to the variable chain length distribution in the C_{30} used in preparation of the SAM.

Wirth and coworkers [140–142] have applied AFM to the characterization of C_{18} monolayers prepared on flat silica gel substrates or silica plates at a typical surface coverage (2.8 μmol/m^2). The AFM technique provided nanometer-scale resolution of surface heterogeneities that are attributable to the earlier silica polishing processes. The authors characterize these surfaces as not homogenous and describe pit regions as deep as 10 nm and up to hundreds of nanometers in width. The enhanced adsorption of a cationic dye to these irregular regions was

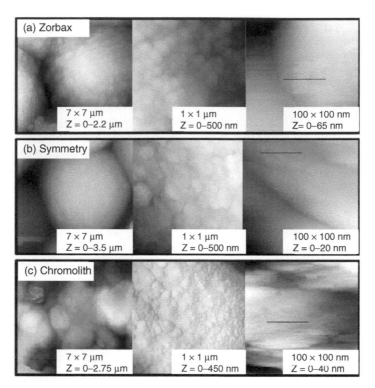

FIGURE 5.17   AFM images of (a) Zorbax SB300, (b) Symmetry 300, and (c) Chromolith chromatographic materials on scales ranging over 3 orders of magnitude. Scan size and Z-scales are indicated in each panel. (Reproduced from Legg, M.A. and Wirth, M.J., et al., *Anal. Chem.*, 78, 6457, 2006. With permission.)

demonstrated using fluorescence spectroscopy [141]; it was suggested that isolated silanols within these regions were responsible for the adsorption since desorption kinetics for the dye molecule were similar to those observed on an unmodified silica gel surface [143].

   More recently, Legg and Wirth [144] have characterized the topography of three commercial monomeric C_{18} stationary-phase materials using AFM. The surfaces were described as relatively smooth, even more so than the chemically modified fused silica materials used in their previous work [141]. Junctions between colloid particles are readily visualized (Figure 5.17). Adhesion force measurements were performed using a chemical force microscopy technique with a silicon contact-mode tip on the flat regions of the material surface. Relatively low adhesion forces together with FTIR spectra data indicate a low abundance of isolated silanols on the investigated surface segment. The authors suggest that the observed nonlinear chromatographic tailing on these materials may stem from a contribution of isolated silanols concentrated in the more irregular surface structures (i.e., pores). The authors also note that adhesion force measurements could not be performed on discontinuous surface regions owing to curvature restrictions imposed by current AFM technology.

5.5.4 NMR Spectroscopy

The application of NMR spectroscopy to the study of the conformational structure of chromatographic stationary phases over the past 25 years is the topic of several reviews [145–147]. The commonly measured nuclei of the stationary-phase are ^1H and ^{13}C, which provide an aggregate measure of the structural and dynamic aspects of chromatographic material. The relative motion of selective deuteration regions within the alkyl chains can also be resolved with ^2H NMR. Furthermore, ^{29}Si NMR can be applied for the characterization of the silica substrate and the silanization chemistry. NMR can be applied to chromatographic materials in solution (i.e., suspended state) as well as to dry bonded phases (i.e., solid state), the latter of which is largely characterized by the presence of anisotropic (directionally dependent) interactions.

^{29}Si cross-polarization magic angle spinning (CP-MAS) NMR has been used for the characterization of the bonding chemistry of chromatographic stationary phases. In a recent study, C_{18} and C_{30} phases prepared via surface and solution polymerization were compared [119]. The authors reported that the degree of cross-linking was similar in the two types of C_{18} phases. In the case of the C_{30} phases, however, the surface-polymerized material exhibited significantly more cross-linking than the solution-polymerized phases; this effect was attributed to differences in bonding density. In a related study [148], it was noted that the degree of cross-linking increased with the density of the alkyl ligands on the surface. These results suggest that the alkyl ligands are bonding on the surface in "islands" with increasing bonding density rather than a more even distribution across the surface.

As with FTIR and Raman spectroscopy, spectral indicators of static and dynamic conformational order have been characterized. These include chemical shifts, line widths, and the relaxation times for the various nuclei under investigation. A common measure of *gauche/trans* conformational order termed the "*γ-gauche* effect" was originally demonstrated for a series of linear alkanes [149]. A chemical shift was attributed to a *gauche* conformation at a γ position from the carbon under examination. Alkyl chains tethered to silica surfaces in all-*trans* conformations result in a signal at 32.6 ppm in the ^{13}C NMR spectra, whereas all-*gauche* conformations result in a signal at 30.0 ppm [150,151].

Several studies have investigated the influence of surface coverage, alkyl chain length, and temperature on alkyl chain conformation in relation to the γ-*gauche* effect. Jinno et al. [151] first demonstrated this effect for monomeric and polymeric C_{18} phases using a ^{13}C CP-MAS NMR technique. Pursch et al. [103,152,153] investigated a series of stationary phases that varied in alkyl chain length, bonding density, and subject to a range of temperatures using ^{13}C CP-MAS NMR spectroscopy. On the basis of the ^{13}C signals at 32.6 and 30.0 ppm (see Figure 5.18), it was apparent that an increase in alkyl chain conformational order results from an increase in surface coverage or chain length, and with reduced temperature. Near-crystalline order was observed for a C_{22} stationary at high surface coverage ($\approx 7.0\,\mu mol/m^2$) and a polymeric phase of extended chain length (C_{34}). Complementary suspended-state ^{13}C NMR studies of extended-length stationary phases (C_{30}) were conducted [154] in the presence of methanol and methyl *tert*-butyl ether (MTBE). NMR spectra in

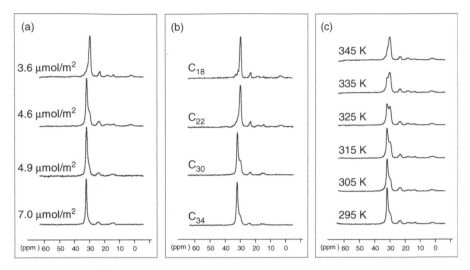

FIGURE 5.18 ^{13}C CP-MAS NMR spectra of alkyl stationary phases illustrating the γ-*gauche* effect for changes in (a) surface coverage, (b) alkyl chain length, and (c) temperature for a C$_{22}$-modified silica. (Reproduced from Pursch, M. Ph.D. Dissertation "Neue Synthesewege zur Darstellung von langkettigen Alklinterphasen-Untersuchungen zur Oberflä—chenmorphologie durch Festkö rper-NMR-Spektroskopie und HPLC," Universität Tübingen, Tübingen, Germany, 1997. With permission.)

methanol were similar to solid-state results; in contrast, the stronger eluent (MTBE) resulted in an increase in alkyl chain *gauche* conformations. Ansarian et al. [82] used ^{13}C CP-MAS NMR with a Sil-ODA$_{25}$ phase to show the dramatic effect of temperature on order. The phase is a liquid crystal with a true transition temperature; below the T_c, the phase is highly ordered and shape-selective; above the T_c, there is a sharp transition to disordered alkyl ligands. Meyer et al. [27] used the technique to demonstrate that there is much more chain mobility in immobilized poly(ethylene-*co*-acrylic acid) phases when there is a higher percentage of acid groups in the polymer chain.

   ^{13}C CP-MAS NMR spectroscopy has also been used to investigate potential differences due to different synthetic routes and substrates. Srinivasan et al. [119] investigated the influence of synthetic routes on the conformational order and mobility of C$_{18}$ and C$_{30}$ stationary phases. The C$_{18}$ phases had significantly different bonding densities, and the higher-density phase showed a higher degree of ordering. However, it is difficult to determine whether the difference is due to the synthetic route or the bonding density of the alkyl ligands. Similarly, the bonding densities of the C$_{30}$ phases were not directly comparable; however, very little difference in the alkyl ligand ordering was observed and suggests a minimal effect of bonding chemistry for these extended length phase materials. In a similar study of C$_{18}$ alkyl phases [120], an increase in order with increasing bonding density was demonstrated for stationary-phase densities of 2–8.2 μmol/m^2. There was one exception in the study: the 2 μmol/m^2 phase consistently showed a higher degree of ordering than the

2.7 μmol/m^2 phase. The authors hypothesize that this effect is due to a difference in synthetic routes where the lower-density phase was made with a stoichiometrically limited portion of trichlorooctadecylsilane in the presence of water, allowing for polymerized "islands" of alkyl ligand, and that the 2.7 μmol/m^2 phase was made under anhydrous conditions, resulting in a more uniform surface. Srinivasan et al. [155] attempted to use this technique to differentiate between the effect of substrate with a low-density zirconia phase (2.5 μmol/m^2), a higher-density zirconia phase (4.0 μmol/m^2), and a silica phase (4.2 μmol/m^2); however, no conclusions could be made as the zirconia phases had a layer of graphitic carbon on the surface, which interfered with the spectroscopic measurement.

High-power decoupling (HPDEC) MAS NMR was used for the determination of the *trans/gauche* ratio in solvated immobilized poly(ethylene-*co*-acrylic acid) phases and a C$_{30}$ phase [27]. In 100% methanol, the aminopropyl-linked poly(ethylene-*co*-acrylic acid) phase, glycidoxypropyl-linked poly(ethylene-*co*-acrylic acid), and C$_{30}$ phase had *trans/gauche* ratios of 40/60, 38/62, and 59/41, respectively, whereas in a volume fraction of 50/50 methanol/water, the *trans/gauche* ratios were 47/53, 44/56, and 60/40. It is interesting to note that the alkyl ligand ordering does change with solvent composition and that this change is also evident in the separation of the β-carotene isomers. In contrast, the C$_{30}$ ligands do not appear to be affected by the solvent composition.

Dynamic motion of the alkyl stationary phases can also be obtained from NMR studies through an analysis of line shapes, comparisons between single-pulse (SP) and CP-MAS spectra, and various relaxation time constants. Zeigler and Maciel [156] have provided a detailed explanation of these spectral indicators for the characterization of a C$_{18}$-modified silica. Sander et al. [9] have provided an overview of these indicators in the context of alkyl-modified surfaces. In general, line widths are narrower for more mobile species, given that the various molecular orientations and effects are more averaged over the sampling timescale. In contrast, line widths for more rigid species tend to be more broadened. Gangoda et al. [157] investigated selectively deuterated shorter-chain (C$_7$–C$_{12}$) alkyl surfaces by wide-line ^{2}H NMR and demonstrated that deuterium atoms near the distal end exhibited much narrower signals (i.e., higher mobility) than those proximal to the silica surface. Motion at the first carbon near the silica surface was considerably more restricted than carbons toward the middle of the alkyl chain. In a similar effort, Zeigler and Marciel [156] described lower-density deuterated C$_{18}$ phases characterized as highly disordered chains interacting strongly with the silica surface, whereas deuterated C$_{18}$ phases with a two- to threefold increase in surface coverage possessed considerable conformational order. Kelusky and Fyfe [158] describe the increased mobility of deuterated alkyloxysilanes immobilized onto silica through ^{2}H line widths with an increase in alkyl chain length (C$_3$–C$_{16}$); however, the surface coverages of these chromatographic materials were not reported, and, therefore, the interpretation of the results remains inconclusive.

NMR relaxation measurements such as cross-polarization time constants (T_{CH}) can also yield useful information regarding alkyl-chain mobility. Sindorf and Maciel [159] have characterized the relative mobility of low-density C$_8$ and C$_{18}$ stationary phases ($\approx$2.6 and 1.7 μmol/m^2, respectively) as a function of carbon position from the silica surface. Carbons closest to the silica surface were more restricted than carbons distal from the surface, with the end methyl groups demonstrating the most mobility. These results are consistent with those of two other related ^{13}C NMR investigations of alkyl stationary phases [160,161], which demonstrated that the mobility of methylene segments along the alkyl chain were further reduced with an increase in bonding density.

The investigation of Pursch et al. [150] employed a selective population inversion (SPI) ^{13}C CP-MAS NMR experiment to characterize the mobility of particular conformation segments within the alkyl chain of stationary-phase materials. Herein, the methylene signal at 32.6 ppm (*trans* conformation) was inverted with short contact times, indicative of relative rigidity, whereas the methylene signal at 30.0 ppm was described as a more mobile moiety with a mixture of *gauche* and *trans* conformations. In support of these data, Pursch et al. [152] report the use of 2D wide-line separation (2D-WISE) NMR to correlate chain conformation (^{13}C signal) with methylene segment mobility (^{1}H line-widths); alkyl chains containing primarily *trans* conformations (32.6 ppm) are significantly more rigid than are chains containing *gauche* conformers.

Spin-diffusion solid-state NMR experiments have been used to probe the alignment of specific domains of molecular order within chromatographic stationary phases [162]. In general, the technique involves the initial magnetization of two domains with differing mobility that results in a combined NMR signal from both domains. Selection of one domain is then achieved through the utilization of a specific relaxation time; herein, only one domain of the sample (i.e., the more mobile component) is magnetized and results in a distinct signal. Magnetization is then transferred to neighboring nuclei through dipolar interactions (i.e., spin diffusion), which results in a reduction in the original signal and an increase in the magnetization of the rigid components. The extent of magnetization during the spin diffusion mixing time is used to determine the size of areas of equal mobility and can be determined in the range of 0.5–200 nm. Raitza et al. [162] determined that the alkyl ligands in polymeric C_{30} stationary phases exist in distinct domains: (1) an ordered all-*trans* region with an average size of 82 nm^2 with an average alkyl chain length of 4.3 nm and (2) a more mobile domain containing *gauche* conformations with an average size of 125 nm^2 with an average alkyl chain length of 3.1 nm. A similar spin diffusion NMR technique was applied for the conformational investigations of glycidoxypropyl-linked 5% poly(ethylene-*co*-acrylic acid) stationary phases by Meyer et al. [163]. The authors have determined a domain size of 16 ± 2 nm for the primarily *gauche* mobile component and a smaller domain size of 3.2 ± 0.4 nm for the rigid all-*trans* component.

Finally, dipolar and chemical shift correlation (DIPSHIFT) NMR was used to investigate the molecular motion and dynamics of the immobilized alkyl ligands of poly(ethylene-*co*-acrylic acid) stationary phases [164]. Through the measurement of the [13]C–[1]H dipolar couplings, it was possible to discern the geometry of motion of the alkyl ligands, and this can be utilized to determine the degree of order within distinct domains of the stationary-phase. When the domain contained ligands primarily in the *trans* configuration, they tended to be relatively ordered and did not exhibit any significant difference with temperature. In contrast, if the domains were composed primarily of ligands in the *gauche* conformations, then the degree of disorder was observed to increase with increasing temperature. Additional experiments were performed to contrast the regional difference between a dry stationary-phase with a stationary-phase that was suspended in 95:5 methanol/water: the *gauche* regions were deemed slightly more mobile in the suspended state, whereas the all-*trans* regions possess comparable rigidity in both the solid state and the suspended states.

5.5.5 FLUORESCENCE SPECTROSCOPY

Fluorescence spectroscopy involves the absorption of a photon of light by the molecular species of interest to produce an excited electronic state, followed by the release of a photon of a different energy as the species returns to the ground electronic state. The majority of alkyl stationary phases do not contain fluorophores; therefore a direct spectroscopic investigation of phase structure is not feasible. Alternatively, an indirect measure of the alkyl phase morphology can be elucidated via the interaction of fluorescent molecular probes with the stationary-phase. Many fluorescence techniques require the use of optically flat surfaces; therefore, tethered alkyl-chain materials for use in fluorescence spectroscopy investigations are commonly prepared on quartz and silica plates. Aspects of stationary-phase morphology, including the heterogeneity and polarity of the phase, the spatial distribution of alkyl chains on the silica substrate, and the lateral diffusion of the adsorbed molecular probes, have been extensively investigated. Fluorescence spectroscopy has also been applied in the investigations of the alkyl chain conformational structure as influenced by bonding density, alkyl chain length, and temperature through its interaction with a suitable fluorescent probe solute.

The polarity of alkyl-based stationary-phase materials has been investigated through the incorporation of the several chemically bonded or adsorbed molecular probes followed by fluorescence spectroscopy. First, Lochmüller et al. [165] incorporated low levels of dansylamide into alkyl-modified silica and monitored the changes in excitation and emission fluorescence wavelengths. Differences were attributed to polarity changes from differing solvent environments as well as to the microheterogeneity present in the silica. Other researchers [166,167] utilized pyrene as an adsorbent probe on alkyl stationary phases of various lengths (C_2 and C_{18}) and synthetic chemistries (monomeric and polymeric). The changes in the fluorescence spectra were attributed to dipole-induced dipole and $\pi-\pi$ interactions between pyrene and the stationary-phase and solvent environments. From the results of the Carr and Harris study [168], they concluded that pyrene was considered "shielded" by the C_{18} polymeric phase when in equilibrium with 70:30 methanol:water; under highly aqueous conditions, the spectral differences suggest that pyrene is more exposed to the solvent [168].

The diffusion-related molecular processes occurring within a C_{18} stationary-phase have also been investigated using pyrene as a fluorescent probe [169]. Particular spectral bands were attributed to pyrene excimers formed in a diffusion-limited reaction. Rate constants for this formation were then used to estimate the microviscosity of the stationary-phase. A similar application of total internal reflection fluorescence spectroscopy [170] determined similar values of microviscosity for a C_{18} chromatographic material. Ståhlberg [171] measured the kinetics of pyrene excimer formation and cautioned that the evaluation of rate constants, and thus viscosity estimates, is inconsistent and dependent greatly on the various microenvironments present within the stationary-phase.

The existence of various adsorption sites within an alkyl chromatographic material has also been explored by the use of fluorescence spectroscopy. Specifically, a technique termed fluorescence recovery after pattern photobleaching (FRAPP) has been applied to the adsorptive interaction of molecular probes with C_{18} chromatographic materials [172–174]. This experiment describes the diffusional relaxation of probe molecules within a concentration gradient present in the alkyl stationary-phase. Wirth et al. [100,172–175] described the first application of FRAPP to chromatographic materials through lateral diffusion investigations of acridine orange within C_{18} monolayers. Further investigations of surface coverage and temperature effects for these monolayers demonstrated a significant increase (500-fold) in acridine orange diffusion coefficients with an increase in C_{18} chain surface coverage (up to 3.5 μmol/m^2) [173] and only a slight increase in diffusion coefficients with an increase in temperature over the range of 10–55°C [174]. Hansen and Harris [100,172] describe FRAPP experiments with a bulkier eight-ring PAH probe molecule, rubrene, on C_4, C_8, and C_{18} monomeric alkyl phases. Diffusion rates were observed to decrease as the chain length was decreased. For the C_{18} phases, the rate of diffusion also decreased with a reduction in surface coverage (as estimated through contact angle measurements). Both of these studies conclude that irregular regions of the surface represent unfavorable adsorption sites that contribute to relatively large energetic (primarily entropic) barriers and result in slow lateral diffusion within the alkyl stationary-phase.

More direct evidence of dissimilar absorption sites within alkyl chromatographic materials has been demonstrated by single-molecule fluorescence spectroscopy conducted by Wirth and coworkers [142,176,177]. Herein, the adsorption of a single molecule of a dye probe (1,1'-dioctadecyl-3,3,3',3'-tetramethyllindocarbocyanine perchlorate; DiI) was observed in dynamic equilibrium at a C_{18} chromatographic interface in contact with water or acetonitrile. The majority of the molecular probes would laterally diffuse; some molecules would stop on the surface and then resume diffusive movement. Three distinct adsorption processes were characterized, with desorption times scales of 70 ms, 7 s, and >2 min. The two longest desorption events were directly observed at nanometer-scale indentations that were present on a C_{18} surface prepared on silica plates (Figure 5.19). The authors suggest that similar nanoscale pores within silica on chromatographically relevant surfaces may promote such adsorptive processes and thus contribute to the peak tailing observed in RPLC separations. This was further supported by Zhong et al. [178] who used single-molecule fluorescence spectroscopy on 10 μm and 10 nm Luna particles (Phenomenex, Torrance, CA) with a C_{18} bonding density of 3.0 μmol/m^2 and rhodamine 6G in acetonitrile as a probe molecule. The average desorption time was found to be 61 ms; however, strong adsorption events were rare, occurring approximately 0.3% of the total observation time.

Fluorescence anisotropy decay measurements, which are based on the excitation of probes with polarized light and subsequent polarized fluorescence emission, can

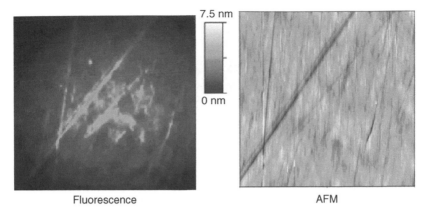

Fluorescence   AFM

FIGURE 5.19 **(See color insert following page 280.)** Fluorescence image for cationic dye molecule (DiI) on a C_{18} surface at high concentration measured using a charge-coupled device (CCD) camera (left). AFM image of the same region ($20 \mu m \times 20 \mu m$) (right). (Reproduced from Wirth, M.J., et al., *J. Phys. Chem. B*, 107, 6258, 2003. With permission.)

also provide indirect information regarding alkyl chain conformation and structure. The relative orientation of the fluorescent probe within the stationary-phase can be resolved through an analysis of the degree of polarization of the emission [179]. The orientation of long hydrophobic probe (1,4-*bis*(*o*-methylstyryl)benzene) on a C_{18} chromatographic phase with moderate bonding density ($2.8 \mu mol/m^2$) in the presence of 100% water is generally centered in the plane of the surface. In the case of experiments conducted with acridine orange [140], the probe was observed to rotate freely in the plane on a similar chromatographic surface but was hindered toward rotation out of the surface plane (i.e., in the direction perpendicular to the silica surface). Pursch et al. [152] employed a nonpolar probe (1,6-diphenylhexatriene; DPH) to investigate the alkyl chain structure of C_{22} polymeric phases over a range of surface coverages and temperatures. Fluorescence anisotropy decay curves were used to determine half-cone angles of the probe's wobble-like motion within the alkyl phase; lower angles are indicative of more restricted movement. An increase in half-cone angle was observed for an increase in temperature and for decrease in surface coverage (Figure 5.20). This observation is attributed to an increase in free volume for diffusional motion of the probe as a result of disordering of the alkyl chains. Likewise, the partitioning of the probe into the alkyl phase is favored at higher surface coverages, but reaches a maximum at a surface coverage of $4.9 \mu mol/m^2$. This is analogous to the maximum partitioning of a naphthalene probe observed for higher surfaces coverages of a monomeric C_{18} phase [97].

5.5.6 OTHER EXPERIMENTAL TECHNIQUES

This section will cover other experimental approaches that are less frequently applied (yet are not insignificant) toward the structural elucidation of stationary-phase surfaces. This includes electron paramagnetic resonance (EPR), small angle neutron

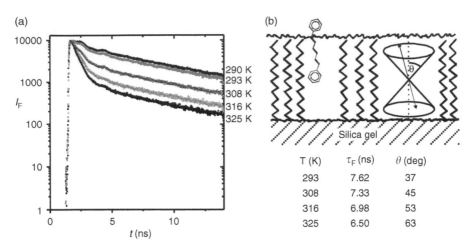

FIGURE 5.20   (See color insert following page 280.) Temperature influence for fluorescence anisotropy measurements on C_{22}-modified silica. (a) Decay curves for DPH at different temperatures, (b) schematic of wobble motion of DPH and resulting fluorescence lifetimes (τ_F) and half-cone angles (θ). (Reproduced from Pursch, M., et al., *J. Am. Chem. Soc.*, 121, 3201, 1999. With permission.)

scattering (SANS), X-ray reflectivity, and ellipsometry techniques. In addition, the application of nonlinear chromatographic techniques to measure various adsorption sites of chromatographic surfaces will also be discussed.

Similar to the approach with fluorescence anisotropy measurements, EPR can be utilized to provide an indirect measure of conformational order through stationary-phase interactions with an appropriate spin molecular probe [180]. Wright et al. [180] has applied EPR to investigate a series of variable bonding density C_{18}-modified surfaces in methanol or acetonitrile solvents with two spin probes of dissimilar molecular shapes. These included the spherical spin probe 2,2,6,6-tetramethyl-1-piperidinyloxy (TEMPO) and the larger and flatter spin probe N-oxyl-4,4-dimethylspiro[oxazolidine-5α-cholestant] (CSL). Changes in EPR line widths and peak heights are considered indicative of probe rotational mobility. Little change in mobility was observed for the TEMPO probe with respect to both bonding density of the stationary-phase and solvent environment. In contrast, the rotational mobility for the bulkier CSL probe was significantly reduced in stationary phases with an increase in surface coverage. Control experiments with unmodified silica were conducted to establish that the restricted CSL mobility was attributable to probe intercalation within denser alkyl chains of higher surface-coverage stationary phases. More recently, Ottaviani et al. [181] applied EPR techniques to measure the adsorption of various molecular spin probes on various C_{18} stationary phases. An increase in adsorption of the spin probes that contain both polar and nonpolar moieties (4-C_n-dimethylammonium-2,2,6,6-tetramethylpiperidine-1-oxyl bromide) is ascribed to an ordered region of the C_{18} chain distal from the surface that forms in mixed organic and water solvent (40–70% acetonitrile or methanol). In addition, an increase in adsorption was observed in mixtures of acetonitrile (compared to methanol), and the molecular probes with longer alkyl chain moieties (C_{16}) were adsorbed less than probes with shorter alkyl chain moieties (C_9) under these conditions.

The geometry of an alkyl chain in motion can be resolved through use of neutron scattering techniques [182,183]. Herein, geometric information of the bulk alkyl phase is based on the rotational isomeric state model that describes the segmental motion of three or five contiguous carbon–carbon bonds in either a *trans* or *gauche* conformational state. For a C_{18}-modified silica surface, alkyl chain motion was dependent on an increase in temperature; three-carbon segments were observed to change conformation at lower temperatures, whereas changes in five-carbon segments were observed for only high temperatures ($>80°C$). The application of SANS [184,185] represented the first direct measurement of thickness and volume fraction for a series of monomeric and polymeric chromatographic stationary phases. Alkyl-modified surfaces were prepared on wide-pore silica with varying alkyl chain length (C_8, C_{18}, and C) and surface coverages. To reduce scattering from the pore structure, experiments were carried out on samples in a methanol/deuterated methanol solvent. Results are summarized in Table 5.2. The polymeric C_{18} phase exhibits a thickness comparable to the length of C_{18} chain in the all-*trans* conformation (≈ 2.3 nm).

Wasserman [186] has described the use of both low-angle X-ray reflectivity and ellipsometry for the determination of thickness of C_{10}–C_{18} SAMs prepared on surface silanol groups of silicon plates. Ellipsometry is based on the reflection of polarized light from a sample and depends on the sample's thickness and refractive index. X-ray reflectivity measures the intensity of X-rays reflected from a surface (or interference pattern) that is characteristic of the distance between interfaces. The thickness of the SAMs was consistent with fully extended alkyl chains with all-*trans* conformations and excellent agreement was observed between the two methods.

Differential scanning calorimetry (DSC) is a common technique for the classification of individual phase transitions in liquid-crystalline materials and has been applied for the phase characterization of alkyl-modified chromatographic surfaces. Hansen and Callis [187] applied DSC to investigate phase changes in C_{18} and C_{22}

TABLE 5.2
Physical Properties of Alkyl-modified Silica Obtained by SANS Measurements

Phase Type	Surface Coverage (μmol/m^2)	Thickness (nm)	Alkyl Chain Volume Fraction
C_8 monomeric	3.7	1.0 ± 0.2	0.65 ± 0.15
C_{18} monomeric	3.7	1.7 ± 0.3	0.65 ± 0.15
C_{18} polymeric	7.5	2.1 ± 0.3	0.65 ± 0.15
C_{30} monomeric	2.8	2.5 ± 0.4	0.88 ± 0.1

Source: Sander, L.C. et al., *Anal. Chem.*, 62, 1099, 1990.

chromatographic sorbents at moderate surface coverages (2.0–2.5 μmol/m^2) and concluded that no direct phase transition was evidenced for dry materials. Jinno et al. [188] observed a weak transition for a polymeric C_{18} phase at $\approx$45°C and an even weaker transition for a monomeric C_{18} phase at $\approx$35°C. The presence of a definitive first-order phase transitions in these materials, however, is questionable. Wheeler et al. [189] concluded that temperature-dependent conformational changes occur within tethered alkanes but it does not resemble a well-defined phase transition typical of a bulk freezing–melting process. In contrast, distinct endothermic phase transitions have been observed with the liquid crystalline comb-shaped polymer (ODA$_n$) chromatographic phases described by Ihara and coworkers [54,78,190] (see Figure 5.6). These materials undergo a dramatic shift from an ordered conformational state (as evidenced by NMR) to an anisotropic configuration within a temperature range of 30–47°C (depending on solvent). In addition, the selectivity differences between a naphthacene/chrysene solute pair is drastically reduced above the critical temperature ($\approx$42°C) for which the conformational state of the polymer (ODA$_{25}$) becomes disordered; in contrast, no sharp decrease in selectivity with temperature was observed for monomeric and polymeric C_{18} stationary phases [80].

Gritti, Guiochon, and coworkers [16,191–207] have characterized the heterogeneity of a wealth of chromatographic materials under a range of operating conditions by adsorption energy distributions (AED) derived from adsorption isotherm data obtained from nonlinear chromatography. Adsorption isotherms are generally determined via a dynamic frontal analysis method, for which relatively high concentrations of probe solutes populate the adsorption sites within the stationary-phase. More complete details regarding this approach and a summary of the results from these prolific studies are provided in a recent review [16]. Numerous C_{18}-bonded monomeric columns on various silica supports were examined in these studies; a more limited number of polymeric-type columns (C_{18} and C_{30}) were also considered. Adsorption site energies were determined for a range of solutes (phenol, caffeine, propranolol, naphthalene sulfonate, nortryptiline, amtryptiline); the effect of pressure, temperature, and mobile phase composition were investigated for the neutral solutes and the effect of buffer, salts, pH, and ionic strength was investigated for the ionizable compounds. Different adsorption energies that range between a few and 30 kJ/mol have been reported for this group of solutes over the range of conditions. Higher-energy sites are populated first at lower concentrations, until saturation is reached. The lower-energy sites then fill and become significant only at the higher concentrations. These results provide an indirect measure of the surface heterogeneity of commonly used C_{18}-bonded chromatographic materials. The surfaces have been described as a heterogeneous mixture of adsorption sites that vary in their saturation capacity and steric availability as well as in the molecular-level interactions that are accessible to the solute probe on retention.

5.5.7 MOLECULAR SIMULATIONS

A number of molecular simulation approaches have also been employed to elucidate the specific structural features of alkyl-modified surfaces. Recognizing that molecular simulations represent an idealized view of a chromatographic surface that may be significantly heterogeneous, these experiments are still valuable in clarifying the importance of a particular parameter (e.g., bonding density and chemistry, temperature, type of solvent) on conformational changes within the model system. It is also important to highlight that the results obtained from one simulation model are inherently influenced by the initial model construction and therefore represent only a fraction of possible chromatographic materials and conditions. Several older reviews [208,209] have been published regarding the molecular dynamic (MD) simulations of alkyl-modified chromatographic surfaces. A more recent review [9] describes the results of these simulations in the context of stationary-phase conformational order indicators.

The majority of simulation efforts have been toward the development of a representative chromatographic model for the more commonly used monomeric-type RPLC phases. The first applications of MD simulations to monomeric chromatographic model systems were conducted over a decade ago [210,211]. Schure [211] simulated C_8- and C_{18}-modified amorphous silica at low to moderate density (2.1 μmol/m^2) and included solvent mixtures of methanol and water. The alkyl chains were described as relatively disordered, and methanol enrichment within the chain environment was observed. Klatte and Beck described MD simulations for C_8-, C_{12}- and C_{18}-modified surfaces over a range of alkyl chain densities in vacuum [210] and later within a range of water/methanol solvent mixtures [212,213]. Approximately 25% *gauche* defects were overall present in the chains at room temperature. At lower densities, the surface topography of the C_{18} phases in vacuum was characterized as rough and disordered. Yarovsky et al. [214,215] studied *n*-alkyldimethylsilyl (C_4, C_8, and C_{18}) modified amorphous silica at surface densities of 1.6–3.7 μmol/m^2 in vacuum and observed a reduction in both alkyl chain *gauche* fractions and overall chain mobility with an increase in surface coverage. Mountain and coworkers [216,217] applied MD simulations to an RPLC model consisting of a slab of solvent (methanol, acetonitrile) surrounded by C_8 chains tethered to a fixed surface with relatively high surface coverage (5.1 μmol/m^2). The C_8 alkyl chain end-to-end distance profile was bimodal in nature, with a small portion of fully extended chains with an all-*trans* configuration and the majority of chains containing a mixture of *gauche* defects (reduced end-to-end distance). The structural features of the alkyl chains (i.e., end-to-end distance, tilt angle) were not significantly influenced by the presence of solvents. Ban et al. [218] described a monomeric-phase model constructed of low-density (1.9 μmol/m^2) octadecyl-dimethylsilyl ligands on amorphous silica subjected to various solvent environments (water, methanol, *n*-hexane). The distribution of solvent molecules within the alkyl phase changed with solvent composition. Dou et al. [219] obtained similar results for a C_{18} monomeric model at 2.6 μmol/m^2 in various compositions of methanol and water. Ashbaugh et al. [220] characterized the interfacial region between a system of *n*-C_{18} alkyl chains tethered to solid surface (at 50% surface coverage) in contact with a water layer at 300 K. The water molecules sufficiently permeate into the alkyl chains, and this infiltration is attributed, in part, to the inherent roughness of the alkyl layer at the relatively low surface coverage employed.

Wick et al. [221,222] have applied Monte Carlo (MC) simulations to RPLC models composed of an *n*-hexadecane retentive phase with a water–methanol mobile

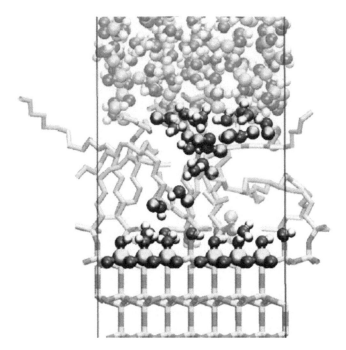

FIGURE 5.21 (See color insert following page 280.) Snapshot of a methanol bridge cluster for a MC simulated monomeric C_{18} chromatographic surfaces. Solutes involved in the cluster are highlighted in darker, opaque coloring. (Reproduced from Zhang, L., et al., *J. Chromatogr. A*, 1126, 219, 2006. With permission.)

phase of varying composition. The partition coefficients of alkane and alcohol solute probes were quantified in an effort to characterize RPLC retention thermodynamics. Zhang et al. [223] investigated the conformation of dimethyloctadecyl-bonded silica model at 2.9 μmol/m^2 in the presence of pure water using MC sampling techniques. The alkyl chains were described as oriented away from the surface, rather than collapsed or shrunken owing to the presence of 100% water mobile phase. More recent studies [224] have simulated monomeric-type chromatographic surfaces under various solvent conditions to improve the description of chain conformation and solvent partitioning in RPLC processes. Clusters of methanol and water were demonstrated to "bridge" between the solvent interface region of the alkyl chains and the silanols of the silica surface (Figure 5.21). These recent results provide invaluable information regarding the molecular-level solvation processes that may promote RPLC separations on commonly used chromatographic materials.

To investigate the controlling features of polymeric-type materials that result in highly shape-selective separations, Lippa et al. [225,226] have conducted MD simulations of polymeric alkylsilane materials to identify structural trends of the chromatographic models with their corresponding shape-recognition capabilities. Models of C_8-, C_{18}-, and C_{30}-modified silicas in vacuum were simulated to examine systematic and localized changes in alkyl phase architecture that results from changes in chain

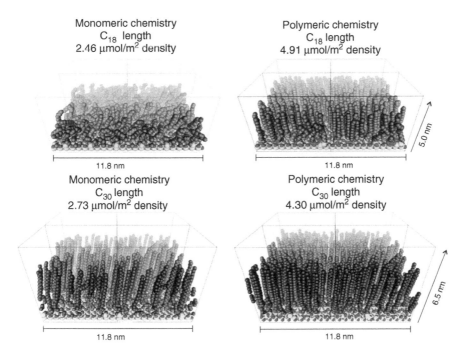

Monomeric chemistry
C_{18} length
2.46 µmol/m² density

Polymeric chemistry
C_{18} length
4.91 µmol/m² density

11.8 nm

5.0 nm

Monomeric chemistry
C_{30} length
2.73 µmol/m² density

Polymeric chemistry
C_{30} length
4.30 µmol/m² density

11.8 nm

11.8 nm

6.5 nm

FIGURE 5.22 **(See color insert following page 280.)** Perspective four-cell view snapshots of various stationary-phase models, which illustrate the contrasting surface topography with bonding density and alkyl chain length. (Adapted from Lippa, K.A., et al., *Anal. Chem.*, 77, 7852, 2005.)

length in addition to variations in surface coverage, bonding chemistry (monomeric vs. polymeric), and temperature. To allow for the observation of significant changes in surface morphology, models of extended dimensions (to include up to 87 individual alkyl chains) were constructed. Surface densities ranged from 1.6–5.9 µmol/m² for C_8, C_{18}, and C_{30} ligands and the polymeric models were constructed of three oligomer units to approximate bonding in these phases.

Simulation results from these polymeric phase models highlighted several trends in molecular structure that are pertinent to the elucidation of shape-recognition processes in RPLC. Both the alkyl chain *trans* fraction and the phase thickness increased with increased surface coverage as individual chains were more constrained by the neighboring chains. Longer-chain-length phases were more ordered than shorter-chain-length phases at identical surface coverages. Consistent with spectroscopic results, an increase in conformational order was observed at reduced temperature. The resultant model snapshots shown in Figure 5.22 exemplified the aggregate structural features observed for chromatography phase models of varying bonding chemistry, density, and length. Chain extension and increased *trans* conformations are observed with models of polymeric C_{18} phases. C_{30} stationary-phase models exhibit significantly increased *trans* conformations, even with lower surface densities (see Figure 5.22). The applicability of solvent-free simulation results for representing

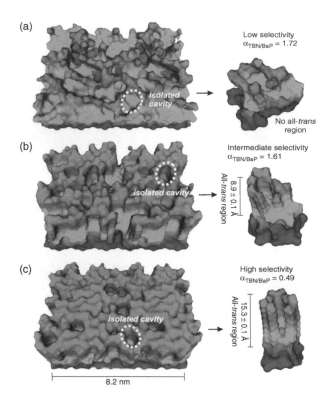

FIGURE 5.23 **(See color insert following page 280.)** Molecular surfaces generated with a 2.0 Å radius probe for MD simulation snapshots of (a) C_{18} monomeric model at 3.28 μmol/m² surface coverage representing an actual RPLC phase with low shape selectivity ($\alpha_{TBN/BaP}$ = 1.72) and two C_{18} polymeric models at (b) 3.89 μmol/m² and (c) 5.94 μmol/m² surface coverages also representing actual phases with intermediate to high shape selectivity ($\alpha_{TBN/BaP}$ = 1.61 and 0.49, respectively). Surfaces are color coded according to the height along the z-axis. Included are the side-view surface areas generated for the isolated cavities with an average diameter of ≈10 Å. (Adapted from Lippa, K.A., et al., *J. Chromatogr. A*, 1128, 79, 2006. With permission.)

RPLC processes has been questioned [203,224]. It is important to recognize that a substitute of solvent for a larger model with a greater number of alkyl chains was deemed necessary to provide adequate resolution of local phase structural changes and thus the characterization of potentially important features within chromatographic phases that represent shape-selective materials.

Analytical shape computation techniques were applied for the detection of cavities and the calculation of molecular surface properties of isolated cavity features and other ordered formations within these resultant alkyl stationary-phase simulation models [227]. Deep cavities (8–10 Å wide) within the alkyl chains were identified for C_{18} polymeric models representing shape selective stationary phases (Figure 5.23). Similar-structure cavities with significant alkyl-chain ordered regions (>11 Å) were isolated from two independent C_{18} models (differing in temperature, bonding chemistry and density) that represent highly shape-selective materials. The size and depth of these ordered regions increased (up to 28 Å) for the extended-length C_{30} alkyl-phase models. The results provide an initial physical representation of polymeric-type alkyl-modified surfaces and identify potential molecular features that may be involved in the shape-selective retentive processes. These molecular representations of chromatographic surfaces are exclusive of the indisputably significant interactions with the shape-constrained solutes that are used to characterize the stationary phase's shape selectivity. The molecular picture of a complete chromatographic retention process on such surfaces may differ (if even subtly) as solutes approach and interact with the surface chains and potential binding sites.

The advantages of MC techniques over MD methods toward the simulation of RPLC materials and processes have been brought forth in the recent literature [224]. Shortcomings of the MD approach are ascribed to the limited time frame (generally nanoseconds) for which simulations are feasible and the requirement of the closed ensemble (i.e., the number of molecules within the model system). Leach [228] provides a comparison between the two techniques and notes that the two methods differ primarily in their ability to sample the model's phase space. MD simulations are restricted to conformational constraints within the model and therefore may be limited in sampling all conformational changes over the given time period; the advantage here is a more detailed description of local phase structure and dynamic behavior. In contrast, MC methods randomly and more efficiently sample the configurational space and can better describe molecular systems for which significant phase changes may occur. In addition, MC methods provide more representative statistical averages and can be better suited to characterize partitioning processes (e.g., lattice model). Hybrid MD/MC methods have been successfully employed for simulating polymers [229] and various larger-scale biological systems [230–234] and may represent a viable tool for more comprehensive modeling of stationary-phase retentive processes. Such a case may include a more representative equilibrium structure of the stationary and solvent phases with time-dependent structural detail regarding the particular phase features that may be contributing to the retentive process.

5.6 MECHANISM OF SHAPE SELECTIVITY

The study of retention processes in RPLC has been the topic of numerous investigations and reviews [10,12,58,59,189,235–240] and remains a predominant topic in the current chromatography literature [11,13,16,241]. The continuing pursuit of a more complete understanding of the retention process in RPLC is anticipated to yield improvements in separation techniques and contribute toward the development of novel sorbent materials. Retention processes in RPLC were initially described [242,243] as the difference between the solute's preference for the mobile phase and the stationary-phase (solvophobic theory), which is largely driven by the formation of the solute cavity in the solvent. Advances in the selectivity evaluation of chromatographic materials demonstrated that solute retention could also be controlled by the surface coverage and chain length of the stationary-phase [7,47,94], thereby minimizing the prevailing influence of the solvent phase. Retention models based on statistical

(a) Methanol/water

(b) Acetonitrile/water

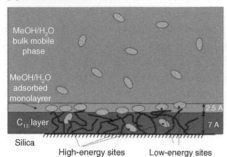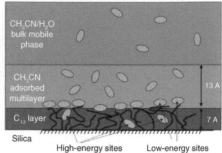

FIGURE 5.24 **(See color insert following page 280.)** Schematic comparison of the adsorption mechanisms of a solute from aqueous solutions of (a) methanol and (b) acetonitrile onto a RPLC material. Three different "phases" (bulk mobile phase, the adsorbed mono- or multilayer of organic modifier molecules, and the C_{18} phase) are involved in the chromatographic system. The solute is represented by small ovals. (Reproduced from Gritti, F. and Guiochon, G., *Anal. Chem.*, 77, 4257, 2005. With permission.)

thermodynamics theory were thus developed [14,97,244,245] that described solute partitioning between a mobile phase and the stationary-phase. Herein, the solute interaction with the chromatographic sorbent resembles bulk-phase partitioning in which the entire solute is fully embedded in the stationary-phase. An alternative adsorption model [14,244] was proposed to account for the observation [246] that the driving force for retention of a solute on a tethered C_{18} stationary-phase was significantly weaker than the driving force for partitioning of the solute into a bulk-phase liquid C_{18} phase. The adsorption model considers that only a fraction of a solute's surface area is involved in a surface adsorption process and thus the free energy of transfer is reduced relative to partitioning into the bulk phase. An "interphase" model [244] that describes solute partitioning into semiordered regions within the stationary-phase was also proposed to account for these differences in energetics.

A more recent review of retention mechanisms in RPLC portrays the partition and adsorption models as extreme positions on a broad spectrum of possible mechanisms [16]. Chromatographic surfaces are viewed as heterogeneous with a range of adsorption sites that differ in inherent energy levels and overall abundance. The apparent retention of a solute may, however, result from a combination of partition and adsorption processes inside and on the surface of the chromatographic stationary-phase. In addition, the overall mechanism can involve a combination of molecular-level interactions (e.g., hydrophobic, polar, hydrogen bonding) occurring within each process. The view can become more complicated when one considers the presence of additional "phases," such as adsorbed monolayers of organic modifiers onto the tethered stationary-phase layer [247] (see Figure 5.24).

The exact mechanism controlling shape-selective retentive processes is not fully understood, although it is clear that the pure partitioning and adsorption models cannot account for differences in retention for isomer separations or the range of selectivityobserved for columns of various surface coverages and alkyl chain lengths.

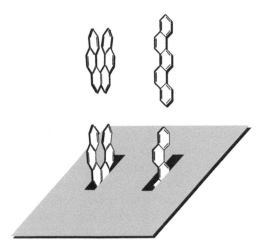

FIGURE 5.25   The "slot model." (Adapted from Wise, S.A. and Sander, L.C., *J. High Resolut. Chromatogr. Chromatogr. Commun.*, 8, 248, 1985.)

A conceptual model termed the "slot" model [47] (Figure 5.25) was originally proposed to account for the molecular geometry-based recognition differences observed with "shape-selective" (typically higher density) polymeric stationary phases. The spaces within these phases are viewed as semirigid "slots" with a cuboid shape of a particular width (i.e., chain spacing) into which solutes penetrate. Planar and more linear molecules with larger L/B ratios preferentially fit within the slots and result in longer retention. Conversely, nonplanar and more square-shaped solutes ($L/B \approx 1$) may be excluded from the slots on the basis of molecular width and thus exhibit less retention. The retention differences for PAH isomers [24] and polyfluorinated PAHs [248] on polymeric bonded alkyl phases, in addition to the separation of carotene and xanthophyll isomers [27] on poly(ethylene-*co*-acrylic) copolymer phases, have been described with the "slot" model.

The "lattice" model of Yan and Martire [38,39,249] describes the equilibrium partitioning of block-like solutes through an extension of a partition model based on statistical thermodynamic theory. The model is composed of 3D lattices representing the solvent and stationary phases with a mean-field approximation (i.e., particles are mixed randomly and uniformly) and steric and attractive energy terms. Retention is described through a solute distribution coefficient (K) defined as the ratio of the equilibrium constant of solute in stationary-phase to that of the solvent phase (at infinite dilution) and is represented by the van der Waal volume (V_{vdW}), effective contact area (A_{eff}), and the minimum cross-sectional area (A_{min}) solute parameters. For isomeric solutes, V_{vdW} and A_{eff} vary slightly, and therefore, most of the partitioning (K) is attributed to A_{min}. In addition to the mean-field approximation, this model makes several assumptions, including a distinct boundary between the mobile and stationary phases, an aligned anisotropic (ordered) stationary-phase, and complete inclusion of the solute molecule into the stationary-phase (bulk-phase partitioning). This model can account for the preferential partitioning of more linear, rod-like PAHs over bulkier, nonplanar isomers through alignment with an anisotropic liquid crystalline stationary-phase and isotropic mobile phase model system. Yan and Martire's theory has been adapted to describe the differences in retention of PAHs on crystalline and noncrystalline forms of a comb-shaped C_{18} polymer chromatographic phase [190]. The lattice model has also been invoked in the development of equilibrium dispersive models to describe the RPLC chromatographic resolution of shape-constrained solutes [250].

On the basis of the lattice model, linear relationships between $\ln k'$ and A_{min} were observed for PAH isomer sets of relative molecular mass 278 and 302 [24]. Similar (yet slightly improved) linear relationships were observed between $\ln k'$ and L/B ratios [24]. The concordance of the linear relationships originating from the more rigorous lattice retention model with the empirically based slot model suggests that the 3D shape of the solute provides a significant basis for the RPLC separation of compounds with constrained molecular structure. But it is not molecular shape alone that controls the shape-recognition process. As detailed within this review, the stationary-phase characteristics that influence overall phase molecular order have also been deemed rather important. For these retention models that can be applied to shape-recognition processes on "ordered" stationary phases, solute retention is predicted to increase with increased bonding density until a maximum value is reached. As first observed by Sentell and Dorsey [97], solute retention decreases for very high density chromatographic materials and was attributed to a reduction in the available interchain space within the alkyl phase. The retention maximum for these materials was also observed to be dependent on the overall shape of the solute (see Figure 5.10) [61].

As with most chromatographic separations, the retention process that governs shape-selective processes is undoubtedly complex and may be the compilation of multiple interactions on various segments of stationary-phase. On the basis of the wealth of conformational information obtained from spectroscopic data, it is likely that the alkyl chains within these shape-selective materials are considerably ordered with a significant amount of *trans* conformations. From the recent NMR domain investigations of shape-selective stationary phases [162,163], it is clear that rigid and ordered domains coexist with more mobile and more disordered regions within these materials. RPLC surfaces have also been characterized as heterogeneous with a range of adsorption sites that differ in inherent energy levels and overall abundance [16]. These data have led to the recent suggestion provided by Gritti and Guiochon [16] that RPLC phases possess high-energy cavities within the alkyl chains that may provide stronger dispersion (i.e., van der Waals) interactions with a solute than it would experience at the interface in the presence of the bulk solvent. The ability of a shape-selective RPLC phase to preferentially retain the planar BaP solute compared with nonplanar 1,2:3,4:5,6:7,8-tetrabenzonaphthalene (TBN) solute ($\alpha_{TBN/BaP} < 0.7$) suggests that BaP can readily fit into these high-energy cavities and interact more strongly with the phase via van der Waals interactions and thus result in longer retention. In contrast, a nonplanar solute (e.g., TBN) may not be of the appropriate size or shape to fit within these cavities, and its chromatographic retention may result from an interaction more on the material surface. A conceptual "slot" model of a potential shape-recognition

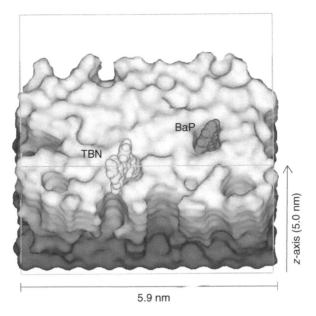

z-axis (5.0 nm)

5.9 nm

FIGURE 5.26 **(See color insert following page 280.)** A representation of the "slot model" illustrating potential constrained-shape solute (BaP) interactions with the conformational ordered cavities of a polymeric C_{18} stationary-phase simulation model. Also included on the chromatographic surface is an identical-scale molecular structure of 1,2:3,4:5,6:7,8-tetrabenzonaphthalene (TBN).

process is illustrated in Figure 5.26. This includes a representation of the polymeric C_{18} stationary-phase at a high surface coverage (5.9 μmol/m^2) from earlier MD simulations [225] together with a depiction of a preferential retention of the planar BaP solute within an isolated cavity compared with a more surface-retained nonplanar TBN solute.

To summarize, a great deal has been learned regarding the factors that influence shape-recognition processes in RPLC through numerous chromatographic and spectroscopic investigations conducted over the past two decades. In general, shape recognition is correlated with alkyl chain conformational order; consequently, any experimental variable or stationary-phase compositional factor that promotes order can promote shape selectivity. However, it is important to acknowledge that several pieces of experimental evidence would be crucial before a rigorous retention model for shape-selective processes could be formulated. This predominantly includes (1) a characterization of the potentially diverse adsorption sites on shape-selective chromatographic materials, (2) an improved surface characterization of shape-selective materials through ever-advancing microscopy techniques, and (3) a systematic chromatographic and spectroscopic characterization of novel and potentially shape-selective materials to build on the recognized conformational order parameters that control the shape-recognition process in RPLC separations.

LIST OF ACRONYMS AND SYMBOLS

2D-WISE	two-dimensional wide-line separation
AED	adsorption energy distribution
A_{eff}	effective contact area
AFM	atomic force microscopy
A_{min}	minimum cross-sectional area
$\alpha_{TBN/BaP}$	selectivity factor (TBN/BaP)
$\alpha_{TRI/o\text{-}TER}$	selectivity factor (TRI/o-TER)
BaP	benzo[a]pyrene
β-CD	β-cyclodextrin
CCD	charge-coupled device
co-ODA$_n$	multianchored poly(octadecyl)acrylate
CP-MAS	cross polarization-magic angle spinning
CSL	N-oxyl-4,4-dimethylspiro[oxazolidine-5α-cholestant]
DBG	dibutyl glutamide
DiI	1,1'-dioctadecyl-3,3,3',3'-tetramethyllindocarbocyanine perchlorate
DIPSHIFT	dipolar and chemical shift correlation
DPH	1,6-diphenylhexatriene
DSC	differential scanning calorimetry
DSG	disteryl glutamide
EPR	electron paramagnetic resonance
FRAPP	fluorescence recovery after pattern photobleaching
FTIR	Fourier transform infrared
gtg	*gauche-trans-gauche*
HPDEC	high-power decoupling
K	thermodynamic partition coefficient
k'	retention factor
L/B	length-to-breadth ratio
LFER	linear free-energy relationship
M	molecular mass
MC	Monte Carlo
MD	molecular dynamics
MPAH	methyl-substituted polycyclic aromatic hydrocarbon
MTBE	methyl *tert*-butyl ether
N	bonding density
n_c	number of carbon atoms
NMR	nuclear magnetic resonance
ODA$_n$	poly(octadecyl) acrylate
o-TER	o-terphenyl
PAH	polycyclic aromatic hydrocarbon
P_c	percent carbon
PCB	polychlorinated biphenyl
PhPh	phenanthro[3,4-c]phenanthrene
RP/WAX	reversed-phase/weak anion exchange

RPLC	reversed-phase liquid chromatography
R_S	peak resolution
S	specific surface area
σ'	steric solute parameter
$\sigma'S$	steric interaction term
S	column steric parameter
SAM	self-assembled monolayer
SANS	small angle neutron scattering
SFG	sum-frequency generation
SP	single-pulse
SPI	selective population inversion
SRM	Standard Reference Material
TBN	1,2:3,4:5,6:7,8-tetrabenzonaphthalene
T_c	critical temperature
T_{CH}	cross polarization time constants
TEMPO	2,2,6,6-tetramethyl-1-piperidinyloxy
TRI	triphenylene
V_{vdW}	van der Waal volume

REFERENCES

1. Katz, E.D. and Ogan, K.L., Selectivity factors for several PAH pairs on C18 bonded phase columns, *J. Liq. Chromatogr.*, 3, 1151, 1980.
2. Ogan, K., Katz, E., and Slavin, W., Determination of polycyclic aromatic hydrocarbons in aqueous samples by reversed-phase liquid chromatography, *Anal. Chem.*, 51, 1315, 1979.
3. Ogan, K. and Katz, E., Retention characteristics of several bonded-phase liquid-chromatography columns for some polycyclic aromatic hydrocarbons, *J. Chromatogr.*, 188, 115, 1980.
4. U.S. EPA test method polynuclear aromatic hydrocarbons—Method 610, Fed. Regist., 44, 233, 1979.
5. Amos, R., Evaluation of bonded phases for the high-performance liquid chromatographic determination of polycyclic aromatic hydrocarbons in effluent waters, *J. Chromatogr.*, 204, 469, 1981.
6. Wise, S.A., Bonnett, W.J., Guenther, F.R., and May, W.E., A relationship between reversed phase C18 liquid chromatographic retention and the shape of polycyclic aromatic hydrocarbons, *J. Chromatogr. Sci.*, 19, 457, 1981.
7. Sander, L.C. and Wise, S.A., Synthesis and characterization of polymeric C18 stationary phases for liquid chromatography, *Anal. Chem.*, 56, 504, 1984.
8. Wise, S.A., Benner, B.A., Jr., Liu, H., Byrd, G.D., and Colmsjö, A., Separation and identification of polycyclic aromatic hydrocarbon isomers of molecular weight 302 in complex mixtures, *Anal. Chem.*, 60, 630, 1988.
9. Sander, L.C., Lippa, K.A., and Wise, S.A., Order and disorder in alkyl stationary phases, *Anal. Bioanal. Chem.*, 382, 646, 2005.
10. Vailaya, A. and Horvath, C., Retention in reversed-phase chromatography: partition or adsorption? *J. Chromatogr. A*, 829, 1, 1998.
11. Vailaya, A., Fundamentals of reversed phase chromatography: thermodynamic and exothermodynamic treatment, *J. Liq. Chromatogr. Relat. Technol.*, 28, 965, 2005.

12. Tchapla, A., Heron, S., Lesellier, E., and Colin, H., General view of molecular interaction mechanisms in reversed-phase liquid-chromatography, *J. Chromatogr. A*, 656, 81, 1993.
13. Molnar, I., Searching for robust HPLC methods—Csaba Horvath and the solvophobic theory, *Chromatographia*, 62, 549, 2005.
14. Dorsey, J.G. and Dill, K.A., The molecular mechanism of retention in reversed-phase liquid-chromatography, *Chem. Rev.*, 89, 331, 1989.
15. Sander, L.C. and Wise, S.A., The influence of column temperature on selectivity in liquid chromatography for shape-constrained solutes, *J. Sep. Sci.*, 24, 910, 2001.
16. Gritti, F. and Guiochon, G., Critical contribution of nonlinear chromatography to the understanding of retention mechanism in reversed-phase liquid chromatography, *J. Chromatogr. A*, 1099, 1, 2005.
17. U.S. Department of Health and Human Services, Report on carcinogens, 11th edn., National Toxicology Program, 2005. Available at http://ntp.niehs.nih.gov/go/16183
18. Yang, S.K. and Silverman, B.D., *Polycyclic Aromatic Hydrocarbon Carcinogenesis: Structure–Activity Relationships*, CRC Press, Inc., Boca Raton, FL, 1988.
19. Zechmeister, L., *cis-trans Isomeric Carotenoids, Vitamins A, and Arylpolyenes*, Academic Press, Inc., New York, 1962.
20. Fetzer, J.C. and Biggs, W.R., The use of large polycyclic aromatic hydrocarbons to study retention in non-aqueous reversed-phase HPLC, *Chromatographia*, 27, 118, 1989.
21. Jinno, K., Molecular planarity recognition for polycyclic aromatic hydrocarbons in liquid chromatography, in Jinno, K. (Ed.), *Chromatographic Separations Based on Molecular Recognition*, Wiley-VCH, Inc., New York, 1997, p. 65.
22. Sander, L.C. and Wise, S.A., Retention and selectivity for polycyclic aromatic hydrocarbons in reversed-phase liquid chromatography, in Smith, R.M. (Ed.), *Retention and Selectivity Studies in HPLC*, Elsevier, Amsterdam, 1995, p. 337.
23. Wise, S.A., Sander, L.C., and May, W.E., Determination of polycyclic aromatic hydrocarbons by liquid chromatography, *J. Chromatogr.*, 642, 329, 1993.
24. Wise, S.A. and Sander, L.C., Molecular shape recognition for polycyclic aromatic hydrocarbons in reversed-phase liquid chromatography, in Jinno, K. (Ed.), *Chromatographic Separations Based on Molecular Recognition*, Wiley-VCH, Inc., New York, 1997, p. 1.
25. Wise, S.A., Sander, L.C., Lapouyade, R., and Garrigues, P., The anomalous behavior of selected methyl-substituted polycyclic aromatic hydrocarbons in reversed-phase liquid chromatography, *J. Chromatogr.*, 514, 111, 1990.
26. Dachtler, M., Kohler, K., and Albert, K., Reversed-phase high-performance liquid chromatographic identification of lutein and zeaxanthin stereoisomers in bovine retina using a C-30 bonded phase, *J. Chromatogr. B*, 720, 211, 1998.
27. Meyer, C., Skogsberg, U., Welsch, N., and Albert, K., Nuclear magnetic resonance and high-performance liquid chromatographic evaluation of polymer-based stationary phases immobilized on silica, *Anal. Bioanal. Chem.*, 382, 679, 2005.
28. Rimmer, C.A., Sander, L.C., and Wise, S.A., Selectivity of long chain stationary phases in reversed phase liquid chromatography, *Anal. Bioanal. Chem.*, 382, 698, 2005.
29. Sander, L.C., Sharpless, K.E., Craft, N.E., and Wise, S.A., Development of engineered stationary phases for the separation of carotenoid isomers, *Anal. Chem.*, 66, 1667, 1994.

30. Ströhschein, S., Pursch, M., Handel, H., and Albert, K., Structure elucidation of beta-carotene isomers by HPLC-NMR coupling using a C-30 bonded phase, *Fresenius J. Anal. Chem.*, 357, 498, 1997.
31. Olsson, M., Sander, L.C., and Wise, S.A., Comparison of the liquid chromatographic behaviour of selected steroid isomers using different reversed phase materials and mobile phase compositions, *J. Chromatogr.*, 537, 73, 1991.
32. Sander, L.C., Parris, R.M., Wise, S.A., and Garrigues, P., Shape discrimination in liquid chromatography using charge transfer phases, *Anal. Chem.*, 63, 2589, 1991.
33. Janini, G.M., Johnston, K., and Zielinski, W.L., Use of a nematic liquid-crystal for gas–liquid-chromatographic separation of polyaromatic hydrocarbons, *Anal. Chem.*, 47, 670, 1975.
34. Janini, G.M., Muschik, G.M., Schroer, J.A., and Zielinski, W.L., Gas–liquid-chromatographic evaluation and gas-chromatography mass spectrometric application of new high-temperature liquid-crystal stationary phases for polycyclic aromatic hydrocarbon separations, *Anal. Chem.*, 48, 1879, 1976.
35. Janini, G.M., Muschik, G.M., and Zielinski, W.L., N,N'-*bis*[para-butoxybenzylidene)-alpha,alpha$'$-bi-para-toluidine—thermally stable liquid-crystal for unique gas–liquid-chromatography separations of polycyclic aromatic-hydrocarbons, *Anal. Chem.*, 48, 809, 1976.
36. Radecki, A., Lamparczyk, H., and Kaliszan, R., A relationship between the retention indices on nematic and isotropic phases and the shape of polycyclic aromatic hydrocarbons, *Chromatographia*, 12, 595, 1979.
37. Sander, L.C. and Wise, S.A., Polycyclic Aromatic Hydrocarbon Structure Index, Natl. Inst. Stand. Technol. Spec. Publ. 922. U.S. Government Printing Office, Washington, 1997.
38. Yan, C. and Martire, D.E., Molecular theory of chromatographic selectivity enhancement for blocklike solutes in anisotropic stationary phases and its application, *Anal. Chem.*, 64, 1246, 1992.
39. Yan, C. and Martire, D.E., Molecular theory of chromatography for blocklike solutes in isotropic stationary phases and its application to supercritical fluid chromatographic retention of PAHs, *J. Phys. Chem.*, 96, 7510, 1992.
40. Garrigues, P., Radke, M., Druez, O., Willsch, H., and Bellocq, J., Reversed-phase liquid chromatographic retention behaviour of dimethylphenanthrene isomers, *J. Chromatogr.*, 473, 207, 1989.
41. Sander, L.C. and Wise, S.A., Determination of column selectivity toward polycyclic aromatic hydrocarbons, *J. High Resolut. Chromatogr. Chromatogr. Commun.*, 11, 383, 1988.
42. Sander, L.C. and Wise, S.A., Evaluation of shape selectivity in liquid chromatography, *LC GC*, 8, 378, 1990.
43. Tanaka, N., Tokuda, Y., Iwaguchi, K., and Araki, M., Effect of stationary-phase structure on retention and selectivity in reversed phase liquid chromatography, *J. Chromatogr.*, 239, 761, 1982.
44. Certificate of Analysis, Standard Reference Material 869a column selectivity test mixture for liquid chromatography (polycyclic aromatic hydrocarbons), National Institute of Standards and Technology (NIST), Gaithersburg, MD, 1998. Available at http://www.nist.gov/SRM
45. Engelhardt, H., Nikolov, M., Arangio, M., and Scherer, M., Studies on shape selectivity of RP C18-columns, *Chromatographia*, 48, 183, 1998.

46. Lippa, K.A., Sander, L.C., and Wise, S.A., Chemometric studies of polycyclic aromatic hydrocarbon shape selectivity in reversed-phase liquid chromatography, *Anal. Bioanal. Chem.*, 378, 365, 2004.

47. Wise, S.A. and Sander, L.C., Factors affecting the reversed-phase liquid chromatographic separation of polycyclic aromatic hydrocarbon isomers, *J. High Resolut. Chromatogr. Chromatogr. Commun.*, 8, 248, 1985.

48. Wilson, N.S., Dolan, J.W., Snyder, L.R., Carr, P.W., and Sander, L.C., Column selectivity in reversed-phase liquid chromatography iii. The physico-chemical basis of selectivity, *J. Chromatogr. A*, 961, 217, 2002.

49. Wilson, N.S., Nelson, M.D., Dolan, J.W., Snyder, L.R., Wolcott, R.G., and Carr, P.W., Column selectivity in reversed-phase liquid chromatography i. A general quantitative relationship, *J. Chromatogr. A*, 961, 171, 2002.

50. Nogueira, R., Lubda, D., Leitner, A., Bicker, W., Maier, N.M., Lammerhofer, M., and Lindner, W., Silica-based monolithic columns with mixed-mode reversed-phase/weak anion-exchange selectivity principle for high-performance liquid chromatography, *J. Sep. Sci.*, 29, 966, 2006.

51. Kayillo, S., Dennis, G.R., and Shalliker, R.A., An assessment of the retention behaviour of polycyclic aromatic hydrocarbons on reversed phase stationary phases: selectivity and retention on C18 and phenyl-type surfaces, *J. Chromatogr. A*, 1126, 283, 2006.

52. Chowdhury, M.A.J., Boysen, R.I., Ihara, H., and Hearn, M.T.W., Binding behavior of crystalline and noncrystalline phases: evaluation of the enthalpic and entropic contributions to the separation selectivity of nonpolar solutes with a novel chromatographic sorbent, *J. Phys. Chem. B*, 106, 11936, 2002.

53. Chowdhury, M.A.J., Ihara, H., Sagawa, T., and Hirayama, C., Recognition of critical pairs of polycyclic aromatics on crystalline, liquid-crystalline and isotropic regions of silica-supported polymer in HPLC, *Chromatographia*, 52, 45, 2000.

54. Chowdhury, M.A.J., Ihara, H., Sagawa, T., and Hirayama, C., Retention versatility of silica-supported comb-shaped crystalline and non-crystalline phases in high-performance liquid chromatography, *J. Chromatogr. A*, 877, 71, 2000.

55. Chowdhury, M.A., Ihara, H., Sagawa, T., and Hirayama, C., Retention behaviors of polycyclic aromatic hydrocarbons on comb-shaped polymer immobilized-silica in RPLC, *J. Liq. Chromatogr. Relat. Technol.*, 23, 2289, 2000.

56. Srinivasan, G., Pursch, M., Sander, L.C., and Müller, K., FTIR studies of C30 self-assembled monolayers on silica, titania and zirconia, *Langmuir*, 20, 1746, 2004.

57. Sander, L.C. and Wise, S.A., in Giddings, J.C., Grushka, E., Cazes, J., Brown, P.R. (Eds.), Adv. Chromatogr. Marcel Dekker, New York, 1986, p. 139.

58. Sander, L.C., Pursch, M., and Wise, S.A., Shape selectivity for constrained solutes in reversed-phase liquid chromatography, *Anal. Chem.*, 71, 4821, 1999.

59. Nawrocki, J., The silanol group and its role in liquid chromatography, *J. Chromatogr. A*, 779, 29, 1997.

60. Berendsen, G.E. and de Galan, L., Preparation and chromatographic properties of some chemically bonded phases for reversed-phase liquid chromatography, *J. Liq. Chromatogr.*, 1, 561, 1978.

61. Sander, L.C. and Wise, S.A., The influence of stationary-phase chemistry on shape recognition in liquid chromatography, *Anal. Chem.*, 67, 3284, 1995.

62. Adrianov, K.A. and Izmaylov, B.A., Hydrolytic poly-condensation of higher alkyltrichlorosilanes, *J. Organomet. Chem.*, 8, 435, 1967.

63. Sander, L.C. and Wise, S.A., Influence of substrate parameters on column selectivity with alkyl bonded phase sorbents, *J. Chromatogr.*, 316, 163, 1984.
64. Fatunmbi, H.O., Bruch, M.D., and Wirth, M.J., Si-29 and C-13 NMR characterization of mixed horizontally polymerized monolayers on silica-gel, *Anal. Chem.*, 65, 2048, 1993.
65. Ulman, A., Self-assembled monolayers of alkyltrichlorosilanes—building-blocks for future organic materials, *Adv. Mater.*, 2, 573, 1990.
66. Orendorff, C.J. and Pemberton, J.E., Alkylsilane-based stationary phases via a displaceable surface template approach: synthesis, characterization, and chromatographic performance, *Anal. Chem.*, 77, 6069, 2005.
67. Gee, M.L., Healy, T.W., and White, L.R., Hydrophobicity effects in the condensation of water films on quartz, *J. Colloid Interface Sci.*, 140, 450, 1990.
68. Fairbank, R.W.P. and Wirth, M.J., Role of surface-adsorbed water in the horizontal polymerization of trichlorosilanes, *J. Chromatogr. A*, 830, 285, 1999.
69. Fairbank, R.W.P., Xiang, Y., and Wirth, M.J., Use of methyl spacers in a mixed horizontally polymerized stationary-phase, *Anal. Chem.*, 67, 3879, 1995.
70. Wirth, M.J. and Fatunmbi, H.O., Horizontal polymerization of mixed trifunctional silanes on silica—a potential chromatographic stationary-phase, *Anal. Chem.*, 64, 2783, 1992.
71. Wirth, M.J. and Fatunmbi, H.O., Horizontal polymerization of mixed trifunctional silanes on silica. 2. Application to chromatographic silica-gel, *Anal. Chem.*, 65, 822, 1993.
72. Wirth, M.J., Fairbank, R.W.P., and Fatunmbi, H.O., Mixed self-assembled monolayers in chemical separations, *Science*, 275, 44, 1997.
73. Parikh, A.N., Schivley, M.A., Koo, E., Seshadri, K., Aurentz, D., Mueller, K., and Allara, D.L., n-Alkylsiloxanes: from single monolayers to layered crystals. The formation of crystalline polymers from the hydrolysis of n-octadecyltrichlorosilane, *J. Am. Chem. Soc.*, 119, 3135, 1997.
74. Schreiber, F., Structure and growth of self-assembling monolayers, *Prog. Surf. Sci.*, 65, 151, 2000.
75. Schreiber, F., Self-assembled monolayers: from "simple" model systems to biofunctionalized interfaces, *J. Phys.: Condens. Matter*, 16, R881–R900, 2004.
76. Ulman, A., Formation and structure of self-assembled monolayers, *Chem. Rev.*, 96, 1533, 1996.
77. Fujimoto, C., Maekawa, A., Murao, Y., Jinno, K., and Takeichi, T., An attempt directed toward enhanced shape selectivity in reversed-phase liquid chromatography: preparation of the dodecylaminated beta-cyclodextrin-bonded phase, *Anal. Sci.*, 18, 65, 2002.
78. Takafuji, M., Fukui, M., Ansarian, H.R., Derakhshan, M., Shundo, A., and Ihara, H., Conformational effect of silica-supported poly(octadecyl acrylate) on molecular-shape selectivity of polycyclic aromatic hydrocarbons in RP-HPLC, *Anal. Sci.*, 20, 1681, 2004.
79. Shundo, A., Nakashima, R., Fukui, M., Takafuji, M., Nagaoka, S., and Ihara, H., Enhancement of molecular-shape selectivity in high-performance liquid chromatography through multi-anchoring of comb-shaped polymer on silica, *J. Chromatogr. A*, 1119, 115, 2006.
80. Rahman, M.M., Takafuji, M., Ansarian, H.R., and Ihara, H., Molecular shape selectivity through multiple carbonyl-pi interactions with noncrystalline solid phase for RP-HPLC, *Anal. Chem.*, 77, 6671, 2005.

81. Takafuji, M., Rahman, A.M., Ansarian, H.R., Derakhshan, M., Sakurai, T., and Ihara, H., Dioctadecyl L-glutamide-derived lipid-grafted silica as a novel organic stationary-phase for RP-HPLC, *J. Chromatogr. A*, 1074, 223, 2005.

82. Ansarian, H.R., Derakhshan, M., Rahman, M.M., Sakurai, T., Takafuji, M., Taniguchi, I., and Ihara, H., Evaluation of microstructural features of a new polymeric organic stationary-phase grafted on silica surface: a paradigm of characterization of HPLC-stationary phases by a combination of suspension-state H-1 NMR and solid-state C-13-CP/MAS-NMR, *Anal. Chim. Acta*, 547, 179, 2005.

83. Horak, J., Maier, N.M., and Lindner, W., Investigations on the chromatographic behavior of hybrid reversed-phase materials containing electron donor-acceptor systems ii. Contribution of pi-pi aromatic interactions, *J. Chromatogr. A*, 1045, 43, 2004.

84. Sandoval, J.E. and Pesek, J.J., Synthesis and characterization of a hydride-modified porous silica material as an intermediate in the preparation of chemically bonded chromatographic stationary phases, *Anal. Chem.*, 61, 2067, 1989.

85. Pesek, J.J., Matyska, M.T., Sandoval, J.E., and Williamsen, E.J., Synthesis, characterization and applications of hydride-based surface materials for HPLC, HPCE and electrochromatography, *J. Liq. Chromatogr. Relat. Technol.*, 19, 2843, 1996.

86. Pesek, J.J., Matyska, M.T., Williamsen, E.J., Evanchic, M., Hazari, V., Konjuh, K., Takhar, S., and Tranchina, R., Synthesis and characterization of alkyl bonded phases from a silica hydride via hydrosilation with free radical initiation, *J. Chromatogr. A*, 786, 219, 1997.

87. Pesek, J.J., Matyska, M.T., and Takhar, S., Synthesis and characterization of long chain alkyl stationary phases on a silica hydride surface, *Chromatographia*, 48, 631, 1998.

88. Pesek, J.J., Matyska, M.T., Oliva, M., and Evanchic, M., Synthesis and characterization of bonded phases made via hydrosilation of alkynes on silica hydride surfaces, *J. Chromatogr. A*, 818, 145, 1998.

89. Sunseri, J.D., Gedris, T.E., Stiegman, A.E., and Dorsey, J.G., Complete methylation of silica surfaces: next generation of reversed-phase liquid chromatography stationary phases, *Langmuir*, 19, 8608, 2003.

90. Stiegman, A.E., Dorsey, J.G., Sunseri, D., and Karpf, J., New synthetic approaches to reverse-phase column preparation, *Am. Chem. Soc.*, 231, 2006, Abstracts.

91. Kobayashi, K., Kitagawa, S., and Ohtani, H., Development of capillary column packed with thiol-modified gold-coated polystyrene particles and its selectivity for aromatic compounds, *J. Chromatogr. A*, 1110, 95, 2006.

92. Liu, X.D., Bordunov, A., Tracy, M., Slingsby, R., Avdalovic, N., and Pohl, C., Development of a polar-embedded stationary-phase with unique properties, *J. Chromatogr. A*, 1119, 120, 2006.

93. Wegmann, J., Albert, K., Pursch, M., and Sander, L.C., Poly(ethylene-co-acrylic acid) stationary phases for the separation of shape-constrained isomers, *Anal. Chem.*, 73, 1814, 2001.

94. Sentell, K.B. and Dorsey, J.G., Retention mechanisms in reversed-phase chromatography. Stationary-phase bonding density and solute selectivity, *J. Chromatogr.*, 461, 193, 1989.

95. Wise, S.A. and May, W.E., Effect of C18 surface coverage on selectivity in reversed phase chromatography of polycyclic aromatic hydrocarbons, *Anal. Chem.*, 55, 1479, 1983.

96. Wirth, M.J. and Fairbank, R.W.P., Preparation of mixed C-18/C-1 horizontally polymerized chromatographic phases, *J. Liq. Chromatogr. Relat. Technol.*, 19, 2799, 1996.

97. Sentell, K.B. and Dorsey, J.G., Retention mechanisms in reversed-phase liquid chromatography. Stationary-phase bonding density and partitioning, *Anal. Chem.*, 61, 930, 1989.

98. Berendsen, G.E. and de Galan, L., Role of the chain length of chemically bonded phases and the retention mechanism in reversed-phase liquid chromatography, *J. Chromatogr.*, 196, 21, 1980.

99. Egelhaaf, H.J., Oelkrug, D., Pursch, M., and Albert, K., Combined fluorescence and NMR studies of reversed HPLC stationary phases with different ligand chain lengths, *J. Fluorescence*, 7, 311, 1997.

100. Hansen, R.H. and Harris, J.M., Lateral diffusion of molecules partitioned into silica-bound alkyl chains: influence of chain length and bonding density, *Anal. Chem.*, 68, 2879, 1996.

101. Jinno, K., Effect of the alkyl chain length of the bonded stationary-phase on solute retention in reversed phase high performance liquid chromatography, *Chromatographia*, 15, 667, 1982.

102. Lochmüller, C.H. and Wilder, D.R., The sorption behavior of alkyl bonded phases in reversed-phase, high performance liquid chromatography, *J. Chromatogr. Sci.*, 17, 574, 1979.

103. Pursch, M., Brindle, R., Ellwanger, A., Sander, L.C., Bell, C.M., Haendel, H., and Albert, K., Stationary interphases with extended alkyl chains: a comparative study on chain order by solid-state NMR spectroscopy, *Solid State Nucl. Magn. Res.*, 9, 191, 1997.

104. Sander, L.C. and Wise, S.A., Effect of phase length on column selectivity for the separation of polycyclic aromatic hydrocarbons by reversed-phase liquid chromatography, *Anal. Chem.*, 59, 2309, 1987.

105. Tanaka, N., Sakagami, K., and Araki, M., Effect of alkyl chain length of stationary-phase on retention and selectivity in reversed-phase liquid chromatography, *Chem. Lett.*, 587, 1980.

106. Bell, C.M., Sander, L.C., Fetzer, J.C., and Wise, S.A., Synthesis and characterization of extended length alkyl stationary phases for liquid chromatography with application to the separation of carotenoid isomers, *J. Chromatogr. A*, 753, 37, 1996.

107. Bell, C.M., Sander, L.C., and Wise, S.A., Temperature dependence of carotenoids on C18, C30, and C34 bonded stationary phases, *J. Chromatogr. A*, 757, 29, 1997.

108. Sander, L.C. and Wise, S.A., Subambient temperature modification of selectivity in reversed-phase liquid chromatography, *Anal. Chem.*, 61, 1749, 1989.

109. Sander, L.C. and Wise, S.A., Shape selectivity in reversed-phase liquid chromatography for the separation of planar and non-planar solutes, *J. Chromatogr. A*, 656, 335, 1993.

110. Saito, Y., Ohta, H., and Jinno, K., Chromatographic separation of fullerenes, *Anal. Chem.*, 76, 266A, 2004.

111. Maroncelli, M., Qi, S.P., Strauss, H.L., and Snyder, R.G., Nonplanar conformers and the phase-behavior of solid normal-alkanes, *J. Am. Chem. Soc.*, 104, 6237, 1982.

112. Snyder, R.G. and Schachtschneider, J.H., Vibrational analysis of the *n*-paraffins. i. Assignments of infrared bands in the spectra of C_3H_8 through *n*-$C_{19}H_{40}$, *Spectrochim. Acta*, 19, 85, 1963.

113. Snyder, R.G., Vibrational study of the chain conformation of the liquid *n*-paraffins and molten polyethylene, *J. Chem. Phys.*, 47, 1316, 1967.

114. Snyder, R.G., Maroncelli, M., Qi, S.P., and Strauss, H.L., Phase-transitions and nonplanar conformers in crystalline normal-alkanes, *Science*, 214, 188, 1981.

115. Sander, L.C., Callis, J.B., and Field, L.R., Fourier transform infrared spectrometric determination of alkyl chain conformation on chemically bonded reversed phase liquid chromatography packings, *Anal. Chem.*, 55, 1068, 1983.

116. Jinno, K., Wu, J., Ichikawa, M., and Takata, I., Characterization of octadecylsilica stationary phases by spectrometric methods, *Chromatographia*, 37, 627, 1993.

117. Singh, S., Wegmann, J., Albert, K., and Müller, K., Variable temperature FT-IR studies on *n*-alkyl modified silicas, *J. Phys. Chem. B*, 106, 878, 2002.

118. Srinivasan, G., Neumann-Singh, S., and Müller, K., Conformational order of *n*-alkyl modified silica gels as evaluated by Fourier transform infrared spectroscopy, *J. Chromatogr. A*, 1074, 31, 2005.

119. Srinivasan, G., Meyer, C., Welsch, N., Albert, K., and Müller, K., Influence of synthetic routes on the conformational order and mobility of C-18 and C-30 stationary phases, *J. Chromatogr. A*, 1113, 45, 2006.

120. Srinivasan, G., Sander, L.C., and Müller, K., Effect of surface coverage on the conformation and mobility of C-18-modified silica gels, *Anal. Bioanal. Chem.*, 384, 514, 2006.

121. Henry, M.C., Wolf, L.K., and Messmer, M.C., *In situ* examination of the structure of model reversed-phase chromatographic interfaces by sum-frequency generation spectroscopy, *J. Phys. Chem. B*, 107, 2765, 2003.

122. Srinivasan, G. and Müller, K., Influence of solvents on the conformational order of C-18 alkyl modified silica gels, *J. Chromatogr. A*, 1110, 102, 2006.

123. Larsson, K. and Rand, R.P., Detection of changes in the environment of hydrocarbon chains by Raman spectroscopy and its application to lipid-protein systems, *Biochim. Biophys. Acta*, 326, 245, 1973.

124. Orendorff, C.J., Ducey, M.W., and Pemberton, J.E., Quantitative correlation of Raman spectral indicators in determining conformational order in alkyl chains, *J. Phys. Chem. A*, 106, 6991, 2002.

125. Liao, Z.H., Orendorff, C.J., Sander, L.C., and Pemberton, J.E., Structure–function relationships in high-density docosylsilane bonded stationary phases by Raman spectroscopy and comparison to octadecylsilane bonded stationary phases, *Anal. Chem.*, 78, 5813, 2006.

126. Orendorff, C.J. and Pemberton, J.E., Raman spectroscopic study of the conformational order of octadecylsilane stationary phases: effects of electrolyte and pH, *Anal. Bioanal. Chem.*, 382, 691, 2005.

127. Orendorff, C.J., Ducey, M.W., Jr., Pemberton, J.E., and Sander, L.C., Structure–function relationships in high density octadecylsilane stationary phases by Raman spectroscopy: 3. Effects of self-associating solvents, *Anal. Chem.*, 3360, 2003.

128. Orendorff, C.J., Ducey, M.W., Jr., Pemberton, J.E., and Sander, L.C., Structure–function relationships in high density octadecylsilane stationary phases by Raman spectroscopy: 4. Effects of neutral and basic aromatic compounds, *Anal. Chem.*, 75, 3369, 2003.

129. Pemberton, J.E., Ho, M.K., Orendorff, C.J., and Ducey, M.W., Raman spectroscopy of octadecylsilane stationary-phase conformational order—effect of solvent, *J. Chromatogr. A*, 913, 243, 2001.

130. Doyle, C.A., Vickers, T.J., Mann, C.K., and Dorsey, J.G., Characterization of liquid chromatographic stationary phases by Raman spectroscopy—effects of ligand type, *J. Chromatogr. A*, 779, 91, 1997.

131. Doyle, C.A., Vickers, T.J., Mann, C.K., and Dorsey, J.G., Characterization of C-18-bonded liquid chromatographic stationary phases by Raman spectroscopy: the effect of temperature, *J. Chromatogr. A*, 877, 41, 2000.

132. Doyle, C.A., Vickers, T.J., Mann, C.K., and Dorsey, J.G., Characterization of C18 bonded liquid chromatographic stationary phases by Raman spectroscopy: the effect of mobile phase composition, *J. Chromatogr. A*, 877, 25, 2000.

133. Ho, M. and Pemberton, J.E., Alkyl chain conformation of octadecylsilane stationary phases by Raman spectroscopy. 1. Temperature dependence, *Anal. Chem.*, 70, 4915, 1998.

134. Ho, M., Cai, M., and Pemberton, J.E., Characterization of octadecylsilane stationary phases on commercially available silica-based packing materials by Raman spectroscopy, *Anal. Chem.*, 69, 2613, 1997.

135. Ducey, M.W., Jr., Orendorff, C.J., Pemberton, J.E., and Sander, L.C., Structure–function relationships in high density octadecylsilane stationary phases by Raman spectroscopy: 1. Effects of temperature, surface coverage and preparation procedure, *Anal. Chem.*, 5576, 2002.

136. Ducey, M.W., Jr., Orendorff, C.J., Pemberton, J.E., and Sander, L.C., Structure–function relationships in high density octadecylsilane stationary phases by Raman spectroscopy: 2. Effect of common mobile phase solvents, *Anal. Chem.*, 5585, 2002

137. Liao, Z., Orendorff, C.J., and Pemberton, J.E., Effect of pressurized solvent environments on the alkyl chain order of octadecylsilane stationary phases by Raman spectroscopy, *Chromatographia*, 64, 139, 2006.

138. Bierbaum, K., Grunze, M., Baski, A.A., Chi, L.F., Schrepp, W., and Fuchs, H., Growth of self-assembled *n*-alkyltrichlorosilane films on Si(100) investigated by atomic-force microscopy, *Langmuir*, 11, 2143, 1995.

139. Pursch, M., Vanderhart, D.L., Sander, L.C., Gu, X., Nguyen, T., and Wise, S.A., C30 self-assembled monolayers on silica, titania and zirconia: HPLC performance, atomic force microscopy and NMR studies of molecular dynamics and uniformity of coverage, *J. Am. Chem. Soc.*, 6997, 2000.

140. Burbage, J.D. and Wirth, M.J., Effect of wetting on the reorientation of acridine-orange at the interface of water and a hydrophobic surface, *J. Phys. Chem.*, 96, 5943, 1992.

141. Wirth, M.J., Ludes, M.D., and Swinton, D.J., Spectroscopic observation of adsorption to active silanols, *Anal. Chem.*, 71, 3911, 1999.

142. Wirth, M.J., Swinton, D.J., and Ludes, M.D., Adsorption and diffusion of single molecules at chromatographic interfaces, *J. Phys. Chem. B*, 107, 6258, 2003.

143. Ludes, M.D., Anthony, S.R., and Wirth, M.J., Fluorescence imaging of the desorption of dye from fused silica versus silica gel, *Anal. Chem.*, 75, 3073, 2003.

144. Legg, M.A. and Wirth, M.J., Probing topography and tailing for commercial stationary phases using AFM, FT-IR, and HPLC, *Anal. Chem.*, 78, 6457, 2006.

145. Sentell, K.B., Nuclear-magnetic-resonance and electron-spin-resonance spectroscopic investigations of reversed-phase liquid-chromatographic retention mechanisms—stationary-phase structure, *J. Chromatogr. A*, 656, 231, 1993.

146. Pursch, M., Sander, L.C., and Albert, K., Understanding reversed-phase liquid chromatography (RPLC) through solid state NMR spectroscopy, *Anal. Chem.*, 71, 733A, 1999.

147. Albert, K., NMR investigations of stationary phases, *J. Sep. Sci.*, 26, 215, 2003.

148. Srinivasan, G., Sander, L.C., and Müller, K., Effect of surface coverage on the conformation and mobility of C-18-modified silica gels, *Anal. Bioanal. Chem.*, 384, 514, 2006.

149. Grant, D.M. and Paul, E.G., Carbon-13 magnetic resonance. ii. Chemical shift data for the alkanes, *J. Am. Chem. Soc.*, 86, 2984, 1964.

298 Advances in Chromatography, Volume 46

150. Pursch, M., Ströhschein, S., Handel, H., and Albert, K., Temperature-dependent behavior of C30 interphases. A solid-state NMR and LC-NMR study, *Anal. Chem.*, 68, 386, 1996.

151. Jinno, K., Ibuki, T., Tanaka, N., Okamoto, M., Fetzer, J.C., Biggs, W.R., Griffiths, P.R., and Olinger, J.M., Retention behaviour of large polycyclic aromatic hydrocarbons in reversed-phase liquid chromatography on a polymeric octadecylsilica stationary phase, *J. Chromatogr.*, 461, 209, 1989.

152. Pursch, M., Sander, L.C., Egelhaaf, H.J., Raitza, M., Wise, S.A., Oelkrug, D., and Albert, K., Architecture and dynamics of C22 bonded interphases, *J. Am. Chem. Soc.*, 121, 3201, 1999.

153. Pursch, M. Ph.D. Dissertation "Neue Synthesewege zur Darstellung von langkettigen Alklinterphasen-Untersuchungen zur Oberflä—chenmorphologie durch Festkö rper-NMR-Spektroskopie und HPLC," Universität Tübingen, Tübingen, Germany, 1997.

154. Ströhschein, S., Pursch, M., Lubda, D., and Albert, K., Shape selectivity of C-30 phases for RP-HPLC separation of tocopherol isomers and correlation with MAS NMR data from suspended stationary phases, *Anal. Chem.*, 70, 13, 1998.

155. Srinivasan, G., Kyrlidis, A., McNeff, C., and Müller, K., Investigation on conformational order and mobility of diamondbond-C18 and C18-alkyl modified silica gels by Fourier transform infrared and solid-state NMR spectroscopy, *J. Chromatogr. A*, 1081, 132, 2005.

156. Zeigler, R.C. and Maciel, G.E., A study of the structure and dynamics of dimethyloctadecylsilyl-modified silica using wide-line ^{2}H NMR techniques, *J. Am. Chem. Soc.*, 113, 6349, 1991.

157. Gangoda, M., Gilpin, R.K., and Figueirinhas, J., Deuterium nuclear magnetic resonance studies of alkyl modified silica, *J. Phys. Chem.*, 93, 4815, 1989.

158. Kelusky, E.C. and Fyfe, C.A., Molecular motions of alkoxysilanes immobilized on silica surfaces: a deuterium NMR study, *J. Am. Chem. Soc.*, 108, 1746, 1986.

159. Sindorf, D.W. and Maciel, G.E., Silicon-29 nuclear magnetic resonance study of hydroxyl sites on dehydrated silica gel surfaces, using silylation as a probe, *J. Phys. Chem.*, 87, 5516, 1983.

160. Gilpin, R.K. and Gangoda, M.E., Nuclear magnetic resonance spectrometry of alkyl ligands immobilized on reversed phase liquid chromatographic surfaces, *Anal. Chem.*, 56, 1470, 1984.

161. Albert, K., Evers, B., and Bayer, E., NMR investigation of the dynamic behavior of alkyl-modified silica, *J. Magn. Reson.*, 62, 428, 1985.

162. Raitza, M., Wegmann, J., Bachmann, S., and Albert, K., Investigating the surface morphology of triacontyl phases with spin-diffusion solid-state NMR spectroscopy, *Angew. Chem., Int. Ed. Engl.*, 39, 3486, 2000.

163. Meyer, C., Busche, S., Welsch, N., Wegmann, J., Gauglitz, G., and Albert, K., Contact-angle, ellipsometric, and spin-diffusion solid-state nuclear magnetic resonance spectroscopic investigations of copolymeric stationary phases immobilized on SiO$_2$ surfaces, *Anal. Bioanal. Chem.*, 382, 1465, 2005.

164. Meyer, C., Pascui, O., Reichert, D., Sander, L.C., Wise, S.A., and Albert, K., Conformational temperature dependence of a poly(ethylene-co-acrylic acid) stationary-phase investigated by nuclear magnetic resonance spectroscopy and liquid chromatography, *J. Sep. Sci.*, 29, 820, 2006.

165. Lochmüller, C.H., Marshall, D.B., and Harris, J.M., Room temperature, excitation wavelength dependent fluorescence at surfaces: a potential method for studying the micro heterogeneity of surface environments, *Anal. Chim. Acta*, 131, 263, 1981.

166. Ståhlberg, J. and Almgren, M., Polarity of chemically modified silica surfaces and its dependence on mobile-phase composition by fluorescence spectrometry, *Anal. Chem.*, 57, 817, 1985.

167. Carr, J.W. and Harris, J.M., Fluorescence studies of the stationary-phase chemical environment in reversed-phase liquid chromatography, *Anal. Chem.*, 58, 626, 1986.

168. Carr, J.W. and Harris, J.M., *In situ* fluorescence detection of polycyclic aromatic hydrocarbons following pre-concentration on alkylated silica adsorbents, *Anal. Chem.*, 60, 698, 1988.

169. Bogar, R.G., Thomas, J.C., and Callis, J.B., Lateral diffusion of solutes bound to the alkyl surface of C18 reversed phase liquid chromatographic packings, *Anal. Chem.*, 56, 1080, 1984.

170. Rangnekar, V.M., Foley, J.T., and Oldham, P.B., Investigation of chromatographic surface viscosity using total internal-reflection fluorescence, *Appl. Spectrosc.*, 46, 827, 1992.

171. Ståhlberg, J., Almgren, M., and Alsins, J., Mobility of pyrene on chemically modified silica surfaces, *Anal. Chem.*, 60, 2487, 1988.

172. Hansen, R.L. and Harris, J.M., Lateral diffusion of molecules partitioned into C-18 ligands on silica surfaces, *Anal. Chem.*, 67, 492, 1995.

173. Kovaleski, J.M. and Wirth, M.J., Lateral diffusion of an adsorbate at chromatographic octadecylsiloxane/water interfaces of varying hydrocarbon density, *J. Phys. Chem. B*, 101, 5545, 1997.

174. Kovaleski, J.M. and Wirth, M.J., Temperature dependence of the lateral diffusion of acridine orange at water/hydrocarbon interfaces, *J. Phys. Chem.*, 100, 10304, 1996.

175. Zulli, S.L., Kovaleski, J.M., Zhu, X.R., Harris, J.M., and Wirth, M.J., Lateral diffusion of an adsorbate at a chromatographic C-18 water interface, *Anal. Chem.*, 66, 1708, 1994.

176. Ludes, M.D. and Wirth, M.J., Single-molecule resolution and fluorescence imaging of mixed-mode sorption of a dye at the interface of C-18 and acetonitrile/water, *Anal. Chem.*, 74, 386, 2002.

177. Wirth, M.J. and Swinton, D.J., Single-molecule probing of mixed-mode adsorption at a chromatographic interface, *Anal. Chem.*, 70, 5264, 1998.

178. Zhong, Z.M., Lowry, M., Wang, G.F., and Geng, L., Probing strong adsorption of solute onto *cis*-silica gel by fluorescence correlation imaging and single-molecule spectroscopy under RPLC conditions, *Anal. Chem.*, 77, 2303, 2005.

179. Wirth, M.J. and Burbage, J.D., Adsorbate reorientation at a water (octadecylsilyl)silica interface, *Anal. Chem.*, 63, 1311, 1991.

180. Wright, P.B., Lamb, E., Dorsey, J.G., and Kooser, R.G., Microscopic order as a function of surface coverage in alkyl-modified silicas—spin probe studies, *Anal. Chem.*, 64, 785, 1992.

181. Ottaviani, M.F., Cangiotti, M., Famiglini, G., and Cappiello, A., Adsorption of pure and mixed solvent solutions of spin probes onto stationary phases, *J. Phys. Chem. B*, 110, 10421, 2006.

182. Beaufils, J.P., Hennion, M.C., and Rosset, R., Neutron scattering study of alkyl chain motion on reversed phase liquid chromatographic packings, *Anal. Chem.*, 57, 2593, 1985.

183. Beaufils, J.P., Hennion, M.C., and Rosset, R., Segmental motion of alkyl chains grafted on silica-gel, studied by neutron-scattering, *J. Phys. (Paris)*, 44, 497, 1983.

184. Glinka, C.J., Sander, L.C., Wise, S.A., and Berk, N.F., Characterization of chemically modified pore surfaces by small angle neutron scattering, *Mat. Res. Soc. Symp. Proc.*, 166, 415, 1990.

185. Sander, L.C., Glinka, C.J., and Wise, S.A., Determination of bonded phase thickness in liquid chromatography by small angle neutron scattering, *Anal. Chem.*, 62, 1099, 1990.
186. Wasserman, S.R., Whitesides, G.M., Tidswell, I.M., Ocko, B.M., Pershan, P.S., and Axe, J.D., The structure of self-assembled monolayers of alkylsiloxanes on silicon: a comparison of results from ellipsometry and low-angle X-ray reflectivity, *J. Am. Chem. Soc.*, 111, 5852, 1989.
187. Hansen, S.J. and Callis, J.B., Differential scanning calorimetry of reversed phase LC packings, *J. Chromatogr. Sci.*, 21, 560, 1983.
188. Jinno, K., Nagoshi, T., Tanaka, N., Okamoto, M., Fetzer, J.C., and Biggs, W.R., Effect of column temperature on the retention of peropyrene-type polycyclic aromatic hydrocarbons on various chemically bonded stationary phases in reversed-phase liquid chromatography, *J. Chromatogr.*, 436, 1, 1988.
189. Wheeler, J.F., Beck, T.L., Klatte, S.J., Cole, L.A., and Dorsey, J.G., Phase-transitions of reversed-phase stationary phases—cause and effects in the mechanism of retention, *J. Chromatogr. A*, 656, 317, 1993.
190. Chowdhury, M.A.J., Boysen, R.I., Ihara, H., and Hearn, M.T.W., Binding behavior of crystalline and noncrystalline phases: evaluation of the enthalpic and entropic contributions to the separation selectivity of nonpolar solutes with a novel chromatographic sorbent, *J. Phys. Chem. B*, 106, 11936, 2002.
191. Gritti, F., Felinger, A., and Guiochon, G., Overloaded gradient elution chromatography on heterogeneous adsorbents in reversed-phase liquid chromatography, *J. Chromatogr. A*, 1017, 45, 2003.
192. Gritti, F. and Guiochon, G., Effect of the mobile phase composition on the isotherm parameters and the high concentration band profiles in reversed-phase liquid chromatography, *J. Chromatogr. A*, 995, 37, 2003.
193. Gritti, F. and Guiochon, G., Effect of the pH, the concentration and the nature of the buffer on the adsorption mechanism of an ionic compound in reversed-phase liquid chromatography. ii. Analytical and overload band profiles on Symmetry-C-18 and Xterra-C-18, *J. Chromatogr. A*, 1041, 63, 2004.
194. Gritti, F. and Guiochon, G., Effect of the ionic strength of salts on retention and overloading behavior of ionizable compounds in reversed-phase liquid chromatography. ii. Symmetry-C-18, *J. Chromatogr. A*, 1033, 57, 2004.
195. Gritti, F. and Guiochon, G., Effect of the ionic strength of salts on retention and overloading behavior of ionizable compounds in reversed-phase liquid chromatography. i. Xterra-C-18, *J. Chromatogr. A*, 1033, 43, 2004.
196. Gritti, F. and Guiochon, G., Influence of the pressure on the properties of chromatographic columns i. Measurement of the compressibility of methanol–water mixtures on a mesoporous silica adsorbent, *J. Chromatogr. A*, 1070, 1, 2005.
197. Gritti, F. and Guiochon, G., Influence of the pressure on the properties of chromatographic columns. iii. Retention volume of thiourea, hold-up volume, and compressibility of the C-18-bonded layer, *J. Chromatogr. A*, 1075, 117, 2005.
198. Gritti, F., Martin, M., and Guiochon, G., Influence of pressure on the properties of chromatographic columns. ii. The column hold-up volume, *J. Chromatogr. A*, 1070, 13, 2005.
199. Gritti, F. and Guiochon, G., Effect of the surface heterogeneity of the stationary-phase on the range of concentrations for linear chromatography, *Anal. Chem.*, 77, 1020, 2005.

200. Gritti, F. and Guiochon, G., Adsorption mechanism in reversed-phase liquid chromatography—effect of the surface coverage of a monomeric C-18-silica stationary-phase, *J. Chromatogr. A*, 1115, 142, 2006.
201. Gritti, F. and Guiochon, G., Influence of the degree of coverage of C-18-bonded stationary phases on the mass transfer mechanism and its kinetics, *J. Chromatogr. A*, 1128, 45, 2006.
202. Gritti, F. and Guiochon, G., Heterogeneity of the adsorption mechanism of low molecular weight compounds in reversed-phase liquid chromatography, *Anal. Chem.*, 78, 5823, 2006.
203. Gritti, F. and Guiochon, G., Adsorption mechanism in reversed-phase liquid chromatography—effect of the surface coverage of a monomeric C-18-silica stationary-phase, *J. Chromatogr. A*, 1115, 142, 2006.
204. Gritti, F. and Guiochon, G., A chromatographic estimate of the degree of heterogeneity of RPLC packing materials 1. Non-endcapped polymeric C-30-bonded stationary-phase, *J. Chromatogr. A*, 1103, 43, 2006.
205. Gritti, F. and Guiochon, G., A chromatographic estimate of the degree of surface heterogeneity of reversed-phase liquid chromatography packing materials. ii. Endcapped monomeric C-18-bonded stationary-phase, *J. Chromatogr. A*, 1103, 57, 2006.
206. Gritti, F. and Guiochon, G., A chromatographic estimate of the degree of surface heterogeneity of RPLC packing materials. iii. Endcapped amido-embedded reversed phase, *J. Chromatogr. A*, 1103, 69, 2006.
207. Kim, H.J., Gritti, F., and Guiochon, G., Effect of the temperature on the isotherm parameters of phenol in reversed-phase liquid chromatography, *J. Chromatogr. A*, 1049, 25, 2004.
208. Schure, M.R., Particle simulation methods in separation science, *Adv. Chromatogr.*, 39, 139, 1998.
209. Beck, T.L. and Klatte, S.J., Computer simulations of interphases and solute transfer in liquid and size exclusion chromatography, *Unified Chromatogr.*, 748, 67, 2000.
210. Klatte, S.J. and Beck, T.L., Molecular-dynamics of tethered alkanes—temperature-dependent behavior in a high-density chromatographic system, *J. Phys. Chem.*, 97, 5727, 1993.
211. Schure, M.R., Molecular dynamics of liquid chromatography: chain and solvent structure visualization, in Pesek, J.J., Leigh, I.E. (Eds.), *Chemically Modified Surfaces*, The Royal Society of Chemistry, Cambridge, 1994, p. 181.
212. Klatte, S.J. and Beck, T.L., Microscopic simulation of solute transfer in reversed phase liquid chromatography, *J. Phys. Chem.*, 100, 5931, 1996.
213. Klatte, S.J. and Beck, T.L., Molecular-dynamics simulations of tethered alkane chromatographic stationary phases, *J. Phys. Chem.*, 99, 16024, 1995.
214. Yarovsky, I., Aguilar, M.I., and Hearn, M.T.W., High-performance liquid-chromatography of amino-acids, peptides and proteins. 125. Molecular-dynamics simulation of *n*-butyl chains chemically bonded to silica-based reversed-phase high-performance liquid-chromatography sorbents, *J. Chromatogr. A*, 660, 75, 1994.
215. Yarovsky, I., Aguilar, M.L., and Hearn, M.T.W., Influence of the chain-length and surface-density on the conformation and mobility of *n*-alkyl ligands chemically immobilized onto a silica surface, *Anal. Chem.*, 67, 2145, 1995.
216. Mountain, R.D., Hubbard, J.B., Meuse, C.W., and Simmons, V., Molecular dynamics study of partial monolayer ordering of chain molecules, *J. Phys. Chem. B*, 105, 9503, 2001.

217. Slusher, J.T. and Mountain, R.D., A molecular dynamics study of a reversed-phase liquid chromatography model, *J. Phys. Chem. B*, 103, 1354, 1999.
218. Ban, K., Saito, Y., and Jinno, K., Characterization of the microscopic surface structure of the octadecylsilica stationary-phase using a molecular-dynamics simulation, *Anal. Sci.*, 20, 1403, 2004.
219. Dou, X., Wang, H., and Han, J.Y., Molecular dynamics simulation of the effects of mobile-phase modification on interactions in reversed-phase liquid chromatography, *J. Liq. Chromatogr. Relat. Technol.*, 29, 2559, 2006.
220. Ashbaugh, H.S., Pratt, L.R., Paulaitis, M.E., Clohecy, J., and Beck, T.L., Deblurred observation of the molecular structure of an oil–water interface, *J. Am. Chem. Soc.*, 127, 2808, 2005.
221. Wick, C.D., Siepmann, J.I., and Schure, M.R., Simulation studies of retention in isotropic or oriented liquid *n*-octadecane, *J. Phys. Chem. B*, 105, 10961, 2001.
222. Wick, C.D., Siepmann, J.I., and Schure, M.R., Simulation studies on the effects of mobile-phase modification on partitioning in liquid chromatography, *Anal. Chem.*, 76, 2886, 2004.
223. Zhang, L., Sun, L., Siepmann, J.I., and Schure, M.R., Molecular simulation study of the bonded-phase structure in reversed-phase liquid chromatography with neat aqueous solvent, *J. Chromatogr. A*, 1079, 127, 2005.
224. Zhang, L., Rafferty, J.L., Siepmann, J.I., Chen, B., and Schure, M.R., Chain conformation and solvent partitioning in reversed-phase liquid chromatography: Monte Carlo simulations for various water/methanol concentrations, *J. Chromatogr. A*, 1126, 219, 2006.
225. Lippa, K.A., Sander, L.C., and Mountain, R.D., Molecular dynamics simulations of alkylsilane stationary-phase order and disorder. 1. Effects of surface coverage and bonding chemistry, *Anal. Chem.*, 77, 7852, 2005.
226. Lippa, K.A., Sander, L.C., and Mountain, R.D., Molecular dynamics simulations of alkylsilane stationary-phase order and disorder. 2. Effects of temperature and chain length, *Anal. Chem.*, 77, 7862, 2005.
227. Lippa, K.A., Sander, L.C., and Mountain, R.D., Identification of isolated cavity features within molecular dynamics simulated chromatographic surfaces, *J. Chromatogr. A*, 1128, 79, 2006.
228. Leach, A.R., *Molecular Modelling: Principles and Applications*, Prentice Hall, London, 2001.
229. Faller, R. and de Pablo, J.J., Constant pressure hybrid molecular dynamics-Monte Carlo simulations, *J. Chem. Phys.*, 116, 55, 2002.
230. Adcock, S.A. and McCammon, J.A., Molecular dynamics: survey of methods for simulating the activity of proteins, *Chem. Rev.*, 106, 1589, 2006.
231. Izaguirre, J.A. and Hampton, S.S., Shadow hybrid Monte Carlo: an efficient propagator in phase space of macromolecules, *J. Comput. Phys.*, 200, 581, 2004.
232. Hampton, S.S. and Izaguirre, J.A., Improved sampling for biological molecules using shadow hybrid Monte Carlo, *Comput. Sci.—ICCS 2004, Pt 2, Proceedings*, 3037, 268, 2004.
233. LaBerge, L.J. and Tully, J.C., A rigorous procedure for combining molecular dynamics and Monte Carlo simulation algorithms, *Chem. Phys.*, 260, 183, 2000.
234. Bernardi, A., Raimondi, L., and Zanferrari, D., Conformational analysis of saccharides with Monte Carlo stochastic dynamics simulations, *Theochem*, 395, 361, 1997.
235. Jaroniec, M., Partition and displacement models in reversed-phase liquid-chromatography with mixed eluents, *J. Chromatogr. A*, 656, 37, 1993.

236. Kaliszan, R., Quantitative structure–retention relationships applied to reversed-phase high-performance liquid-chromatography, *J. Chromatogr. A*, 656, 417, 1993.

237. Lochmüller, C.H., Jiang, C., Liu, Q., and Antonucci, V., High-performance liquid chromatography of polymers: retention mechanisms and recent advances, *Crit. Rev. Anal. Chem.*, 26, 29, 1996.

238. Ranatunga, R.P.J. and Carr, P.W., A study of the enthalpy and entropy contributions of the stationary-phase in reversed-phase liquid chromatography, *Anal. Chem.*, 72, 5679, 2000.

239. Tan, L.C., Carr, P.W., and Abraham, M.H., Study of retention in reversed-phase liquid chromatography using linear solvation energy relationships. 1. The stationary-phase, *J. Chromatogr. A*, 752, 1, 1996.

240. Tijssen, R., Schoenmakers, P.J., Bohmer, M.R., Koopal, L.K., and Billiet, H.A.H., Lattice models for the description of partitioning adsorption and retention in reversed-phase liquid-chromatography, including surface and shape effects, *J. Chromatogr. A*, 656, 135, 1993.

241. Vitha, M. and Carr, P.W., The chemical interpretation and practice of linear solvation energy relationships in chromatography, *J. Chromatogr. A*, 1126, 143, 2006.

242. Horvath, C.S., Melander, W.R., and Molnar, I., Solvophobic interactions in liquid chromatography with nonpolar stationary phases, *J. Chromatogr.*, 125, 129, 1976.

243. Horvath, C.S. and Melander, W.R., Liquid chromatography with hydrocarbonaceous bonded phases; theory and practice of reversed phase chromatography, *J. Chromatogr. Sci.*, 15, 393, 1977.

244. Dill, K.A., The mechanism of solute retention in reversed-phase liquid chromato-graphy, *J. Phys. Chem.*, 91, 1980, 1987.

245. Martire, D.E. and Boehm, R.E., Unified theory of retention and selectivity in liquid chromatography. 2. Reversed phase liquid chromatography with chemically bonded phases, *J. Phys. Chem.*, 87, 1045, 1983.

246. Colin, H. and Guiochon, G., Comparison of some packings for reversed phase high performance liquid solid chromatography. ii. Some theoretical considerations, *J. Chromatogr.*, 158, 183, 1978.

247. Gritti, F. and Guiochon, G., Adsorption mechanism in RPLC. Effect of the nature of the organic modifier, *Anal. Chem.*, 77, 4257, 2005.

248. Luthe, G., Ariese, F., and Brinkman, U.A.T., Retention behaviour of higher fluor-inated polycyclic aromatic hydrocarbons in reversed-phase liquid chromatography, *Chromatographia*, 59, 37, 2004.

249. Yan, C. and Martire, D.E., Molecular theory of chromatography for blocklike solutes in anisotropic stationary phases and its application, *J. Phys. Chem.*, 96, 3489, 1992.

250. Goto, M. and Mccoy, B.J., A distribution kinetics model for chromatography of block-like molecules, *Chem. Eng. Sci.*, 59, 2105, 2004.

6 Liquid Chromatographic Separations of Basic Compounds

David V. McCalley

CONTENTS

6.1 INTRODUCTION

More than 20 years ago, the author was attempting the separation of the cinchona alkaloids, a group of naturally occurring bases, by reversed-phase high-performance liquid chromatography (RP-HPLC). Quinine, used for the treatment of malaria, and quinidine, used as a cardiac antiarrhythmic drug are important members of the group. A published method [1] had achieved some separation, but with rather broad peaks, using a 10 μm μ-Bondapak column and a methanolic mobile phase containing acetic acid. Surely, a much higher resolution separation would be obtained with a 5 μm column. However, the same mobile phase with a 5 μm column from a different manufacturer (Hypersil ODS) showed excessive or complete retention of the alkaloids (even with increase in the methanol concentration) and serious tailing. Other authors had also failed to reproduce the alkaloid separation on a variety of RP columns [2]. Here was an early example of the difficulties posed by the analysis of basic compounds. Tailing peaks with low efficiency and the poor reproducibility of separations on columns made from different silica are typical consequences of undesirable interactions of bases with ionized silanol groups on RP columns. These interactions are most serious on the relatively impure silica phases in use at the time and lead not only to problems of low resolution but also to poor quantitation. While more modern high-purity silica phases are now available, which give far superior results at low pH, another problem that leads to broad peaks and tailing is overloading, which occurs even on these modern phases. Further, the poor retention of hydrophilic bases in RP chromatography is yet another difficulty. Use of higher pH, where the ionization of bases is partially suppressed, is an attractive possibility and can give rise to useful separation selectivity. However, interaction with ionized silanols is still an almost inevitable consequence of working at neutral or higher pH, and thus different strategies may need to be employed. Many workers have attempted the use of alternative materials to silica such as polymeric-, zirconia-, and carbon-based phases, but so far these have not produced a generally accepted substitute for silica in the analysis of basic compounds.

The aim of this review is to summarize the difficulties likely to be encountered in the LC separation of basic solutes, which include the majority of pharmaceutical and also many biomedically important compounds. An answer to the problem of the separation of the cinchona alkaloids, fit for purpose, was obtained on the Hypersil column by adding the silanol blocking agent hexylamine to the mobile phase, which allowed the extra separation power of the smaller particle column to be exploited [3]. However, alternative solutions to the problem, which will be explored in this review, are more appropriate in particular circumstances; there is no universal solution that is applicable in all cases. The present review will concentrate on the most recent developments in this subject for the past few years. Further background information can be found in earlier reviews by the present author [4,5] and by Snyder [6].

6.2 RP CHROMATOGRAPHY OF BASES

6.2.1 CONVENTIONAL C18 COLUMNS-DETRIMENTAL INTERACTIONS WITH SILANOL GROUPS

Silica is by far the most popular column material for the separation of bases, despite its problems, generally for the same reasons as it is for neutral compounds. It has high mechanical strength; spherical particles of narrow size distribution can be manufactured and packed relatively reproducibly; it is easily modified with various reagents to give different functionality; and it is compatible with water and organic modifiers with no swelling. Furthermore, bonded (most commonly octadecylsilyl) silica used in the RP mode is by far the most popular phase used for the separation of bases owing to the following:

- Numerous advantages including simplicity of use and considerable experience with the technique
- Compatibility with aqueous samples
- Rapid equilibration of the column with the mobile phase (allowing gradient elution)
- Versatility of the mechanism, which can be altered by changing the mobile phase or additives [7]

Many of the problems in the analysis of bases with bonded silica originate from residual underivatized silanol groups. Silanols were discovered on the surface of silica as long ago as 1936 by Kiselev. In a detailed review, Nawrocki reports the surface concentration of hydroxyl groups on bare silica as $8.0 \pm 1.0 \ \mu\text{mol m}^{-2}$, that corresponds to about 5 hydroxyls nm^{-2} [8]. Values of 3–4 hydroxyls nm^{-2} found in more recent experimental work, for example, using deuterium exchange combined with infrared spectroscopy, continue to support these older literature figures [9]. When bonding alkyl ligands to silica via silylation reactions, maximal coverage of about 4–4.5 $\mu\text{mol m}^{-2}$ can be obtained with monomeric bonding. These figures represent the densest coverage of the smallest groups (trimethylsilyl), which are less subject to steric hindrance. For instance, Kirkland et al. [10] achieved a surface coverage of 4.1 $\mu\text{mol m}^{-2}$ for pure fully rehydroxylated silica bonded exclusively with trimethylchlorosilane (TMS). With larger reagents (e.g., octadecylsilylanes) the maximum coverage is less. The number of unreacted silanols can be reduced by endcapping with TMS. It appears that only about 50% of silanols can be reacted, even in the most favorable cases. Some additional internal (nonsurface) silanols may be buried in the silica structure but are inaccessible to solutes for the same steric reasons that hinder reaction with the reagents. Figure 6.1 illustrates the various groups that are present on the surface of a typical endcapped bonded phase (C8 in this example).

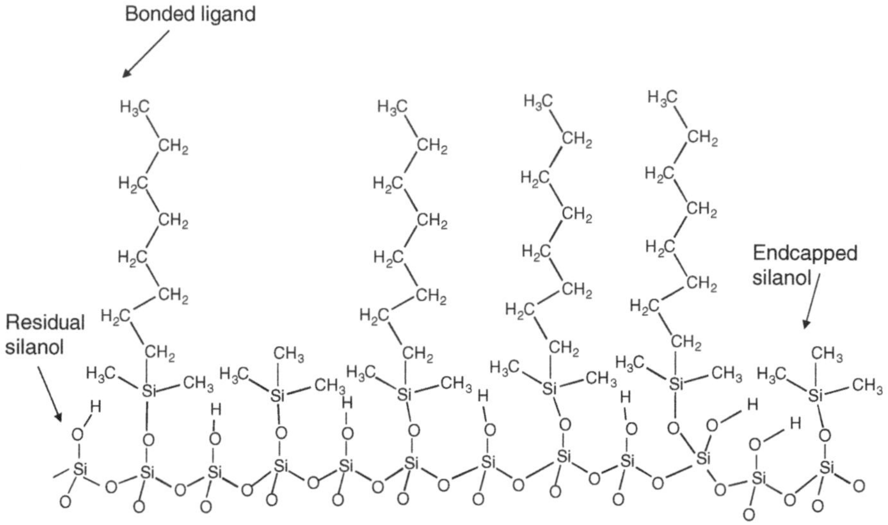

FIGURE 6.1   Groups present on the surface of a typical endcapped alkyl bonded phase (C8).

Silanol groups can be classified as

- *Isolated*, that is, where no interaction takes place with other surface groups
- *Vicinal*, where two silanol groups are close together on the surface and thus can interact by hydrogen-bonding
- *Geminal*, where two –OH groups are attached to the same silicon atom

Köhler and Kirkland [11,12] showed that rehydroxylation of the silica surface, for example, by treatment with dilute hydrofluoric acid (HF), increases the number of homogenously distributed hydrogen-bonded silanols, leading to better chromatographic performance with bases. The silanol concentration of $8.0\,\mu\mathrm{mol\,m^{-2}}$ represents a maximum on rehydroxylated silica; lower values occur if rehydroxylation is not performed especially considering the thermal treatments involved in manufacture. The proportion of the various types of silanols and their activity has been investigated at some length, although conflicting results have been obtained. Nevertheless, there now seems general agreement that the most active are isolated silanols [7,8].

Underivatized silanols can interact with bases by hydrogen-bonding or ionic interactions, although the latter are stronger energetically. Interaction with bases can be described by the following equation:

$$BH^+ + SiO^-M^+ \rightarrow SiO^-BH^+ + M^+$$

where BH^+ is the protonated form of a base B and M^+ represents the mobile phase buffer cation.

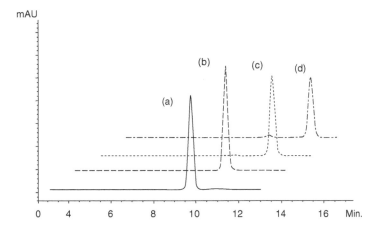

FIGURE 6.2   Chromatograms for 0.32 μg (a) propranolol, (b) nortriptyline, (c) amitriptyl-ine and (d) amphetamine on Primesep 200 (15 × 0.46 cm, 5 μm particles), mobile phase acetonitrile–TFA buffer ^S_WpH 1.96 (34:66, v/v), ^S_WpH 1.95 (41:59, v/v), ^S_WpH 1.95 (42:58, v/v) and ^S_WpH 2.04 (13:87, v/v), respectively. Temperature 30°C, injection volume 5 μL, flow rate 1 mL min^{-1}. Column efficiencies and asymmetry factors 9800, 1.0; 10800, 1.0; 10600, 1.0; 11,400, 1.0 for (a), (b), (c), and (d).

The potential exists for a mixed interaction mechanism for bases on silica RP columns, which involves both ion-exchange and hydrophobic retention. This mixed interaction mechanism has popularly been considered to be in itself the cause of the problem with the analysis of bases. However, the situation is not simple: basic solutes can give excellent peak shapes with mixed-mode phases containing an acetate group within a hydrophobic chain (see Figure 6.2) [13]. It could be that the position of the ionized silanols in conventional phases, buried deep beneath the ODS ligands, provides a dissimilar situation. The kinetics of interaction with silanols might also be very different from those with phases deliberately prepared to give mixed mode interactions.

Neue and Carr [14] suggested that the overall retention factor k for bases on conventional silica RP columns could be described by the relationship

$$k = k_{RP} + k_{IX} + k^*_{RP}k^*_{IX}$$

where k is the overall retention factor, k_{RP} is the contribution of the reversed-phase process, k_{IX} is the contribution of ion-exchange with surface silanols, and $k^*_{RP}k^*_{IX}$ is a multiplicative combination of both processes. They concluded that the multiplicative process involving ionized silanols was dominant in the retention of charged bases with reversed-phase packings. These multiplicative sites might correspond to the subset of very high-energy sites, which have long been suspected to be the cause of poor peak shapes for bases [8]. Wirth et al. [15] have presented direct spectroscopic evidence for the existence of strong specific sites for the adsorption of organic bases. Using cationic dyes in conjunction with fluorescence imaging, they found three specific adsorption sites on fused silica cover slips bonded with C18 ligands with desorption times of

0.07, 2.6, and 26 s, suggesting that tailing should become significant for the latter two sites. Whatever the exact nature of these detrimental interactions involving silanols really is, exponentially tailing peaks can often result. These lead to low efficiency, reduced resolution, and poorer detection limits due to decreased signal/noise ratio. In addition, there may be problems with measurement of peak area and quantitation, because it may be difficult to determine the end of a peak.

The pK_a of silanols is fundamental in determining their chromatographic influence. The *average* pK_a of silanol groups was reported as 7.1 [8]. Trace metals seem to increase the acidic nature of the surface; they may increase the activity of neighboring silanols rather than interacting directly with analytes. Metals especially Fe, Al, Ni, Zn also have chelating properties. Thus, the use of highly pure silicas free of metal contamination was strongly recommended by Köhler and Kirkland [11,12]. They classified silicas as "Type A," which included many of the older impure materials that could contain tens or even hundreds of $mg\,kg^{-1}$ of some metals, and "Type B," which included silicas of much higher purity. The widespread introduction of these Type B silicas by column manufacturers has probably been the most significant factor in the improvement of the chromatography of basic solutes in recent years. Tailing of bases, even for small solute loads, can still be found even on some Type B silica columns at pH 3.0, which is well below the average pK_a of silanols [16]. It is possible that this finding is linked to residual trace metal contamination that might persist on the earlier Type B phases studied, increasing the acidity of the silanols. The tailing of the same solutes on the same stationary phases operated at pH 7.0, where about half of the residual silanols would now be expected to be ionized was considerably higher [17]. This result again suggests the particular involvement of *ionized* silanols in poor peak shape.

The question remains as to the exact mechanism of the tailing that originates from the presence of ionized silanols on the phase, for example, at intermediate pH. Accessible ionized silanols constitute a relatively small number of strongly adsorptive sites. It is often supposed that the tailing is of kinetic origin, owing to the slow kinetics of desorption from ionized silanol groups. According to Guiochon [18], the overloading of such sites even by quite small amounts of ionized base is more likely to be responsible for the poor peak shape, that is, the tailing has a thermodynamic rather than a kinetic origin. This hypothesis was based on work with the basic drug amitriptyline at a mobile phase pH of 6.9. Kinetic tailing was proposed to be effective only with extremely small concentrations of injected solute. Certainly, overloading phenomena may be more pervasive than previously thought in all aspects of peak shape problems and is considered in detail in the next section. However, it is the author's experience that tailing associated with ionized silanols does not reduce even with the smallest detectable amounts of basic solute. Clearly, further studies are necessary to elucidate the detailed mechanisms involved in peak tailing of bases.

6.2.2 OVERLOADING

Overloading is a term that describes the decrease in efficiency or change in retention that occurs with increased quantities of sample; it can be caused by excessive sample volume or sample mass [7]. It results from the nonlinearity of the distribution of the solute between the stationary and mobile phase as the solute mass is increased. Sample volume is generally fixed by the use of a small volume injection loop, appropriate for the column dimensions and analysis conditions, leaving the problem of mass overload, which can be considerably more difficult to overcome. Clearly, overloading is likely to be problematic in preparative chromatography but may also cause difficulties in analytical situations. For example, in the pharmaceutical industry, there are usually legal requirements to report quantities of those impurities that are above the 0.1% (or in some cases, an even lower) level. Such determinations must be performed with sufficient amounts of the sample to allow detection and quantitation of the impurities. The peaks of major constituents may then become broadened sufficiently through overloading to interfere with or obscure these impurity peaks. In low ionic strength mobile phases, overload of ionized bases can occur even near their ultraviolet (UV) detection limit; thus, peak shape and retention become a function of solute mass over the entire quantitation range, which is highly undesirable. Figure 6.3 shows the variation of column efficiency with solute mass for a 15 cm × 0.46 cm (i.e., conventional dimensions) XTerra MS column using acetonitrile-formic acid at pH 2.7. Three solutes are uncharged under these conditions (3-phenylpropanol, caffeine, and phenol) while the bases (propranolol, nortriptyline) and the acid (2-naphthalenesulfonic acid) are fully charged. There is a striking distinction between the overloading behavior of the charged and uncharged compounds (note the logarithmic scale of the sample

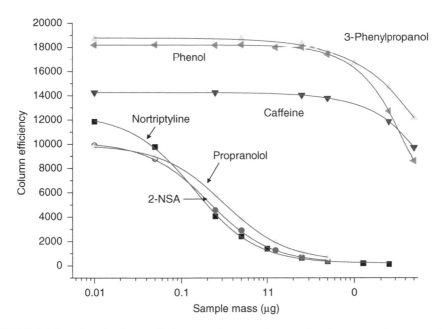

FIGURE 6.3 Plot of column efficiency against sample mass for three neutral compounds (3-phenylpropanol, caffeine, phenol), and three charged compounds (propranolol, nortriptyline, 2-NSA [2-naphthalenesulfonic acid]) on XTerra MS (15 × 0.46 cm, 3.5 μm particles). Mobile phase; acetonitrile–formic acid (overall concentration 0.02 M) $_w^w$pH 2.7 (28:72, v/v) except for caffeine (12.5:77.5, v/v). Flow rate 1 mL min^{-1}. Column temperature 30°C. Injection volume 5 μL.

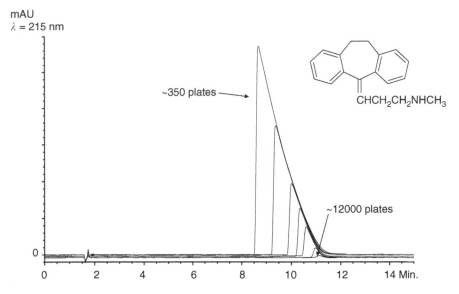

FIGURE 6.4   Overlaid chromatograms for 5, 2.5, 1, 0.5, 0.25, 0.05, 0.01 μg (Loading factor, L_f = 1.3-0.003%) nortriptyline. Conditions as Figure 6.3.

mass axis) [19], which suggests either ionic sites of opposite charge to the solute are present on the column, or that ionic repulsion of the solute is taking place. It is possible even that ionized solutes can only access a proportion of the stationary phase area that is accessible to neutral solutes. However, the fact that rapid overload of both acids and bases occurs argues somewhat against the existence of charged sites on this column, as both negatively and positively charged sites would be required. Independent studies [20] have suggested that the silanols on this phase are not ionized at pH 2.7 (see also below), consistent with the good peak shapes obtained on this phase with small solute mass [19]. Considerable deterioration in efficiency results even by use of 0.1 μg of the charged compared with a mass a hundred times greater (10 μg) for the uncharged solutes. Figure 6.4 shows overlaid chromatograms for 0.01–5 μg of nortriptyline using the same conditions. While peaks for the smallest sample mass are symmetrical, broadening takes places with increasing sample mass leading to right-angled triangle peak shapes; retention decreases with increasing solute mass. The shape of the peaks is different from that which results from interactions of small amounts of base with a column that contains ionized silanols; such peaks in contrast usually give rise to exponential tailing. Furthermore, the same overloading behavior for charged solutes was found at low pH on PRP-1, a polymeric column prepared from polystyrene-divinyl benzene. No negatively charged sites were found on this phase, at least not at low pH [21]. This result is a further argument against the involvement of ionic groups in overloading on these phases. Clearly, PRP-1 has no silanols, and XTerra MS is a highly inert hybrid silica phase. Thus, silanols cannot be directly responsible for this overloading behavior. The results point to the possibility of repulsion of ionic solutes as being the cause of overloading.

Figure 6.3 gives an idea of the practical effects of the overloading, which occurs for each solute in a specific mobile phase. However, overloading depends on k, as it governs the fraction of the sample present in the stationary phase. The loading capacity of a column is usually expressed as its saturation capacity, which can be loosely defined in frontal analysis as the maximum uptake of solute by the column before breakthrough occurs in a given mobile phase. Guiochon [22] gave an approximate expression for the thermodynamic contribution to column efficiency (equivalent to mass overload induced peak broadening):

$$N_{th} = \frac{4}{L_f}\left(\frac{1+k_0}{k_0}\right)^2 \tag{6.1}$$

k_0 is the retention factor for small sample mass, L_f is the loading factor (the sample mass w_x divided by the saturation capacity w_s), and N is measured using the base peak width. Knox and Pyper [23] and Guiochon assumed that the column HETP is the sum of two independent contributions, which are due to the nonlinear behavior of the isotherm (H_{th}) and mass transfer kinetics (H_0)

$$H = H_{th} + H_0 \tag{6.2}$$

Hence,

$$L/N_{tot} = L/N_{th} + L/N_0 \tag{6.3}$$

$$1/N_{tot} - 1/N_{th} + 1/N_{kin} \tag{6.4}$$

$$1/N_{tot} = \frac{N_0 + N_{th}}{N_{th}N_0} \quad \text{or} \quad N_{tot} = \frac{N_0 N_{th}}{N_{th} + N_0} \tag{6.5}$$

$$N_{tot}/N_0 = \frac{1}{1 + (N_0/N_{th})} \tag{6.6}$$

$$N_{tot} = \frac{N_0}{1 + (N_0/N_{th})}. \tag{6.7}$$

Combining Equations 6.1 and 6.7,

$$16\,t_r^2/W^2 = \frac{N_0}{1 + (1/4)N_0 L_f\left[k_0^2/(1+k_0)^2\right]} \tag{6.8}$$

$$W^2 = 16\,t_0^2/N_0(1+k_0)^2 + 4t_0^2 k_0^2 L_f \quad \text{or} \quad W^2 = 16\,t_0^2/N_0(1+k_0)^2 + 4t_0^2 k_0^2 w_x/w_s \tag{6.9}$$

t_0 is the void volume of the column and W is the peak width at base of the overloaded peak.

Snyder used a form of this equation with a minor empirical modification to calculate w_s [7,24]. w_s can be obtained using Equation 6.9 in principle with two injections, one using a small solute mass to yield N_0 and k_0, the other using a moderately overloaded injection (e.g., to decrease the efficiency by half its original value) to give the value of W at a given value of the loading factor. This method gives an approximate

TABLE 6.1

Retention Factor and Saturation Capacity of Nortriptyline and 2-Naphthalenesulfonic Acid in Various Buffers

Solute	Buffer	$_w^w$pH	I (mM L^{-1})	k	w_S (mg)
Nortriptyline	0.02 M Formic	2.7	1.9	10.3	0.4
	0.2 M Formic/NH$_3$	2.7	18.3	10.9	2.2
	0.02 M Phosphate	2.7	22.0	11.2	3.0
	0.008 M TFA	2.3	7.9	16.0	2.8
2-NSA	0.02 M Formic	2.7	1.9	0.7	0.2
	0.02 M Phosphate	2.7	22.0	0.6	2.6

Column: XTerra MS (pyridine) 3.5 μm, 15 × 0.46 cm ID. Mobile phase acetonitrile: buffer 28–72 (v/v). The calculations of ionic strength ignore the effect of the organic solvent.

estimate of w_s, which becomes increasingly incorrect as the loading factor increases, due partially to the approximation involved in Equation 6.1. w_s measured in this way may also correspond to the saturation only of a subset of high energy sites, as proposed by Guiochon (see below), and thus the values differ substantially for ionized compounds from those measured by frontal analysis. However, if this subset of sites indeed exists, they completely dominate the column performance, rendering the substantial extra capacity of the column redundant. Thus, this measurement has perhaps even greater significance for basic compounds that the measurement of total saturation capacity by frontal analysis. Measurement of w_s from column efficiency is less time consuming than from frontal analysis, and is an acceptable method considering the large differences shown, for instance, between charged and uncharged solutes. Table 6.1 gives some values in different mobile phases of the saturation capacity w_s for the charged base nortriptyline, calculated from the Snyder equation, using a hybrid phase prepared with a classical pyridine catalyst. We believe that this phase has a negligible number of charged groups (negatively charged silanols or positively charged catalyst residues that occur on some phases [25]) on its surface under the mobile phase conditions specified. Changing from 0.02 to 0.2 M formic acid, adjusting the pH of the latter to the same as the former with ammonia, increases ionic strength by a factor of almost 10 times. k for nortriptyline is only slightly increased from 10.3 to 10.9, although the saturation capacity is increased more than five times from 0.4 to 2.2 mg. If strong ionic retention sites for cations were present on XTerra, the increased concentration of competitive cations (NH_4^+) in the mobile phase would be expected to *decrease* the retention of nortriptyline. Similar results were obtained with 0.02 M potassium phosphate buffer, which has slightly higher ionic strength than (pH adjusted) 0.2 M formic acid, and increases w_s of nortriptyline considerably, without increasing its k. Thus, higher ionic strength alone can lead to higher saturation capacity, possibly by diminishing the effect of mutual repulsion of similarly charged ions held on the phase surface. Table 6.1 additionally shows the similar effect of ionic strength on an anionic compound (2-naphthalenesulfonic acid); note that anionic compounds also have low saturation capacities, which can interpreted similarly in terms of mutual repulsion of these ions. Ion pair effects seem to be very

small with these mobile phases. While solute cations must be accompanied by counterions of opposite charge to preserve the neutrality of the system, we do not believe
that solute cations are held exclusively on the surface as neutral or "contact" ion pairs.
Carr and coworkers [26,27] have demonstrated a very low incidence of ion pairing
with similar hydrophilic anions, at least in the mobile phase, although the situation
in the stationary phase could be different. In contrast, 0.008 M trifluoroacetic acid
has a lower ionic strength than phosphate but still gives a similar saturation capacity.
This result could be attributed to the additional recognized ion pair effects of TFA,
as shown by the considerable increase in k for nortriptyline, from $k = 10$ in 0.02 M
formic acid to $k = 16$ in 0.008 M TFA. Ion-pairing reduces the number of charged
species held on the surface and thus the effects of mutual repulsion. Very similar
results were obtained for overloading of basic peptides using gradient elution, con-
firming a more universal nature to these findings. Figure 6.5 shows the separation of
the Alberta basic peptide mixture, which contains peptides P1–P4 with one to four
basic lysine residues. When injected at normal strength (1.3–2.5 µg each peptide),
the peak capacity for P3 was 131, 216, and 232 using an acetonitrile buffer gradient
containing 0.02 M formic acid, 0.008 M TFA, and 0.02 M phosphate, respectively,
in line with the ionic strength and ion-pair ability of the various additives. Narrower
peaks can be clearly observed for the normal strength mixture when phosphate or TFA
is used. In contrast, the diluted mixture shows similar peak capacities for all three
mobile phases. With TFA, the increased retention of all peptides due to ion pairing is
also apparent [28,29].

Snyder originally showed that the total saturation capacity of a RP column for
ionized species measured by frontal analysis was similar for charged and uncharged
compounds [24]. He attributed the initial rapid overload and deterioration in peak
shape at low pH to overload of (relatively scarce) ionized silanols. This could at
least be a contributory mechanism on the older Type A columns available at the time.
(The explanation of overload of ionized silanols as a cause of poor peak shape has also
been given by Guiochon—see above, [18]). According to Snyder, filling of plentiful
hydrophobic sites would then explain the high total saturation capacity. Gritti and
Guiochon [30,31] determined the saturation capacities of a large number of charged
and uncharged compounds, mostly on Type B RP columns. They deduced that the
surface of silica RP columns is heterogeneous, consisting of a number of weak, intermediate, and strong sites. Overloading by very small amounts of ionized bases could
be explained by overloading of the scarce strong sites that are completely filled, prior
to the filling of plentiful weaker sites. However, the measured adsorption energies
of the strongest sites on many of these Type B columns were insufficient to indicate
they were ionized silanols, and instead, gave adsorption energies consistent with them
being different types of hydrophobic site. For example, solutes may be adsorbed at the
interface between the stationary and mobile phase (weaker sites) while stronger sites
could correspond with the penetration of solutes deeper into the hydrophobic layer
[32]. Some supporting evidence for these multiple hydrophobic sites was obtained by
McCalley [19], who showed that a sharp peak above the void volume with a long tail
was obtained by injection of very large amounts of the basic drug nortriptyline (see
Figure 6.6) using a highly inert XTerra column at low pH. As mentioned earlier, this
column appears to have no ionized silanols at the pH of operation (2.7). It appears from

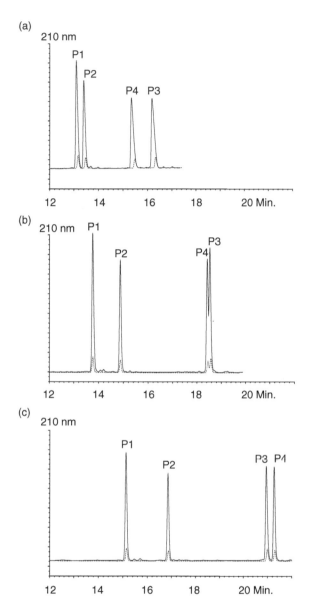

FIGURE 6.5 (a) Analysis of Alberta basic peptides on Discovery C18 (25 × 0.46 cm, 5 μm particles), at normal working strength (approx 1.3–2.5 μg each peptide) and diluted 10 times in mobile phase (dotted line). Solvent A 0.9 g/l formic acid (0.02 M) in water (pH 2.7); solvent B 0.9 g/l formic acid (0.02 M) in acetonitrile (pH 2.7).Gradient 5%B to 42.5% B in 30 min (1.25% acetonitrile/min) (b) As (a) but solvent A 0.02 M phosphate buffer in water (pH 2.7); solvent B acetonitrile: 0.04 M phosphate buffer in water (pH 2.7) 50:50 v/v. Gradient 10%B to 85% B in 30 min (1.25% acetonitrile/min) (c) As (a) but Solvent A 0.9 g/l TFA (0.0079 M) in water (pH 2.3); Solvent B 0.9/l TFA in acetonitrile.

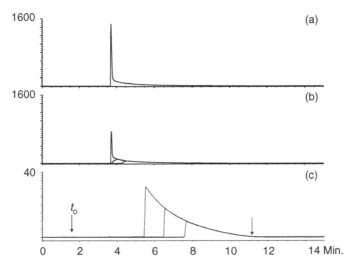

FIGURE 6.6 Chromatograms for nortriptyline on XTerra MS for (a) 250 μg ($L_f = 63\%$) (b) 100 (25%), 150 (38%) and 200 μg (50%). (c) 12.5 (3%), 25 (6%) and 50 μg (13%). t_o indicates the column void time. Other conditions as in Figure 6.3.

Figure 6.6 that right-angled triangle peaks are obtained until a sample mass of 150 μg is introduced, which could be attributed to the gradual filling of strong hydrophobic sites. Increased loading after this point (200 μg and above) could correspond to the saturation of these sites, and the filling subsequently of weaker sites, giving rise to the superimposed sharp peak that does not move further toward the position of the void volume of the column, indicating the greater abundance of these sites. This result cannot be explained using the hypothesis of mutual solute repulsion alone. The theory of Gritti and Guiochon [32] considers that ionized bases are exclusively ion-paired when held on the surface of RP columns even in the presence of additives such as formic acid, which accounts for the similar total saturation capacities shown by ionized and neutral compounds. However, other interpretations of Figure 6.6 are possible; note also that the highest solute concentrations used clearly overload the buffer concentration [19]. It is evident that the explanations of the overloading of ionized solutes differ in some aspects, and that further studies are necessary in this area to shed more light on the mechanism.

Overloading effects seem even more complex at intermediate pH because silanols are now partially ionized and involved in the retention of bases. As mentioned previously, the *overload* of now-ionized silanols could at least be part of the cause of tailing peaks, even when very small amounts of ionized base are used [18,24]. However, it has been observed that as solute mass increases in experiments at pH 7, column efficiency may *improve* from an initially low value to a maximum, afterward declining in the usual way [33]. This observation could be due to the "blocking" or saturation of ionized silanols by a portion of the sample, such that the rest interacts mainly by hydrophobic processes, resulting in better efficiency. At higher pH still, the solute should not be ionized if appreciably above its pK_a and therefore should

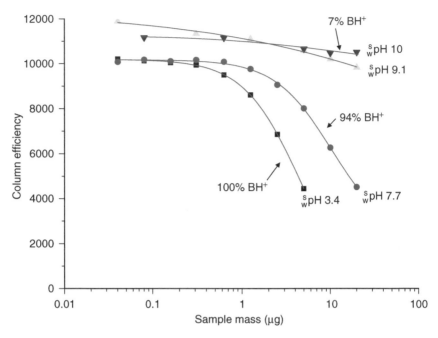

FIGURE 6.7 Plots of efficiency against mass amitriptyline ($^s_w pK_a$ 8.9) at various pH values on XTerra RP18 (15 × 0.46 cm, 5 µm particles). Mobile phase acetonitrile- buffer (35:65 v/v) using phosphate buffers $^w_w pH$ 2.7 ($^s_w pH$ 3.4), $^w_w pH$ 7.0 ($^s_w pH$ 7.7), and ammonium carbonate buffers $^w_w pH$ 8.5 ($^s_w pH$ 9.1) and $^s_w pH$ 10.0. Percentage ionization of the base (% BH$^+$) is indicated.

behave similar to a neutral compound, giving much higher saturation capacities (and no ionic interactions, even if silanols are ionized). Indeed this appears to be true, as shown for amitriptyline ($^s_w pK_a$ 8.9) analyzed on an XTerra RP18 column using a mobile phase at $^s_w pH$ 10 as indicated in Figure 6.7 [33]. Clearly, the decline in efficiency with increasing solute mass is serious at pH 2.7, but this decline diminishes considerably as the solute becomes deprotonated as the pH is raised. However, this strategy does not necessarily work for solutes of higher pK_a. At the maximum pH of long-term stability for this column (around 10), the continued appreciable ionization of such solutes appears to be sufficient to cause the usual detrimental interactions with silanols, which become ionized at this high pH. This factor may not be problematic for amitriptyline (Figure 6.7) which is only 7% ionized in the 35% ACN-phosphate buffer ($^s_w pH$ 10) used for this study. The column efficiency remains high even with sample loads in excess of 10 µg. However, the column efficiency of the stronger base amphetamine ($^s_w pK_a$ 9.8) is poor, even for small solute mass, when analyzed at $^s_w pH$ 10 (see Figure 6.8), and even increases somewhat with sample load. Amphetamine is almost 40% protonated under these conditions [33], and the results are consistent with interaction of the charged base with ionized silanols and the blocking effect described earlier.

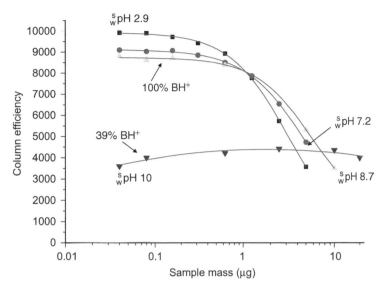

FIGURE 6.8 Plots of efficiency against mass amphetamine ($^S_w pK_a$ 9.8) at various pH values on XTerra RP18. Mobile phase acetonitrile-buffer (10:90, v/v), using phosphate-buffers $^W_w pH$ 2.7 ($^S_w pH$ 2.9), $^W_w pH$ 7.0 ($^S_w pH$ 7.2), and ammonium carbonate buffers $^W_w pH$ 8.5 ($^S_w pII$ 8.7) and $^S_w pH$ 10.0.

6.2.3 ANALYSIS OF BASES ON DIFFERENT TYPES OF RP COLUMN BASED ON SILICA

It has been shown earlier that the problems of the analysis of bases on C18 columns based on silica include interaction with or overload of ionized silanols and overloading of the hydrophobic phase surface. Both processes can both lead to broad, tailing peaks, although the tailing may be of different origin. Furthermore, the low mobile phase pH necessary to suppress silanol ionization and the pH limit of stability of approximately 7–8 for conventional columns, can lead to low retention of protonated (especially hydrophilic) bases. Researchers have continually attempted to improve the speed and selectivity of analysis of bases, for instance, by the use of different bonded ligands, very small particle columns, or monoliths consisting of single rods of silica. In this section, the variety of types of bonded silica columns is discussed with respect to their efficacy in addressing the problems of analysis of bases.

Conventional C18 columns are usually prepared with monofunctional silylating reagents such as $ClSi(CH_3)_2 R$, where $R = -C_{18}H_{37}$, leading to monomeric-bonded phases. Difunctional or trifunctional silanes using reagents such as $Cl_2Si(CH_3)$ R or Cl_3SiR are also used. However, the use of difunctional or trifunctional silanes does not lead to phases with unbonded silanol concentrations that are much different from those of monomeric phases [34]. Using difunctional or trifunctional reagents can lead to phases with coverage greater than $4.0 \,\mu mol\, m^{-2}$, which may have some polymeric character, leading to somewhat greater stability of the phases, useful particularly in

HPLC-MS. Although no systematic studies have been reported, it does not appear that there is much difference in the activity of these different phases toward basic compounds. Furthermore, manufacturers do not always reveal the class of reagent from which their phases are prepared. Therefore, in the following discussion, no particular distinction is made between phases prepared from these different silylating agents.

6.2.3.1 RP Columns with Shorter Alkyl Chains or Other Functional Groups

The overwhelming majority of separations of bases have been carried out on RP columns with alkyl chain lengths of C18 or similar. McCalley [35] investigated the effect of chain length on peak shape and retention of basic compounds, comparing C4, C8, C18, and cyanopropyl phases synthesized on the same base silica (Inertsil-3), a Type B material of low silanophilic activity. Direct comparison of the phases was difficult, however, owing to the different coverage of these phases; the shorter alkyl ligands generated higher coverage, presumably because of reduced steric hindrance. The hydrophobicity of the columns decreased as expected, shown by decreased retention of benzene along the series C18, C8, C4, and CN. This decrease in retention was obtained also for the nine bases, which consisted of a series of weak to moderately strong bases (pyridine, pK_a 5.2 to nortriptyline, pK_a 10.3) in acetonitrile-buffer pH 3. All the test probes were completely protonated under these conditions. These relatively inert Type B columns should produce separations that are largely based on hydrophobic interactions. Little difference in selectivity for these bases was found between any of the columns at pH 3. With acetonitrile–phosphate buffer pH 7.0, retention of the bases increased, as shown by the greater acetonitrile concentrations necessary for their elution at similar k to that shown at pH 3. Some of the increase in retention can be attributed to decreased protonation. The hydrophobic retention of the unprotonated species is generally much greater than that of the ionized form and the overall k will increase even if only a small proportion of the solute is ionized. However, silanophilic retention processes make a larger contribution to the retention of bases at pH 7.0, reducing the relative influence of hydrophobic retention. This may explain why only small decreases in mean k for the bases were obtained with decreasing ligand chain length at pH 7.0. The cyanopropyl column gave the highest mean k for the bases of all columns at pH 7.0, attributable to additional polar interactions between the solutes and surface group. The cyanopropyl column showed marked selectivity differences for the individual bases at pH 7.0 compared with the alkyl-bonded phases, showing that such columns may be useful when alternative selectivity is required to achieve a separation. Both at pH 7.0 and 3.0, decreased peak asymmetry was obtained for the bases along the series C18 (less symmetrical) > C8 > C4 ~ CN (more symmetrical). The improvement in peak shape for C8 compared with C18 phases was confirmed for Discovery and Kromasil columns, although C4 and cyano versions of these phases were not available/not tested. The improved peak shape obtained with shorter chain phases might be due to the greater surface coverage achievable. However, other factors such as freer access of buffer and solute ions to the surface of short chain phases may be involved.

A detailed examination of the selectivity of different phases can be obtained by considering the data from the classification system developed by Snyder and coworkers [36] (see later sections).

6.2.3.2 Columns with Sterically Hindered Ligands

Kirkland et al. [10] introduced sterically hindered bonded phases such as diisopropyl-n-octyl bonded silica in 1989, developed to facilitate the analysis of bases at low pH ($\leq$3) where at least the majority of silanols are not ionized. The same authors had shown previously that typical bonded phases (at least those available at the time) were not stable at low pH, even over relatively short periods of time [37,38]. They highlighted the serious loss of short-chain silanes, particularly the endcapping, from RP columns used at low pH; nevertheless, slow loss of longer chain bonded phases such as C8 and C18 was also demonstrated. The same authors also produced a diisobutyl n-octadecyl phase showing greater stability [39,40]. While it was claimed that sterically hindered packings gave better efficiency for the strongly basic tricyclic antidepressant drugs than conventional bonded phases prepared on the same silica, such columns have not been favored by other researchers when considered especially for the separation of bases. Nevertheless, these phases seem to have considerable advantages in their extremely high stability at *low* pH (although they are not especially stable at *high* pH) [41], even at elevated temperature. It is believed that these columns are not endcapped to preserve their high stability at low pH; furthermore, the surface coverage achievable with these phases is rather lower than other phases, presumably because of steric hindrance. Both factors may be detrimental to peak shape of basic solutes, in addition to the normal factor of the activity of the silica from which they are prepared. This author believes that the stability of RP columns at low pH, particularly with regard to loss of endcapping and its effect on the column performance for basic compounds, is an important but so far little-researched topic.

6.2.3.3 Columns with Embedded Groups

6.2.3.3.1 Embedded (Neutral) Polar Groups
In 1990, Supelco introduced the first commercial column of this type (pK_b-100), with an embedded polar amide group within a long alkyl chain [42], prepared on relatively impure (Type A) silica. It seems this phase was prepared in a two-step process, involving the acylation of a preformed aminopropylsilylated silica. Some aminopropyl groups remained unreacted and become charged at low pH. While this property could be beneficial in repelling positively charged bases from deleterious interactions with the surface, excess retention and poor peak shapes of acidic compounds contained within the sample could result, due to anion exchange interactions. In the mid-1990s, Supelco introduced the ABZ and ABZ$^+$ phases that were similar but used in addition, a smaller acid chloride with less steric hindrance than the main bonding ligand to derivatize more of the free amino groups; this procedure is somewhat analogous to endcapping, although has a different purpose. The ABZ$^+$ phase was claimed to give better peak shapes for codeine than a number of competitive phases while maintaining good peak shapes for acidic compounds such as benzoic acid [43].

The asymmetry factor of codeine at pH 7.0 was shown to be independent of addition of triethylamine (TEA), a common silanol blocking agent (see later text), whereas on competitive columns, peak shape improved on the addition of TEA, indicating greater silanol activity. The authors proposed that a layer of water is adsorbed by the surface amide functionalities that may shield the residual silanols of the silica from the solute. The improved wetting of embedded polar group (EPG) phases by aqueous solutions, and thus their ability to function in mobile phases with low or no organic modifier content, necessary to obtain sufficient retention for the separation of hydrophilic basic solutes, is a further advantage of this type of phase. A later development by Supelco was Discovery Amide, prepared on a pure (Type B) silica using a one-step synthesis using an appropriate embedded amide silane. It has the structure

$$-O-Si-(CH_3)_2-(CH_2)_3NHCO(CH_2)_{14}-CH_3.$$

Coverage is $3.0\,\mu mol\,m^{-2}$ on $200\,m^2\,g^{-1}$ silica (12% C). The phase is endcapped with a small silylating reagent in the conventional way. Ascentis RP-amide columns are the latest version of the Supelco amido column bonded on higher surface area $450\,m^2\,g^{-1}$ pure silica (25% C). A polymeric bonding is used to increase the stability and reduce bleed, especially for MS work.

O'Gara et al. [44] from Waters Corporation reported the preparation of an embedded octyl carbamate phase (12% C) using a single-step synthesis in 1995, having the structure

$$-O-Si-(CH_3)_2-(CH_2)_3O-CONHC_8H_{17}.$$

Another reported advantage of the single-step procedure was improved reproducibility of phase preparation. Polar and basic analytes had lower k on this phase compared with a classical endcapped C8 phase, which was attributed to a weakening of the interaction with unbonded silanols owing to the presence of the embedded group. The tailing factor of the basic drug amitriptyline was improved at pH 7.0 as compared with the classical C8 phase, prepared on the same base silica. No appreciable difference in stability was noted between the conventional and EPG phases at pH 3.0 or 7.0. These phases are now also available in RP18 format on a variety of Waters packings. For example, SymmetryShield RP18 has surface coverage $3.2\,\mu mol\,m^{-2}$, the same as Symmetry C18, made on the same silica without an EPG [45]. The tailing factor of the basic drug amitriptyline measured in methanol–phosphate buffer pH 7.0 was only 1.1 on SymmetryShield as compared with 2.1 on Symmetry. A reduction in methylene group selectivity was observed for the EPG phases, attributed as before to the water layer bound by the hydrophilic carbamate groups. Decreased hydrophobic as well as decreased silanophilic interactions may thus contribute to the reduced retention of bases shown by EPG phases. It was also noted that EPG phases preferentially retain phenolic compounds and that EPG phases in general have different selectivity to conventional bonded phases. The improved peak shape shown by EPG phases might also be influenced by hydrogen-bonding of surface silanols with the embedded group instead of with basic solutes [34].

Euerby and Petersson [46] investigated the selectivity and peak shape for a number of EPG phases. They emphasized the differences in EPG phases prepared by 1- and

2-step procedures and demonstrated tests that could easily distinguish between these types of EPG phase.

McCalley [35] independently evaluated the EPG columns Discovery Amide and SymmetryShield, comparing their performance with the equivalent non-EPG phases. The study involved measurements of k, column efficiency, and asymmetry factor for nine basic solutes in methanol–phosphate buffer at pH 3.0 and 7.0, and acetonitrile–phosphate buffer at pH 3.0. On average, considering all conditions, the EPG columns showed better peak shape than the conventional columns, in line with the results obtained by the manufacturers. Reduced retention of the bases on the EPG column was noted, which could be due to reduced hydrophobicity of the phase (shown by the reduced retention of benzene) as well as reduced retention on silanols. Some differences in selectivity from that on the equivalent conventional phases for the nine test bases were also noted on the EPG columns. However, for the Discovery and Discovery Amide columns, relatively little difference in performance was obtained for the tests using acetonitrile at pH 7; acetonitrile appears to emphasize interactions with ionized silanols at this pH, and it is possible that silanol interactions are relatively large in comparison to the effects of the embedded group. It also seems possible that the EPG technology may be relatively more beneficial when applied to more strongly acidic Type A, rather than pure Type B silicas. The selectivity differences that have been shown between EPG and C18 phases may be a more useful aspect of these phases, although the differences seem to relate largely to the hydrogen-bond acceptor properties of the phases, and thus are more relevant to the separation of mixtures containing also compounds such as phenols, rather than those containing exclusively basic solutes [35,45,47].

The bleed from EPG phases detected in HPLC-MS is greater than that for conventional phases. For older EPG phases, increased bleed may have indeed been related to increased loss of stationary phase. However, on modern EPG phases, the increased efficiency of ionization of the bleed from EPG columns may be the major cause of the problem rather than excessive phase loss. Nevertheless, the presence of a water layer close to the surface in EPG phases could potentially promote bleed from EPG phases, and this subject warrants further study [47]. Concern over phase bleed has somewhat restricted use of EPG phases in HPLC-MS.

6.2.3.3.2 Columns with Embedded Ionic Groups
Recently, phases having embedded acetate groups within an alkyl chain have become commercially available (Primesep, SIELC technologies), which have a mixed mode cation exchange/hydrophobic interaction with basic solutes [48]. Primesep A, Primesep 100, and Primesep 200, have cationic groups of decreasing acidity, which can exchange with basic solutes. The group in Primesep A is reported by the manufacturer to be ionized over its entire range of operation, whereas the pK_a of the groups in Primesep 100 and 200 is reported to be 1 and 2, respectively. Thus, changing the pH by varying the nature and concentration of acid additives like TFA enables the state of ionization of the column groups to be changed, affecting the retention of the basic compounds relative to acidic and neutral compounds. The hydrophobic retention of the column can be controlled in the normal way by changing the concentration of organic modifier (e.g., acetonitrile) in the mobile phase. With these columns, both

functional groups, ion exchange and RP, are part of a single ligand bonded to silica, so it is perhaps easier to maintain exact ratio of IE vs. RP; this may be more challenging in columns containing physical mixtures of discrete ion exchange and RP particles.

Our independent evaluations of these RP/cation exchange phases have indeed shown that they produce excellent peak shapes for small sample mass of basic solutes (see Figure 6.2) [13]. It is possible that silanols may be shielded from interaction with protonated bases by the presence of the charged ligands. In any case, as mentioned earlier, the popular assumption that a mixed mode mechanism leads inevitably to tailing is shown to be unfounded. The retention of the quaternary amine benzyltriethylammonium chloride, together with the bases amphetamine, benzylamine, and codeine, increases from pH 2 leveling off at $\sim^s_w$pH 3–4 on Primesep 100 (see Figure 6.9). If the column groups indeed have a pK_a of 1.0, as claimed by their manufacturer, a sharper leveling of retention as the pH was raised might be expected.

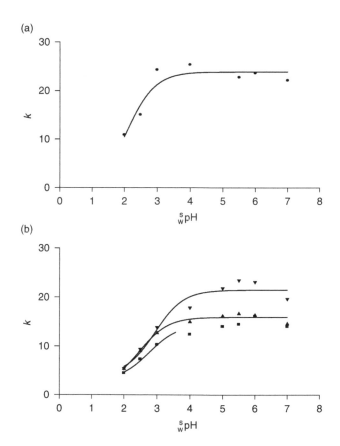

FIGURE 6.9 Plot of retention factor (k) against mobile phase pH for Primesep 100 (15 × 0.46 cm, 5 μm particles). Temperature 30°C, injection volume 5 μL, flow rate 1 mL min^{-1} (a) BTEAC (●), mobile phase acetonitrile–phosphate buffer (45:55, v/v), (b) benzylamine (■), codeine (▲) and amphetamine (▼), mobile phase acetonitrile–phosphate buffer (16:84, v/v), (29:71, v/v) and (32:68, v/v), respectively.

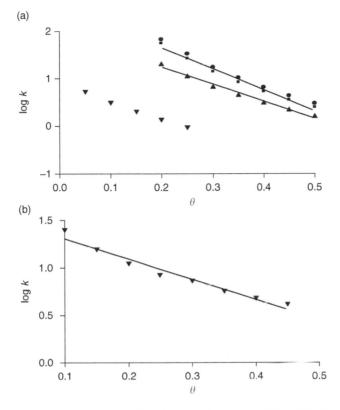

FIGURE 6.10 Plots of log k vs. the volume fraction of organic modifier (θ) in the mobile phase for (a) nortriptyline (■, $r^2 = .98$), diphenhydramine (▲, $r^2 = .99$), amitriptyline (●, $r^2 = .99$) and amphetamine (▼, $r^2 = .99$) on Primesep 200 and (b) for amphetamine (▼, $r^2 = .96$) on Primesep 100 using ^s_wpH 2.0 phosphate buffers. Other conditions as Figure 6.9.

It is possible that this result reflects a greater spread of pK_a of column groups, as might be expected when ionic groups are in close proximity to one another. An alternative explanation is that the pK_a of column groups may be affected by the presence of the organic solvent. Comparable results were obtained on the Primesep 200 phase. This phase, however, has weaker embedded acidic groups and allows the elution of more hydrophobic bases like amitriptyline and nortriptyline, which have excessive retention on Primesep 100. Plots of log k vs. the volume fraction of organic modifier in the mobile phase for four basic drugs on Primesep 200, and for amphetamine on Primesep 100 are shown in Figure 6.10. The plots are reasonably linear ($r^2 = .98$–.99) showing behavior similar to that of conventional RP columns. A useful feature of these columns was found to be increased loadability for protonated bases; in some cases, saturation capacities were 10 times higher or more than that found for conventional columns operated under similar mobile phase conditions (see Table 6.2). The increased loadability might be explained by the provision of extra retention sites for protonated bases, or the contribution of the ligands to neutralization of the charge on the column surface, if ionic repulsion is responsible for column overload (see following text).

TABLE 6.2
Saturation Capacities in mg on Primesep 100, Primesep 200, and XTerra RP18 Columns of the Same Dimensions (15 cm × 0.46 cm, 5 μm Particle Size)

Solute	XTerra RP18 (Neutral EP gp)	Primesep 100 (Strong acid gp)	Primesep 200 (Weak Acid gp)
Amphetamine	3	22	19 mg
Benzylamine	0.5	23	(Insufficient retention)
Amitriptyline	5	(Excessive retention)	10
Diphenhydramine	5	(Excessive retention)	15
Propranolol	0.5	(Excessive retention)	16
Nortriptyline	5	(Excessive retention)	17
Codeine	12	38	21

Mobile phase: acetonitrile–phosphate buffer (overall 0.05 M), w^spH 2.0. Note that saturation capacities do not correspond to the total saturation capacity determined by frontal analysis (see text).

6.2.3.4 Hybrid Inorganic–Organic Polymer Phases

Waters introduced porous spherical hybrid inorganic–organic polymer packings in order to overcome some of the limitations of silica without losing its desirable qualities. The XTerra range of packings are synthesized using a mixture of two organosilanes, one that forms silica units [e.g., tetraethoxysilane (SiOEt)$_4$] and one that forms methylsiloxane units [e.g., $CH_3(SiOEt)_3$], typically in a 2:1 molar ratio. As a consequence, some of the silanol groups on the phase are replaced with methyl groups [49]. Bonded phases synthesized from this substrate include two trifunctional materials, XTerra MS-C8 and MS-C18, which are optimized for maximum stability, and two monofunctional embedded carbamate materials (RP8 and RP18), which are especially designed for the separation of basic analytes. Test solutes were slightly less retained on the hybrid materials, attributed to their lower surface area and carbon content compared with similar phases produced on pure silica. The tailing factor for amitriptyline at pH 7 was significantly lower on the hybrid C18 column compared with a similar phase based on pure silica (1.47 vs. 2.16). Tailing factors on the RP18 version were reported as 0.95–1.32 for a variety of basic pharmaceuticals at pH, 2.5, 7.0, and 9.5. The good performance of the phase was attributed to the reduced concentration of residual silanols. The hybrid columns exhibit enhanced stability at high pH and are recommended for general use up to a pH ~ 10. This higher pH gives the potential to analyze some bases above their pK_a such that they are deprotonated, and could lead to elimination of silanol effects (see following text). Rosés and Neue [20] studied the retention of Li$^+$ ions as a function of mobile phase pH as a test of the residual silanol activity of a variety of Waters phases, including XTerra. Retention increased continuously as the pH was raised from pH 3 using Resolve C18 (a Type A silica); from about pH 7 upward for Symmetry C18 (a Type B silica); no increase in retention was noted even at pH 10 for XTerra MS (see Figure 6.11). The onset of the increase in retention of Li$^+$ was delayed to higher pH values for these bonded phases compared with the corresponding bare silica/hybrid (results not shown), indicating

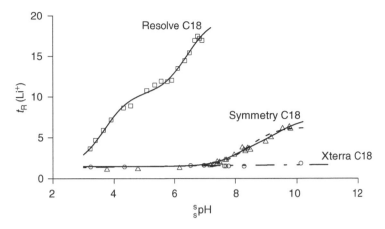

FIGURE 6.11   Plot of retention of Li^+ against pH on different Waters columns (based on [20]).

that C18 ligands have a shielding effect on the silanol groups. Residual-ionized silanols were concluded to be absent on the hybrid phase even at pH 10. However, this evaluation may be optimistic for XTerra MS, (as indeed also may be the evaluation for the all the columns). It is unlikely, for instance, that a hydrophilic inorganic ion such as Li^+ could undergo the strong multiplicative RP/IX interaction postulated for organic bases (see previous text); alternatively, it might also not penetrate the hydrophobic structure of a bonded phase and interact with the surface in the same way as an organic base.

Davies and McCalley [33] used the quaternary amines benzyltriethylammonium chloride, berberine, and some basic compounds as probes, measuring retention and efficiency on XTerra RP18 as a function of pH over the range 2.7–10. They found little evidence of ionized silanols up to pH 7.0, but some deterioration in peak shape of some bases and increase in retention of the quats at pH 8.5; more pronounced effects were found at pH 10. Thus, it is indeed possible that Li^+ does not emulate exactly the interactions of organic bases. However, it is also conceivable that the RP18 version and the MSC18 version of XTerra studied previously [20] might perform somewhat differently in these tests, accounting for the difference between the results of the two research groups.

Waters developed an ethyl-bridged form (BEH) of the hybrid phase [50], which uses an internal difunctional bridging silane, on the basis of a previous report by Unger [51]. 1,2-*bis* (triethoxysilylethane) was used in place of $CH_3(OEt)_3$ in the synthesis procedure with tetraethoxysilane. Co-condensation of 1 and 4 equivalents of these reagents, respectively, gave a hybrid of empirical formula $SiO_2(O_{1.5}SiCH_2CH_2SiO_{1.5})_{0.25}$. An octadecyl surface coverage of 3.2–$3.4\,\mu mol\,m^{-2}$ was achieved on this material after bonding, comparable to conventional silica-ODS phases. Superior peak shape for basic compounds in a test using methanol-phosphate buffer pH 7.0 was found for the BEH compared with a similarly prepared C8 phase prepared on pure silica. This improvement was attributed to a lower acidity of surface silanol groups in the BEH compared with silica. However, no particular suggestion was made of increased inertness of the BEH phase compared with the original methyl hybrid phase (XTerra). The main advantage of the new BEH material appears to be further increased stability of the phase at high pH. An efficiency loss of only 20% was demonstrated by exposure to 50 mM TEA (pH 10) for around 150 h in an accelerated aging test at 50°C.

A disadvantage of the hybrid materials is that they give somewhat lower efficiency (even for neutral compounds) than purely silica based materials [5]. This factor is hardly surprising, considering they have some polymeric character, and that polymeric materials are known to give lower efficiency (see later sections).

6.2.3.5 Monolithic Silica RP Columns

Silica monoliths, which are a single piece of porous material, were developed by Tanaka and Nakanishi, and have been commercialized by Merck in a format where the phase is clad within a PEEK casing of conventional dimensions (e.g., 10×0.46 cm ID) [52,53]. Their advantage is the much lower backpressures they generate compared with particulate columns of comparable efficiency. The Merck monolith was suggested to have the efficiency of a 3 μm particulate column but the backpressure equivalent to a 15 μm column [54]. Desmet and coworkers suggested that the unique advantage of commercial clad monoliths may lie with the very high plate numbers, which are accessible with long (or coupled) columns. Alternatively, the sub-2-μm particulate columns, which are competitive with monoliths, produce faster analysis when a limited plate number is required, especially when used with systems that are capable of delivering pressures in excess of 400 bar [55,56]. However, a new generation of superior monolith columns with reduced domain size (skeleton size + through pore) may increase the competitiveness of silica monoliths [57].

For specific application to basic compounds, the current clad commercial silica monoliths produced by Merck have been shown to give reasonable performance for neutrals and strong bases at acid pH. Asymmetry factors can be greater than those on conventional phases, even for neutral compounds. However, their performance with strong bases at pH 7 was found to be inferior to that shown by many modern particulate columns. Severe tailing was found for a number of basic solutes [58,59]. Recently, Tanaka and coworkers have demonstrated that endcapping of commercial bonded silica monoliths *in situ* with trimethylsilylimidazole can give improved performance with basic solutes. A solution of the reagent in toluene was passed through the column at low flow rate and elevated temperature [60].

We believe that new developments (also including the commercial availability of smaller diameter columns that reduce solvent consumption and are more suitable for MS work) will also increase the interest in monolith columns.

6.2.3.6 Columns with Sub-2 μm Particles

HPLC systems have recently become commercially available, which allow the use of pressures up to 1000 bar. Columns of particle size around 1.7 μm and up to approximately 10-cm long can be operated at their optimum flow rate within the pressure capability of the system. The main advantage of these columns is that they can generate the same plate count as longer columns of larger particles but in a shorter analysis time. Independent evaluations of the possibilities and limitations of these new systems have begun to appear [61,62].

To date, insufficient work has been done to ascertain whether the general advantages of very small particle columns for neutral compounds also can be demonstrated for basic compounds, and that no unexpected differences in selectivity or peak shape (other than a reduction in the plate height) occur as a result of the reduction in particle size.

6.2.4 OTHER TYPES OF COLUMN (e.g., POLYMERS, ZIRCONIA)

While silica-based materials are by far the most popular in RP chromatography, other materials have also been used for particular analyses, especially where column stability may be an issue. Carr and coworkers have developed RP zirconia columns, which are stable over the pH range 1–13 and in some variants up to 200°C. An initial problem in column manufacture was that $Zr-O-Si$ bonds are not as stable as $Si-O-Si$ bonds, so silanization is not useful to produce RP packings. However, zirconia particles can be coated with organic polymers such as polybutadiene [63]. While silanol groups are a complication on silica-based phases, Zr(IV) sites (which are hard Lewis acids) cause very strong adsorption of hard Lewis base analytes (e.g., $RCOO^-$, R-SO_3^-). These sites can be blocked by adding a strongly competing Lewis base (e.g., F^-, PO_4^{3-}) to the mobile phase at sufficiently high concentration. These strong Lewis bases are small anions with a high electron density and low polarizability, which form strong complexes with the electropositive Zr(IV) atoms. The adsorption of strong Lewis bases from the mobile phase in turn confers cation exchange properties on the PBD-zirconia phase [64]. The phases exhibit very significant selectivity differences from silica ODS phases in phosphate buffers at pH 7. For cationic solutes, ion-exchange interactions can provide the majority of the retention on PBD-zirconia phases, whereas RP interactions dominate on high quality ODS phases. Snyder and coworkers using their test procedure (see following text) also found that bonded zirconia phases show much increased retention of cations and reduced retention of hydrogen-bond acceptors compared with Type B silica columns [65]. Results for the analysis of bases appear promising for zirconia columns, but few independent studies of their performance compared with silica-based columns have been published.

Buckenmaier et al. used polystyrene-divinylbenzene phases, and a polyvinyl alcohol phase, principally to study the overloading behavior of basic solutes on a surface free of silanol groups and their effects [21]. However, cation exchange sites were discovered on these purely polymeric phases at pH 7, giving rise to peak tailing of bases at intermediate pH as is also found for silica-based columns. Nevertheless, these polymers can be used at high pH (e.g., pH 12) where peak shapes improve again. An advantage of using pH 12 is that overloading occurs much less readily, and thus column efficiency is maintained at much higher mass load. The efficiencies obtained in general were considerably lower than is found typically for silica-based phases when used at low pH, and this is a continuing general disadvantage of the use of polymeric materials.

6.2.5 Influence of the Solute

Snyder and coworkers [6] reviewed the dependence of silanol interactions on the chemical structure of the interacting base. Deductions were based on data of the relative efficacy of amine silanol blocking agents that were more popular at the time; the stronger the deactivation effect of the modifier, presumably the stronger its binding to silanols. Stronger interaction of an amine or quaternary ammonium compound was said to be favored by

- Greater basicity (larger pK_a)
- Reduced steric hindrance around the nitrogen atom, for example, dimethyl derivatives give stronger interactions than those containing branched substituents
- Increased substitution on the nitrogen; tertiary amines interact more strongly than secondary amines, which interact more strongly than primary amines, presumably because of the electron donating ability of alkyl groups
- One long alkyl group substituted onto the N atom, which makes the amine more hydrophobic without greatly increasing the steric hindrance around the nitrogen atom

Few systematic studies have been published since.

McCalley [66,67] studied the peak shapes produced by a series of closely related alkyl pyridine compounds, using methanol, acetonitrile, and THC as modifiers in combination with phosphate buffers at pH 7.0, using a Type B phase. The compounds had rather similar pK_a over the range 5.1–6.7, and in combination with organic solvent effects (see following text), it seems likely that solutes were mostly uncharged in these studies. Little correlation was found between solute pK_a and peak shape, which seems hardly surprising given these conditions. However, marked influence of solute stereochemistry was observed. Peak tailing increased along the series 2-, 3-, and 4-methyl pyridine as the alkyl substituent is moved further away from the nitrogen atom. Peak tailing decreased for the series 2-methyl, 2-ethyl, and 2-propyl pyridine as the size of the substituent next to the nitrogen atom increased. Peak asymmetry decreased for the series 4-methyl, 4-ethyl, 4-isopropyl, and 4-*tert*-butylpyridine where the substituents are positioned more remotely from the basic center. These results indicate a possibility that the overall size of the molecule may influence its ability to penetrate the bonded ligands and interact with the surface. The source of the detrimental interactions observed in these experiments is uncertain. Hydrogen bonding of nonionized solutes with silanols could play a part.

In the case of stronger bases that are partially or totally ionized under the analysis conditions, there is clearly a relationship between the pK_a of the base (and thus its degree of ionization) and the degree of detrimental interaction [35]. For nine basic solutes analyzed on nine different columns at $_w^w$ pH 7 using acetonitrile, the lowest average A_s (1.4) was obtained for codeine, which had the lowest $_w^w pK_a$ (8.5) and the highest for nortriptyline (3.0), which has the highest $_w^w pK_a$ (10.2), with other solutes showing a reasonable correlation of tailing with pK_a. It is not surprising that if ionic interactions can be responsible for peak tailing, the degree of solute ionization should correlate with peak tailing; however, it seems unlikely that this is the only factor involved. Unfortunately, unlike the aforementioned study with the substituted pyridines, there was insufficient structural relationship between the solutes in this second study to make any firm deductions on the effect of stereochemistry on peak tailing under conditions where the solutes were ionized. Further studies are necessary in this area.

6.2.6 Influence of the Mobile Phase

6.2.6.1 Mobile Phase pH and Its Effect on Solute pK_a

Before assessing the effect of mobile phase pH on the chromatography of bases, some consideration must be given to the measurement itself. Conventionally, the pH is measured in the aqueous buffer before mixing with the organic modifier ($_w^w$pH scale) [68]. However, measuring the pH in the mobile phase after mixing with the organic modifier has been recommended [69]. For these measurements, the pH electrode can be calibrated with buffers of known pH made up in the same aqueous–organic mixture as used for the mobile phase (giving $_s^s$pH) or more conveniently by calibration in the usual aqueous standards (giving $_w^s$pH). Conversion between the $_w^s$pH and $_s^s$pH scales is simply achieved from the relationship

$$_w^s pH - _s^s pH = \delta$$

Values of δ are easily obtained from the literature. Rosés and Bosch, who have researched extensively in this area, have discussed the shortcomings of $_w^w$pH measurements. The pH variation when adding methanol or acetonitrile to the aqueous buffer depends on the particular buffer, on its concentration, and on the fraction of organic solvent in the mixture. Buffer solutions prepared from anionic and neutral (uncharged) acids (e.g., HAc/Ac$^-$, H$_2$PO$_4^-$/HPO$_4^{2-}$) increase their pH when acetonitrile or methanol is added, whereas buffers from cationic acids (e.g., NH$_4^+$/NH$_3$) show the reverse trend. The pK_a variation of analytes follows the same pattern as that of buffer components. Much better fits of retention time as a function of pH are obtained for basic (and acidic) solutes when pH is measured in the aqueous–organic mobile phase. These results are of importance in the interpretation of the chromatography of basic solutes. For example, in the commonly used phosphate buffers at intermediate pH, the $_w^s$pH in the aqueous–organic mixture is higher than $_w^w$pH, while the $_w^s pK_a$ of basic solutes BH$^+$ is lower. Thus, while a basic solute of $_w^w pK_a$ 9 would be expected to be almost completely charged in a mobile phase of $_w^w$pH 7, a proper consideration of appropriate mobile phase values shows this to be not true. The pK_a of organic bases in aqueous–organic solvents can be estimated by a variety of methods; some values in methanol–water and acetonitrile water at various combinations were determined using capillary electrophoresis by Buckenmaier et al. [70,71] and are shown in Table 6.3. The advantages of CE include the good sensitivity of the technique, and its separation power, which obviates the need for the use of highly pure compound. At high buffer pH, the unprotonated, uncharged base has no electrophoretic mobility (μ_{base}) and thus migrates with the electroosmotic flow (EOF). Alternatively, a fully protonated,

TABLE 6.3

$_w^s pK_a$ **Values of Bases in Methanol–Water and Acetonitrile–Water as Determined by Capillary Electrophoresis**

Compound	Water	20% MeOH	30% MeOH	40% MeOH	60% MeOH	70% MeOH
Nortriptyline	10.2	10.0	9.8	9.7	9.4	9.3
Diphenhydramine	9.2	9.0	8.9	8.8	8.5	8.2
Codeine	8.2	8.1	7.9	7.8	7.6	ND
Procainamide	9.3	9.2	9.1	8.9	8.7	8.5
Benzylamine	9.5	9.3	9.2	9.2	9.0	8.9
Protriptyline	10.7	10.4	10.3	10.2	9.9	9.7
Amitriptyline	9.3	9.3	9.1	8.9	8.5	8.3

Compound	Water	20% ACN	40% ACN	60% ACN
Nortriptyline	10.2	10.1	9.9	9.7
Diphenhydramine	9.2	9.2	8.9	8.7
Codeine	8.2	8.1	7.8	7.5
Procainamide	9.3	9.3	9.2	9.0
Benzylamine	9.5	9.4	9.1	8.8

ND = not determined.

that is, positively charged base exhibits maximum mobility and elutes faster than the EOF because of electrostatic solute attraction by the cathode. Intermediate mobility is a function of the dissociation equilibrium of the base. The μ_{base} is obtained from Equation 6.10 by a straightforward measurement of migration time of the base (t_{base}) and EOF (t_{EOF}); the latter can be obtained from any convenient neutral marker (e.g., acetone). L_{cap} and l_{eff} are capillary length (inlet to outlet) and effective capillary length (inlet to detection window), respectively, and V is the voltage applied across the capillary

$$\mu_{base} = \frac{L_{cap} l_{eff}}{V} \left[\frac{1}{t_{base}} - \frac{1}{t_{EOF}} \right] \tag{6.10}$$

$$pK_a = pK_a' - \frac{Az^2 \sqrt{I}}{1 + Ba_0 \sqrt{I}}. \tag{6.11}$$

Plotting μ_{base} vs. pH gives a sigmoidal curve, whose inflection point reflects the apparent base-pK_a', which may be corrected for ionic strength, I, using Equation 6.11 in order to obtain the thermodynamic pK_a value in the respective solvent composition. Parameters A and B are Debye-Hückel parameters, which are functions of temperature (T) and dielectric constant (ε) of the solvent medium. For the buffers used, $z = 1$ for all ions; a_0 expresses the distance of closest approach of the ions, that is, the sum of their effective radii in solution (solvated radii). Examples of the plots are shown in Figure 6.12.

Analysis of bases has been carried out with low, intermediate, or high pH mobile phases, which are discussed in the following sections.

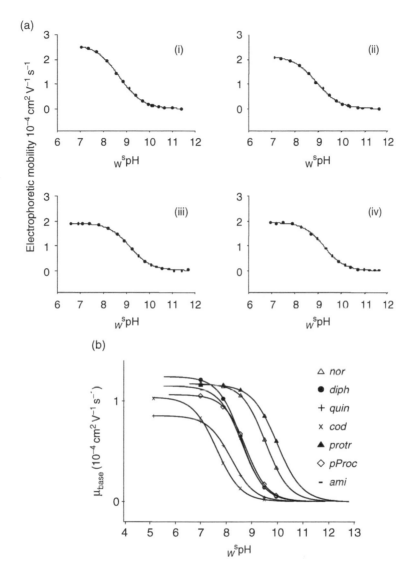

FIGURE 6.12 (a) Electrophoretic mobility of diphenhydramine (μ_{base}) *vs.* $_w^s pH$ in (i)ACN/aqueous buffer 60:40 v/v ($_w^s pH$ = 8.69), 40:60 ($_w^s pH$ = 8.92), 20:80 ($_w^s pH$ = 9.16), 0:100 ($_w^s pH$ = 9.25). Sigmoidal curves through data points calculated by nonlinear regression (Sigma Plot 5.0). Detection wavelength: 214 nm, T = 25°C, V = 20 kV. (b) Plots of electrophoretic mobility (μ_{base}) vs. $_w^s pH$ in methanol–water–buffer (60:20:20, v/v/v). Sigmoidal curves through data points calculated by non-linear regression (Sigma Plot 5.0). Detection wavelength (bases): 214 nm; T = ambient 25°C; Vrun = 10 kV except for cod and quin at $_w^s pH$ 5.18, where Vrun = 20 kV; typical electrical currents were about 10 and 20 μA for 10 and 20 kV, respectively.

6.2.6.1.1 Use of Low pH

The advantages of using low pH to inhibit ionization of unbonded silanol groups have already been mentioned. McCalley evaluated eight early Type B silica phases with nine basic solutes at $_w^w pH$ 7 and with eight basic solutes at $_w^w pH$ 3 [16,17]. For the seven solutes common to both studies, the average efficiency of 25 cm columns using methanol–phosphate buffer was 6000 plates with average asymmetry factor, $A_s = 3.4$ at pH 7, compared with $N = 9600$ and $A_s = 1.9$ at pH 3. In a later study [35] using nine more recent Type B columns and eight to nine basic solutes, the average N using acetonitrile–phosphate buffer was 14,500, $A_s = 2.0$ at $_w^w pH$ 7 compared with $N = 18,600$, $A_s = 1.3$ at $_w^w pH$ 3.0. In each case, small amounts of solute were used ($0.2\,\mu g$) to avoid overloading effects. In both studies, a clear advantage can be seen in the use of low pH to obtain the best peak shape, consistent with the suppression of ionization of silanols. For the later study, peak tailing was hardly a problem at low pH, at least for the columns used. Indeed, they were selected as likely to be some of the best available phases at the time (1999). On the other hand, the results at pH 7 were considerably poorer, especially for the strongest base, nortriptyline that had an average $A_s \sim 3$ over all the phases using acetonitrile. These results show clearly, why low pH has remained the most popular for the analysis of bases. Nevertheless, low retention of hydrophilic bases can be a severe restriction in work at low pH. At pH 7, the combined effects of decreasing solute ionization (especially for bases of low pK_a), which increases hydrophobic interactions and increasing silanol ionization, can both contribute to higher overall retention.

McCalley [72] subsequently studied the effects of different buffers at low pH on the retention and peak shape of bases, principally with the aim of finding any differences between phosphate buffers, traditionally used with UV detection, and volatile buffers such as formic acid or trifluoroacetic acid used in mass spectrometry. At low pH, formic acid gave slightly lower efficiencies on average than phosphate when using 200 ng of solute. This result is at least partially attributable to overloading that takes place to some extent even with these low solute levels in this low ionic strength mobile phase (see earlier text). TFA was found to be a useful volatile buffer at low pH, giving increased retention compared with phosphate, owing to ion-pair effects and good column capacity (as mentioned earlier). However, suppression of MS sensitivity may be a disadvantage of its use [73]. In one case (Waters Symmetry column), large differences in selectivity for acids and bases were noted in formic acid compared with phosphate at the same pH and molar concentration (see Figure 6.13) [25]. This result is due to positively charged catalyst residues on this phase, which are ionized at low pH, and provide anionic exchange sites, which are influential in low ionic strength mobile phases (e.g., formic acid). Thus the protonated bases diphenhydramine and nortriptyline have low retention in formic acid owing to repulsion from the phase, whereas benzenesulfonic and 2-naphthalenesulfonic acids have high retention because of retention on the positively charged sites.

6.2.6.1.2 Intermediate pH

Clearly, there are disadvantages in the use of intermediate pH in terms of poorer peak shape for bases compared with low pH, owing to silanol ionization [16,17]. An exception to these results was found later for the hybrid inorganic/organic phase

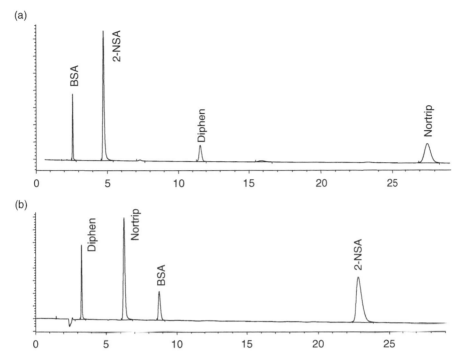

FIGURE 6.13 Separation of acidic and basic test compounds on a Symmetry 100 column (25 × 0.46 cm, 5 μm particles). Mobile phase (a), acetonitrile-0.02 M phosphate buffer, pH 2.7 (28:72 v/v). Mobile phase (b), acetonitrile-0.02 M formic acid, pH 2.7 (28:72 v/v). Flow rate, 1 mL min^{-1}. Detection, UV at 215 nm. Column temperature, 30°C.

XTerra, which gave peak shapes at pH 7 similar to those obtained at acid pH, owing to the apparent lack of silanol ionization [33]. However, useful selectivity differences from those found at low pH can be obtained even on conventional silica phases, despite the inferior peak shapes [35]. A further disadvantage of use of intermediate pH is that both silanols and solutes are partially ionized. Changes in ionization of both will occur if the pH is not carefully controlled, which may compromise the reproducibility of the separation [5]. Nevertheless, at pH 7, some weak bases will be virtually deprotonated if the pH is above their pK_a, especially when considering the effect of the organic solvent (see earlier text). In theory, therefore, the solute, being essentially neutral, should not be able to undergo coulombic interaction with silanols even if the latter are ionized, therefore yielding good peak shapes. The enhanced hydrophobic retention of the neutral compared with the ionized form of the solute, leads to usefully higher retention of hydrophilic bases even if the solute is only partially deprotonated (see earlier text) and may also give useful selectivity effects.

At pH 7.0, ammonium phosphate buffer was found to give similar retention but somewhat improved peak shape as compared with potassium phosphate buffers, presumably because of the superior deactivating effect of the ammonium ion on ionized silanols [72]. In contrast, this effect was hardly observed at low pH, probably because there was little silanol ionization under these conditions on the inert modern phases studied. Ammonium acetate is not at all a good buffer at pH 7, but is often used as a mobile phase in HPLC-MS owing to its volatility. Nevertheless, it gave reasonable peak shape and reproducibility of retention for small quantities of many basic solutes. However, for solutes whose pK_a was close to the mobile phase pH, poor efficiency was obtained, presumably because of poor buffering ability of the mobile phase. Note that we evaluated ammonium acetate mobile phases only with standard solutions containing low solute concentrations. More serious effects of poor buffering might occur with higher concentrations of solute, or with the presence of matrix compounds, which themselves could alter the pH.

6.2.6.1.3 High pH

A few RP columns are now available for higher pH work, which allow chromatography of higher pK_a basic solutes when they are substantially deprotonated. Further increases in retention are obtained compared with low and intermediate pH owing to increased hydrophobic retention of the neutral species. Overloading effects are considerably reduced for the neutral compound compared with the charged species and different selectivity of separations can be obtained [33]. A problem with this approach is the stability of the underlying silica and bonded ligands in conventional bonded phases, a constraint that often prevents analysis at sufficiently high pH.

McCalley [35] evaluated the Zorbax Extend phase, developed by Kirkland and coworkers [74], which has a bidentate ligand bonded to the silica surface. This column was reported to be especially stable at high pH, and indeed was designed for the routine separation of free uncharged bases. With methanol–phosphate buffers, low pK_a solutes such as pyridine and codeine, which would be expected to be completely deprotonated at pH 11, gave asymmetry factors comparable to values obtained at pH 3 and 7. The values for the strongest base nortriptyline were clearly worse at high pH. Perhaps even a small degree of ionization of the solute at high pH can still give rise to serious peak tailing because of ionization of residual column silanols. The embedded polar group hybrid inorganic–organic polymer XTerra RP18 was evaluated by Davies and McCalley [33] and found to give the highest efficiency for compounds of moderate pK_a (e.g., codeine and quinine) at the highest pH studied ($_w^s pH$ 10.0). However, compounds of higher basicity, which remain significantly protonated even at $_w^s pH$ 10, gave worse peak shape than they did at low pH. For example, the efficiency for small sample mass (<0.1 μg) of amitriptyline ($_w^s pK_a = 8.9$) improves over the range $_w^s pH$ 3.4–10.0 (see Figure 6.7) presumably because the solute becomes mostly neutral at high pH (only 7% ionized, calculated using the Henderson–Hasselbach Equation), and the peak shape is unaffected even if silanols are ionized. Alternatively, the peak shape for small sample mass of amphetamine ($_w^s pK_a = 9.8$, ~40% ionized at $_w^s pH = 10$) is much poorer at high pH, presumably because its ionization is still sufficient to cause substantial detrimental interaction with the now ionized silanols (Figure 6.8).

Fornal et al. [75] determined selectivity differences for bases in RP-HPLC under high pH conditions. They used quantitative structure retention relationships (QSRR) to model retention behavior. They reported that the stability of the columns they used (Waters XTerra MS, Zorbax Extend, Thermo BetaBasic) was limited with deterioration occurring for XTerra (apparently the most stable phase) just below 200 injections when using a mobile phase of acetonitrile–water or methanol–water both containing 0.1% ammonia ($^{w}_{w}$pH 10.4). Selectivity changes could be obtained both by interchange of the stationary phases and by change of the organic modifier.

More work is necessary to discover more fully the potential and limitations of high pH work.

6.2.6.2 Nature of the Organic Modifier

Claessens and coworkers made a systematic study of the differences between isoeluotropic aqueous methanol, acetonitrile, and tetrahydrofuran for the separation of a mixture that included the bases hexylpyridine and heptylpyridine [76]. They proposed that silanophilic interactions, which differed widely between the different modifiers, were the cause of differences in peak shapes for the bases. The bases gave worst asymmetry factors when using acetonitrile, leading them to suggest this modifier as best for testing as it would emphasize differences between the columns. McCalley [16,17] studied the peak shape of basic solutes using isoeluotropic mixtures of methanol, acetonitrile, and THF in combination with phosphate buffers at pH 3 and 7. At pH 7, the best peak shapes were obtained using THF and the worst with acetonitrile. At pH 3, THF also gave the best peak shapes but acetonitrile generated higher efficiencies than methanol. Despite these favorable results, THF seems to be seldom used as organic modifier, probably as a result of stability and safety concerns with this solvent. A second study [35] confirmed that lower asymmetry factors and higher column efficiency were obtained using methanol at pH 7 rather than acetonitrile, but that higher efficiencies were obtained with acetonitrile than methanol at pH 3. It is possible that at pH 3, when silanol effects are minimized on modern Type B columns (as used in [35]) that the dominant factor is the lower viscosity of acetonitrile. At flow rates somewhat above the optimum, which are very commonly employed in HPLC, increased solute diffusivity results in higher efficiency since increased mass transfer dominates at these flow rates. Results at pH 7 are more difficult to interpret. In these studies [16,17,35], isoeluotropic eluents were used for practical purposes, since it was felt that peak shapes should be compared under conditions of similar retention of the solutes. However, the higher concentration of methanol used (65%, compared with 40% acetonitrile) results in a higher mobile phase $^{s}_{w}$pH when anionic buffers such as phosphate are used; a lower solute $^{s}_{w}$pK_a also occurs because of the concentration of organic solvent. These effects work in the same direction and result in less protonation of the analytes in the methanolic mobile phase. It is likely that this factor contributes considerably to these observed differences, although other possible interpretations exist. For example, methanol is capable of hydrogen bonding with silanol groups whereas acetonitrile is not, and peak symmetry might also be affected by different ligand wetting in different modifiers of different concentration, resulting in different accessibility to silanol groups on the column surface.

6.2.6.3 Use of Silanol Blocking Agents

These are mobile phase additives (usually amines), which by competitive interaction, block the detrimental effect of silanol groups. As noted in a previous review [4], their use has declined, principally because the availability of inert, high purity silica columns has made them less necessary. Also, these additives may give rise to slow equilibration, be difficult to remove from the column, cause interference with detection, and generally add to the complexity of the method. In some cases, they may even undergo chemical reaction with sample components. Nevertheless, such reagents (e.g., TEA) are still quite frequently used, for example, in the pharmaceutical industry, where routine methods established on older Type A phases may not have been updated because of the expense and inconvenience of revalidation of the procedure.

Carr and coworkers [77] studied the retention of different amines on PBD-Zirconia phases. On PBD zirconia, they found that steric hindrance has a much greater effect on retention of bases in comparison to silica ODS phases; thus, quaternary amines have substantially less retention than the corresponding less hydrophobic primary amine. This result agreed with previous findings of Sokolowski and Wahlund [78]. Thus, secondary and tertiary amines such as dimethylbutylamine and TEA were less effective silanol blockers. Only ammonia and primary amines were able to improve the chromatographic properties of basic solutes on PBD-zirconia. It would be interesting to check the application of these results also to silica–ODS phases.

6.2.7 TEMPERATURE

Lundanes and Greibrokk [79] reviewed the general principles of the use of elevated temperature as a parameter in LC. Sandra and coworkers [80] summarized the advantages of elevated temperature as

- Varied selectivity compared with ambient temperature.
- Shorter analysis times.
- The reduction of the mobile phase viscosity allowing the use of longer columns and the generation of higher efficiency. At high linear velocities, where interphase mass transfer broadening dominates, the increased analyte diffusion coefficient (which results also from the decrease in eluent viscosity) gives increased efficiency [81].

It is important to note that, at least for neutral compounds, temperature has only a small effect on efficiency when the column is operated at its optimum flow velocity. Above the optimum velocity, temperature has a beneficial effect on efficiency, whereas below the optimum velocity, increased solute diffusion has a relatively large negative effect on efficiency [32,82,83].

McCalley [82,83] showed that the retention of some bases at neutral pH *increased* with temperature over the range ambient to 60°C, in contrast to the usual effect of decrease in retention for neutral compounds. This observation accounted for some of the marked selectivity differences that can be observed in the separation of mixtures containing different classes of compound with increasing temperature. Pronounced increases in efficiency with increasing temperature were demonstrated for basic compounds at a mobile phase pH of 7. These increases were over and above any expected due to decreased mobile phase viscosity and increased solute diffusivity, which were shown for the same compounds at pH 3. It was later demonstrated that the increase

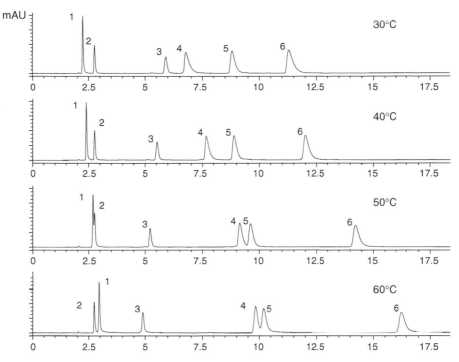

FIGURE 6.14 Chromatograms obtained for sample mix using acetonitrile–phosphate buffer (40:60 v/v) $_w^s$pH 7.8 using Inertsil ODS 3 V (25 × 0.46 cm, 5 μm particles), in the temperature range 30–60°C. Peaks: (1) benzylamine, (2) bteN (3) berberine chloride, (4) quinine, (5) protriptyline, (6) nortriptyline. Flow rate: 1 mL min^{-1}. Sample volume 5 μL, sample mass 0.1 μg, UV detection at 210 nm. Mobile phase acetonitrile–phosphate buffer pH (40:60 v/v).

in retention and efficiency for bases could be explained by reduction in the pK_a of solutes with increasing temperature [83]. As the base becomes deprotonated, hydrophobic retention of the solute increases. Quaternary amine compounds, which are ionized across the temperature range, do not show these anomalous effects, providing a rationale for the selectivity differences obtained as a function of temperature in Figure 6.14. It can be seen that as the temperature is increased from 30°C to 60°C, the retention of the quaternary compound benzyltriethylammonium chloride (peak 2) decreases, while the retention of the base benzylamine (peak 1) increases, leading to a reversal in the order of elution of the peaks. Note that the increased retention of bases with temperature should be expected when the mobile phase $_w^s$pH is close to the $_w^s$pK_a of the base. Even nortriptyline (peak 6 in Figure 6.13), the strongest base studied, shows a considerable increase in retention time from ~11.5 min at 30°C to ~16.5 min at 60°C; its $_w^s$ pK_a may be reduced to ~9 at 60°C with 40% ACN. A contributing factor is the much higher k of the neutral compound compared with the protonated base.

Difficulties of work at high temperatures include stationary phase stability and analyte stability. However, it has been shown that the relatively short residence times in the column at high temperatures tend to minimize the effects of analyte instability

[81]. Although most classical silica–ODS columns are not stable at temperatures much above 60°C (dependent on eluent conditions), columns made from other materials such as polymeric and zirconia-based phases have been used for high temperature work (see earlier text).

6.2.8 TESTING METHODS FOR RP COLUMNS

The evaluation of RP columns has long posed a problem for analysts who wish to classify both the separation selectivity and the peak shape likely to be obtained from a particular phase; McCalley [84] and Dorsey [85] have reviewed these tests. The work of Kele and Guiochon [86] has been fundamental to the philosophy of column testing. These authors have shown that the reproducibility of columns (at least those produced by the major manufacturers), both within batch and batch to batch, is excellent. In this case, the testing of single columns (a strategy used by the majority of workers in this field for reasons of time and economy) is a valid approximation because the results for a single column are likely to be reasonably representative of columns of this type produced by the manufacturer, at least at the time of testing. However, some particular stationary phases have been in popular use for decades. It seems inconceivable that *some* changes in the phases have not occurred over this time, for example, owing to changes in the specification of reagents supplied to the column manufacturer by outside parties or improvements in column hardware, which might be made at least to improve the reproducibility of the product. Improvements in packing procedures may allow the production of columns of higher efficiency even if the stationary phase itself remains unchanged. Thus, a little caution is necessary in the examination of results of tests performed many years previously. Alternatively, the pressure from users who wish to maintain a particular separation constant over a period of many years has acted as an influence in the opposite direction. Some manufacturers record definite advances or changes in their products by designations such as ODS-1, ODS-2, ODS-3, and so forth.

The first generally column-accepted test was the Engelhardt test [87], which has been considered in detail in an earlier review [4]. Briefly, this test involves the analysis of a mixture of neutral compounds, an acidic compound (phenol), and some weak bases (including aniline) in an unbuffered methanol–water mixture. While this test (as far as bases are concerned) seems to provide a convenient means of distinguishing between highly active Type A columns and the more inert Type B columns, it does not distinguish well between individual Type B phases [16,17,35], all of which perform well in the test. In contrast, tests with stronger bases in buffered mobile phases can show clear differences between these Type B phases. There is also some uncertainty as to what this test is measuring; it seems clear that the major current problems are with the analysis of ionized bases, whereas weak bases are likely to be mostly present as neutral compounds under the conditions of the Englehardt test, although of course the pH of an unbuffered mobile phase is rather indeterminate.

Sander and Wise [88] reported a new standard reference material SRM 870, which contains the neutral compounds ethylbenzene and toluene, to monitor hydrophobicity and methylene selectivity; uracil (void volume marker) quinizarin to measure activity toward metal chelators; and amitriptyline to measure activity toward bases. The mobile phase recommended was 80:20 methanol–phosphate buffer (20 mM potassium phosphate pH 7), giving a final phosphate concentration in the mixture of 4 mM. They noted that retention and peak shape for amitriptyline were strongly influenced by mobile phase ionic strength at lower levels, but there was much less influence at higher levels (above ~2 mM). Also, they noted that peak asymmetry and retention of amitriptyline were not correlated. The combined effect of injection volume and mass overload was assessed. It was found that efficiency decreased with increasing sample volume up to 100 μL. It was recommended that 2 μL injections of the mixture were used, resulting in the introduction of less than 5 μg of each solute. However, no detailed studies on the effect of solute mass for amitriptyline were carried out, despite indications from the published chromatograms that the peak shape of the basic solute was the most affected of those in the mixture by solute mass. Other studies [89] have shown that at pH 7, efficiency may sometimes *increase* with increasing sample mass, followed by the usual decrease as active sites on the column are blocked by the sample itself, allowing a greater proportion of the sample to travel down the column relatively free of these detrimental interactions. There was broad correlation between the results for the NIST test and the Englehardt test; however, in some cases, relatively good peak shape was obtained for components of the Engelhardt test, but asymmetric peak shape for amitriptyline, supporting the discussion earlier.

Neue and coworkers [90] at Waters have worked on the characterization of RP packings for many years. The mobile phase in their test consists of phosphate buffer pH 7 (overall concentration 20 mM) mixed with methanol in the volume ratio 35:65. Uracil is used as the void volume marker, naphthalene and acenapthene are the purely hydrophobic compounds, propranolol (400 mg/L) and amitriptyline (100 mg/L) are the basic probes, and butylparaben and diisopropyl phthalate are the markers for polar selectivity of the packings. Originally, the relative retention of amitriptyline and acenapthene was calculated as a measure of silanophilic interaction. However, it was later realized that this measurement was not free from hydrophobic interaction (presumably a high value for this relative retention could be caused not only by high retention of amitriptyline caused by silanophilic interaction, but also by low retention of acenapthene caused by low hydrophobicity). Thus, an empirical subtraction method was used to remove the hydrophobic contribution from the retention factor of amitriptyline, leaving exclusively the silanophilic contribution of the amino group. A plot of this "silanophilic interaction" against hydrophobicity (retention of acenapthene) reveals packings of high silanophilic acitivity (e.g., Waters Spherisorb ODS-2) at the top of the chart. Packings with low silanophilic activity at the bottom of the chart included all 10 of the EPG packings tested (see Figure 6.15). A problem with the test is that amitriptyline is only partially ionized and thus its retention depends on accurate adjustment of the mobile phase pH. It appeared that there was reasonable correlation between the Waters and NIST silanophilicity tests, although this is hardly surprising since the test compound (amitriptyline) is the same and the test conditions similar (methanol-phosphate buffer pH 7). It was noted that the peak shape of amitriptyline was always asymmetric, and that therefore a comparison of the two tests should ideally use the same amount of solute; the large amount of amitriptyline in the NIST test was remarked on. Overloading depends on the retention factor that can vary from column to column, dependent for example, also on the hydrophobic retention

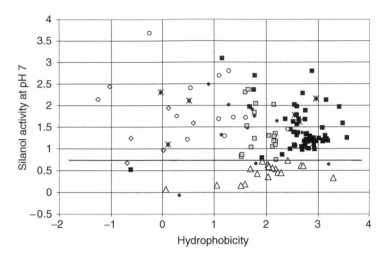

FIGURE 6.15 Evaluation of silanophilic activity of various groups of phases according to reference [90]. Designations in (a): black squares: C18, gray squares: C8, circles: phenyl, diamonds:cyano, stars: fluorinated packings, triangles: packings with embedded polar groups, small back diamonds: uncategorized packings.

properties of the column. Thus, the amount of solute that partitions into the stationary phase varies between columns. The best procedure is to try to minimize overloading so that effects are small whatever the value of k. A direct comparison between the Waters and Tanaka test (see later text) was more difficult owing to differences in the mobile phase conditions and test solutes.

Euerby and coworkers [91–93] have continued to use and adapt the Tanaka test for column characterization. The test includes pentylbenzene and butylbenzene for hydrophobic interactions, relative retention of triphenylene and o-terphenyl for shape selectivity, relative retention of caffeine and phenol for hydrogen bonding capacity— a measure of the number of available silanol groups/degree of endcapping, and the ion-exchange capacity, measured as the relative retention of benzylamine (2.5 μg injected) and phenol at pH 7.6 and 2.7. The mobile phase used for the last test is 20 mM phosphate buffer (pH 7.6 or 2.7) in 30:70 methanol–water. Benzylamine can give very poor peak shape even at pH 2.7 on some Type A columns. It is possible again that this amount of basic compound could give rise to some overloading effects; however, the effect of overloading on retention time (which is measured in the Tanaka test rather than any peak shape parameter) is generally less severe than the effect on column efficiency.

McCalley and Brereton [84] used principal components analysis (PCA) to investigate the data sets produced previously [16,17] for column assessments for bases at pH 3 and 7. A range of five solutes was necessary to cover possible interactions between the column and base at either pH. For example, there was little correlation between asymmetry values for codeine and nortriptyline when using methanol– phosphate buffers at pH 3, showing that different column evaluations would be obtained depending on the choice of solute. Furthermore, the rankings of columns according to mean peak asymmetry for all test compounds at pH 3 using a particular modifier (e.g., methanol) was different from the ranking at pH 7 using the same modifier. The ranking of the columns using acetonitrile at pH 7 was different from that using methanol at pH 7. It was concluded that not only a variety of solutes but also different conditions were needed to test columns effectively, and that PCA could be used for the judicious selection of probes. These results emphasized the difficulties that are likely to be encountered for those trying to develop a column test that contains a single basic solute analyzed with a single mobile phase, and why little agreement has been found between different test procedures that use different probes and conditions.

Vervoort and coworkers [94] compared the results of the Engelhardt, Tanaka, and Galushko tests with the results of tests with basic compounds published by McCalley [16,17,35]. They wished to predict the suitability of seven Type B stationary phases for the successful analysis (i.e., good peak shape) of a suite of their own proprietary pharmaceutical compounds. The Engelhardt, Tanaka, and Galushko tests resulted in different ranking of the Type B phases. Furthermore, these three tests were in disagreement with rankings using the range of basic compounds of the McCalley test. They pointed out that the amount and nature of the modifier influence the eluent pH and solute pK_a, and thus the state of ionization of the test solute. Furthermore, they showed differences in the rankings dependent on the nature of the buffer constituents and the test compounds used in agreement with the findings of McCalley. It is hardly surprising that tests produce different results, considering all these variables that can affect the analysis of basic compounds, as discussed earlier. According to Vervoort, only the procedures described by McCalley using a range of probes and conditions produced comparable column rankings to those found when using their own pharmaceutical compounds.

Little has been said so far on test procedures aimed primarily at the evaluation of selectivity for different RP columns. One of the most comprehensive of these is the hydrophobic subtraction model, developed by Snyder and coworkers; this method is applicable to all types of solutes including bases [36]. The model recognizes that hydrophobic interactions dominate the RP retention process; thus in the absence of any other mechanism, plots of retention ($\log k$) for one column vs. another should yield straight line with roughly unit slope and little scatter, which indeed can be shown to be a reasonable approximation. However, clearly, other contributions to retention exist, which lead to scatter in these $\log k$ plots. In particular, for basic compounds, ionized silanols can retain protonated bases by ion exchange, whereas neutral silanols can hydrogen bond with proton-acceptor solutes. Shape selectivity interactions and hydrogen bonding of acidic solutes with a basic column group are also incorporated into the model, which is so called because it assumes that the major contribution of hydrophobic retention is first subtracted in order to envisage the remaining contributions from other interactions. It leads to a general equation for RP-HPLC retention and column selectivity.

$\text{Log } \alpha = \log k / \log k \text{ (ethylbenzene)}$

$$= \eta'\mathbf{H} \quad - \quad \sigma'\mathbf{S*} \quad + \quad \beta'\mathbf{A} \quad + \quad \alpha'\mathbf{B} \quad + \quad \kappa'\mathbf{C}$$

$\eta'\mathbf{H}$	$\sigma'\mathbf{S*}$	$\beta'\mathbf{A}$	$\alpha'\mathbf{B}$	$\kappa'\mathbf{C}$
hydrophobic	steric resistance	column H-bond acidity	H-bond basicity	ion interaction
	(to bulky interactions)	(non-ionized silanols)	(from sorbed water?)	(ionized silanols)

Ethylbenzene is used as a nonpolar reference solute. Greek letters represent empirical, eluent, and temperature-dependent properties of the *solute*, which are relative to the values for ethylbenzene, for which all solute parameters are identically zero. Detailed studies have allowed the selection of optimum solutes for evaluation of each interaction mode. Bold capitals represent eluent and temperature-independent properties of the *column*; these column values are relative to a hypothetical average Type B C18 column. Any column that behaves identically as this hypothetical reference column will have $H = 1$ and all other values S^*, A, B, $C = 0$. The ion exchange term is of particular importance for the analysis of basic solutes; columns are tested with the quaternary amine berberine at pH 2.8 and 7.0 using phosphate buffers. Quaternary amines remain ionized independent of pH. The principal role of the model is to allow the choice of different phases of very different selectivity (e.g., to resolve compounds that coelute on one phase) or similar selectivity (e.g., to replace one type of column with a similar substitute). The magnitude of the C term gives an idea of the silanophilic activity of the phase. Newer phases made from Type B silica are less acidic than older Type A columns and therefore should have lower values of C, thus enabling suitable columns for the analysis of bases to be selected. The model can be used to predict suitable columns to achieve a desired selectivity between bases and other types of compound. The USP has recently suggested using the Snyder evaluation alongside the NIST test, the latter being a "column performance" test, involving measurement of peak shape, to allow a comprehensive evaluation of properties [95].

6.3 OTHER SEPARATION MECHANISMS FOR THE ANALYSIS OF BASES

RP chromatography has been by far the most dominant mechanism used for the separation of basic compounds. Normal-phase (NP) chromatography is applicable for the separation of bases, for example, using silica columns with largely nonpolar eluents containing small amounts of polar modifiers such as alcohols and amines. For example, NP chromatography was shown to give peak shapes similar to those obtained in RP separations of the cinchona alkaloids [96] and with useful alternative selectivity. However, the general problems of NP chromatography for all compounds remain. These include the powerful elution properties of water, which can have a serious effect even when only small quantities are present in nonpolar mobile phases. Consequently, the technique is incompatible with the injection of water and the water content of mobile phases needs to be carefully controlled to achieve reproducible separations.

Hydrophilic interaction chromatography (HILIC) or aqueous normal-phase chromatography (ANP) refers to the use of polar stationary phases (e.g., bare silica, silica, or polymeric phases with bonded zwitterionic ligands, diol phases) in combination with a mobile phase rich in organic solvent but containing a significant amount of water (typically at least 3%). Bell [97] summarized the advantages of this technique as follows:

- Gives an alternative selectivity of separation to RP-HPLC
- Retention of polar compounds is greater than in RP-HPLC

- Compatibility with preparative separations owing to volatility of the mobile phase and ease of recovery of solutes
- Higher flow rates are possible because of the low mobile phase viscosity
- Analysis of polar and ionic compounds is possible from complex matrices
- Good compatibility of the mobile phase with MS and evaporative light scattering detectors

Furthermore, HILIC separations (even with silica columns) do not suffer the problems of classical NP separations. An excellent review of the technique has recently been published by Irgum [98]. The good peak shape of bases has been a constant feature of HILIC separations. It appears that the tailing mechanism shown in RP separations does indeed also depend on the presence of the hydrophobic alkyl ligands as well as on silanol groups. Clearly, these ligands are not present on bare silica phases. Figure 6.16 shows the separation of a variety of acidic and basic solutes using a silica HILIC phase. Column efficiencies for nortriptyline, diphenhydramine, benzylamine, and procainamide ranged from 22,000 to 25,000 plates with asymmetry factors 1.00–1.05; using the same conditions, loss of efficiency for injection of 5 μg of diphenhydramine was around 10% compared with small mass injection when using acetonitrile–ammonium formate buffer pH 3.0, which compares favorably with the results in RP chromatography (see earlier text).

It has been considered that HILIC stationary phases absorb or imbibe water and that partition of analytes occurs between this layer of water and water in the bulk mobile phase. This mechanism occurs alongside ionic retention, as many of the commonly used HILIC stationary phases (as is the case with bare silica) have ion-exchange properties [34]. Partitioning into the water layer may explain the retention

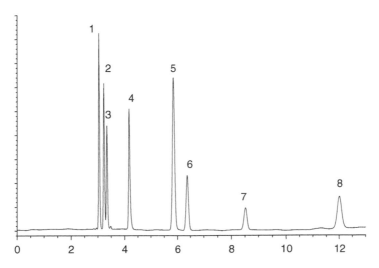

FIGURE 6.16 Separation of 1 = phenol; 2 = 2-naphthalene sulfonic acid; 3 = p-xylenesulfonic acid; 4 = caffeine; 5 = nortriptyline; 6 = diphenhydramine; 7 = benzylamine; 8 = procainamide on Atlantis silica column (25 × 0.46 cm, 5 μm particles). Mobile phase acetonitrile-0.1 M ammonium formate pH 3.0 (85:15, v/v), 1 mL min^{-1}.

of acidic species that cannot be retained on silica columns by ion exchange. As the water content of the mobile phase increases, retention usually decreases. Clearly, other retention mechanisms may occur as well as IX and this partition mechanism, such as polar interactions between column ligands and the solute, and the hydrophobic interactions described by Bidlingmeyer [99], although the latter effect is much weaker than on C18 bonded phases. This combination of mechanisms explains the unique selectivity of the HILIC technique [98].

Grumbach et al. [100] recommended the use of acetonitrile with bare silica columns, with concentration not greater than 95% or less than 70%. At least 5% of the mobile phase should be water to allow for the formation of the aqueous layer and to allow solubility of buffer, if one is used. In some cases, methanol can be used to form the polar layer. It was noted that while bare silica can be used at pH < 1 (no bonded ligands to hydrolyze, as in RP-HPLC) it is more susceptible to dissolution at intermediate pH (presumably since it not protected by a C18 layer), and should not be used above pH 6. Buffers such as ammonium acetate at pH 5 and ammonium formate at pH 3 were recommended at 5–20 mM concentrations. They reported the elution strength of various solvents using silica and HILIC conditions as

THF < acetone < acetonitrile < isopropanol < ethanol < methanol < water.

Interchange of these solvents can promote selectivity differences.

It appears that the use of HILIC for the separation of basic compounds is increasing and it can provide a useful alternative selectivity to RP, with polar compounds being retained more than nonpolar compounds. The compatibility of HILIC eluents with MS detection seems to be a particular advantageous feature of the technique. Improved understanding of the separation mechanism may lead to its increased use.

6.4 CONCLUDING REMARKS

Much progress has been made with the analysis of basic compounds using liquid chromatography over the last decade. Probably, the major advance has been in the availability of high purity silica as the base material for the preparation of bonded phases. While only about half of the silanol groups can react with the bonding reagent or with the endcapping material, because of steric effects, it appears that the acidity of residual silanols in some of these new phases is sufficiently low that their ionization can be suppressed at low pH values where the bonded ligand remains reasonably stable. Thus, the tailing that was often experienced for bases, even when small quantities of solute were used and low pH conditions were employed, is now a less serious problem. However, the mechanism by which this tailing occurs on phases with ionized silanols is still not completely understood. It is popularly thought that the mere presence of a mixed mode mechanism, involving both hydrophobic and ionic interactions, is the cause of tailing. However, phases, which have been deliberately prepared to incorporate anionic (negatively charged) groups within a hydrophobic chain, do not show these problems. The problem may be caused by the situation of ionized silanols buried deep underneath the hydrophobic ligands. Strong sites may be caused by the synergistic or multiplicative effect of hydrophobic and ionic retention; evidence for the existence of very strong sites with slow kinetics has come from fundamental spectroscopic studies. It seems possible that the tailing peaks on phases that have ionized silanols may be caused by the overloading of such sites, although kinetic effects are also possible, especially with smaller solute amounts.

The problem of poor peak shape cannot be entirely avoided even with the use of the most modern pure silica RP columns that seem to have few, if any, ionized silanols on their surface at low pH, unless very small amounts of solute are used. Broad peaks can result from a different mechanism (i.e., overloading of the hydrophobic surface of the phase). It is possible that this overloading is caused by the mutual repulsion of ionized basic solutes on the surface; alternatively, the phase consists of a relatively small number of high-energy hydrophobic sites that are easily overloaded, together with a much larger number of lower energy sites. Whatever the cause of the overloading mechanism, it can be a serious problem, especially when low ionic strength mobile phases are used, such as those favored for HPLC-MS. Overloading can still be observed for basic solutes in such mobile phases, even if only a fraction of a microgram of solute is injected. Thus, peak shape (and retention) can be a function of sample mass right down to the detection limit of UV spectroscopy.

Hybrid silica organic phases are stable at higher pH than conventional silica-based phases; they have shown some promise for the analysis of bases because they offer the possibility of analysis of the solute as a free base. In this case, any interactions with ionized silanols could be avoided, and the overloading properties of the solute should resemble those of a neutral compound. A further advantage of these phases appears to be the higher pK_a of the silanol groups, which do not appear to be ionized at intermediate pH as they are on conventional silica phases. However, the strongest organic bases remain protonated at the limit of stability of hybrid phases (around pH 10). Ionization of silanols at high pH occurs even on these phases, leading to the usual problems. Newer versions of the phases, which are stable at even higher pH, have recently become available but so far have not been fully evaluated.

Despite the problems with silica, it has remained dominant as a stationary phase for the analysis of bases for the same reasons as it has for the separation of other classes of solute. Polymeric phases still give lower efficiency than silica phases, and at low pH seem to suffer the same overloading effect as silica-based phases. Ionogenic groups seem to be introduced into polystyrene–divinyl-based phases during their manufacture, and these can lead to tailing of bases at intermediate pH where these groups become ionized. Other phases, such as those made from zirconia, show some promise for the analysis of bases but have not been fully evaluated as yet.

As with other types of solutes, chromatographers have attempted to improve the speed and efficiency of analysis of bases by the use of smaller particle (e.g., sub-2 μm) or monolithic columns. Small particle columns have not yet been fully evaluated for the analysis of bases, to determine whether they give equivalent selectivity, and reduction in plate height commensurate with the reduction in particle size, as has been demonstrated for neutral compounds. Commercial silica monolith columns give reasonable performance for the analysis of bases at low pH, but show evidence of the presence of ionized silanols at intermediate pH, giving tailing peaks. Temperature is another parameter that has been investigated as a means of either speeding up the analysis (since the optimum mobile phase flow velocity moves to higher values) or by increasing column efficiency (e.g., by allowing the use of longer columns at the same backpressure). Temperature has been shown also to have other beneficial effects, such as the reduction of the pK_a of bases, reducing their ionization at a given pH and thus reducing the detrimental effects of unfavorable column interactions.

Bases remain the most problematic class of compounds analyzed by HPLC. While much progress has been made, clearly many aspects of their chromatography would benefit from further study.

ACKNOWLEDGMENTS

The author thanks Dr. Uwe Neue of Waters Associates, Milford, MA, USA for the provision of Figures 6.1 and 6.15, and Dr David Bell of Supelco, Bellefonte, PA, USA for helpful information on EPG phases.

REFERENCES

1. Johnston, M.A., et al., *J. Chromatogr.*, 189, 241, 1980.
2. Hobson-Frohock, A., and Edwards, W.T., *J. Chromatogr.*, 249, 369, 1982.
3. McCalley, D.V., *Chromatographia*, 17, 264, 1983.
4. McCalley, D.V., *LC.GC*, 17, 440, 1999.
5. McCalley, D.V., *J. Sep. Sci.*, 26, 187, 2003.
6. Stadalius, M.A., Berus, J.S., and Snyder, L.R., *LC×GC*, 6, 494, 1988.
7. Snyder, L.R., Kirkland, J.J, and Glajch, J.L., *Practical HPLC Method Development*, Wiley, New York, 1997.
8. Nawrocki, J., *J. Chromatogr. A*, 779, 29, 1997.
9. Christie, A.A., and Egeberg, P.K., *Analyst*, 130, 738, 2005.
10. Kirkland, J.J., Glajch, J.L., and Farlee, R.D., *Anal. Chem.*, 6, 2, 1989.
11. Köhler, J., et al., *J. Chromatogr.*, 352, 275, 1986.
12. Köhler, J., and Kirkland, J.J., *J. Chromatogr.*, 385, 125, 1987.
13. Davies, N.H., Euerby, M.R., and McCalley, D.V., *J. Chromatogr. A*, 1138, 65, 2007.
14. Neue, U.D., and Carr, P.W., *J. Chromatogr. A*, 1063, 35, 2005.
15. Ludes, M.D., and Wirth, M.J., *Anal. Chem.*, 74, 386, 2002.
16. McCalley, D.V., *J. Chromatogr. A*, 769, 169, 1997.
17. McCalley, D.V., *J. Chromatogr. A*, 738, 169, 1996.
18. Gritti, F., and Guiochon, G., *J. Chromatogr. A*, 1132, 51, 2006.
19. McCalley, D.V., *Anal. Chem.*, 78, 2532, 2006.
20. Méndez, A., et al. *J. Chromatogr. A*, 986, 33, 2003.
21. Buckenmaier, S.M.C., McCalley, D.V., and Euerby, M.R., *Anal. Chem.*, 74, 4672, 2002.
22. Golshan-Shirazi, S., and Guiochon, G., *Anal. Chem.*, 60, 2364, 1988.
23. Knox, J.H., and Pyper, H.M., *J. Chromatogr.*, 363, 1, 1986.
24. Eble, J.E., et al., *J. Chromatogr.*, 384, 45, 1987.
25. McCalley, D.V., *Anal. Chem.*, 75, 3404, 2003.
26. Dai, J., et al., *J. Chromatogr. A*, 1069, 225, 2005.

27. Dai, J. and Carr, P.W. *J. Chromatogr. A*, 1072, 169, 2005.
28. McCalley, D.V., *J. Chromatogr. A*, 1038, 77, 2004.
29. McCalley, D.V., *LC.GC (Int.)*, 18, 290, 2005.
30. Gritti, F., and Guiochon, G., *Anal. Chem.*, 77, 1020, 2005.
31. Gritti, F., and Guiochon, G., *J. Chromatogr. A*, 1095, 27, 2005.
32. Guiochon, G., *J. Chromatogr. A*, 1126, 6, 2006.
33. Davies, N.H., Euerby, M.R., and McCalley, D.V., *J. Chromatogr. A*, 1119, 11, 2006.
34. Neue, U.D., in *HPLC Columns*, Wiley-VCH, 1997
35. McCalley, D.V., *J. Chromatogr. A*, 844, 23, 1999.
36. Snyder, L.R., Dolan, J.W., and Carr, P.W., *J. Chromatogr. A*, 1060, 77, 2004.
37. Glajch, J.L., et al., *J. Chromatogr.*, 318, 23, 1985.
38. Glajch, J.L., Kirkland, J.J., and Köhler, J., *J. Chromatogr.*, 384, 81, 1987.
39. Kirkland, J.J., Dilks, C.H., Henderson, J.E., *LC×GC*, 11, 290, 1993.
40. Kirkland, J.J., and Henderson, J.W., *J. Chromatogr. Sci.*, 32, 473, 1994.
41. Claessens, H.A., and Kirkland, J.J., *J. Chromatogr. A*, 691, 3, 1995.
42. Ascah, T.L., and Feibush, B., *J. Chromatogr.*, 506, 357, 1990.
43. Ascah, T.L., et al., *J. Liq. Chrom. Rel. Technol.*, 19, 3049, 1996.
44. O'Gara, J.E., et al., *Anal. Chem.*, 67, 3809, 1995.
45. Neue, U.D., et al., *Chromatographia*, 54, 169, 2001.
46. Euerby, M.R., and Petersson, P., *J. Chromatogr. A*, 1088, 1, 2005.
47. Bell, D.S., et al. *Improved LC-MS Analysis Using EPG Stationary Phases* (Supelco Presentation) 2006.
48. Manufacturers' Literature, SIELC Technologies, 2006.
49. Cheng, Y.F., et al., *LC.GC*, 18, 1162, 2000.
50. Wyndham, K.D., et al., *Anal. Chem.*, 75, 6781, 2003.
51. Unger, K.K., *Porous Silica*, Elsevier, Amsterdam, 1979.
52. Nakanishi, J.K., and Soga, N., *J. Am. Ceram. Soc.*, 74, 2518, 1991.
53. Ishizuka, N., et al., *J.Chromatogr. A*, 797, 133, 1998.
54. Leinweber, F.C., et al., *Anal. Chem.*, 74, 2470, 2002.
55. Desmet, G., et al., *Anal. Chem.*, 78, 2150, 2006.
56. Billen, J., Gzil, P., and Desmet, G., *Anal. Chem.*, 78, 6191, 2006.
57. Tanaka, N., *Presentation at HPLC 2006*, San Francisco, USA, June 2006.
58. McCalley, D.V., *J. Chromatogr. A*, 965, 51, 2002.
59. Kele, M., and Guiochon, G., *J. Chromatogr. A*, 960, 19, 2002.
60. Yang, C., et al.,*J. Chromatogr. A*, 1130, 175, 2006.
61. De Villiers, A., et al., *J. Chromatogr. A*, 1127, 60, 2006.
62. Nguyen, D.T., et al., *J. Chromatogr. A*, 1128, 105, 2006.
63. Li, J., and Carr, P.W., *Anal. Chem.*, 69, 2202, 1997.
64. Yang, X., Dai, J., and Carr, P.W., *J. Chromatogr. A*, 996, 13, 2003.
65. Gilroy, J.J., et al., *J. Chromatogr. A*, 1026, 77, 2004.
66. McCalley, D.V., *J. Chromatogr. A*, 664, 139, 1994.
67. McCalley, D.V., *J. Chromatogr. A*, 708, 185, 1995.
68. *IUPAC Compendium of Analytical Nomenclature*, Blackwell, Oxford, 1998.
69. Subirats, X., Bosch, E., and Rosés, M., *J. Chromatogr. A*, 1121, 170, 2006.
70. Buckenmaier, S.M.C., McCalley, D.V., and Euerby, M.R., *J. Chromatogr. A*,1004, 71, 2003.
71. Buckenmaier, S.M.C., McCalley, D.V., and Euerby, M.R., *J. Chromatogr. A*, 1026, 251, 2004.
72. McCalley, D.V., *J. Chromatogr. A*, 987, 17, 2003.
73. Temesi, D., and Law, B., *J. Chromatogr. B*, 748, 21, 2000.

74. Kirkland, J.J., et al., *Anal. Chem.*, 70, 4344, 1998.
75. Fornal, E., Borman, P., and Luscombe, C., *Anal. Chim. Acta*, 570, 267, 2006.
76. Claessens, H.A., Vermeer, E.A., and Cramers, C.A., *LC×GC (Int.)*, 6, 692, 1993.
77. Yang, X., Dai, J., and Carr, P.W., *Anal. Chem.*, 75, 3153, 2003.
78. Sokolowski, A., and Wahlund, K.G., *J. Chromatogr.*, 189, 299, 1980.
79. Lundanes, E., and Greibrokk, T., *Adv. Chromatogr.*, 44, 45, 2006.
80. Lestremau, F., et al., *J. Chromatogr. A*, 1109, 191, 2006.
81. Thompson, J.D., and Carr, P.W., *Anal. Chem.*, 74, 1017, 2002.
82. McCalley, D.V., *J. Chromatogr. A*, 902, 311, 2000.
83. Buckenmaier, S.M.C., McCalley, D.V., and Euerby, M.R., *J. Chromatogr. A*, 1060, 117, 2004.
84. McCalley, D.V., and Brereton, R.G., *J. Chromatogr. A*, 828, 407, 1998.
85. Rogers, S.D., and Dorsey, J.G., *J. Chromatogr. A*, 892, 57, 2000.
86. Kele, M., and Guiochon, G., *J. Chromatogr. A*, 869, 181, 2000.
87. Engelhardt, H., and Jungheim, M., *Chromatographia*, 29, 59, 1990.
88. Sander, L.C., and Wise, S.A., *J. Sep. Sci.*, 26, 283, 2003.
89. McCalley, D.V., *J. Chromatogr. A*, 793, 31, 1998.
90. Neue, U.D., et al., *J. Sep. Sci.*, 26, 174, 2003.
91. Euerby, M.R., McKeown, A.P., and Petersson, P., *J. Sep. Sci.*, 26, 295, 2003.
92. Kimata, K., et al., *J. Chromatogr. Sci.*, 27, 721, 1989.
93. Euerby, M.R., and Petersson, P., *LC. GC (Eur)*, 13, 665, 2000.
94. Vervoort, R.J.M., et al., *J. Chromatogr. A*, 931, 67, 2001.
95. USP www.usp.org/USPNF/columnsDB.html
96. McCalley, D.V., *J. Chromatogr.*, 260, 184, 1983.
97. Bell, D.S., *Lecture Presented at HPLC 2006*, San Francisco, June 2006.
98. Hemstrom, P., and Irgum, K., *J. Sep. Sci.*, 29, 1784, 2006.
99. Bidlingmeyer, B.A., Del Rios, J.K., and Korpi, J., *Anal. Chem.*, 54, 442, 1982.
100. Grumbach, E.S., et al., *LC.GC*, 22, 1010, 2004.

7 Hyphenated Chromatographic Techniques in Nuclear Magnetic Resonance Spectroscopy

Stacie L. Eldridge, Albert K. Korir,
Christiana E. Merrywell, and Cynthia K. Larive

CONTENTS

7.1 INTRODUCTION

The coupling of chromatographic techniques to nuclear magnetic resonance (NMR) detection has become increasingly important over the last few decades. In particular, the hyphenation of high-performance liquid chromatography (HPLC) and capillary electrophoresis (CE) to NMR have been at the forefront of these developments. For unknown or complex substances, it is not uncommon to have overlapping signals in the NMR spectrum complicating the structure elucidation of unknown analytes. For this reason, the experimental protocol typically requires that sample components undergo a separation or extraction step prior to NMR using HPLC, supercritical fluid chromatography (SFC), or CE. While it is possible to collect fractions of interest and measure the NMR spectrum offline, coupling of the separation directly with online NMR detection is a practical way to eliminate unnecessary transfer steps and minimize sample loss or degradation. This review describes recent advances in hyphenation of chromatographic techniques with NMR detection, presents the advantages and limitations of the various approaches, and summarizes selected applications of these methods. The recent review by Jayawickrama and Sweedler on the hyphenation of capillary separations with NMR also touches on many of the topics discussed herein [1].

NMR spectroscopy is a powerful method for chemical analysis. NMR detection is often used to identify the presence of impurities or degradation products in a sample. It can be used to investigate the dynamics of molecular interactions and report on the properties of the bulk solvent. NMR is an excellent tool for quantitative analysis since a pure standard of the analyte is not required. However, because of its ability to unambiguously determine both the connectivity and spatial arrangements of atoms in a molecule, the primary use of NMR in analytical chemistry is for structure elucidation. An additional benefit of NMR is its nondestructive nature, facilitating subsequent analysis by other analytical techniques, especially methods like mass spectrometry (MS) that provide complementary structural information. Despite these benefits, NMR suffers from two significant disadvantages. Although direct analysis of mixtures is possible where resonances of the species of interest can be resolved, in many cases because of the complexity of the sample, analytes must be separated prior to detection. A further limitation is that NMR has inherently poor sensitivity when compared to other spectroscopic detection methods, hindering the analysis of samples that are mass-limited or extremely low in concentration. In order to better understand the way that different NMR parameters affect sensitivity, it is instructive to review some of the basic principles of NMR spectroscopy.

7.1.1 Fundamentals of NMR Detection

It is understood that in most spectroscopic techniques, a spectrum is generated by absorption of a quantized unit of energy that causes a transition from a lower energy ground state to a higher energy excited state. The difference in energy of these two states is proportional to the frequency, v, and Planck's constant, h, as shown in the following equation:

$$\Delta E = hv. \tag{7.1}$$

In NMR spectroscopy, this energy difference is in the radiofrequency (RF) region of the electromagnetic spectrum and is extremely small, such that the populations of the ground and excited spin states are almost equal at room temperature. The energy difference between the ground and excited electronic states in ultraviolet (UV)-visible absorption spectroscopy is much greater than in NMR, therefore the frequencies are also much larger. For example, visible absorption at a wavelength of 500 nm corresponds to a frequency of 600×10^{12} Hz, whereas ^{1}H NMR frequencies range from 60 to 900 MHz, corresponding to wavelengths on the order of 5 m to 30 cm, depending on the strength of the applied magnetic field. In NMR, the frequency of the absorbed electromagnetic radiation is proportional to the applied magnetic field

$$v = \left(\frac{\gamma}{2\pi}\right) \cdot B_0, \tag{7.2}$$

where γ is the magnetogyric ratio, a constant for a given type of nucleus, and B_0 is the static magnetic field in units of Tesla. The sensitivity of NMR is proportional to the difference in the populations of the upper and lower energy states as given by the Boltzmann distribution

$$\frac{N_l}{N_u} = e^{(\Delta E/k_B T)} = e^{(hv/k_B T)} = e^{(\gamma \hbar B_0/k_B T)}, \tag{7.3}$$

where N_l and N_u are the populations of the lower and upper energy states, respectively, ΔE is the energy difference between the spin states, k_B is the Boltzmann constant $(1.38 \times 10^{-23}$ J/K), $\hbar$ is Planck's constant divided by 2π $(6.63 \times 10^{-34}/2\pi$ J $\cdot$ s), and T is temperature in units of Kelvin. As shown in Equation 7.3, NMR is more sensitive at higher magnetic field strengths owing to a larger difference in the populations of the upper and lower energy states. However, even at the highest magnetic field strengths currently available, this difference in energy is small. Therefore, the populations of the ground and excited states are nearly equal, making NMR a spectroscopic technique of inherently low sensitivity. Even with this fundamental limitation, there are many experimental parameters and hardware options that contribute to the sensitivity of an NMR measurement, as described in the following section.

7.1.2 NMR Sensitivity

As Beer's law in absorption spectroscopy has a path length dependence, the observe volume, V_{obs}, or active volume of an NMR probe is an important determinant of the sensitivity of NMR measurements. The observe volume is the fraction of the total sample volume, V_{tot}, that returns a signal when a sample is inserted in an NMR tube or is injected into a flow system. The relationship between chromatographic peak shape, peak volume and flow rate, and sensitivity in hyphenated NMR measurements is complex and is discussed in greater detail in Section 7.2. For the purpose of this discussion, the sample is assumed to be present at a uniform concentration in a sample volume, V_{tot}. The probe observe factor, f_0, is calculated as shown in the following equation:

$$f_0 = \frac{V_{obs}}{V_{tot}}. \tag{7.4}$$

From this equation one might assume that an f_0 of 1 ($V_{obs} = V_{tot}$) would provide maximum sensitivity. However, due to magnetic susceptibility effects, probe observe factors significantly less than 1 are typically required to obtain good line shape. For example, a 5 mm tube probe may require a volume of 600 or 750 μL for optimum sample homogeneity although the probe V_{obs} is only 250 μL. In tube-based experiments, special tubes, such as those produced by Shigemi Inc., with susceptibility-matched plugs can be used to confine the sample to V_{obs}. A related parameter, the fill factor, f_{fill}, is the fraction of volume inside the observe coil that can be occupied by the sample

$$f_{fill} = \frac{V_{obs}}{V_{coil}}, \tag{7.5}$$

where V_{coil} is the internal volume of the coil. Again, intuitively one would expect that a f_{fill} of 1 would provide the best sensitivity, however, magnetic susceptibility effects generally require that the sample be maintained at some distance from the coil to achieve sufficiently narrow and homogeneous line shape. In many probes, there is typically an air gap between the coil and the sample tube such that f_{fill} is significantly less than 1. The walls of the sample container further reduce f_{fill} as they increase the distance of the sample from the coil. For example, users of tube-based NMR probes can choose to use thin-walled tubes, which can increase f_{fill} by about 15% [2].

In analytical chemistry, sensitivity is traditionally defined as the slope of the calibration curve produced by plotting the measured signal versus the sample concentration. Numerous strategies have been introduced in order to heighten the sensitivity of NMR and lower the limits of detection. Although the traditional approach of signal averaging can provide significant gains, with spectral S/N increasing as the square root of the number of transients coadded, there are practical limits on the gains that can be obtained, in our experience, leveling off at around 24–48 h. In addition, signal averaging is not useful for real-time, online detection of separations. Another strategy for improving S/N is the use of higher magnetic fields, such as 21.1 Tesla

(900 MHz), since sensitivity increases as $B_0^{7/4}$ [3]. Increasing the magnetic field has the advantage that in addition to improving sensitivity, it also increases the dispersion of NMR resonances and produces spectra with less second order coupling. In addition to the specific determinants of S/N defined by Equation 7.6, NMR measurements are also affected by a variety of data acquisition and processing parameters, including the repetition time, digital resolution, pulse width, post-acquisition apodization, and other experimental considerations beyond the scope of this chapter. Readers interested in additional details about optimizing data acquisition are encouraged to refer to a text that provides greater detail on the subject of NMR spectroscopy [4,5].

Because sensitivity depends on so many different experimental factors, NMR spectroscopists generally use the signal-to-noise ratio, S/N, as a figure of merit for sensitivity comparisons. For example, in a comparison between NMR probes or spectrometers from two vendors, the spectral S/N measured for a standard sample acquired with specified acquisition parameters and probe geometry would provide a direct indication of relative sensitivity. The S/N is calculated for an NMR experiment as the peak signal divided by the root mean square (RMS) noise, given by Equation 7.6, and is directly related to the performance of the radiofrequency coil [3,6]

$$S/N = \frac{[(k_0(B_1/i)V_sN_s\gamma\hbar^2 I(I+1)\omega_0^2)/3k_BT\sqrt{2}]}{V_{\text{noise}}}. \tag{7.6}$$

The S/N defined by Equation 7.6 for a given NMR resonance is proportional to the square of the nuclear precession frequency (ω_0, rad/s), the magnitude of the transverse magnetic field (B_1) induced in the RF coil per unit current (i), the number of spins per unit volume (N_S), the sample volume (V_s), and k_0 a scaling constant that accounts for magnetic field inhomogeneities. The S/N is inversely proportional to the noise generated in the RF receiver and by the sample (V_{noise}) as defined by the Nyquist theorem,

$$V_{\text{noise}} = \sqrt{4k_BTR_{\text{noise}}\Delta f}, \tag{7.7}$$

where R_{noise}, measured over a specified spectral width (Δf, Hz), is a representation of conductive losses from the coil itself, magnetic (eddy current), and dielectric losses in the sample and surroundings.

As with other absorption spectroscopies, NMR S/N is directly dependent on sample concentration expressed as N_S, the number of spins per unit volume, in Equation 7.6. Because spectral S/N can be increased through signal averaging, in making comparisons between probes it is instructive to calculate the concentration sensitivity, S_C, by dividing S/N by the square root of the acquisition time and the molar concentration

$$S_C = \frac{S/N}{C\sqrt{t_{\text{acq}}}}. \tag{7.8}$$

As shown in Equation 7.8, for a fixed acquisition time, S_C is the S/N per unit concentration, for example, mM. It is important to note from the sample volume term, V_S, in Equation 7.6, that S_C is directly proportional to V_{obs}. Therefore, if a 4 mm diameter sample cell is replaced with a cell with a diameter of 3 mm, keeping the same coil length, both the concentration sensitivity and the observe volume will decrease by a factor of 1.78 (16/9).

7.1.3 CRYOPROBES AND MICROCOIL PROBES

7.1.3.1 Cryoprobes

Another strategy for improving the S/N is through RF coil design. Thermal noise in the coil and in the associated circuitry for the first stage of signal amplification is the dominant noise contribution to NMR spectra. Therefore, the introduction of probes with cryogenically cooled detection coils and preamplifiers has greatly advanced NMR sensitivity by reducing V_{noise}, as shown by Equation 7.7 [7–9]. Although the detection coil is cooled to 20–25 K, these cold probes are engineered such that the sample is maintained at room temperature. Therefore, increases in sensitivity result only from noise reduction and are not due to the effect of sample temperature (T) on S/N. Cryogenically cooled detection systems can provide sensitivity that is four times greater than conventional probes, although for high salt or "lossy" samples, a gain of a factor of 2 in sensitivity is more typical. A sensitivity increase by a factor of 4 means that the same spectral S/N can be obtained using a quarter of the sample required with a conventional probe. Alternatively, since S/N gains by signal averaging accrue as the square root of the number of coadded scans, for an equal amount of sample, a similar quality cold probe spectrum can be measured in 1/16th the time required using a conventional NMR probe. The sensitivity of flow NMR experiments is intrinsically limited in on-flow analysis when spectra are acquired on the fly or because only a limited amount of sample can be injected in the chromatographic step without compromising the quality of the separation. Therefore, the commercial introduction of cryogenically cooled flow probes has been a significant advance in hyphenated NMR measurements [9].

7.1.3.2 Microcoil Probes

For samples available in limited quantities, such as natural products, metabolites, or pharmaceutical impurities, miniaturized or microcoil probes can provide a useful mechanism to improve NMR sensitivity at a fraction of the cost of a cryogenically cooled detection system [10,11]. There is no firm definition of the term "microcoil," but it is generally used to refer to coils with V_{obs} on the order of a few microliters to as small as 1 nanoliter. The detailed theory of the design of solenoidal microcoils can be found elsewhere and is beyond the scope of this chapter [12,13]. An exciting area of recent innovation is the incorporation of multiple microcoils within a single NMR probehead for multiplexed analysis [14]. These types of probes are not yet available commercially, although they do show promise for the future of hyphenated NMR technology. One of the biggest challenges in the development of multiplexed probes is the fact that each coil occupies a different spatial region within the probe and therefore, each will experience a different magnetic field environment [15].

To better assess the sensitivity of NMR probes for mass-limited samples, the mass sensitivity, S_M, should be considered instead of the concentration sensitivity, S_C. The

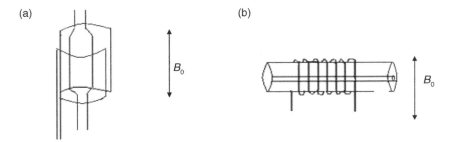

FIGURE 7.1   RF coil geometries and their orientation with respect to the magnetic field, B_0; (a) a saddle or Helmoltz RF coil and (b) a solenoidal coil, shown here wrapped directly around a separation capillary.

S_M is calculated by normalizing S/N to the number of moles of analyte, mol, within V_{obs} during the acquisition time, t_{acq}, as shown in the following equation [10,12]:

$$S_M = \frac{S/N}{mol\sqrt{t_{acq}}}. \tag{7.9}$$

In addition to their smaller V_{obs}, NMR microcoils generally have a geometry that is different from the normal saddle or Helmholtz design, shown in Figure 7.1a. Instead, microcoil probes are usually constructed in a solenoidal geometry (Figure 7.1b), oriented perpendicular to the applied magnetic field. There are several reasons that the solenoidal geometry is favored over the Helmholtz design in the construction of microscale NMR coils, including both practical and theoretical benefits. From a practical standpoint, it is easier to construct a small solenoidal coil with good homogeneity properties and it can be wrapped directly onto the separation capillary or flow cell, offering an improved fill factor. From a theoretical standpoint, it is useful to consider how the RF coil sensitivity, B_1/i, scales with coil size. The coil sensitivity relates the magnitude of the B_1 or transverse magnetic field produced per unit current flowing through the windings of the coil [3,10]. In addition to providing the excitation pulses, the NMR signal is also detected through this coil; therefore, the coil sensitivity is directly related to S/N as shown in Equation 7.6. For a saddle coil, the coil sensitivity corresponds to

$$\frac{B_1}{i} = \frac{\eta\mu_0\sqrt{3}}{\pi}\left[\frac{2dh}{\left(d^2+h^2\right)^{3/2}} + \frac{2h}{d\sqrt{d^2+h^2}}\right], \tag{7.10}$$

where η defines the number of turns in the coil, μ_0 is the permeability of free space, and h and d are the coil height and diameter, respectively. For a solenoidal geometry, the coil sensitivity can be expressed as shown in Equation 7.11, under the provision that $h \gg r$, where r is the coil radius

$$\frac{B_1}{i} = \frac{\mu_0\eta}{d\sqrt{1+(h/d)^2}}. \tag{7.11}$$

Providing that h/d remains constant, the RF coil sensitivity is inversely proportional to the coil diameter for solenoidal coils of diameter greater than about $100\,\mu m$. It can be shown theoretically and has been demonstrated experimentally that the sensitivity of a solenoidal coil is about three times that of a saddle coil of the same size [3]. In order to translate the coil sensitivity into measurement of S/N defined by Equation 7.6, we also have to consider the noise characteristics of NMR microcoils. For coils of diameter less than 3 mm, the major noise source is from the resistance of the coil itself, and the noise from even lossy biological samples can be neglected [10].

For microcoil probes, mass sensitivity is a more appropriate figure of merit than concentration sensitivity because of their small coil volumes and generally high values of f_{fill}. However, in making sensitivity comparisons between probes it is important to consider the total sample volume, V_{tot}, required for the analysis, and not simply V_{obs}. The reader should also recognize that high mass sensitivity is typically obtained by analyzing a limited amount of sample in a small volume, a strategy that is only useful for compounds of high solubility. Although this chapter is focused on the use of NMR as a detector for separations, it is also important to consider the utility of microcoil NMR probes for direct analysis of samples in cases where the sample is available in a small volume. For example a recent report describes automated acquisition of NMR spectra for quality control in combinatorial libraries by direct injection and analysis of 5 μL samples from 384 well-plates into a commercially available CapNMR™ (Protasis/MRM Corp., Savoy, IL) flow probe [16]. As described in the following section, when NMR is used as a chromatographic detector, regardless of the coil geometry, an important consideration is how the size of a chromatographic or electrophoretic peak compares with the volume of the NMR detection cell.

7.2 NMR AS AN ONLINE DETECTOR FOR SEPARATIONS

There are many factors to take into consideration when designing a hyphenated NMR experiment, including how the sample is delivered to the detection cell and what volume fraction of the analyte is interrogated by the coil. Chromatographic resolution and spectroscopic sensitivity are two aspects of flow-NMR that must be managed in a careful balance. The resolution of a chromatographic separation can often be improved by using higher flow rates and short NMR acquisition times, but this may lead to a decrease in detection sensitivity in on-flow experiments. In addition, the available hardware options, that is, loop or fraction collection in LC-NMR and special characteristics of the probe (e.g., standard format or cryogenically cooled detection coils), will also affect the outcome.

The advantages of coupling a separation method with NMR detection are maximized when the volume of the separated analyte peak closely matches the total volume of the NMR flow cell. As shown diagrammatically in Figure 7.2, the best sensitivity would be achieved in (a) with the most intense portion of the chromatographic peak centered in the NMR flow cell. Ideally, the detection volume should be about one-third of the eluting peak volume, to avoid peak distortion [1]. There are a variety of commercial flow cell designs ranging from the CapNMR probe with

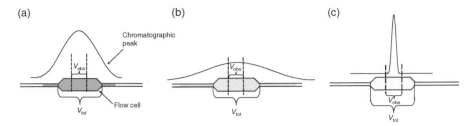

FIGURE 7.2 (a) Representation of a chromatographic peak eluted in a volume that closely matches the flow cell of the NMR probe. (b) A chromatographic peak in a volume that is much larger than the NMR flow cell. (c) A chromatographic peak in a volume much smaller than the flow cell of the NMR probe.

detection volume of about 5 µL to a conventional flow probe with V_{tot} of 120 µL and a V_{obs} of 60 µL. Assuming a chromatographic peak with a width of 0.5 min eluting at a flow rate of 1.0 mL/min, the peak volume would be 500 µL; therefore, only a fraction of this peak would be analyzed, even in a probe with a V_{tot} of 120 µL. As shown in Figure 7.2b, this is a less sensitive arrangement than in (a) because a smaller fraction of the peak is detected. Figure 7.2c shows the consequence of having a peak much smaller than the detection volume. Because of mixing and diffusion within the flow cell, a detection volume much larger than the volume of the separated peak results in peak dispersion and reduced sensitivity. Analyte peaks in CE or capillary electrochromatography (CEC) are typically less than 500 nL while chromatographic peaks in capillary HPLC (CapLC) range from approximately 100 nL to 2 µL. Although the CapLC chromatographic peak volume closely matches the CapNMR probe active volume (V_{obs}) of 1.5 µL, it is the total volume of the detection cell that determines how well NMR sensitivity is maximized. The first CE-NMR and CEC-NMR experiments involved the use of solenoidal coils with volumes ranging from 240 to 400 nL. This represented a reduction of the detection volume by a factor of ~1000.

One way of compensating for the relatively poor sensitivity of NMR is by injection of more concentrated samples or larger sample volumes, however, column overload can lead to a dramatic loss of separation efficiency resulting in peak broadening and possibly peak overlap. Another source of peak broadening and tailing is long transfer lines from the separation column to the NMR flow cell as well as broadening induced by mixing within the bubble cell used for detection. These factors can adversely affect the quality of the NMR signal obtained from an otherwise symmetric chromatographic peak. Lewis et al. [17] performed experiments to compare UV and NMR profiles obtained in CapLC-NMR and LC-NMR, with the results shown in Figure 7.3. These results show that the CapLC peak volume is well matched to the flow cell volume and that more tailing occurred in conventional LC-NMR than in CapLC-NMR, probably because of greater mixing due to the larger diameter of the LC-NMR tubing and flow cell.

Another factor that needs to be considered is the dilution of the chromatographic peak during the separation process. The extent of chromatographic dilution can be

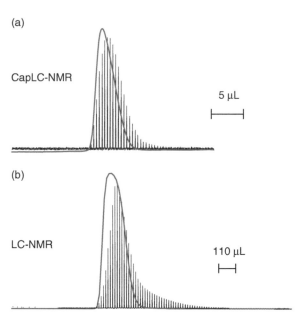

FIGURE 7.3 Peak tailing comparison illustrated by the UV profile (solid line) overlaid on the NMR flow cell profile. In (a) the HPLC peak volume is well matched to the CapNMR flow cell whereas in (b) considerably more tailing occurs on the analytical scale LC-NMR. (Reprinted from Lewis, R. J. et al. *Magn. Reson. Chem.* 2005, 43, 783–789, with permission from John Wiley & Sons.)

estimated by the following equation:

$$D = \frac{C_0}{C_{\max}} = \frac{r^2 \varepsilon \pi \, (1 + k) \, \sqrt{2LH\pi}}{V_{\text{inj}}}, \tag{7.12}$$

where C_0 is the initial analyte concentration, $C_{\max}$ is the final analyte concentration at peak maximum, r is the column radius, ε is the column porosity, k is the retention factor, L is the column length, H is column plate height, and V_{inj} is volume of sample injected. Equation 7.12 indicates that the column radius significantly influences the dilution of a chromatographic peak, the dilution effect being smaller for smaller diameter columns, another benefit of capillary HPLC separations.

When using NMR as a detector for online separations, additional consideration must be given to how the sensitivity is affected by the movement of nuclei past the detector cell. Aside from the physical hardware setup, the chromatographic and spectroscopic parameters also play a role in the quality of the resulting data. Flow rate, solvent composition, and residence and acquisition times can be optimized to provide optimal results. NMR sensitivity and chromatographic resolution tend to have an inverse relationship with respect to online LC-NMR experiments. By slowing the flow rate, more scans can be acquired for a particular analyte in the flow cell, but chromatographic resolution suffers owing to diffusion of peaks in the transfer tubing between the LC column and the NMR probe [18].

The signal intensity and line shape can be significantly affected in continuous-flow detection and, in the case of CE, additional broadening is produced by the electrophoretic current. The amount of time that a nucleus spends in the NMR flow cell or the residence time, τ, is related to the observe volume and the applied flow rate, F,

$$\tau = \frac{V_{obs}}{F}. \tag{7.13}$$

For example, for a V_{obs} of 1 μL and an F of 6 μL/min, the residence time for a pulsed spin is 10 seconds. By increasing the flow rate to 12 μL/min, the residence time is decreased to 5 seconds. This decreases the number of NMR acquisitions that can be recorded for a particular spin, reducing the S/N of the resulting NMR spectrum.

The residence time of flowing nuclei is also related to both the spin-lattice (T_1) and spin–spin (T_2) relaxation times

$$\frac{1}{T_{1\,flow}} = \frac{1}{T_{1\,static}} + \frac{1}{\tau}, \tag{7.14}$$

$$\frac{1}{T_{2\,flow}} = \frac{1}{T_{2\,static}} + \frac{1}{\tau}, \tag{7.15}$$

where $T_{1\,flow}$ and $T_{2\,flow}$ are the relaxation times in a flowing system and $T_{1\,static}$ and $T_{2\,static}$ are the relaxation times for spins in a stationary system. In a flowing system, the T_1 relaxation time of the sample nuclei is reduced owing to constant replacement of pulsed spins in the active volume of the NMR detector. This allows for the application of a faster pulse repetition time, resulting in a greater S/N. However, the T_2 relaxation time of a sample spin may also be reduced. If the residence time (τ) of the observed spin is too short to give sufficient time for full signal decay before it leaves the detection region, line broadening will occur. Since the NMR line width is inversely related to both T_2 and τ in a flow system, faster flow rates result in broader line shapes and ultimately, a decrease in S/N [19].

In electrophoretic separations, the linewidth of an NMR resonance is also affected by current-induced magnetic field gradients accompanied by thermal effects. Solenoidal microcoils that are placed perpendicular to the static magnetic field, B_0, generate an additional local magnetic field gradient owing to the electrophoretic current (Figure 7.4) [1,20,21]. This induced magnetic field, B_i, can be described by the following equation:

$$B_i = \frac{\mu_0 i r}{2\pi R^2}, \tag{7.16}$$

where i is the electrophoretic current, r is the radial distance from the coil to the center of the capillary, and R is the internal diameter of the capillary. This local magnetic field adds to the disruption of the homogeneity in the overall magnetic field and can rarely be recovered by shimming. Since the local magnetic field produced by the electrophoretic current is directly proportional to the magnitude of the current, performing separations at higher voltages increases the magnetic field inhomogeneity

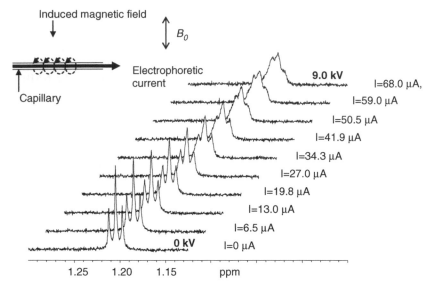

FIGURE 7.4   Microcoil CE-NMR spectra of triethylamine methyl peak in 1 M borate buffer with increasing applied voltage 0.0–9.0 kV by increments of 1.0 kV. Inset is a schematic of electrophoretic current-induced magnetic field. (Reproduced from Jayawickrama, D. A., Sweedler, J. V. *J. Chromatogr. A* 2003, 1000, 819–840, with permission from Elsevier.)

resulting in increased line broadening. The spectra in Figure 7.4 show the resulting line broadening of the NMR signal when the applied voltage is increased from 0.0 to 9.0 kV by increments of 1.0 kV.

7.3 SOLVENT SUPPRESSION

Compared with other types of chromatographic detectors, a unique challenge faced with NMR detection is suppressing the resonance(s) of the solvents used in the separation. Although a variety of effective solvent suppression methods are available, these methods also suppress signals from compounds in the same chemical shift region as the solvent resonance, complicating spectral analysis, particularly for unknown compounds. Even when deuterated solvents are used, the residual proton resonances (especially HOD) are usually still much more intense than those of the analytes. Although the smaller residual proton NMR resonance of deuterated solvents reduces the size of the spectral region removed by solvent suppression, the expense associated with the use of deuterated solvents in chromatographic separations can be prohibitive, especially with flow rates in the mL/min range. Therefore, an advantage of capillary separation methods, such as CapLC or CE for online NMR detection is their lower flow rates and reduced solvent consumption.

The development of simple, multiple-frequency solvent suppression techniques has greatly improved the quality of data that can be obtained using LC-NMR. One of the most useful methods for multiple resonance solvent suppression in LC-NMR is based on the WET (Water suppression Enhanced through T_1 effects) pulse sequence in which the solvent signals are selectively excited by a series of variable tip-angle selective RF pulses, with each pulse followed by a dephasing magnetic field gradient pulse [22,23]. After elimination of the water resonance and the main (^{1}H-^{12}C) resonance of an organic solvent such as acetonitrile or methanol, the ^{13}C satellite peaks are typically the most intense peaks remaining. These satellite peaks can be easily suppressed in WET through the use of ^{13}C decoupling to collapse them into a singlet resonance at the frequency of the ^{12}CH$_3$CN or ^{12}CH$_3$OH resonance. One problem typically encountered in gradient elution LC-NMR is the change in resonance frequency of the solvent peaks as the solvent composition changes. This hampers some solvent suppression techniques, and can make isocratic elution preferable in analysis of complex solutions [24]. However, because of the importance of gradient elution to reverse-phase chromatographic separations, a strategy has been developed with WET suppression that involves acquiring a scout scan prior to each acquisition period. This scout scan is acquired with a low tip-angle nonselective RF pulse without solvent suppression to measure the solvent frequencies at the current mobile phase composition and adjust the frequency of the selective RF pulses of the suppression scheme accordingly [23]. In this way, the NMR acquisition software can automatically seek, identify, and effectively suppress solvent resonances in real time during on-flow NMR detection in a gradient elution separation.

7.4 LC-NMR

As liquid chromatography plays a dominant role in chemical separations, advancements in the field of LC-NMR and the availability of commercial LC-NMR instrumentation in several formats has contributed to the widespread acceptance of hyphenated NMR techniques. The different methods for sampling and data acquisition, as well as selected applications will be discussed in this section. LC-NMR has found a wide range of applications including structure elucidation of natural products, studies of drug metabolism, transformation of environmental contaminants, structure determination of pharmaceutical impurities, and analysis of biofluids such as urine and blood plasma. Readers interested in an in-depth treatment of this topic are referred to the recent book on this subject [25].

7.4.1 EARLY LC-NMR EXPERIMENTS

One of the biggest challenges in the development of the first LC-NMR systems was the design of a probe, which allowed flow from the LC and contained the transmittance/receiver coil. Among the problems experienced with early attempts at coupling LC and NMR was minimization of band broadening, peak tailing, and turbulent flow to reduce "memory effects." Memory effects arise when some analyte remains in the flow cell as a new peak enters the flow cell. This problem has been reduced by improvements to probe design, although problems may still be encountered especially when closely spaced peaks are analyzed. Other problems such as excessive band broadening resulting from placement of the HPLC instrument at long distances from the magnet have been reduced by the introduction of actively shielded magnets with greatly reduced fringe fields. Another challenge that had to be overcome in early experiments, but is no longer a consideration, was the large amount of data storage space required for even simple LC-NMR experiments. Initial setups typically used an external cell for the lock solvent, usually D_2O [26]. One of the first applications of LC-NMR was for the determination of structures and relative amounts of components of jet fuel mixtures [27,28]. LC-NMR detection limits at the time were comparable to that of the traditional means of this type of analysis, refractive index detection [27].

7.4.2 Recent LC-NMR Probe Developments

7.4.2.1 Cryogenically Cooled Flow Probes

The advent of cryogenically cooled probes has significantly improved NMR detection limits. In these cryoprobes, or cold probes, the radiofrequency coils and electronics are cooled to about 20 K, thus greatly reducing electronic noise. This can result in an improvement of the S/N ratio by up to a factor of 4 compared with that obtained by conventional probes. The first results from a cryoprobe with built-in flow configuration were reported by Spraul et al. [29] To test the novel cryoflow probe, human urine samples obtained from a healthy female volunteer 4 hours after a single dose of 500 mg of acetaminophen were analyzed by LC-NMR MS. The results obtained demonstrated the characterization of acetaminophen metabolites using 40% less material than previously reported with a conventional LC-NMR flow probe. A new metabolite was identified using 16 scans per time slice in the on-flow mode. Cryogenically cooled flow probes have also found application in conjunction with solid phase extraction, described in greater detail in Section 7.4.3.2.

7.4.2.2 Microcoil Flow Probes

Another strategy for improving NMR sensitivity for directly coupled separations is the use of solenoidal microcoils, for example, the CapNMR probe. This probe yields excellent line shapes and has less interference from the residual solvent background signal because of the small flow cell volume of $5 \mu L$ [2]. These features allow the CapNMR probe to be coupled to capillary HPLC, thus offering unique advantages over conventional LC-NMR. Steep solvent gradients in conventional LC-NMR often require long equilibration times. However, the CapNMR probe can quickly recover from a chromatographic solvent gradient owing to its small flow cell volume. An additional advantage is that the small flow cell and capillary HPLC format afford the economical use of deuterated solvents under flow conditions.

As described in Section 7.4.2, one of the challenges of online LC-NMR is the need to match the chromatographic peak to the active volume of the CapNMR flow cell. An excellent discussion of the comparison of CapLC-NMR with other NMR probe types has been provided by Lewis et al. [17] Table 7.1 shows the sensitivity comparison of the CapNMR probe with larger volume probes. It is important to note that the experimental design used by these authors adjusted the concentration of analyte such

TABLE 7.1

A Comparison of the Sensitivity of the CapNMR Probe with Large Volume Probes

Probe	Tube	Volume (μL)	Relative S/N
Gas	10^{-3}	0.1	10^{-4}
Supercritical fluid	0.1–1	10^{-3}–10^{-4}	10^{-3}–10^{-4}
Liquid	1	$<10^{-5}$	10^{-2}
5 mm ^{1}H	5 mm	500	1.0
3 mm ^{1}H	3 mm	150	1.7
3 mm ^{1}H	3 mm Shigemi	100	2.9
Flow ^{1}H	Flow cell	110	2.5
5 mm Cryo ^{13}C/^{1}H	2.5 mm	120	5.3
5 mm Cryo ^{13}C/^{1}H	3 mm Shigemi	110	8.0
CapNMR	Flow cell	5	4.9

Source: Lewis, R. J. et al. *Magn. Reson. Chem.* 2005, 43, 783–789.

that 3 μg of the compound was present in the total volume of the flow cell, V_{tot}, rather than V_{obs}. The results in Table 7.1 show that the CapNMR probe has approximately two-fold enhancement in S/N over the standard LC-probe.

Because of its greater sensitivity, smaller volume flow cell, and reduced band broadening, CapLC-NMR has increasingly been used for the identification of analytes in complex mixtures. For example, Krucker et al. [30] have demonstrated the use of this technique in the on-flow separation and identification of tocopherol homologues using only 1.33 μg of each compound. For this analysis the authors reported a 200-fold reduction in the consumption of mobile phase solvents and amount of sample used compared with the previous analysis of the same analytes by conventional LC-NMR [31]. Figure 7.5 is a pseudo 2D NMR spectrum showing the ^{1}H NMR spectra of the analytes as a function of retention times. Figure 7.6 shows the extracted ^{1}H NMR spectra of the tocopherol homologues. The homologues can clearly be distinguished even though only 1.33 μg of each compound was injected for analysis. Since these compounds are easily oxidizable by air and light, this approach clearly illustrates the suitability of hyphenated NMR techniques for analysis of unstable compounds. For complete structure elucidation, 2D NMR experiments were performed by stopping the chromatographic run when the peak of interest reached the NMR detection cell. Figure 7.7 is a COSY spectrum of one of the homologues acquired at 600 MHz using the CapNMR probe in a stopped-flow mode. Once data acquisition is completed, separation can be resumed to facilitate the analysis of the next chromatographic peak.

7.4.3 SAMPLING STRATEGIES

Depending on the information required from the sample components, there are a variety of different LC-NMR analyte introduction schemes that can be employed.

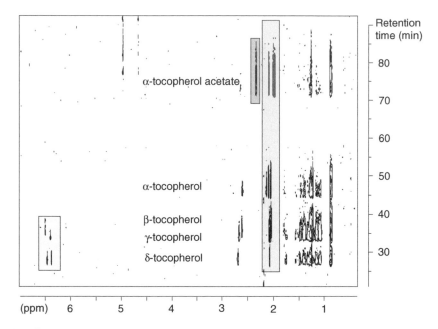

FIGURE 7.5   An on-flow CapLC-NMR spectrum showing the separation and identification of tocopherol derivatives. The boxes in the figure show the resonances of aromatic (~6.5 ppm), methyl (2.0–2.2 ppm) and methylene protons of the acetate group (~2.3 ppm), which are used to distinguish the tocopherol homologues. (Reproduced with permission from Ref. 30. Copyright (2004) American Chemical Society.)

The process for introducing the HPLC-separated analyte peaks into the NMR probe can be configured using two different modes of operation: on-flow (continuous flow) or stopped-flow.

7.4.3.1   On-Flow NMR

The earliest and simplest sampling strategy is the on-flow method. In on-flow measurements, the separation is performed continuously and the outlet of the column is connected via tubing, usually polyetheretherketone (PEEK), to the NMR probe. The flow is continuous, limiting the residence time of the analyte in the probe, and consequently, reducing the apparent sensitivity. The flow can also cause a decrease in the magnetic field homogeneity, which can have a dramatic impact on the resolution of the NMR spectrum, as can flow broadening as discussed in Section 7.2. In on-flow experiments, illustrated in pathway A of Figure 7.8, the separated analytes are eluted from the chromatographic column, travel through a photodiode array detector (PDA), and are delivered directly to the NMR detection cell, where spectra are acquired as the eluent flows through the NMR probe. This mode of operation is ideal for investigating concentrated samples for which NMR spectra can be measured without extensive signal averaging.

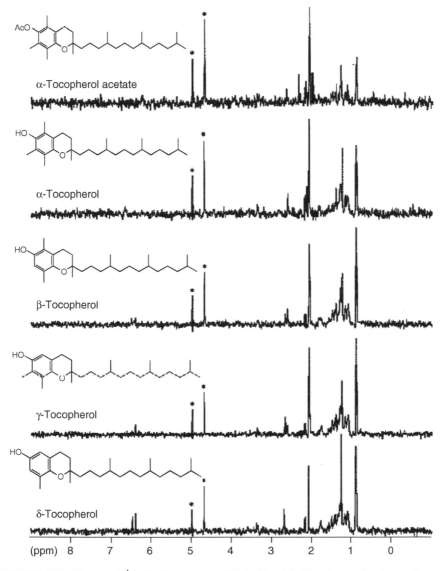

FIGURE 7.6 Extracted ^{1}H NMR spectra (600 MHz, 1.5 µL solenoidal microprobe) of the tocopherol homologues at the corresponding peak maxima; * residual solvent signals. (Reproduced with permission from Ref. 30. Copyright (2004) American Chemical Society.)

7.4.3.2 Stopped-Flow NMR

Stopped-flow experiments, illustrated by pathway A of Figure 7.8, are better suited for sample components present at low concentrations since a separated analyte can be selected and held in the NMR probe allowing extensive signal averaging. Execution of stopped-flow acquisition requires calibration of the exact length of time it takes

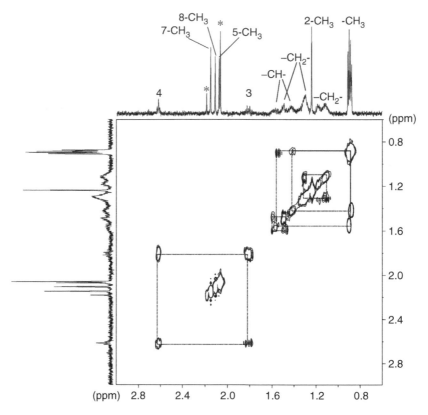

FIGURE 7.7 Stopped-flow COSY spectrum of 1.33 μg of α-tocopherol obtained on a 600 MHz NMR spectrometer using the CapNMR probe. (Reproduced from Krucker, M. et al. *Anal. Chem.* 2004, 76, 2623–2628. Copyright (2004). Used with the permission of the American Chemical Society.)

for a chromatographic peak to travel between the UV detector and the flow cell of the NMR probe. This allows the use of automated acquisition routines that trigger off the UV signal to stop the pump at the appropriate time, parking the analyte peak in the flow cell of the NMR probe for data acquisition. An advantage of stopped-flow over on-flow is that residence time of the analyte is not limited. However, during the acquisition time, peaks of interest still in the transfer tubing or on the column will diffuse, resulting in poorer chromatographic resolution and reduced *S/N* in NMR spectra measured for later eluting peaks. Another advantage of CapLC-NMR experiments is reduced band broadening in stopped-flow experiments because of the narrower bore capillaries and tubing employed [17]. In addition to enabling the analysis of components present at low concentrations through signal averaging, stopped-flow experiments allow the acquisition of lengthy 2D NMR spectra that are usually required for structure elucidation. However, advances in the rapid acquisition of 2D NMR data sets promise to facilitate measurement of real-time LC-2D NMR experiments [32].

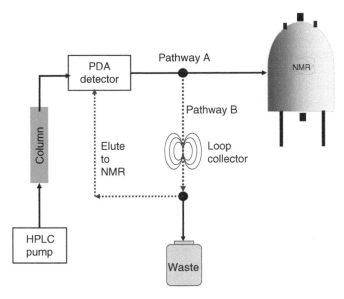

FIGURE 7.8 Schematic of HPLC-NMR setup illustrating the pathways used for on-flow and stopped flow modes (pathway A) and loop collection mode (pathway B), shown by the dashed line.

An alternative approach involves the use of loop collection, illustrated by pathway B of Figure 7.8. In this approach, the analytes eluting from the chromatographic column are first detected by the PDA, and then individually transferred to collection loops where the fractions are stored until analyzed. The contents of the loops are flushed one at a time through the PDA detector to the NMR probe for analysis. An advantage of this configuration is that it permits the online collection of many analytes from a single chromatographic run. It is also possible to perform loop collection from multiple chromatographic separations followed by NMR analysis of all the collected fractions, essentially decoupling the separation from the NMR analysis step. However, the analyte in the collected fraction will undergo dilution by diffusion during loop storage, which may reduce the detection sensitivity compared with stopped-flow analysis.

7.4.3.3 Solid-Phase Extraction-NMR

An alternative sample introduction method uses solid phase extraction (SPE) in lieu of loops to trap chromatographic peaks. In this method, the eluent from the chromatographic separation is directed into SPE cartridges where the analyte of interest is trapped. These cartridges may then be dried using a suitable gas, such as nitrogen, to remove mobile phase solvents before eluting the analyte with a small volume of deuterated organic solvent. This approach allows the use of protonated solvents during the chromatographic separation, saving valuable deuterated solvents, of which only small volumes are needed for transfer from the SPE cartridge to the flow cell of the probe [33]. It is also possible to further increase the concentration of the analyte by performing multiple separations and trapping the same analyte repeatedly on an SPE cartridge. However, one of the major drawbacks of using SPE sample trapping is loss of the most polar compounds due to poor retention by the SPE packing material and difficulty in eluting the very hydrophobic compounds from the SPE cartridge. Even so, many compounds suitable for LC-NMR analysis will have physicochemical properties that make them amenable to analyses using SPE [34].

SPE sample trapping can also increase sensitivity by more closely matching the chromatographic peak to the active volume of the NMR probe. For example, Djukovic et al. [35] used this approach to separate a standard mixture containing 1 mM of the anti-inflammatory drugs ibuprofen, naproxen, and phenylbutazone. These authors reported a significant improvement in S/N by up to a factor of approximately 15 using this approach compared with conventional LC-NMR. Among the advantages reported by the authors are low detection limits ($\sim$4 μg of sample/injection) and rapid analysis because of the simultaneous separation, preconcentration, and NMR analysis. A similar approach involving LC-SPE and CapNMR has been used for the rapid iden-tification of complex sesquiterpene lactones and esterified phenylpropanoids present in a crude extract of *Thapsia garganica* [36]. The HPLC peaks with volumes of 0.5–1 mL were concentrated to approximately 10 μL by SPE, a volume that more closely matches that of the CapNMR flow cell (5 μL). In this work, however, the LC-SPE was performed offline prior to analysis by CapNMR to facilitate acquisition of 2D NMR spectra. Nevertheless, the setup allowed structure determination of complex natural products constituting less than 0.5% of the extract. Figure 7.9 shows 2D spectra obtained from some of the components of the plant extracts, including those from less-sensitive NMR experiments such as NOESY and HMBC. These results demonstrate the potential of LC-SPE-CapNMR for the analysis of complex mixtures allowing structural elucidation of complex natural products.

For example, Miliauskas et al. [37] used LC-SPE-NMR for the identification of radical scavenging compounds in extracts of *Rhaponticum carthamoides*. Figure 7.10 is an online LC-UV chromatogram of the water-butanol fraction of the plant extract. The authors reported that only the structures of peaks 1 and 2 in the chromatogram have previously been isolated and identified. Using LC-SPE-NMR, it was possible to selectively detect and identify relatively rare radical scavenging compounds without the need for offline chromatographic separation. The 1D NMR spectra of all the com-pounds (flavonoid glycosides and chlorogenic acid) were recorded after flushing the compounds from the SPE cartridges to the NMR probe. Figure 7.11 is a representative spectrum of one of the compounds obtained in that fashion. The spectrum was recor-ded online in deuterated methanol and is representative of a new radical scavenging compound in an extract of *Rhaponticum carthamoides*. Because of the nondestructive nature of NMR, the pure compound was recovered and infused into a mass spectro-meter for molecular weight determination. Because the amount of material trapped on one cartridge was not sufficient for 2D NMR experiments using a 400 MHz LC-NMR probe, analytes obtained from multiple runs were combined for offline analysis. The number of chromatographic runs necessary for multiple trapping on an SPE cartridge can be reduced by using higher field magnets and cryoflow probe technology. For example, automated LC-SPE-NMR analysis with detection using a 20 μL cryoflow probe has been used by Exarchou et al. [38] to identify natural product components

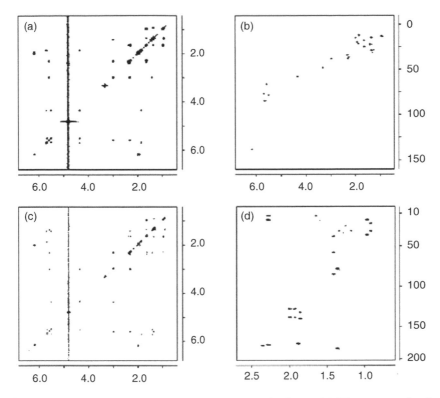

FIGURE 7.9 Results obtained using the CapNMR probe for (a) COSY spectrum of a chromatographic peak containing thapsigargicin, (b) HSQC, (c) NOESY and (d) HMBC of a chromatographic peak containing thapsigargin. (Reproduced from Lambert, M. et al. *Anal. Chem.* 2006, 79, 727–735. Copyright (2006). Used with the permission of the American Chemical Society.)

in Greek oregano extracts. These authors compared LC-SPE-NMR analysis with two different systems, a 400 MHz spectrometer with a conventional (120 μL V_{tot}) flow probe and a 600 MHz spectrometer with a lower volume (20 μL V_{tot}) cryoflow probe. The sensitivity of spectra measured with these systems for a 2 mM standard solution manually injected revealed an improvement in S/N by about a factor of 8 using the 600 MHz system with the cryoprobe after normalizing the S/N per microliter contained in the flow cell. Since sensitivity scales as $B_o^{7/4}$, we can account for about a factor of 2 gain in sensitivity simply from the increase in magnetic field strength. For online LC-SPE-NMR analysis, the authors saw additional practical improvements in sensitivity attributed to the fact that the 20 μL flow cell more closely matches the elution volume of the SPE cartridge (~30 μL), avoiding the dilution that occurred when the SPE peaks were analyzed in the 120 μL V_{tot}. The sensitivity accrued by online SPE and a 20-μL cryoflow probe enabled acquisition of 2D NMR experiments (HMQC and HMBC) as well as direct observation of ^{13}C NMR spectra, an important asset for natural product structure elucidation. An alternative approach taken by

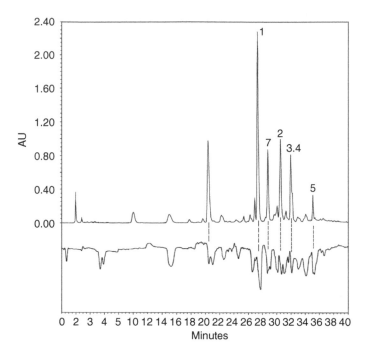

FIGURE 7.10 LC-UV chromatogram for the separation of the water-butanol fraction of the *Rhaponticum carthamoides* extract. The upper profile is the UV signal of the radical scavenger at 254 nm while the lower profile is the signal obtained when a model radical, added postcolumn, is reduced as the scavenger radical elutes leading to a reduction in absorbance. This approach is a means of rapidly probing the activity of an individual peak without the need of isolation. (Reproduced from Miliauskas, G. et al. *J. Nat. Prod.* 2005, 68, 168–172. Copyright (2005). Used with the permission of the American Chemical Society.)

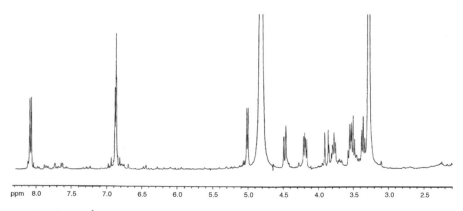

FIGURE 7.11 ^{1}H NMR spectrum of one of the five relatively rare flavonoid glycosides in an extract of *Rhaponticum carthamoides*, obtained by LC-SPE-NMR technique. (Reproduced from Miliauskas, G. et al. *J. Nat. Prod.* 2005, 68, 168–172. Copyright (2005). With permission American Chemical Society.)

Xu et al. [39] used a semi-preparative LC-SPE-NMR system providing a sensitivity enhancement more than 30-fold greater than standard LC-NMR.

7.4.4 LC-NMR-MS

Owing to the speed, superior sensitivity, and complementary structural information provided by MS, hyphenating LC-NMR methods with MS detection offers many benefits [40,41]. The manner in which these techniques are coupled can be varied according the desired results. One option would be to set up the LC, NMR, and MS in series where the sample would be analyzed first by NMR and then transferred to the MS for subsequent analysis. This configuration has several disadvantages in that NMR experiments are both less sensitive and more time-consuming than the acquisition of MS data, and that the interface to the MS would exceed the backpressure tolerated by most NMR flow probes. A more practical option is to set up the NMR and MS detection methods in parallel following chromatographic separation. In this case, a fraction of the sample, typically about 5%, is diverted to the MS while the rest is directed to the NMR flow cell. Most analyses that use MS detection in parallel with NMR use the MS data independently to complement the NMR data for purposes of structure elucidation. However, it is also possible to use MS results in lieu of UV detection to initiate stopped-flow NMR spectral acquisition. This is an especially useful strategy for compounds that do not contain a good UV chromophore. By introduction of a short time delay between MS and NMR data acquisition, the MS results can be matched against a database to decide whether a compound is already known, in which case there is no need for NMR data acquisition. This process of dereplication avoids time-consuming analysis of known products and is especially important in natural product analysis or in the identification of drug metabolites [38,42,43]. LC-NMR-MS has also been used for the direct identification of carbohydrates in beer [44]. However, for the most part, directly coupled LC-NMR-MS remains a niche technique. An alternative strategy effective for many laboratories is to perform LC-NMR or LC-MS analyses in parallel or to collect fractions following LC-NMR analysis that are then subjected to offline MS analysis [45].

7.4.5 LC-NMR Applications

7.4.5.1 Phytochemical and Natural Products Analysis

The time-consuming and tedious process of isolating novel bioactive compounds from natural products has benefited greatly from the introduction of LC-NMR and several of the examples already discussed relate to the analysis of phytochemical natural products [33,36,38,45–47]. More recently, there has been increased attention placed on analysis of marine and microbial organisms [48,49]. Traditionally, activity-guided fractionation is used to focus attention on determining the structures of components with interesting biological activity [50]. In the past, this has involved tube-based analysis of purified bioactive fractions, but more recently, LC-NMR has been used to improve the analysis throughput and to avoid contamination and decomposition commonly associated with offline fraction collection and analysis.

An alterative approach to traditional bioactivity-guided fractionation was described in a study to investigate compounds in the bark of *Erythrina vogelii*, a plant that has long been used in the treatment of illness on the Ivory Coast. In this study, crude extracts were analyzed by low-flow (0.1 mL/min) on-flow LC-NMR allowing acquisition of 256 scans per increment over the 19 hr chromatographic run. The peaks were collected after LC-NMR analysis and subjected to a secondary LC microfractionation and bioassay [51]. After LC-MS/MS and UV analysis, a novel compound with strong antifungal activity was identified as 5,7,4'-trihydroxy-2'-methoxy-5'-(3-methylbut-2-enyl)-isoflavanone. Using a tube-based system, Schroeder et al. [52] have recently described a similar strategy performing differential 2D NMR analyses for spectra of crude fungal extracts to identify novel bioactive compounds, which are isolated and further characterized downstream following additional fractionation steps.

7.4.5.2 Drug Discovery and Development

In the development of new pharmaceuticals, LC-NMR is used primarily for the identification of drug metabolites and in stability testing for the structure elucidation of degradation products. Biofluids, such as blood or urine, are commonly used in studies to characterize drug metabolites. In one such study, urine from rats dosed with 2-chloro-4-trifluoromethylaniline (CTFMA) was analyzed by ^{19}F-NMR to identify the metabolic products of the compound *in vivo*. It was shown that the major metabolic product of CTFMA is 2-amino-3-chloro-5-trifluoromethylphenylsulfate, of which 33.5% of the original dose was excreted within the first 48 hours following administration [53]. In another study, the fluorinated metabolites of the HIV-1 reverse transcriptase inhibitor 5-chloro-1-(2',3'-dideoxy-3'-fluoro-*erythro*-pentofuranosyl)uracil were identified using HPLC coupled with ^{19}F-NMR and MS. The major metabolites were identified as the glucuronide conjugate of the original compound and 3-fluoro-ribolactone [54]. LC-NMR was also used by Sohda et al. [55] to characterize rat urinary metabolites of Zonampanel monohydrate (YM872), a highly water-soluble agent that has antagonist effects at the AMPA subtype of glutamate receptors. These authors administered ^{14}C-labeled YM872 to rats to follow urinary excretion and for metabolite profiling. The structures of two metabolites isolated by preparative HPLC were determined using LC-NMR and several MS methods (including MALDI-TOF, LC-MS, and LC-MS/MS) [56].

Regulatory bodies such as the Food and Drug Administration (FDA) in the United States require the identification of all impurities above the 0.1% level in formulated pharmaceuticals. Once identified, the structure of the impurity is typically confirmed through synthesis to provide absolute structure identification and for use as standards in subsequent quality assurance analyses. Together, LC-MS and LC-NMR play important roles in stability testing. For example, parallel analysis by LC-NMR and LC-MS was used for the rapid structure elucidation of an unknown impurity in 5-aminosalicylic acid, which is marketed for the treatment of acute ulcerative colitis and Chron's disease [57]. In another study, Fukutsu et al. [58] used a combination of LC-NMR, LC-MS, and LC-IR to identify degradation products of cefpodozime prozetil.

7.4.5.3 Chemical Analysis

LC-NMR has also found application in studies addressing a variety of problems in chemical analysis. Some studies make use of the ability of NMR to serve as a general detector for organic compounds, especially in applications involving quantification of different structural isomers or the analysis of compounds that do not ionize well and therefore are ill-suited for MS. For example, LC-NMR has been used to quantify α and γ-linolenic acids and for the characterization of regioisomers of the food flavoring agent ethyldimethylpyrazine [59,60]. LC-NMR has also found application in environmental analyses. Preiss et al. [61] used LC-NMR and LC-MS to characterize polar explosives and their transformation products in groundwater samples collected from ammunition waste sites. The power of LC-NMR for structure elucidation was demonstrated through its use in structural studies of natural organic matter isolated from freshwater and soil [62]. LC-NMR studies can also take advantage of the ability of NMR to serve as a highly specific detector for compounds that contain a high abundance spin $1/2$ nucleus such as ^{19}F, ^{29}Si, or ^{31}P or have incorporated into their structure a nucleus that is normally at low abundance, for example, ^{13}C or ^{15}N. Blechta et al. [63] used LC-NMR experiments with ^{1}H-^{29}Si indirect detection to analyze mixtures of siloxan polymers. Other studies take advantage of the unique ability of NMR to study dynamic processes like isomerization, for example, the interconversion of rotational isomers or enol–keto tautomers [64,65].

7.5 SUPERCRITICAL FLUID CHROMATOGRAPHY-NMR

Over the past three decades, supercritical fluid chromatography (SFC) has become increasingly popular as a separation technique. Most supercritical fluids are inexpensive and can be allowed to evaporate into the atmosphere without harming the environment. Supercritical fluids are widely used in industrial applications, especially for extractions, because of the ease and low cost of analyte recovery. Many of the physical properties of supercritical fluids are intermediate between those of a substance in its liquid and gaseous states. Therefore, SFC combines many of the characteristics of liquid and gas chromatography. Particularly important properties of supercritical fluids are their high densities (0.2–0.5 g/cm^3) allowing them to dissolve large, nonvolatile molecules, and the ease with which the density can be varied through changes in temperature and pressure. Another important physical property of supercritical fluids is their low viscosity allowing SFC to be conducted at higher flow rates, providing faster elution times than liquid chromatography. The most commonly used mobile phase for SFC is supercritical carbon dioxide (CO_2). It readily dissolves large, nonpolar organic molecules, such as n-alkanes containing between 5 and 30 carbons and polycyclic aromatic compounds. Supercritical CO_2 has the advantage of having relatively low critical point parameters ($T_c = 31°C$ and $P_c = 1072$ psi) and because of the absence of protons, is ideal for use with online NMR detection.

7.5.1 SFC-NMR INSTRUMENTATION

In order to utilize NMR detection for online SFC separations, a flow system capable of operating under high temperatures and pressures is required. The first report on the hyphenation of SFC and NMR by Dorn and coworkers describes a custom-built NMR probe capable of operating at temperatures of 100°C and 3500 psi [66]. Building on this initial work, high-pressure NMR probes were constructed by other researchers that proved successful for use with supercritical fluids [67,68]. In the past decade, probe designers have concentrated on developing high-pressure cells that fit within commercially available probes in order to take advantage of their built-in electronic circuitry and temperature capabilities [69–72]. Various nonmagnetic materials have been utilized in probe modification, which are capable of withstanding high pressures. Some examples include single-crystal sapphire with a titanium valve, sealed quartz tubing, fused silica capillary, and high-strength plastics [73]. Many recent studies reporting improved NMR detection cells have used NMR to make measurements of solutes or solvent components in supercritical fluids and are not focused on using NMR as a detector for SFC. Readers interested in this topic are directed to a thorough review of NMR studies involving supercritical fluids by Yonker and Linehan [73].

7.5.2 SFC-NMR APPLICATIONS

SFC-NMR was used for the separation and identification of a mixture of five vitamin A acetate isomers using supercritical CO_2 as the eluent [74]. An advantage pointed out in this report is the lack of a ^1H solvent resonance, eliminating the need for solvent suppression and allowing unrestricted observation of the entire ^1H chemical shift range. SFC-NMR experiments have also demonstrated the use of supercritical CO_2 for the separation of metoprolol, without any problems due to degradation of this compound [75]. Because amine-containing compounds can react with CO_2 to form carbamic acids, these authors used SFC-NMR to examine the stability of several alkyl-substituted secondary benzyl amines and some primary aromatic amines. It was determined from this study that sterically crowded groups like isopropyl or tert-butyl groups can shield the amine proton and prevent its reaction with CO_2. Although the decreased viscosity of supercritical solvents can be an advantage for their use in separations, it results in ^1H T_1 relaxation times roughly 3 to 10 times longer than in liquid solvents, causing difficulty in the quantification of on-flow SFC-NMR spectra. Fischer et al. [76] addressed this problem through the use of immobilized free radicals to enhance the rates of nuclear relaxation. In another study Braumann et al. [77] demonstrated the use of ^1H NMR for the direct online monitoring of the supercritical fluid extraction of roasted coffee and black pepper.

An area of study related to this topic is the use of subcritical, but superheated water as a mobile phase for chromatographic separations [78]. These separations use water heated to 100–220°C and pressures up to 50 bar, avoiding problems due to hydrolysis and oxidation, which is common when supercritical water is used. Although this is a new area of investigation, several reports on the hyphenation of HPLC using superheated water and NMR detection have appeared. The separation of a mixture of model drug compounds (paracetamol, caffeine, and phenacetin) was carried out using superheated D_2O as the mobile phase demonstrating both LC-NMR and LC-NMR-MS detection [79]. LC-NMR using superheated D_2O has also been used for the analysis of a methanolic extract of powdered ginger and ecdysteroids in crude plant extracts [80,81].

7.6 CAPILLARY ELECTROPHORESIS-NMR

The hyphenation of CE and NMR combines a powerful separation technique with an information-rich detection method. Although compared with LC-NMR, CE-NMR is still in its infancy; it has the potential to impact a variety of applications in pharmaceutical, food chemistry, forensics, environmental, and natural products analysis because of the high information content and low sample requirements of this method [82–84]. In addition to standard capillary electrophoresis separations, two CE variants have become increasingly important in CE-NMR, capillary electrochromatography and capillary isotachophoresis, both of which will be described later in this section.

7.6.1 INTRODUCTION TO CE

Capillary electrophoresis was first introduced in the early 1980s when small, fused-silica capillaries ($<100\,\mu$m ID) protected by an external layer of polyimide coating became available [85,86]. Although the first commercial CE instrument was not introduced until 1989, over 36,000 publications have appeared on this topic encompassing a wide range of disciplines. Electrophoretic separation methods are based on differences in electrophoretic mobility, μ_e, which depends on the charge-to-mass ratio of each species and the frictional retarding factors encountered as ions migrate in solution through the electric field, E. In electrophoresis, cations migrate toward the cathode and anions migrate toward the anode. In addition to movement of ions based on electrophoretic mobilities, CE separations have an additional contribution from electroosmotic flow (EOF), giving an overall migration velocity, v, defined by the following equation:

$$v = (\mu_e + \mu_{eof})E, \qquad (7.17)$$

where μ_{eof} is the mobility due to electroosmotic flow. Capillary electrophoresis is generally carried out using a continuous buffer system in small diameter capillaries ($<100\,\mu$m ID), which increase the cooling efficiency and minimize problems associated with thermal effects. The fused-silica capillaries commonly used in CE have at their surface (Si—OH) groups that are ionized at pH > 3. As illustrated in Figure 7.12, owing to the presence of these surface SiO^- groups, a fixed layer of positive charges forms along the inner wall of the capillary. Because this fixed layer is not sufficiently dense to completely neutralize the negative charge of the capillary wall, a second more mobile layer of cations is also formed. Both layers together are termed the diffuse double layer (Figure 7.12). Since the outer layer of cations

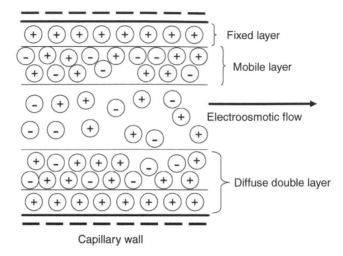

FIGURE 7.12   Representation of the diffuse double layer responsible for electroosmotic flow in capillary electrophoresis.

is not bound tightly to the SiO^- groups, in the presence of an electric field, the hydrated cations migrate toward the cathode, dragging bulk solvent with them and generating EOF. Under conditions of normal CE operation, all ions migrate toward the cathode, eluting in the order: cations, neutral molecules, and finally, anions. Although electrophoretic velocity of an anion is toward the anode, the greater velocity generated by the EOF carries it toward the cathode and the detector. For neutral molecules $\mu_e = 0$, therefore, they elute as a group migrating at a velocity determined by the electroosmotic flow.

An advantage of CE is its extremely high separation efficiency, largely owing to the absence of band broadening mechanisms typical of chromatographic methods [87]. If the radius of a CE capillary is greater than seven times the double layer thickness, then a flat flow profile is expected [88]. This flat flow profile causes all of the solute molecules to experience the same velocity, regardless of their position inside the capillary. This means that each separated solute elutes as a narrow band, producing narrow peaks of high efficiency and giving CE a significant advantage over other separation techniques that rely on pressure-driven flow, which produces parabolic flow profiles.

7.6.2 EARLY CE-NMR EXPERIMENTS

Studies incorporating NMR detection in electrophoresis began in 1972 when Packer and coworkers examined Me_4N^+ ions under the influence of an electric current [89]. They determined that electrical current density in the sample and the radial motion of ions under the influence of the magnetic field induced undesirable magnetic field gradients. This work was followed up by Holz and Muller who suggested that the undesirable field gradients and their effects can be minimized if the electric field is oriented parallel to the static magnetic field [90]. Through the incorporation of this geometry, they were able to measure the electrophoretic mobilities of small ions [91,92]. Because of the low magnetic field used in these studies, high concentrations of ions were required for NMR detection and the effects of Joule heating required the use of a gelling agent for stabilization. In 1988, Sarrinen and Johnson reported a high-resolution electrophoretic NMR (E-NMR) experiment using free electrophoresis in a vertical U-tube cell [93]. Samples of low ionic strength and concentration (10 mM) were used and current densities were kept below 0.5 A/cm^2 to avoid adverse effects due to heating. In the following year, a comprehensive review was published exploring the theory and limitations of combining electrophoresis with NMR [94]. The authors provided a detailed discussion on the ability of NMR to determine electrophoretic mobilities and diffusion coefficients. Emphasis was placed on the current limitations and possible mechanisms for improvement.

In 1994, the first online coupling of CE and NMR was reported using NMR microcoil detection cells with 5–200 nL volumes [95,96]. In these seminal papers, Sweedler and coworkers described the design and optimization of RF microcoils constructed by hand-wrapping copper wire around the separation capillary as well as experimental results obtained for both static and on-flow NMR measurements using a variety of capillary sizes. NMR acquisition times <1 minute yielded well-resolved spectra and limits-of-detection in the nanogram range for a detection cell with ~5 nL active volume. An in-depth explanation of how NMR band intensities can vary under the effects of electrophoretic current, applied potential, and flow rate were also given.

In 1998, a report by Bayer and coworkers provided details of an alternative configuration in which a single NMR probe could be used with three different types of miniaturized separation systems: CapLC, CE, and CEC [97]. Instead of wrapping the radiofrequency coil around the capillary, it was attached to a glass tube (2 mm OD and 1.7 mm ID) allowing the capillary to be exchanged without having to remove the coil. The capillary contained an enlarged 240 nL detection cell in order to increase V_{obs} and the residence time of the analyte (Figure 7.13). Increasing the residence time increases the number of scans that can be coadded, thereby increasing S/N. For measurements conducted under CE conditions, 20 kV was applied across the NMR detection capillary. This was sufficient to perform a clean separation of a lysine and histidine mixture. The CapLC separation used a packed capillary outside of the magnet that was connected to the NMR detection capillary with Teflon tubing. Both the injection and separation of analytes was performed via a pressure driven HPLC pump. The stereoisomers of hop bitter acids were separated and identified despite loading large sample quantities onto the column. The CEC setup utilized the CE power supply together with the HPLC pump. The electric field was applied across the length of the column and capillary and the combined effects of pressure and EOF shortened the elution time when compared with CapLC. In addition, the plug profile that developed because of EOF decreased band broadening, improving the overall resolution. A complete separation of five alkyl benzoates was achieved despite the column overloading that was necessary for NMR detection.

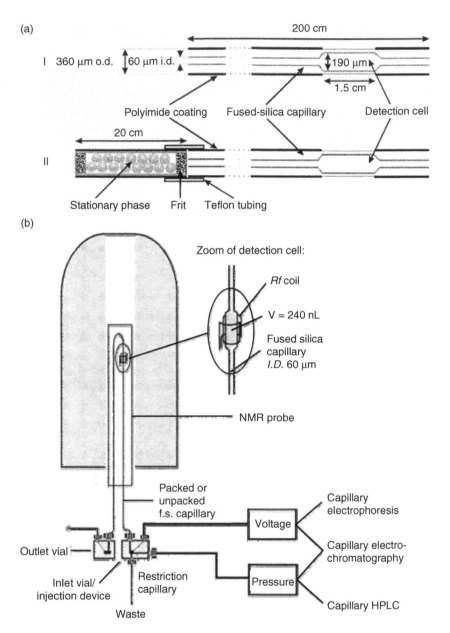

FIGURE 7.13   (a) I. Capillary with enlarged detection cell for NMR detection. The capillary is suitable for CE separations. II. By connecting a packed capillary in front of the NMR detection capillary, the system is adapted to use with CEC and CapLC. (b) Scheme of the instrumentation for the coupling of capillary separation techniques with NMR. (Reproduced from Pusecker, K. et al. *Anal. Chem.* 1998, 70, 3280–3285. Copyright (1998). Used with the permission of the American Chemical Society.)

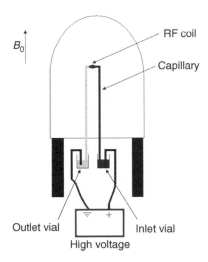

FIGURE 7.14   Instrumental setup for CE-NMR with online NMR detection using a solenoidal microcoil probe.

7.6.3 CE-NMR Instrumentation

A typical setup for CE-NMR is given in Figure 7.14. Because small diameter capillaries are used in CE-NMR, detection is typically accomplished using low-volume microcoil NMR probes. A solenoidal microcoil is either wrapped around the capillary itself or wrapped around a sleeve through which the capillary is inserted. As mentioned in Section 7.1.2, the closer the detection coil is to the sample, the greater the probe fill factor. For this reason, the former design is preferred in which the capillary acts as both as the sample container and the coil form [98]. CE-NMR can also be performed using small saddle coils, although because of the difficulty in fabrication, they tend to be larger than solenoidal coils.

In CE-NMR experiments, a voltage is applied across the capillary maintaining the flow of charged species. The analyte ions are separated on the basis of their electrophoretic mobilities and detected as they flow through the NMR coil. Buffers are usually contained in plastic vials and Pt electrodes can be used safely near the magnet. A high power voltage (up to 30 kV) source is needed and in some cases, a syringe pump is used. These and other magnetic objects must be kept at a safe distance from the magnet. There are typically two modes of injection used in CE-NMR: hydrodynamic and electrokinetic. Hydrodynamic injection uses either pressure or vacuum at one end of the capillary while the other end is submersed in the sample solution. This injection mode is nonselective as the composition of the plug injected is the same as the sample solution. Electrokinetic injection is induced by the application of potential with the injection end of the capillary immersed in the sample. Using this method, the composition of the sample plug depends on the mobility and charge of the ions injected. In either case, the volume of the sample loaded into the capillary must be kept small (1–10 nL) to maintain the high separation efficiency of CE.

One of the problems associated with CE-NMR is that application of the separation voltage can cause a large perturbation of the NMR signals due to a magnetic field gradient induced by the electric field. Olson et al. [21] addressed this issue using periodic stopped-flow CE in which the separation voltage is applied for 15 s and then switched off for 1 min to acquire the NMR data. This method eliminates the effect of magnetic field gradients, improving NMR sensitivity because of the improved line shape and the longer acquisition time. In these experiments, the authors employed field-amplified sample stacking to preconcentrate the analyte several fold. In another approach to solving this problem, Wolters et al. [99] used an NMR probe constructed with multiple solenoidal microcoils for continuous-flow capillary electrophoresis. In this approach, the electrophoretic flow from a single separation capillary is split into multiple outlets, each with its own detection coil. The outlets are cycled between electrophoretic flow and NMR detection to allow for continuous CE with stopped-flow detection.

7.6.4 Capillary Electrochromatography-NMR

Instrumentation for CEC-NMR is similar to CE-NMR except that instead of an open capillary, a packed LC column is used. The separation is electroosmotically driven rather than pressure-driven and the higher sample loading capacity compared with CE increases the sensitivity of NMR detection. Because electro-osmotic pumping leads to plug rather than hydrodynamic flow, greater separation efficiencies are achieved as compared with conventional LC. Two mechanisms for separation can occur in CEC, either in combination or separately: (1) development of partition equilibria between the stationary phase of the column and the mobile phase and (2) differences in charge. A disadvantage of CEC separation is the potential for bubble formation, although this can be minimized by pressurizing the buffer vials (~500 psi) or using low conductivity buffers like 2-(N-morpholino) ethanesulfonic acid (MES) and *tris*-(hydroxymethyl)aminomethane (TRIS).

7.6.5 Capillary-Isotachophoresis-NMR

Capillary-isotachophoresis (cITP) is a variant of CE that utilizes a discontinuous buffer system and a capillary modified to minimize electroosmotic flow to separate analytes into discrete zones on the basis of their electrophoretic mobilities. The sample is sandwiched between two different ionic buffers: the leading electrolyte (LE) with the highest mobility and the trailing electrolyte (TE) with the lowest mobility. Upon application of a high voltage (e.g., 15 kV) across the capillary [100], a steady state condition is reached in which each ion type is separated into its own discrete zone flanked by ions possessing a higher mobility on one side and a lower mobility on the other according to the Kohlrausch regulating function [101,102]. After the formation of these distinct analyte zones, an adjustment of ion concentration occurs within each zone [103]. By judicious selection of the leading and trailing buffers, analyte concentration enhancements of 2–3 orders of magnitude can be achieved. Therefore, cITP results in narrow separated sample zones of high concentration, ideal for online microcoil NMR detection [104].

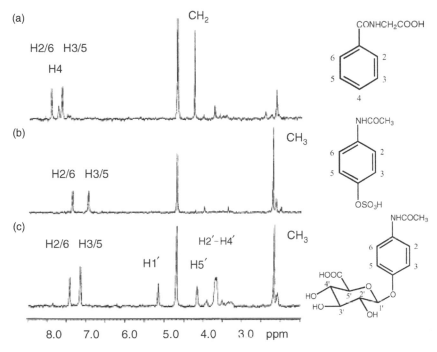

FIGURE 7.15 Continuous-flow CEC-NMR spectra (a) Paracetamol glucuronide; (b) paracetamol sulfate; and (c) endogenous hippurate. (Reproduced from Pusecker, K. et al. *Anal. Commun.* 1998, 35, 213–215. With permission from RSC publishers.)

7.6.6 Applications of CE-NMR and Related Techniques

Early CE-NMR experiments were aimed at understanding the effects of electrophoresis on NMR spectral properties and determining physicochemical properties such as electrophoretic mobilities and diffusion coefficients. Since then, CE-NMR has garnered attention as an effective hyphenated technique with a variety of applications

Pusecker et al. demonstrated the use of CE- and CEC-NMR for drug metabolite identification [105]. They were able to successfully identify the main metabolites of the drug paracetamol (4-hydroxyacetanilide), the glucuronic acid and sulfate conjugates, in a human urine sample (Figure 7.15a,b). Endogenous hippurate was also detected (Figure 7.15c), and is known to be present in urine at millimolar concentrations [106]. The *S/N* they obtained using CE was around 3 and it was estimated that approximately 10 ng of each metabolite was detected. They reported slightly better *S/N* with the CEC separation due to the higher sample loading capacity and larger NMR observe volume.

In an effort to improve separation efficiency, a novel configuration was reported by Albert and coworkers [107] describing a splitless capillary setup capable of separating components of a mixture of unsaturated fatty acid methyl esters. The chromatographic setup consisted of a versatile splitless connection enabling easy exchange between three types of separation methods: CapLC, CEC, and pressurized CEC (pCEC). In the CEC experiments, the electric field was applied solely across the separation column instead of the whole capillary system, allowing the use of higher voltages and shorter analysis times. In addition, a larger NMR detection cell was used in the experiments (~750 nL), increasing the observe volume and enhancing peak shape, resolution, and sensitivity. The results obtained for each CapLC, CEC, and pCEC separation were compared. The spectral resolution achieved from CapLC-NMR was sufficient to provide the assignments for cis and trans isomers of the unsaturated fatty acid methyl esters.

An example of the compromise between better separation efficiency and improved NMR spectral resolution is demonstrated in a publication by Bayer and coworkers [108]. Improvement in the separation was achieved by optimizing the pH of the buffers used. Continuous CE-NMR was utilized to separate caffeine and aspartame, ingredients commonly found in soft drink beverages. Although it provided a good separation, the use of a glycine buffer at pH 10 caused problems due to spectral overlap between the glycine and analyte resonances. To remedy this problem, a formic acid buffer at pH 5 was used. Under these conditions, the order of elution was reversed and the intensity of the aspartame resonance was increased. However, migration times were longer and the caffeine and aspartame peaks were not as well resolved. In the same report, a comparison between gradient and isocratic CEC-NMR was performed by analysis of the analgesic Thomapyrin®containing acetaminophen, caffeine, and acetylsalicylic acid. Isocratic conditions yielded poorly resolved peaks with a solvent peak eluting very close to acetaminophen. These results were attributed to the large volume injected because of the relatively high concentration of the sample required for NMR detection. Gradient elution demonstrated superior separation efficiency by applying a solvent gradient of 0% to 30% CD_3CN over 25 min. In addition, the overall separation time was decreased to less than half the time required under isocratic conditions.

The applicability of cITP-NMR for the analysis of trace impurities was demonstrated by the selective detection of 1.9 nmol of atenolol injected in a sample containing a 1000-fold excess of sucrose [100]. cITP-NMR has also been used for the analysis of a cationic neurotoxin present in a homogenate of the hypobranchial gland of the marine snail *Calliostoma canaliculatum* [109]. Korir et al. [110] used an anionic cITP separation with online NMR detection to separate and identify nanomole quantities of heparin oligosaccharides. Although only a few cITP-NMR applications have appeared, the ability to selectively separate, concentrate, and detect charged analytes makes cITP-NMR a potentially powerful method for trace analysis.

Another important application of hyphenated NMR methods is to provide insights into processes that affect the separation. For example, online NMR detection of the water chemical shift was used to noninvasively probe intracapillary temperatures in CE separations with subsecond temporal resolution and spatial resolution on the order of 1 mm [111]. Lacey et al. [112] followed up this report with a second NMR study using a novel 2-turn vertical solenoidal coil to measure temperature increases of more than 50°C in a chromatographic frit of the type used in CEC. Insights into the mechanisms underlying cITP have also been investigated utilizing online NMR measurements to monitor sample stacking and measure the acidity of different zones by following the chemical shift of the methyl protons of the acetate buffer [113]. In this study, online measurements of the acetate chemical shift were used to determine the pD (pD is the equivalent of pH for D_2O solutions) of the LE and TE on either side of the analyte band as 5.18 ± 0.04 and 3.5 ± 0.1, respectively. The most interesting pD measurements are those across the LE/sample and sample/TE boundaries. However, broadening of the NMR spectra at these interfaces, attributed to magnetic susceptibility differences, precluded accurate pD determination for these regions. A subsequent report comparing the electrophoretic migration behavior of ions in anionic cITP resolved some of these problems and allowed the accurate measurement of intracapillary pD across the analyte band, including the LE/sample and sample/TE interfaces [110]. These authors observed changes in the chemical shift of imidazole, a sensitive NMR pH indicator, at the analyte zone boundaries suggesting stacking of hydroxyl ions at the zone interfaces. In this report, magnetic susceptibility differences were observed across the analyte zones in cationic cITP, but not in the corresponding anionic cITP-NMR runs, leading the authors to suggest that the broadening of resonances at the interfaces in cationic cITP-NMR experiments arises from the inadvertent focusing of trace paramagnetic metal ions present in the buffer solutions. The interaction of analytes and chiral selectors like β-cyclodextrin has also been studied by online cITP-NMR [114,115]. For example, with a binding constant of 200 M^{-1}, β-cyclodextrin has a relatively high affinity for propranolol. As a result of inclusion complex formation, the neutral β-cyclodextrin stacks along with propranolol as indicated by increases in the intensity of the β-cyclodextrin resonances in the analyte zone shown in Figure 7.16. Changes in the β-cyclodextrin resonance chemical shifts in the focused analyte band also reflect the formation of the propranolol inclusion complex.

7.6.7 Future Directions of CE-NMR

CE is well recognized as a powerful separation method and when coupled to an information-rich detection method such as NMR, one might expect the application of this technology to a large number of challenging problems in chemical analysis rather than the few applications reviewed in this article. Unlike LC-NMR, there are currently no commercially available CE-NMR systems, a factor that has limited its widespread acceptance and more universal application. Improvements in sensitivity will continue to advance the use of NMR as a detector for electrophoretic separations. These improvements may come from a combination of advances in probe design, for example, NMR as a detector for microfluidic devices [116], and through more extensive use of stacking techniques such as isotachophoresis, field amplified sample stacking, or pH-mediated sample stacking to enhance sensitivity.

7.7 CONCLUSIONS

Hyphenation of chromatographic separations with NMR detection allows the acquisition of data sets with high information content. This is especially useful for

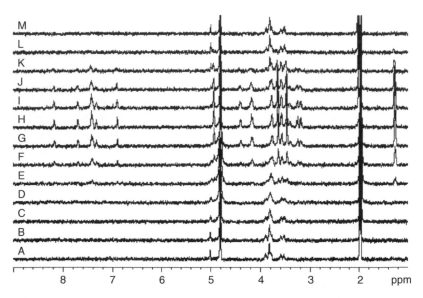

FIGURE 7.16   The results of a cITP-NMR experiment to study the interaction between prop-ranolol (1.9 nmol injected) and β-CD, initially present at 10 mM in all solutions. Each spectrum was measured by coaddition of 8 FIDs with a time resolution of 11.28 s per spectrum. Spectra A–D contain only the resonances of the leading electrolyte (160 mM acetate, pD 4.76, and 10 mM β cyclodextrin). The leading and trailing edges of the analyte band are detected in spectra E and K, respectively, with spectra F–J giving a time-slice profile of the focused pro-pranolol band. Note the increased intensity of the cyclodextrin resonances, for example, the anomeric proton resonance at 5 ppm, in the spectra containing the analyte. Spectra L and M contain only the resonances of the TE, 160 mM acetic acid, pD 2.45, and 10 mM β cyclodextrin. (Reproduced from Almeida, V. K., Larive, C. K. *Magn. Reson. Chem* 2005, 43, 755–761, with permission from John Wiley & Sons.)

structure determination or identification of new or unknown compounds present in complex sample matrices, for example, natural products, drug metabolites, or degradation products of formulated pharmaceuticals. NMR detection is also the method of choice for compounds that are difficult to analyze by MS, for example, mixtures of isomers. Although LC-NMR has become a relatively mature field, recent advances in the development of cryogenically cooled flow probes and the use of SPE as a sample trapping method have led to significant increases in the sensitivity of these measurements and the application of this technology to increasingly challenging analytical problems is anticipated. By comparison, CE-NMR is an emerging technique, but one that holds significant promise for future applications.

REFERENCES

1. Jayawickrama, D. A., and Sweedler, J. V. *J. Chromatogr. A* 2003, 1000, 819–840.
2. Olson, D. L., Norcross, J. A., O'Neill-Johnson, M., Molitor, P. F., and Detlefsen, D. J. *Anal. Chem.* 2004, 76, 2966–2974.

3. Hoult, D. I., and Richards, R. E. *J. Magn. Reson.* 1976, 24, 71–85.
4. Claridge, T. D. W. *High-Resolution NMR Techniques in Organic Chemistry*; Elsevier Ltd.: Oxford, 1999.
5. Keeler, J. *Understanding NMR Spectroscopy*; John Wiley and Sons, 2005.
6. Webb, A. G. *Prog. Nucl. Magn. Reson. Spectrosc.* 1997, 31, 1–42.
7. Black, R. D., Early, T. A., Roemer, P. B., Mueller, O. M., Mogro-Campero, A., Turner, L. G., and Johnson, G. A. *Science* 1993, 259, 793–795.
8. Anderson, W. A., Brey, W. W., Brooke, A. L., Cole, B., Delin, K. A., Fuks, L. F., Hill, H. D. W., et al. *Bull. Mag. Res.* 1995, 17, 98–102.
9. Kovacs, H., Moskau, D., and Spraul, M. *Prog. Nucl. Magn. Reson. Spectrosc.* 2005, 46, 131–155.
10. Lacey, M. E., Subramanian, R., Olson, D. L., Webb, A. G., and Sweedler, J. V. *Chem. Rev.* 1999, 99, 3133–3152.
11. Schroeder, F. C. and Gronquist, M. *Angew. Chem. Int. Ed.* 2006, 45, 7122–7131.
12. Peck, T. L., Magin, R. L., and Lauterbur, P. *J. Magn. Reson., Ser. B* 1995, 108, 114–124.
13. Wu, N., Webb, A. G., Peck, T. L., and Sweedler, J. V. *Anal. Chem.* 1995, 67, 3101–3107.
14. Hou, T., Smith, J., MacNamara, E., Macnaughtan, M., and Raftery, D. *Anal. Chem.* 2001, 31, 2541–2546.
15. MacNamara, E., Hou, T., Fisher, G., Williams, S., and Raftery, D. *Anal. Chim. Acta* 1999, 397, 9–16.
16. Jansma, A., Chuan, T., Albrecht, R. W., Olson, D. L., Peck, T. L., and Geierstanger, B. H. *Anal. Chem.* 2005, 77, 6509–6515.
17. Lewis, R. J., Bernstein, M. A., Duncan, S. J., and Sleigh, C. J. *Magn. Reson. Chem.* 2005, 43, 783–789.
18. Elipe, M. V. S. *Anal. Chim. Acta* 2003, 497, 1–25.
19. Laude, D. A. and Wilkins, C. L. *Anal. Chem.* 1984, 56, 2471–2475.
20. Wolters, A. M., Jayawickrama, D. A., Webb, A. G., and Sweedler, J. V. *Anal. Chem.* 2002, 74, 5550–5555.
21. Olson, D. L., Lacey, M. E., Webb, A. G., and Sweedler, J. V. *Anal. Chem.* 1999, 71, 3070–3076.
22. Ogg, R. J., Kingsley, P. B., and Taylor, J. S. *J. Magn. Reson., Ser. B* 1994, 104, 1–10.
23. Smallcombe, S. H., Patt, S. L., and Keifer, P. A. *J. Magn. Reson.* 1995, 117, 295–303.
24. Wolfender, J.-L., Ndjoko, K., and Hostettmann, K. *J. Chromatogr. A* 2003, 1000, 437–455.
25. Albert, K., Ed. *On-Line LC-NMR and Related Techniques*; John Wiley and Sons Ltd.: West Sussex, England, 2002.
26. Bayer, E., Albert, K., Nieder, M., and Grom, E. *J. Chromatogr.* 1979, 186, 497–507.
27. Haw, J. F., Glass, T. E., and Dorn, H. C. *Anal. Chem.* 1981, 53, 2327–2332.
28. Haw, J. F., Glass, T. E., Hausler, D. W., Motell, E., and Dorn, H. C. *Anal. Chem.* 1980, 52, 1135–1140.
29. Spraul, M., Freund, A. S., Nast, R. E., Withers, R. S., Mass, W. E., and Corcoran, O. *Anal. Chem.* 2003, 75, 1546–1551.
30. Krucker, M., Lienau, A., Putzbach, K., Grynbaum, M. D., Schuler, P., and Albert, K. *Anal. Chem.* 2004, 76, 2623–2628.
31. Strohschein, S., Rentel, C., Lacker, T., Bayer, E., and Albert, K. *Anal. Chem.* 1999, 71, 1780–1785.
32. Shapira, B., Karton, A., Aronzon, D., and Frydman, L. *J. Am. Chem. Soc.* 2004, 126, 1261–1265.

Advances in Chromatography, Volume 46

33. Exarchou, V., Krucker, M., van Beek, T. A., Vervoort, J., Gerothanassis, I. P., and Albert, K. *Magn. Reson. Chem.* 2005, 43, 681–687.
34. Jaroszewski, J. W. *Planta Med.* 2005, 71, 795–802.
35. Djukovic, D., Liu, S., Henry, I., Tobias, B., and Raftery, D. *Anal. Chem.* 2006, 78, 7154–7160.
36. Lambert, M., Wolfender, J.-L., Stærk, D., Christensen, S. B., Hostettmann, K., and Jaroszewski, J. W. *Anal. Chem.* 2006, 79, 727–735.
37. Miliauskas, G., van Beek, T. A., de Waard, P., Venskutonis, R. P., and Sudholter, E. J. R. *J. Nat. Prod.* 2005, 68, 168–172.
38. Exarchou, V., Godejohann, M., van Beek, T. A., Gerothanassis, I. P., and Vervoort, J. *Anal. Chem.* 2003, 75, 6288–6294.
39. Xu, F. and Alexander, A. J. *Magn. Reson. Chem.* 2005, 43, 776–782.
40. Corcoran, O. and Spraul, M. *Drug Discovery Today* 2003, 8, 624–631.
41. Wilson, I. D., Lindon, J. C., and Nicholson, J. K. *Anal. Chem.* 2000, 72, 534A–542A.
42. Dear, G. J., Plumb, R. S., Sweatman, B. C., Ayrton, J., Lindon, J. C., Nicholson, J. K., and Ismail, I. M. *J. Chromatogr. B* 2000, 748, 281–293.
43. Yang, Z. *J. Pharm. Biomed. Anal.* 2006, 40, 516–527.
44. Duarte, I., Godejohann, M., Braumann, U., Spraul, M., and Gil, M. *J. Agric. Food Chem.* 2003, 51, 4847–4852.
45. Iwasa, K., Cui, W., Sugiura, M., Takeuchi, A., Moriyasu, M., and Takeda, K. *J. Nat. Prod.* 2005, 68, 992–1000.
46. Lambert, M., Staerk, D., Hansen, S. H., Sairafianpour, M., and Jaroszewski, J. W. *J. Nat. Prod.* 2005, 68, 1500–1509.
47. Wolfender, J.-L., Queiroz, E. F., and Hostettmann, K. *Magn. Reson. Chem.* 2005, 43, 697–709.
48. Bobzin, S. C., Yang, S., and Kasten, T. P. *J. Chromatogr. B* 2000, 748, 259–267.
49. Pham, L. H., Vater, J., Rotard, W., and Mügge, C. *Magn. Reson. Chem.* 2005, 43, 710–723.
50. Kraus, W. *J. Toxicology* 2003, 22, 495–508.
51. Queiroz, E. F., Wolfender, J.-L., Atindehou, K. K., Traore, D., and Hostettmann, K. *J. Chromatogr. A* 2002, 974, 123–134.
52. Schroeder, F. C., Gibson, D. M., Churchill, A. C. L., Sojikul, P., Wursthorn, E. J., Krasnoff, S. B., and Clardy, J. *Angew. Chem.* 2006, preprint ahead of publication, 45, 7122–7131.
53. Scarfe, G. B., Wright, B., Clayton, E., Taylor, S., Wilson, I. D., Lindon, J. C., and Nicholson, J. K. *Xenobiotica* 1999, 29, 77–91.
54. Schockor, J. P., Unger, S. E., Savina, P., Nicholson, J. K., and Lindon, J. C. *J. Chromatogr. B* 2000, 748, 267–279.
55. Sohda, K.-Y., Minematsu, T., Hashimoto, T., Suzumura, K.-I., Funatsu, M., Suzuki, K., Imai, H., Usui, T., and Kamimura, H. *Chem. Pharm. Bull.* 2004, 52, 1322–1325.
56. Pan, C., Liu, F., Wang, W., Drinkwater, D., and Vivilecchia, R. *J. Pharm. Biomed. Anal.* 2006, 40, 581–590.
57. Novak, P., Tepeš, P., Fistrić, I., Bratoš, I., and Gabelica, V. *J. Pharm. Biomed. Anal.* 2006, 40, 1268–1272.
58. Fukutsu, N., Kawasaki, T., Saito, K., and Nakazawa, H. *J. Chromatogr. A* 2006, 1129, 153–159.
59. Sýkora, J., Bernášek, P., Zarevúká, M., Kurfürst, M., Sovová, H., and Schraml, J. *J. Chromatogr. A* 2007, 1139, 152–155.
60. Sugimoto, N., Yomota, C., Furusho, N., Sato, K., Yamazaki, T., and Tanamoto, K. *Food. Addit. Contam.* 2006, 23, 1253–1259.

61. Preiss, A., Elend, M., Gerling, S., and Tränckner, S. *Magn. Reson. Chem* 2005, 43, 736–746.
62. Simpson, A. J., Tseng, L.-H., Simpson, M. J., Spraul, M., Braumann, U., Kingery, W. L., Kellcher, B. P., and Hayes, M. H. B. *Analyst* 2004, 129, 1216–1222.
63. Blechta, V., Sýkora, J., Hetflejš, J., Šabata, S., and Schraml, J. *Magn. Reson. Chem.* 2006, 44, 7–10.
64. Cardoza, L. A., Cutak, B. J., Ketter, J., and Larive, C. K. *J. Chromatogr. A* 2004, 1022, 131–137.
65. Zhou, C. C. and Hill, D. R. *Magn. Reson. Chem.* 2007, 49, 128–132.
66. Allen, L. A., Glass, T. E., and Dorn, H. C. *Anal. Chem.* 1988, 60, 390–394.
67. Ballard, L., Yu, A., Reiner, C., and Jonas, J. *J. Magn. Reson.* 1998, 133, 190–193.
68. Woelk, K., Rathke, J., and Klingler, R. *J. Magn. Reson., Ser. A* 1994, 109, 137–146.
69. Bai, S., Taylor, C. M., Mayne, C. L., Rugmire, R. J., and Grant, D. M. *Rev. Sci. Instrum.* 1996, 67, 240–243.
70. Pravica, M. G., and Silvera, I. F. *Rev. Sci. Instrum.* 1998, 69, 479–484.
71. Wallen, S. L., Schoenbachler, L. K., Dawson, E. D., and Blatchford, M. A. *Anal. Chem.* 2000, 72, 4230–4234.
72. Yamada, H., Nishikawa, K., Honda, M., Shimura, T., Akasaka, K., and Tabayashi, K. *Rev. Sci. Instrum.* 2001, 72, 1463–1471.
73. Yonker, C. and Linehan, J. *Prog. Nucl. Magn. Reson. Spectrosc.* 2005, 47, 95–109.
74. Braumann, U., Händel, H., Strohschein, S., Spraul, M., Krack, G., Ecker, R., and Albert, K. *J. Chromatogr. A* 1997, 761, 336–340.
75. Fischer, H., Gyllenhaal, O., Vessman, J., and Albert, K. *Anal. Chem.* 2003, 75, 622–626.
76. Fischer, H., Tseng, L.-H., and Raitza, M., Albert, K. *Mag. Reson. Chem.* 2000, 38, 336–342.
77. Braumann, U., Händel, H., and Albert, K. *Anal. Chem.* 1995, 67, 930–935.
78. Coym, J. W. and Dorsey, J. G. *Anal. Lett.* 2004, 37, 1013–1023.
79. Smith, R. M., Chienthavorn, O., Wilson, I. D., Wright, B., and Taylor, S. D. *Anal. Chem.* 1999, 71, 4493–4497.
80. Saha, S., Smith, R. M., Lenz, E., and Wilson, I. D. *J. Chromatogr. A* 2003, 991, 143–150.
81. Louden, D., Handley, A., Lafont, R., Taylor, S. D., Sinclair, I., Lenz, E., Orton, T., and Wilson, I. D. *Anal. Chem.* 2002, 74, 288–294.
82. Cardoza, L. A., Almeida, V. K., Carr, A., Larive, C. K., and Graham, D. W. *Trends Anal. Chem.* 2003, 22, 766–775.
83. Tagliaro, F., Manetto, G., Crivellente, F., and Smith, F. P. *Forensic Sci. Int.* 1998, 92, 75–88.
84. Altria, K. D. and Elder, D. *J. Chromatogr. A* 2004, 1023, 1–14.
85. Jorgenson, J. W. and Lukacs, K. D. *Anal. Chem.* 1981, 53, 1298–1302.
86. Jorgenson, J. W. and Lukacs, K. D. *Science* 1983, 222, 266–272.
87. Baker, D. R. *Capillary Electrophoresis*; John Wiley & Sons, Inc.: New York, 1995.
88. Ewing, A. G., Wallingford, R. A., and Olefirowicz, T. M. *Anal. Chem.* 1989, 61, 292A–303A.
89. Packer, J., Rees, C., and Tomlinson, D. *J. Advan. Mol. Relaxation Processes* 1972, 3, 119–131.
90. Holz, H. and Muller, C. *J. Magn. Reson.* 1980, 40, 595–599.
91. Holz, M. and Muller, C. *Ber. Bunsenges. Phys. Chem.* 1982, 86, 141–147.
92. Holz, H., Lucas, O., and Muller, C. *J. Magn. Reson.* 1984, 58, 294–305.

93. Saarinen, T. R., and Johnson, C. S. *J. Am. Chem. Soc.* 1988, 110, 3332–3333.

94. Johnson, C. S. and He, Q. *Electrophoretic Nuclear Magnetic Resonance*; Academic Press: New York, 1989.

95. Wu, N., Peck, T. L., Webb, A. G., Magin, R. L., and Sweedler, J. V. *J. Am. Chem. Soc.* 1994, 116, 7929–7930.

96. Wu, N., Peck, T. L., Webb, A. G., Magin, R. L., and Sweedler, J. V. *Anal. Chem.* 1994, 66, 3849–3857.

97. Pusecker, K., Schewitz, J., Gforer, P., Tseng, L.-H., Albert, K., and Bayer, E. *Anal. Chem.* 1998, 70, 3280–3285.

98. Gfrörer, P., Schewitz, J., Pusecker, K., and Bayer, E. *Anal. Chem.* 1999, 315A–321A.

99. Wolters, A. M., Jayawickrama, D. A., Webb, A. G., and Sweedler, J. V. *Anal. Chem.* 2002, 74, 5550–5555.

100. Wolters, A. M., Jayawickrama, D. A., Larive, C. K., and Sweedler, J. V. *Anal. Chem.* 2002, 74, 2306–2313.

101. Kohlrausch, F. *Ann. Phys. Chem.* 1897, 62, 209–239.

102. Dismukes, E. B. and Alberty, R. A. *J. Am. Chem. Soc.* 1954, 76, 191–197.

103. Radola, B. J., Ed. *Analytical Isotachophoresis*; VCH Verlagsgesellschaft: New York, 1988.

104. Kautz, R. A., Lacey, M. E., Wolters, A. M., Foret, F., Webb, A. G., Karger, B. L., and Sweedler, J. V. *J. Am. Chem. Soc.* 2001, 123, 3159–3160.

105. Pusecker, K., Schewitz, J., Gfrörer, P., Tseng, L.-H., Albert, K., Bayer, E., Wilson, I. D., Bailey, N. J., Scarfe, G. B., Nicholson, J. K., and Lindon, J. C. *Anal. Commun.* 1998, 35, 213–215.

106. Bales, J. R., Nicholson, J. K., and Sadler, P. J. *Clin. Chem.* 1985, 31, 757 763.

107. Rapp, E., Jakob, A., Schefer, A. B., Bayer, E., and Albert, K. *Anal. Bioanal. Chem.* 2003, 376, 1053–1061.

108. Gfrörer, P., Schewitz, J., Pusecker, K., Tseng, L.-H., Albert, K., and Bayer, E. *Electrophoresis* 1999, 20, 3–8.

109. Wolters, A. M., Jayawickrama, D. A., and Sweedler, J. V. *J. Nat. Prod.* 2005, 68, 162–167.

110. Korir, A. K., Almeida, V. K., and Larive, C. K. *Anal. Chem.* 2006, 112, 7078–7087.

111. Lacey, M. E., Webb, A. G., and Sweedler, J. V. *Anal. Chem.* 2000, 72, 4991–4998.

112. Lacey, M. E., Webb, A. G., and Sweedler, J. V. *Anal. Chem.* 2002, 74, 4583–4587.

113. Wolters, A. M., Jayawickrama, D. A., Larive, C. K., and Sweedler, J. V. *Anal. Chem.* 2002, 74, 4191–4197.

114. Jayawickrama, D. A. and Sweedler, J. V. *Anal. Bioanal. Chem.* 2004, 378, 1528–1535.

115. Almeida, V. K. and Larive, C. K. *Magn. Reson. Chem* 2005, 43, 755–761.

116. McDonnell, E. E., Han, S., Hilty, C., Pierce, K. L., and Pines, A. *Anal. Chem.* 2005, 77, 8109–8114.

8 Organo-Silica Hybrid Monolithic Columns for Liquid Chromatography

Luis A. Colón and Li Li

CONTENTS

8.1 INTRODUCTION

Liquid chromatography (LC), in its broader sense, continues to be an indispensable technique in the analytical laboratory with a remarkable impact on chemical analysis. This is the result of numerous theoretical and technological advances, particularly in the area of high-performance liquid chromatography (HPLC). Column technology is one of the most investigated topics among separation scientists. Consequently, new products in HPLC columns are frequently introduced into the market. This is not surprising since it is within the column that the chromatographic processes take place, making the column the "heart" of the separation system. One aspect in the development of HPLC column technology has focused on new separation media to achieve more efficient and faster separations than those encountered in conventional LC. HPLC separations are typically performed using one of three column formats: open tubes, columns packed with small particles, and very recently, in monolithic

391

columns. For the most part, open tubular columns are fused-silica capillaries in which the stationary phase is attached to the inner walls of the capillary. The fabrication of this column format is quite simple, and very efficient separations are achieved; however, the amount of stationary phase available to achieve separation is low because of the limited amount of surface area available, leading to low mass loadability. Further, open tubular columns are limited to separations in the analytical scale. Columns packed with small beads provide a significantly larger surface area than that found in the open tubular columns, and hence, a larger amount of stationary phase is available. Packed columns exist for analytical and preparative scale purposes. It is very well established that decreasing the particle diameter brings an increase in separation efficiency and resolution. The small packing materials(<2 μm), however, come at the price of an increased column backpressure that is necessary to drive the mobile phase through the column, which in most cases, requires the use of ultrahigh pressure liquid chromatography (UHPLC) [1–6].

Monolithic columns have been introduced as an alternative to particle packed columns and consist of a chromatographic support material manufactured as a continuous, rod-like, interconnected skeletal structure with flow-through paths that provides for very high permeable columns. Although the idea had been planted previously, the modern and practical concept of a monolith as the chromatographic column in HPLC was realized in the late 1980s and early 1990s [7–14] and has attracted considerable attention since [10–19]. A historical perspective and development of the monolith was recently discussed by Svec and Huber [20]. The unique properties of a monolith lie in its bimodal pore structure, namely the micrometer-sized through-pores and the nanometer-sized mesopores. The size of the skeleton and through pores can be controlled separately. A higher permeability and shorter diffusion path lengths can be obtained with a large through-pore/skeleton size ratio in a range of 1–4, much greater than the 0.25–0.4 found for conventional columns packed with particles [21,22]. The large through-pore/skeleton size ratio is responsible for the significant reduction of the pressure requirements to drive the mobile phase through the column. In addition, the short diffusion path lengths provides for a decrease in the C-term of the van Deemter equation at high linear velocities. This provides a route for fast separations while maintaining chromatographic performance for both small and large molecules [14,18,23–32]. The reduced inlet pressures have also allowed the possibility of exploring the use of peristaltic pumps to push the mobile phase through the chromatographic bed for simple and portable HPLC system [33,34]. The monolithic structure can also simplify column technology since a confine of a mold eliminates difficulties encountered with frit fabrication in packed capillary columns for CEC and capillary liquid chromatography (CLC) [35,36]. The absence of retaining frits in CEC, in particular, eliminates a host of problems that have been associated with the frits in packed capillary columns [e.g., nonspecific adsorption, bubble formation, and electro-osmotic flow (EOF) discontinuity in CEC]. Another characteristic of the monolith is that properties such as porosity, surface area, and functionality can be tailored, which can directly affect the flow rate and chromatographic efficiency of the system [37,38]. In principle, a consistent fabrication of monolithic columns should provide a very reliable means of producing HPLC columns since they are prepared by a chemical rather than the mechanical process used for packed particle columns.

Monoliths have been prepared in both small (50–500 μm) and large (3–4.6 mm) inner diameter columns, as well as in other attractive formats that include traditional SPE and 96-well disk arrays [39–45].

The two most common support materials used for monolithic columns are based on organic polymers and silica [7–19,20–34,38,46,47]. The silica-based columns, however, are the most widely used because of their mechanical stability in organic solvents. These columns are typically prepared through well-known sol-gel chemistry starting with a silane precursor (e.g., tetramethoxysilane, TMOS) [13,15,18,19,23] and once the inorganic silica network is formed, its surface is then modified by conventional silane chemistry to impart a desired chromatographic selectivity. For reasons that will become clear later, there are situations in which the fabrication of the silica monolith involves the addition of more than one silane precursor. The additional precursor contains at least one organic moiety attached to the silicon center (e.g., methyltrimethoxysilane), which does not participate in the hydrolysis/condensation reactions that lead to the silica monolith. The structure formed is then known as an organo-silica hybrid, in which the inorganic silica component provides mechanical and structural stability while the organic feature provides bulk modification for the specific application. The term organically modified silica (ORMOSIL) is frequently found in the literature dealing with silica sol-gel chemistry, referring to the organo-silica hybrids just mentioned. In chromatography, ORMOSIL most frequently refer to silica materials that have been modified by grafting an organic moiety on the surface of already synthesize inorganic silica. To avoid any confusion, we will not use the term ORMOSIL here, rather we will use organo-silica hybrids or just organo-silica to refer to the material that is synthesized from precursors that lead to the silica material containing an organic functionality, instead of grafting the functional group on the inorganic silica surface. Organo-silica hybrids have been implemented in chromatography as thin films for open tubular LC or CEC and GC [48–53] and as particulates [54–59]. The use of organo-silica materials in the monolithic format had not attracted considerable attention until recently. Therefore, we present herein recent advances in the area of organo-silica hybrid monolithic columns. This overview is not exhaustive or inclusive; rather, it provides a flavor of the latest progress in a fast-growing subfield of monolithic column technology for LC.

8.2 INORGANIC-SILICA MONOLITHS

It is important to briefly mention the process involved in the preparation of the inorganic-silica monolith (i.e., nonhybrid) before considering the organo-silica counterpart. This will establish the framework that leads to the synthesis of the organo-silica hybrid monolith for LC. The most common approach to fabricate silica-based monolithic columns is by sol-gel processing, which can occur under mild conditions (often at ambient condition). The sol-gel processing used to fabricate silica-based monoliths consists of hydrolysis and condensation reactions of alkoxysilanes under a given set of experimental conditions [13,15,18,23,60,61]; TMOS and/or tetraethoxysilane (TEOS) are the most common. The chemical reactions involved are schematically shown in Figure 8.1. The hydrolysis and condensation are performed

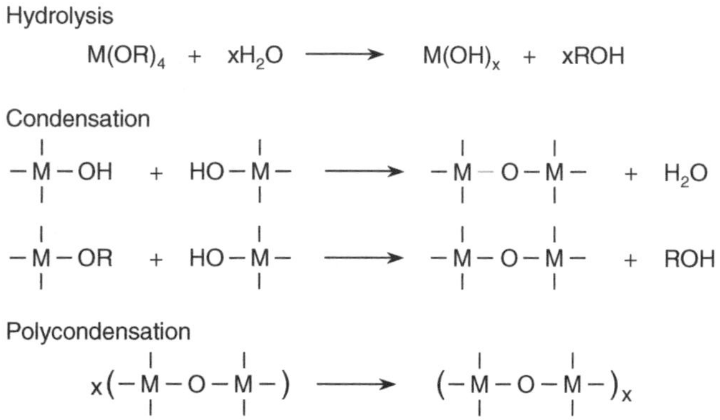

Hydrolysis

$$M(OR)_4 + xH_2O \longrightarrow M(OH)_x + xROH$$

Condensation

$$-\overset{|}{\underset{|}{M}}-OH + HO-\overset{|}{\underset{|}{M}}- \longrightarrow -\overset{|}{\underset{|}{M}}-O-\overset{|}{\underset{|}{M}}- + H_2O$$

$$-\overset{|}{\underset{|}{M}}-OR + HO-\overset{|}{\underset{|}{M}}- \longrightarrow -\overset{|}{\underset{|}{M}}-O-\overset{|}{\underset{|}{M}}- + ROH$$

Polycondensation

$$x\left(-\overset{|}{\underset{|}{M}}-O-\overset{|}{\underset{|}{M}}-\right) \longrightarrow \left(-\overset{|}{\underset{|}{M}}-O-\overset{|}{\underset{|}{M}}-\right)_x$$

FIGURE 8.1   General scheme of reactions involved in the sol-gel processing.

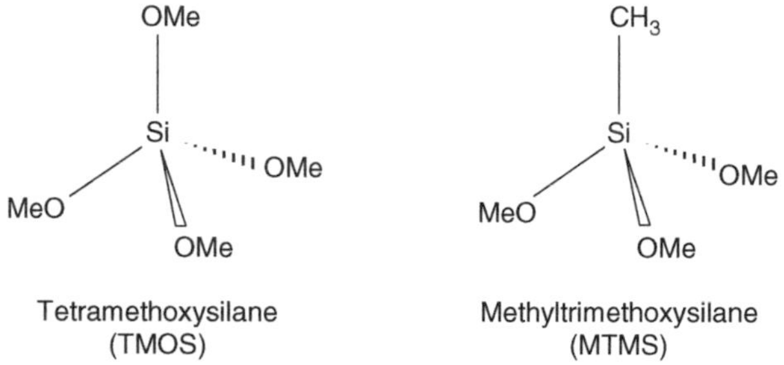

Tetramethoxysilane
(TMOS)

Methyltrimethoxysilane
(MTMS)

FIGURE 8.2   Structure of two silane precursors used to synthesize organo-silica monoliths.

under acidic, basic, or two-step catalysis, initially forming hydroxyl derivatives that consequently undergo condensation liberating alcohol and/or water while Si–O–Si linkages are produced. Polycondensation occurs with multiple linkages to form cyclic oligomers, which eventually lead to the three-dimensional silicate network. To fabricate a porous monolith for chromatographic applications, most typically the silane precursor TMOS (see Figure 8.2) is reacted with water under acidic conditions in the presence of a porogen (e.g., polyethylene glycol, PEG; or polyethylene oxide, PEO). The porogenic agent is used to induce phase separation between the silica- and the water-rich phase, which leads to the porous structure. For example, the water-soluble polymer PEG works as a through-pore template and solubilizer of the silane reagent [15]. The glycol forms strong hydrogen-bonded PEG–silica oligomers, thereby decreasing the solubility of oligomers in solution. The size of the through pore can be controlled by adjusting the PEG/silica ratio. The formation of the through pores in the silica sol-gel system is a competitive process involving phase separation and the sol-gel transition. When phase separation and gel formation occur competitively, various transient structures are permanently frozen in the network. After gelation, the initially fluid reaction system turns into a continuous solid gel phase, with fluid filling the interstices of the gel domains. Open pores are left behind when the fluid is removed [15]. The overall kinetics of sol-gel process depends strongly on a number of factors: the amount of water, the kind of alkoxy groups, the catalyst and its concentration, additives, and reaction temperature. It must be realized that the reaction mixture is submitted to well-controlled heating routines to promote condensation and evaporation of the solvent, leading to the monolithic structure. More specific details on the process can be found in the literature [15,62–64].

A uniform porous silica monolithic column with surface modification suitable for reversed-phase LC was reported in 1996 by Tanaka and coworkers [13]. Since then, considerable investigations have been pursued in an effort to use silica-based monolithic columns in LC and CEC. Presently, monolithic silica columns are commercialized either in a tube or in a fused-silica capillary. Most research laboratories investigating monolithic columns fabricate their own columns, mainly in the capillary format. In many cases, the final step in preparing a silica monolith involves the treatment at very high temperatures (>300°C), which eliminates solvent, unreacted reagents, and most of the organic matter present. Once the inorganic silica monolith is formed, conventional silane chemistry is used to modify the surface of the material to impart the desired selectivity of a stationary phase. In principle, all of the established technologies on surface modification for conventional silica particles can be transferred to the monolithic silica structure. For instance, a toluene solution of octadecyldimethyl-N,N-diethylaminosilane can be introduced into the monolithic structure to produce a C_{18} phase [13,65,66]. For large columns (i.e., 3–4 mm ID), the rod-like monolith is produced individually in a mold, the monolithic structure is then clad in an appropriate tube size. For capillary columns, the monolith is prepared *in situ* and the silica skeleton is covalently attached to the capillary walls, then surface modification can take place. It must be realized that each monolithic column is prepared individually, one-by-one, which can lead to problems with reproducibility; therefore, strict adherence to preparation protocols must be followed during column fabrication.

One drawback of using conventional silane chemistry to bond the stationary phase to the monolithic structure is the limited usable pH range, which in general constrains the applicability of the current phases to the separation of solutes within a pH range of approximately 2–8 [67–69]; although it has been stated that the actual long-term stability is limited to a pH range between 3 and 7 [70]. The instability of the bonded phases at low pH results from the fact that the siloxane bond, by which the phase is attached to the surface of the support material, is prone to nucleophilic cleavage; therefore, the stationary phases become hydrolytically unstable as the pH is lowered. An increase in temperature also leads to the loss of bonded phase [70–72]. At high pH values, on the other hand, the silica dissolves owing to siloxane backbone hydrolysis and this occurs very quickly beyond temperatures of 40°C [73]. Even within the pH range between 3 and 7, an increase in operating temperature increases hydrolysis and solubility of the stationary phase. The stability improvements typically used in conventional HPLC packing materials (i.e., silica beads) can potentially be implemented in monolithic structures, such as the use of sterically protected linkages [74,75], using bidentate ligands [76], synthesizing horizontally polymerized stationary phases [77,78], and bonding the stationary phase through hydride surface modification [79–81]. These approaches, however, have not been explored in detail for silica monolithic columns. The use of hybrid organic–inorganic silica packing support materials, on the other

hand, has shown increased pH stability [48,54–57,59,82]. It becomes clear, therefore, that the route of synthesizing organo-silica hybrids would increase pH stability
of silica-based monolithic columns.

It is well known that during the preparation of the monolithic structure, the silica
network undergoes shrinkage, typically to about 70% of the mold size [13,15,65,66],
which requires the preparation of the monolith in a mold and then clad it in a column
format. In the capillary column, shrinkage can lead to the formation of voids around
the internal surface of the capillary, which results in poor separation efficiency. To
some extent, the difficulties with the *in situ* preparation of monolithic silica capillary
columns beyond 100 μm ID can be attributed to the shrinkage of the silica skeletons
[22,44,83]. An organo-silica hybrid monolithic structure shows less shrinkage during
fabrication than does the pure inorganic silica counterpart [22,44,62,83–85]. From the
fabrication stand-point, therefore, it is also advantageous to synthesize organo-silica
hybrid monoliths.

8.3   ORGANO-SILICA MONOLITHS

Efforts toward the fabrication of more hydrolytically stable and/or to minimize the
shrinking of the silica monolithic material have been pursued by fabricating organo-
silica monoliths. The approach also takes advantage of the sol-gel processing but
using a mixture of silane precursors instead of one. In the mixture of two silanes, for
example, one of them has four functional alkoxides (e.g., TMOS), while the second
contains at most three functional alkoxides (e.g., methyltrimethoxisilane, MTMS);
the nonalkoxide functional group(s) in the second precursor is an organic moiety of
interest, typically introduced for a particular purpose. Figure 8.2 shows the structures of two alkoxysilanes precursors that can be used to fabricate an organo-silica
monolith. The alkoxy moieties can undergo hydrolysis and condensation to form the
silica matrix by sol-gel processing, while the organic one does not participate in
the reactions and provides for a bulk modification of the material. In this fashion,
the organo functional group of interest is introduced by, for example, a trialkoxide,
while the tetralkoxide provides for the silica backbone of the network. The organic
component of the structure is attached by silica–carbon (Si-C) linkage. This attachment is more hydrolytically stable than the monomeric siloxane bonding typically
used in attaching chromatographic stationary phases onto silica surfaces [79,86,87].
A portion of the organic moiety is expected to be buried within the bulk of the
material but another portion is accessible at the surface and available to participate in chromatographic interactions, or it can be modified further if appropriately
chosen.

In a tetralkoxysilane/alkyltrialkoxysilane system, the hydrolysis and condensation reactions proceed not only between similar precursor molecules but also between
the two different precursor species. Figure 8.3 illustrates a simplified scheme of
hydrolysis and condensation reactions in a hybrid system containing TMOS and an
alkyltrimethoxysilane. Notice the difference from the scheme in Figure 8.1; the alkyl
group, represented by R, does not participate in the hydrolysis and condensation reactions that lead to the silica monolith. The sol-gel process kinetics in a hybrid sol-gel

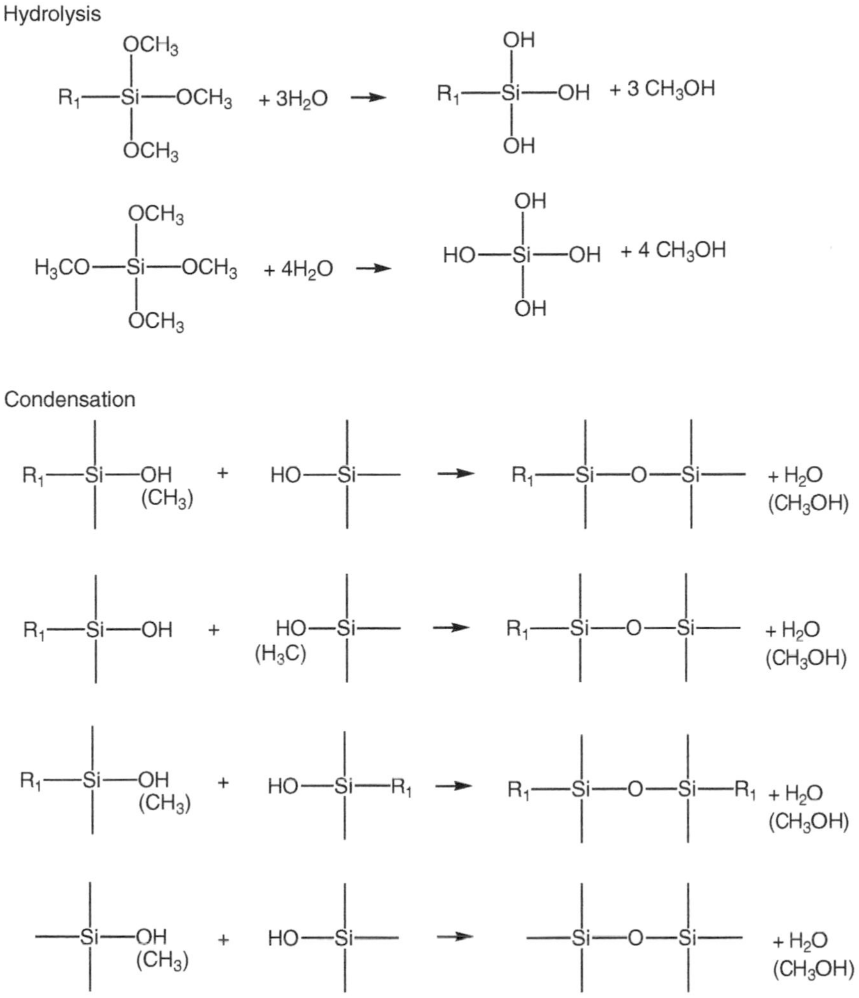

Hydrolysis

Condensation

FIGURE 8.3 Simplified hydrolysis and condensation reactions of TMOS with alkyl-trialkoxysilane during sol-gel processing.

system are different than in a system with a single precursor [52,88]. For example, the rate of hydrolysis for TEOS and its alkyl-monosubstituted alkoxysilanes under acidic conditions follows the trend C_1-TEOS > C_2-TEOS > TEOS > C_8-TEOS [52]. Steric effects of the octyl group in C_8-TEOS lead to a slower rate of hydrolysis than TEOS alone. In a hybrid system, on the other hand, the concomitant condensation reactions appear to influence the hydrolysis of the tetralkoxide. The hydrolyzed species of TEOS appear later and the hydrolyzed species of the monosubstituted alkoxide appear earlier in the hybrid system than if they were reacted by themselves [52]. It is clear that the hybrid system is very complex; but in general, such a system is dependent on the properties of the generated oligomers (e.g., hydrophobicity and

TABLE 8.1
Various Silane Precursors Used to Synthesize
Organo-Silica Hybrid Monolithic Columns

Silane System	Reference
TMOS, C_{18}-TMSPAC, PheDMS	96
TMOS, MTMS	83
TEOS, C_8-TEOS	84
TEOS, APTES	97
TEOS, PTES	100
TEOS, C_8-TEOS	101
TMOS, allyl-TrMOS	95
TMOS, MTMS	123–125
DGS, APTES	108,109,128,129
TMOS, C_8-TMOS, APTES	130

Abbreviations: TMOS: tetramethoxysilane, MTMS: methyl-
triethoxysilane, C_{18}-TMSPAC: octadecyldimethyl [3-
(trimethoxysilyl)propyl] ammonium chloride, PheDMS:
phenyldimethylmethoxysilane, TEOS: tetraethoxysilane,
C_8-TEOS: *n*-octyltriethoxysilane, APTES: aminopropyltri-
ethoxysilane, PTES: phenyltriethoxysilane, allyl-TrMOS:
allyltriethoxysilane, DGS: diglycerylsilane, C_8-TMOS:
n-octyltrimethoxysilane.

polarity), which are determined by the type and amount of the inherent alkyl group used in the alkyltriethoxysilane. A controlled gelation of alkyltrialkoxysilane is much more difficult than TMOS and/or TEOS owing to their extensive cyclization under acid condition [87], premature phase separation over a broad pH range [89], and fewer functional groups for cross-linking. Considering the hybrid system TEOS and its C_8-monosubstituted alkoxysilanes, for example, the highest degree of condensation is obtained when the unsubstituted alkoxysilane is reacted alone compared to the hybrid reaction [52,88]. The hybrid systems also have the ability to self-induced phase separation as a result of immiscibility between the growing oligomers and the solvent. As a result, the morphology of the monolithic matrix largely depends on the composition and type of alkylalkoxysilane used; this is in addition to the other parameters involved in the processing.

8.3.1 Monolithic Structure and the Second Precursor

In principle, many organosilanes can be used as a second precursor to form the organo-silica monolith. Various coprecursors reported for the fabrication of hybrid organo-silica monolithic columns for LC or CEC are listed in Table 8.1. The structure of the organo-silica material is affected by the ratio of the tetralkoxysilane to the alkyltrialkoxysilane in the sol-gel reactions, the type of precursor used, and the processing parameters. This can be seen, for example, in the phase separation behavior of a gelling system containing a mixture of TMOS and vinyltrimethoxysilane (VTMS) [90], where various molar ratios of TMOS to VTMS have been considered (i.e., TMOS:VTMS = 0.2:0.8, 0.5:0.5, 0.8:0.2, and 0:1.0) under acidic conditions. In this system, the hydrolysis was catalyzed by 1 M aqueous HNO_3 and formamide was added as a polar solvent promoting the phase separation and accelerating gelation; all reagents were stirred at $0°C$ before pouring them in a polystyrene vessel for gelation at $40°C$ and then drying at $60°C$. Up to the molar ratio of 0.5 TMOS:0.5 VTMS, the phase separation remained almost unchanged from that of a system composed of TMOS only, which exhibited morphology with well-defined co-continuous micron-sized pores in a very limited concentration region (around the mole ratio of $TMOS:H_2O$ = 1.0:1.5). A graphical representation is provided in Figure 8.4, which illustrates the relationship between the starting composition of the TMOS:VTMS ratio and the resultant macroporous (i.e., micron size) morphology for the cases of TMOS:VTMS molar rations of 0:1 and 0.2:0.8. In a system composed of purely VTMS, owing to the hydrophobicity of vinyl group, gels with co-continuous skeletons and macropores were observed in a narrow composition range. Only 20% incorporation of TMOS results in a broader composition range of the co-continuous macroporous morphology. This is because of the more favorable compatibility with the solvent mixture by incorporating TMOS [64]. As a result, the phase-separated morphology with more coarse structure is frozen in the gelling network at high water concentration, increasing the phase separation tendency [90]. Very similar modification of the phase separating tendency can be performed in the allyltrimethoxysilane system by incorporating 30% TMOS [64].

A silica hybrid macroporous monolith derived from TEOS and n-octyltriethoxysilane (C_8-TEOS) has been proposed by Constantin and Freitag [84] using a two-step catalytic processing. First, the hydrolysis of the silane precursors in a molar ration of 3:1 (TEOS:C_8-TEOS) are hydrolyzed in the solvent (i.e., ethanol or ethanol/N-methylformamide mixtures), catalyzed by 0.1 M HCl/0.001 M KF in the solution; the solution is heated to $80°C$ for 1 h under permanent agitation (1000 rpm) and then cooled to room temperature. The gelation is promoted in a second step using 1% (v/v) N,N-diethylamine as a basic catalyst, then drying at room temperature first, followed by drying in a vacuum oven at $35°C$, 20 mbar. Under such conditions, the morphology of the monolithic structure is mainly determined by the amount of water in the initial reaction system; this is illustrated in images shown in Figure 8.5. A desired, co-continuous, porous structure (similar to the co-continuous structure in Figure 8.4) was observed with an optimum water content of approximately twice the stoichiometric quantity required for hydrolysis (i.e., 200%).

Despite of which precursor is used in the organo-silica hybrid, it appears that each organo-silica hybrid system is completely different from each other. One single set of experimental processing conditions does not necessarily apply to the fabrication of two different systems; therefore, it is difficult to have one single recipe to be applicable to the fabrication of organo-silica monoliths in which the only change would be the alkyl substitute silane. Each particular organo-silica system must be optimized individually in order to obtain desired structures and characteristics that are suitable for LC; this can be a time-consuming task that must be outweighed by the potential benefits of the hybrid system.

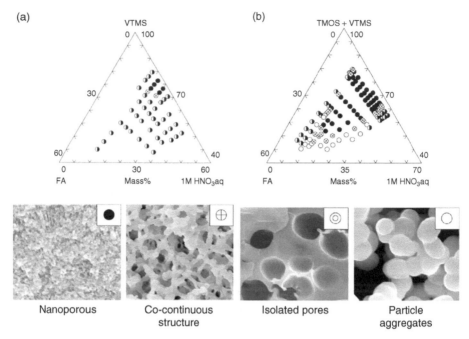

Nanoporous   Co-continuous   Isolated pores   Particle
structure   aggregates

FIGURE 8.4 Relationship between starting material composition and resultant gel morphology for the hybrid system (a) TMOS:VTMS = 0:1.0 and (b) TMOS:VTMS = 0.2:0.8. Closed circles are for nanoporous; crossed circles are for co-continuous structure; open circles are for particle aggregates; double circles are for isolated pores; half-open circles are for macroscopic double phase. The typical resulting morphologies are shown in the SEM images. FA stands for formamide. (Adapted from K. Nakanishi, K. Kanamori, *J. Mater. Chem.*, 15: 3776 (2005), Copyright Royal Society of Chemistry 2005. With permission; A. Itagaki et al., *J. Sol-Gel Sci. Technol.*, 26: 153 (2003). Copyright Kluwer Academic Publishers 2003. With permission).

8.3.2 STATIONARY PHASE AND SILICA BACKBONE IN ONE "POT" REACTION

One of the advantages of organic–inorganic hybrid silica materials is that the physical and/or chemical characteristics can be manipulated by varying the combination of coprecursors in their synthesis. In principle, the introduction of the organic moiety during the fabrication of the silica material can eliminate the postsynthetic grafting procedure used to bind organic groups at the surface of the material to impart a desired surface functionality. An alkyltrialkoxide-derived material, and generally cured more slowly than the tetraalkoxysilane counterparts, depending on the specific kinetics at the pH used for the hydrolysis/condensation sequences [85,91–93]. As a result, the nonhydrolyzed group on the trifunctional silane is preferentially oriented toward the surface of pores and microchannels in the material [85,94,95], altering the properties of the silica surface (e.g., hydrophobicity, hydrophilicity, and binding to guest molecules). If well controlled, the bulk properties (e.g., morphology, mechanical strength) of a monolithic material can be similar to those of inorganic silica. This

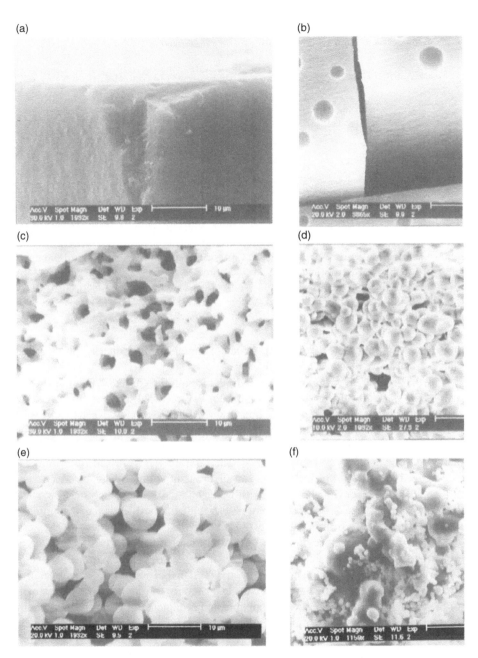

FIGURE 8.5   SEM of xerogels prepared with different amounts of water (Hr). Hr = 100% represents the stiochiometric amount of water required to completely hydrolyze the silanes in the sol-gel reactions. (a) Hr = 100% (magnification × 1932); (b) Hr = 150% (magnification × 3865, the holes at the top surface of the gel result from the entrapment of air bubbles between the gel and its mold); (c) Hr = 200% (magnification × 1932); (d) Hr = 250% (magnification × 1932); (e) Hr = 300% (magnification × 1932); and (f) Hr = 350% (magnification × 1159). (Reprinted from S. Constantin, R. Freitag, *J. Sol-Gel Sci. Technol.*, 28: 71 (2003). With permission. Copyright Kluwer Academic Publishers 2003.)

(a)   (b)

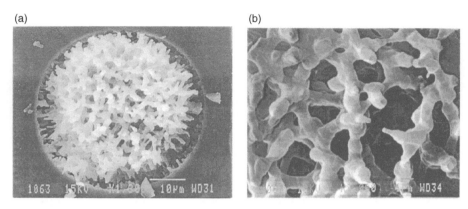

FIGURE 8.6 SEM of an organo-silica hybrid monolith prepared with the precursor N-octadecyldimethyl[3-(trimethoxysilyl)propyl] ammonium chloride to provide reversed phase stationary phase and anodic EOF. (a) Cross-sectional view (magnification 1800 times) and (b) longitudinal view (magnification 7000 times). (Reprinted from J. D. Hayes, A. Malik, *Anal. Chem.*, 72: 4090 (2000). With permission. Copyright American Chemical Society 2000.)

approach would produce a monolithic structure containing the silica support material and the stationary phase in one single step.

In one of the first organo-silica hybrid monolithic capillary columns reported, Malik's group [96] prepared a column for reversed-phase column with anodic EOF for CEC. They fabricated the hybrid monolith by mixing $100\,\mu L$ of TMOS with $100\,\mu L$ of N-octadecyldimethyl[3-(trimethoxysilyl)propyl] ammonium chloride (C_{18}-TMSPAC), $10\,\mu L$ of phenyldimethylsilane (PheDMS), and $100\,\mu L$ of 99% trifluoroacetic acid (TFA) as the catalyst. The mixture was vortexed for 5 min, and centrifuged at 13,000 rpm for 5 min. The supernatant solution was introduced into the capillary tube previously rinsed with water and dried under helium flow while heating from 40°C to 250°C. The column was sealed at both ends and submitted to thermal conditioning from 35°C to 150°C at a rate of 0.2°C/min. The C_{18}-TMS precursor provided the octadecyl moiety for the stationary phase; the positively charged quaternary ammonium moiety provided a positive surface charge to generate anodic EOF. The PheDMS was assumed to be incorporated in the monolithic structure during the sol-gel process reacting with residual silanol groups during the thermal treatment of the column. The monolithic structure obtained is depicted in Figure 8.6a,b. Examination of the structure by scanning electron microscopy (SEM), indicate an average skeleton size of $\sim 1.5\,\mu m$ with macropore sizes between 2 and $4\,\mu m$. The chromatographic performance of this column was evaluated by the separation of a mixture of polycyclic aromatic hydrocarbons (PAHs), mixture of benzene derivatives, and a mixture of aldehydes and ketones (see Figure 8.7). The column efficiencies were reported to be on the order of 150,000–175,000 plates/m.

Anodic EOF for CEC has also been obtained on a silica hybrid synthesized with the two coprecursors TEOS and 3-aminopropyltriethoxysilane (APTES) [97]. Here, no catalyst was added to accelerate the hydrolysis and condensation reactions;

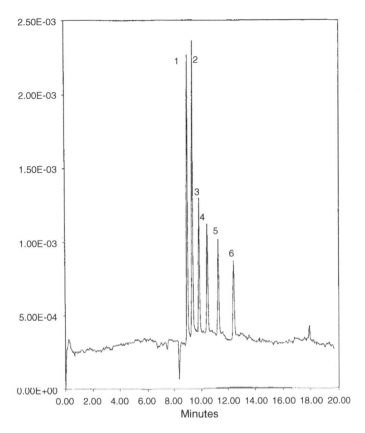

FIGURE 8.7 Separation of aldehydes and ketones by CEC in the organo-silica monolith shown in Figure 8.6. Column: 50 cm × 50 μm ID (46 cm effective length), separation at −25 kV; mobile phase: 70% acetonitrile-30% 5 mM Tris-HCL (pH = 2.3); analytes: (1) benzaldehyde, (2) o-tolualdehyde, (3) butyrophenone, (4) valerophenone, (5) hexaphenone, and (6) heptaphenone. (Reprinted from J. D. Hayes, A. Malik, *Anal. Chem.*, 72: 4090 (2000). With permission. Copyright American Chemical Society 2000.)

instead, the basic amino groups of APTES served as an "internal catalyst" for silica condensation [98]. The monolith was prepared by mixing 112 μL TEOS, 118 μL APTES, 215 μL ethanol, 32 μL water, and 8 mg cetyltrimethyl ammonium bromide (CTAB), added as mesopore template, in a 1.5 mL eppendorf vial. After vortexing for 30 s, the mixture was introduced into a capillary, sealed, and reacted at room temperature for 24 h; the monolithic structure was then rinsed with ethanol and then with water. The CTAB is removed by extraction. When the TEOS and APTES monomers are added in the system just mentioned, gels form within a few minutes owing to the basic properties of the functional group in the APTES monomer. Hydrogen bonds between the amino groups of APTES and silanol groups of hydrolyzed species possibly play a major role in the build-up of the gel network [97,99]; a structure of the aminated organo-silica monoliths structure is presented in Figure 8.8a, which is very different from the co-continuous structure shown in the organo-silica

(a) (b)

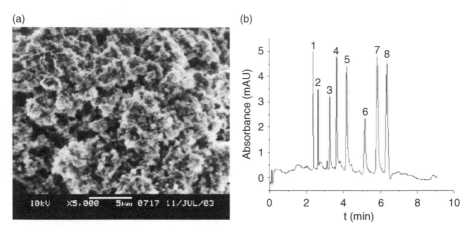

FIGURE 8.8 (a) SEM image of organo-silica monolith prepared with the APTES and TEOS; CTAB used as mesopore templating. (b) Separation of organic acids in the aminated monolith by CEC. Mobile phase: 10 mM citrate with 40% acetonitrile; separation at 10 kV; analytes: (1) *p*-toluenesulfonic acid, (2) *p*-aminobenzosulfonic acid (3) *p*-nitrobenzoic acid, (4) *o*-chlorobenzoic acid, (5) *m*-bromobenzoic acid, (6) benzoic acid, (7) α-naphthyl acetic acid, and (8) *p*-hydroxybenzoic acid. (Reprinted from L. Yan, Q. Zhang, J. Zhang, L. Zhang, T. Li, Y. Feng, L. Zhang, W. Zhang, Y. Zhang, *J. Chromatogr. A*, 1046: 255 (2004). With permission. Copyright Elsevier 2003.)

materials shown in Figures 8.4 and 8.5c. The average mesopore diameter of the monolithic material was 14.2 nm in the 1:1 TEOS:APTES molar ratio, and a surface area of 105.3 m^2/g. The column showed to be effective in the separation of organic acids (see Figure 8.8b) with column efficiency up to 267,000 plates/m and was used to separate triterpenoids in Ganoderma Licidum (used in Chinese medicine).

Other organo-silica monoliths can also be synthesized combining TEOS and silane precursors containing a moiety that can act as the stationary phase for reversed phase LC. Organo-silica monoliths having phenyl or octyl functionality have been synthesized by reacting TEOS with phenyltriethoxysilane (PTES) [100] or C_8-TEOS [101], using a two-step catalytic approach. Initially, the silanes are hydrolyzed in a water/methanol solution using 0.5 M HCl as a catalyst and reacting at 60°C for 4 h. In a second step, dodecylamine is added to the solution as a second catalyst, increasing the pH to accelerate the condensation reactions, and to reduce the gelation time. The dodecylamine is also claimed to act as a supramolecular template to form desired pores in the monolithic matrix [100,101]. Increasing the alkyltriethoxysilane/TEOS ratio in the starting composition results in a larger skeleton size and larger through pores owing to the accelerated phase separation resulting from the addition of the alkyltriethoxysilane. The TEOS/PTES volume ratio of 1.3 has produced materials with a median pore diameter around 5.0 μm, pore volume of 2.9 cm^3/g, mesopores size around 4 nm, and a surface area of 394 m^2/g [100]. On the other hand, an organo-silica monolith prepared with the TEOS/C_8–TEOS volume ratio of 1.80 resulted in a material with a median pore diameter of 1.06 μm

TABLE 8.2
TEOS/C$_8$-TEOS Volume Ratios and Properties of the Resultant Organo-Silica Monoliths[a]

TEOS (µL)	C$_8$-TEOS (µL)	TEOS/C$_8$-TEOS volume ratio	Median pore diameter (µm)	Pore volume (cm^3/g)
88	50	1.76	2.58	3.12
90	50	1.80	1.06	3.25
92	50	1.84	0.70	3.41

[a] Columns were prepared as follows: 180 µL of methanol, 10 µL H$_2$O, 10 µL of 0.5 M HCl, 90 µL of TEOS, and 50 µL of C$_8$-TEOS were mixed and stirred for 3 min, and hydrolyzed at 60°C for 6 h. The mixture was cooled to room before adding 5 mg of dodecylamine. After filling a capillary column, it was allowed to react further for 12 h at 40°C, then rinsed with ethanol and finally dried at 60°C for 48 h.

Source: Reprinted from L.-J. Yan et al., *J. Chromatogr. A,* 1121: 92 (2006). With permission. Copyright Elsevier 2006.

and a pore volume of 3.25 cm^3/g [101]. Small variations in the water concentration used in the sol-gel reaction results in a significant change of the through-pores, as can be seen in Table 8.2 for the TEOS/C$_8$-TEOS system. Both columns have been tested for CEC showing efficiencies of 180,000 plates/m with satisfactory reproducibility on the separation of alkylbenzenes, basic compounds, and PAHs (see Figure 8.9).

8.3.3 ORGANO-SILICA PLATFORM WITH DERIVATIZABLE FUNCTIONALITY

Various organosiloxanes could be used to produce organo-silica monolithic structures containing the stationary phase in one single step. It is clear, however, that the synthetic conditions for the silica-hybrid monolithic structures containing different functionalities must be optimized independently. This can impose difficulties since the addition of the coprecursor can affect the integrity and characteristics of the monolithic structure and at the same time may compromise the amount of stationary phase that can be added for chromatographic interaction. It is extremely difficult to optimize stationary phase content and the appropriate structural properties of the monolith structure in one "pot" reaction synthetic approach. Ideally, one would like to have an optimized synthetic methodology to obtain the hybrid monolithic structure with a given set of desired physical characteristics and, on the other hand, an effective means to provide the stationary phase. This can be achieved by decoupling the hybrid structure formation from the attachment of the stationary phase, an approach used with organo-silica particulates [54–59]. Further, if the organo-silica monolith contains a derivatizable surface functionality incorporated through the bulk modification, it would be attached at the surface through a silica-carbon bond, which provides for hydrolytic stability. Then, chromatographic selectivity can be varied

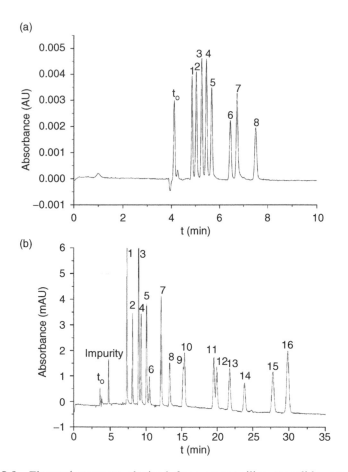

FIGURE 8.9 Electropherograms obtained from organo-silica monoliths prepared with (a) TEOS/PTES volume ratio of 1.3 and (b) TEOS/C$_8$-TEOS volume ratio of 1.80. In (a) anilines are separated at 13 kV and in (b) 16 priority pollutant PAHs are separated. Column: 75 mm ID × 27 cm (20 cm effective length); mobile phase, 5 mM Tris-HCl buffer (pH 8.2) containing 70% v/v acetonitrile. Solutes in (a): (1) acetanilide, (2) aniline, (3) o-toluidine, (4) 3,4-dimethylaniline, (5) 1-naphthylamine, (6) N,N-dimethylaniline, (7) diphenylamine, and (8) N,N-diethylaniline. Solutes in (b): (1) naphthalene, (2) acenaphthylene, (3) fluorine, (4) acenaphthene, (5) phenanthrene, (6) anthracene, (7) fluoranthene, (8) pyrene, (9) benz[a]anthracene, (10) chrysene, (11) benzo[b]fluoranthene, (12) benzo[k]fluoranthene, (13) benzo[a]pyrene, (14) dibenz[a,h]anthracene, (15) indeno[1,2,3-cd]pyrene, and (16) benzo[g,h,i]perylene. (Reprinted from L. Yan et al., *Electrophoresis*, 26: 2935 (2005). With permission. Copyright Wiley-VCH 2005; L.-J. Yan et al., *J. Chromatogr. A*, 1121: 92 (2006). With permission. Copyright Elsevier 2006.)

by reaction of the surface functionality with a desired moiety that will serve as the stationary phase.

The concept of establishing an optimized platform to construct the organo-silica monolith has been demonstrated using allyl functionality [95], although any other

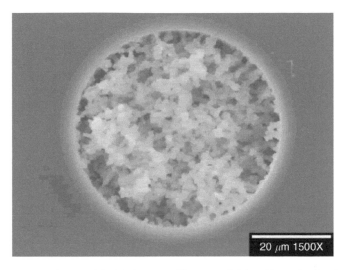

20 μm 1500X

FIGURE 8.10   SEM image of the allylorgano-silica monolith in a 50-μm ID fused-silica capillary column. The monolith was fabricated using allyl-TMS:TMOS (1:4 mol ratio). (Reprinted from H. Colon et al., *Chem. Comm.*, 2826 (2005). With permission. Copyright Royal Society of Chemistry 2005.)

suitable functionality can be used instead. In this synthetic methodology, the allyl hybrid monolithic structure is first synthesized to obtain a given set of physical characteristics (e.g., surface area, porosity, etc.) and then the organo-reactive allyl functionality can be used to bond other desirable moieties. The allyl group is incorporated through the bulk modification process providing the Si-C bonding at the material's surface, while the reactive allyl group provides versatility that allows tailoring of the surface properties of the final material (i.e., chromatographic selectivity). Figure 8.10 shows an SEM image of the hybrid allyl-silica monolith synthesized inside a 50 μm ID fused-silica capillary column. The monolith is fabricated by mixing allyl-trimethoxy silane (allyl-TMS) and TMOS in a molar ratio of 1:4, in a solution containing 1 mL aqueous acetic acid (0.01 M), and 108 mg PEG (10,000 MW) as porogen. Urea (90 mg) is used to promote mesopore uniformity [102]. The solution is stirred for 1 h at 0°C, introduced into the capillary column, and heated at 50°C for 40 h, after sealing the column's ends. The column is then rinsed and reheated at 50°C for 24 h, after which the monolith is heated from 50°C to 150°C at a rate of 1.0°C/min. The monolith is allowed to cool to room temperature and rinsed with ethanol before wash and equilibrate with the mobile phase. This hybrid allyl-silica monolith prepared with an allyl-TMS:TMOS ratio of 1:4, produced through pore sizes around 2 μm, mesopores of about 8 nm, and a surface area of 209 m^2/g. The mesopores can be adjusted by controlling the amount of porogen (i.e., PEG) added to the reaction mixture, as indicated in Table 8.3. About 2.1 mmol quantity of allyl functionality per gram of material was incorporated in the organo-silica hybrid monolith, of which, about 68% was found to be accessible at the surface (i.e., 1.4 mmol/g of material) [95].

TABLE 8.3
Physical Characteristics of Organo-Silica Monolith
as a Function of PEG in the Reaction[a]

Amount of PEG (mg)	Surface area by BET (m^2/g)	Average pore diameter (nm)
108	209	8.4
90	225	6.5
50	372	2.8

[a] An allyl-TMS:TMOS more ratio of 1:4 was used to prepare the organo-monolith. The values for surface area and pore diameter are average of duplicates.
Source: Reprinted from H. Colon et al., *Chem. Comm.*, 2826 (2005). With permission. Copyright Royal Society of Chemistry 2005.

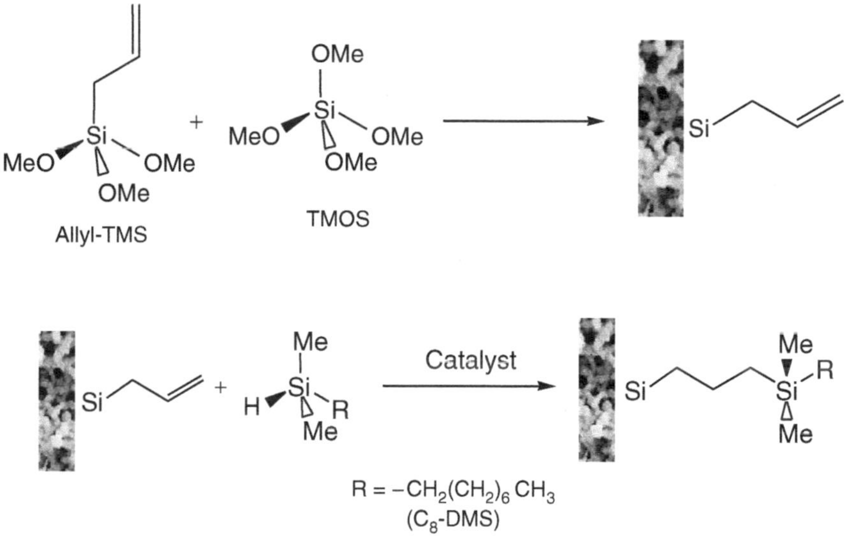

FIGURE 8.11 Overview of hybrid allyl-silica monolith synthesis and allyl and further modification with C$_8$-DMS.

Having all of the physical characteristics of the monolith already established, the allyl functionality of the organo-silica hybrid can be used as a means of introducing other chemistries to the surface. For example, the allyl group can be modified through hydrosilylation, similar to the approach used by Pesek and coworkers [79–81,103,104] to attach terminal olefins to hydride silica surfaces. Figure 8.11 shows a scheme that illustrates the approach by reacting octyldimethylsilane (C$_8$-DMS) with the hybrid allyl-monolith. Using chloroplatinic acid as the catalyst, C$_8$-DMS has been reacted with the monolithic structure. The reaction consists of

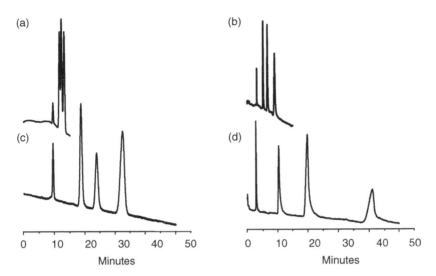

FIGURE 8.12 Separation of model compounds using hybrid monolithic columns. (a) and (c) hybrid allyl-monolith before hydrosilylation reaction, (b) and (d) hybrid allyl-monolith after modification with C_8-DMS. (a) and (b) show separations by capillary LC at 90 psi—mobile phase of acetonitrile/4 mM borate buffer (pH = 9.3) (40/60)—while (c) and (d) show separations by CEC at 25 kV—mobile mobile phase of acetonitrile/4 mM borate buffer (pH = 9.3) (20/80). Solutes are (in order of elution) DMSO, benzene, toluene, and ethylbenzene.

filling the allyl-monolithic column with a solution of isopropanol containing 0.12 mg/mL chloroplatinic acid and 300 mM C_8-DMS and reacting for 24 h at 80°C. It is estimated that 0.44 mmol C_8-DMS is incorporated per gram of monolithic material, which corresponds to a C_8 surface coverage of 2.1 μmol/m^2, under the experimental conditions used. Figure 8.12 shows chromatograms of alkylbenzenes separated via CEC and capillary LC in the allyl-monolith before (a and c) and after (b and d) hydrosilylation with C_8-DMS. It can be noticed that the hydrophobicity of the allyl-monolith is sufficient to be used as a separation media by itself. However, it is also clearly shown that the retention of the model compounds increase significantly after reaction with C_8-DMS. A myriad of possibilities exists, including reaction with other silane species and/or preparing the organo-silica monolith with a different functionality. Nevertheless, it is evident that this is a feasible approach to obtain an organo-silica monolith with desired physical properties and then tune the surface chemistry to obtain chromatographic selectivity.

8.3.4 REDUCTION OF SHRINKAGE AND LARGER BORE COLUMNS (>100 μm ID)

Shrinkage of the monolithic structure has been a considerable challenge when fabricating silica-based monolithic columns. In general, shrinkage is induced by the capillary pressure within the pores of the silica matrix as it is formed and condensation between neighboring silanol or alkoxide groups during the aging and drying processes; this is also called "syneresis" [60]. In many cases, shrinkage causes the monolithic structure to detach from the column walls and or cracking of the monolith, rendering an unusable column. A variety of suggestions have been proposed to reduce shrinkage in sol-gel derived materials, such as removing the unreacted water from the monoliths before gelation, exchanging the solvent in wet gel with low surface tension solvents (e.g., hexane, ethanol), and modifying the surface with trimethyl-chlorosilane (TMCS) [104–106]. Usually, when preparing a silica-based monolith inside of a fused silica capillary column the capillary is pretreated to increase the number of silanol groups at the silica surface. The silanol groups on the walls can be incorporated into the silica network through condensation reactions, thus the silica monolith is anchored onto the walls of the capillary more efficiently. In the majority of cases, the treatment involves rinsing the capillary column with a basic solution (e.g., NaOH) and then water; an acidic rinse is often included also. Other secondary treatments have been postulated to improve the mechanical strength and shrinkage, for example, attaching a film of the monolithic gel to the capillary wall before the formation of the monolith [107]. A secondary treatment using 2% APTES at 110°C has been used claiming that the amino group of APTES provides electrostatic bonding between the anionic silica monolith and the cationically modified capillary surface [108,109]. It has been reported that in such columns can be operated at flow rates as high as 500 μL min^{-1} with no occurrences of monolith detachment from the capillary column [108,109].

The main problem with shrinkage is with the *in situ* fabrication of monolithic columns having an internal diameter greater than 100 μm, particularly when using the single precursor TMOS or TEOS. For practical reasons, it is important to optimize the preparation of silica monoliths in columns having bores greater than 100 μm. Hybrid monolithic structures show less unfavorable shrinkage than their nonhybrid counter parts [22,44,62,83–85]. One possible reason is the steric hindrance offered by the alkyl group introduced in an alkylalkoxysilane, for example, which can hinder the condensation between neighbor silanol or alkoxide groups. In addition, there is more "flexibility" in the hybrid matrix when compared with the TMOS-derived one. This flexibility enables the gel to endure the stress generated by shrinkage. Furthermore, the decreased concentration of the silanol groups in the hybrid sol decreases the extent of syneresis since there is less connectivity.

Using an alkylsilane as a second precursor in combination with TMOS (i.e., forming an organo-silica monolith), the *in situ* formation of monoliths in 530-μm ID capillary columns has been made possible [22,44,62]. Hybrid organo-silica columns with diameters of 100, 200, and 530 μm have been synthesized from mixtures of TMOS and MTMS [22,44,83]. Shrinking of the monolithic structure was less than that observed on silica monoliths prepared from TMOS alone. These larger column diameters provide for more flexible operating conditions. Figure 8.13 illustrates SEM images of organo-silica monoliths prepared from mixtures of TMOS and MTMS in capillaries columns with various inner diameters. The hybrid monolithic columns were prepared from a mixture of TMOS and MTMS in a 3:1 ratio (9 mL volume), PEG (1.05 g), and urea (2.03 g) in 0.01 M acetic acid (20 mL), which had been stir-ring for 45 min at 0°C before filling the capillary columns. The reaction proceeded inside the column overnight at 40°C and then the column was treated at 120°C for

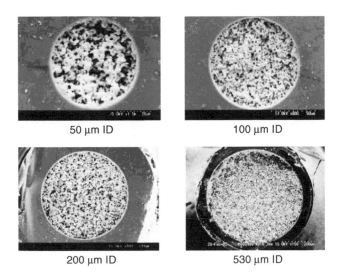

| 50 µm ID | 100 µm ID |
| 200 µm ID | 530 µm ID |

FIGURE 8.13 SEM images of macroporous methylorgano-silica monoliths prepared in capillaries with various inner diameters. (Adapted from K. Nakanishi, *Bull. Chem. Soc. Jpn.*, 79: 673 (2006). With permission. Copyright The Chemical Society of Japan 2006; M. Motokawa et al., *J. Chromatogr. A*, 961: 53 (2002). Copyright Elsevier 1992.)

3 h. The hybrid monolithic columns are then reacted with an octadecyldimethyl-N,N-diethylaminosilane to produce the stationary phase bearing C_{18} groups for reverse phase chromatography. The 500-µm ID organo-silica monolithic columns showed smaller through pores and skeleton sizes, which in turn results in higher column efficiency, as compared to capillary columns of smaller diameters; this however, comes at the expense of permeability [44]. The permeability still showed to be higher (2.5–4 times) than capillary columns packed with 3-µm particles having similar efficiency. The 530-µm ID columns provided good peak shape for gradient elution of proteins at a flow rate up to 100 µL/min. Furthermore, as depicted in Figure 8.14, the larger diameter column also afforded an increase in sample loadability when compared with 200-µm ID monolithic columns, accommodating injection volumes of 10 µL.

8.3.5 ORGANO-SILICA MONOLITHS FOR AFFINITY LIGAND ENCAPSULATION (ENTRAPMENT)

Affinity monolith chromatography makes use of immobilized ligands in the monolithic support to retain species by means of specific and reversible interactions, many of biological origin [110]; a classical example is the interaction of an enzyme immobilized on a monolithic structure and its substrate. Affinity ligands on silica-based monoliths are typically attached to the surface of the monolithic structure by bonding an organo-reactive silane at the surface of inorganic silica that is then linked to the affinity ligand of interest. Another very attractive approach that can be used to immobilize affinity ligands in monolithic structures is through entrapment. The ligand

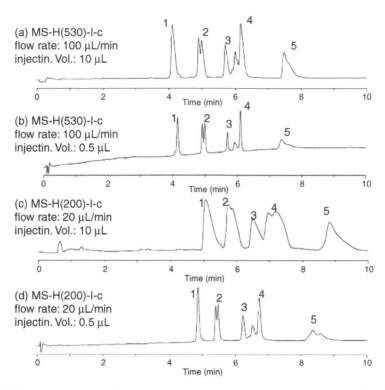

FIGURE 8.14 Separation of a standard protein mixture (1: ribonuclease A, 2: insulin, 3: cytochrome C, 4: lysozyme, 5: BSA), concentration 0.1 mg/mL for each protein) on organo-silica monolithic capillary columns that were further modified with ODS. (a, b) are from 530 μm ID × 50 mm length columns. (c, d) are from 200 μm ID × 50 mm length columns. Mobile phase: (A) acetonitrile containing 0.1% TFA, (B) H$_2$O containing 0.1% TFA. Gradient condition: the linear gradient A 10–50% (0–10 min). (Reprinted from M. Motokawa et al., *J. Sep. Sci.*, 29: 2471 (2006). With permission. Copyright Wiley-VCH 2006.)

is incorporated in the monolithic structure at the time the network matrix is forming. As the silica material polymerizes, it grows around the ligand, thereby entrapping the ligand within the silica matrix. The process of entrapment or encapsulation in the silica matrix is schematically represented in Figure 8.15. Since the sol-gel processing is performed in aqueous conditions, silica entrapment allows the ligand to be in a more compatible solvent, minimizing denaturation of the biological ligand (e.g., protein). The entrapment approach provides for a fabrication route in which the affinity ligand and the support material are formed in a single-step reaction, avoiding chemical attachment that can potentially alter the recognition nature of the ligand. A disadvantage of the process, however, is that during the hydrolysis and condensation reactions, alcohols byproducts can exist in large concentrations in the local environment of the encapsulation, which in turn can denature some desired ligands (e.g., proteins); nonspecific interactions with surface silanols can also affect the entrapped recognition element. This problem is minimized by entrapment in organo-silica hybrid

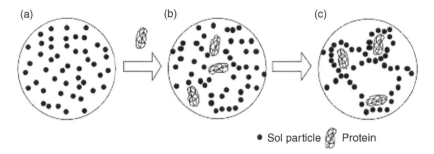

• Sol particle 🔬 Protein

FIGURE 8.15 Protein encapsulation in a silica matrix via sol-gel processing; the process include (a) formation of the sol particles during hydrolysis and initial condensation, (b) addition of the protein into the sol solution, and (c) the growing silica network entraps the protein molecule. (Reprinted from M. Kato et al, *Anal. Chem.*, 74: 1915 (2002). With permission. Copyright American Chemical Society 2002.)

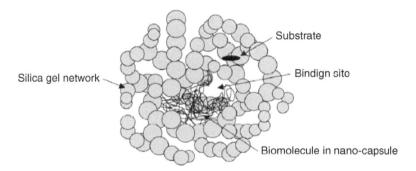

FIGURE 8.16 Representation of a biomolecule nanoencapsulated in the silica network. (Reprinted from M. Kato et al., *J. Sep. Sci.*, 28: 1893 (2005). With permission. Copyright Wiley-VCH 2005.)

monoliths, instead of purely inorganic silica encapsulation, providing for improved biocompatibility.

Protein encapsulation by sol-gel chemistry, in particular, has attracted much attention for the development of sensing platforms, and the support of enzyme bioreactors and bioaffinity chromatographic columns [111–119]. The monolith must contain appropriate pore sizes that are small enough to prevent leaching of the entrapped molecules, yet large enough to allow smaller analytes to enter the porous structure and to provide sufficient space to accommodate any conformational changes required during the affinity binding process. The mesopores of the monolithic structure form a "nanocapsule" that entraps the biomolecule of interests, as illustrated in Figure 8.16 [120]. The organo-silica hybrid monolithic materials can be prepared from a variety of organo silane monomers combine with selected alkoxysilanes to avoid shrinkage, while providing the required structural and chemical characteristics, and preserving biological activity for a prolonged period. To avoid denaturing of the entrapped proteins, a very mild and biocompatible sol-gel processing method must be used.

For example, a two-step processing method can be used in which the hydrolysis is performed under acidic condition and then a buffered solution containing the protein of interest is added to the hydrolyzed silica sol to initiate gelation under conditions that are protein compatible [121].

It is important to realize that the selection of the proper organosilanes is essential in order to preserve protein activity while maintaining proper structural characteristics (e.g., pore size) [33,50–52,85,108,109,122–125]. Using a hybrid system containing the mixture of MTMS or dimethoxydimethylsilane (DMDMS) and TMOS in acidic conditions (0.04 M HCl), organo-silica monoliths have been synthesized to encapsulate proteins to perform enantiomeric separations by CEC [123–125]. For example, an organo-silica monolith containing bovine serum albumin (BSA) has been prepared by hydrolyzing the monomeric solution ($\sim$20 min) and then adding a BSA solution ($\sim$5% w/v) in 50 mM phosphate buffer, before filling the column. After aging the column at room temperature for 3–4 days, enantiometric separations were performed by CEC. The resolution of D,L-tryptophan (Trp) was achieved using BSA-encapsulated columns containing MTMS–TMOS gel, but not with columns using TMOS only. Different composition of the gel matrix can influence the enantioselectivity of the proteins as the physical and chemical properties of the monolithic matrix vary. In the organo-silica monolith prepared from MTMS and TMOS to encapsulate BSA, an increase in the MTMS/TMOS ratio gives rise to an increase in the retention of D,L-Trp, as well as an increased enantioselectivity, as shown in Figure 8.17 [125]. The same organo-silica monolithic system has also been used to encapsulate trypsin and integrated with a column for capillary electrophoresis (CE) to provide protein digestion and separation in one single system [126]. The entrapped trypsin showed an activity toward [Tyr8]-bradykinin that was higher than that of free trypsin and an increased stability under continuous flow use compared to free solution.

Several new silane precursors have been reported by Brennan's group [85,127] designed to reduce shrinkage and to increase biocompatibility when entrapping proteins in sol-gel derived monolithic materials; three are shown in Figure 8.18. Diglycerylsilane (DGS) and APTES, for example, have been used to prepare organo-silica monolithic columns entrapping biomolecules for frontal affinity chromatography [108,109,128,129]. APTES is added to provide cationic sites that counterbalance the anionic charge of the surface, reducing nonselective interactions. A typical column preparation involves the hydrolysis of DGS in aqueous HCL (1 M HCl) for 15–25 min before adding a mixture of PEO, APTES, and the biomolecule in HEPES buffer. The solution is loaded inside the column and aged for at least 5 days at 4°C, leading to a hydrated macroporous structure encapsulating the protein. SEM images of a typical column containing the protein dihydrofolate reductase (DHFR) are shown in Figure 8.19 [108]. The size and proportion of the macropores is highly dependent on the molecular weight of the PEO used; homogeneous and reproducible columns have been found to be formed with an optimal level of 8% w/v for MW 10,000 PEO [108]. The organo-silica monolithic column shows a bimodal pore distribution with pores around 0.5 μm for eluent flow and around 3–5 nm in which the protein is entrapped. The columns are suitable for pressure-driven LC and can be operated at relatively high flow rates with low backpressure. The monolithic column containing

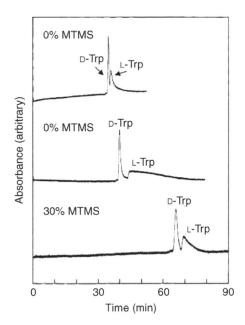

FIGURE 8.17 Separation of D,L-Trp by BSA-encapsulated column prepared using different ratios of MTMS. Conditions: Gel composition: 5% (w/v) BSA in 50 mM phosphate buffer (pH 5); TMOS (containing 0–30% MTMS)-hydrolyzed solution = 5:1. (Reprinted from K. Sakai-Kato, M. Kato et al., *J. Pharm. Biomed. Anal.*, 31: 299 (2003). With permission. Copyright Elsevier 2003.)

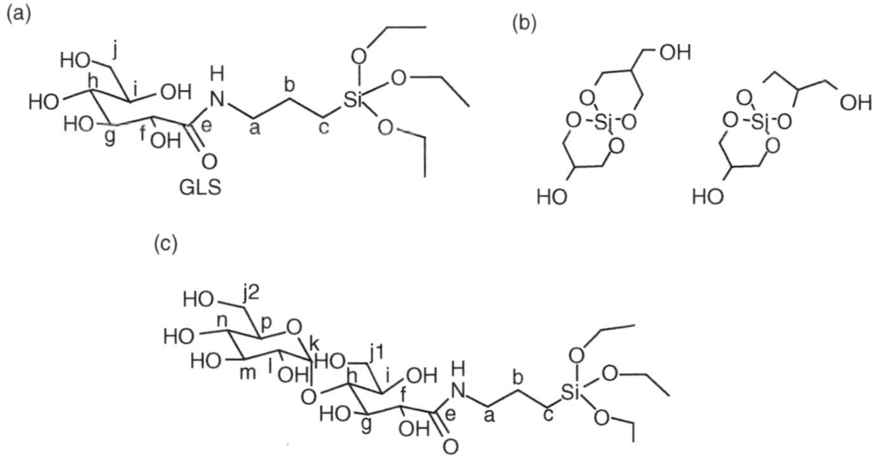

(a)

HO—j
h
HO
HO
g
i—OH
f OH
e
H
N
a
b
c
Si
O
O
O
O
GLS

(b)

OH
O
O-Si-O
O
HO

O
O-Si-O
O
HO
OH

(c)

HO—j2
n
HO
HO
m
p
O
OH
k
HO
h
f
g
j1
i—OH
OH
O
e
H
N
a
b
c
Si
O
O
O

FIGURE 8.18 Chemical structures of silanes precursors used to synthesize organo-silica monolith to entrap biomolecules: (a) gluconamidylsilane (GLS), (b) maltonamidylsilane (MLS), and (c) possible structures of diglycerylsilane (DGS). (Reprinted from Y. Chen et al., *J. Mater. Chem.*, 15: 3132 (2005). Copyright Royal Society of Chemistry 2005; M. A. Brook et al., *J. Sol-Gel Sci. Technol.*, 31: 343 (2004). Copyright Kluwer Academic Publishers 2004. With permission.)

(a) (b)

FIGURE 8.19 SEM images of organo-silica monolith prepared with a solution of DGS/PEO/APTES, after 5 days of aging. (a) Image of monoliths formed in 1-mm capillaries that pulled away from the capillary wall and were removed from the capillary under flow. (b) Image of a monolith in a 250-µm ID capillary column. (Reprinted from R. J. Hodgson et al., *Anal. Chem.*, 76: 2780 (2004). With permission. Copyright American Chemical Society 2004.)

entrapped DHFR has been used to study the binding of various inhibitors of DHFR using frontal affinity chromatography (see Figure 8.20) [108]. Similar columns have also been used in combination with MALDI-MS/MS for high-throughput screening of compound mixtures [127] and for immunoextraction [109].

8.4 FINAL REMARKS

Organo-silica hybrid monolithic columns are a second generation of silica-based monoliths for chromatographic applications. They offer improved characteristics over the inorganic silica-based monoliths. There are, however, challenges with this technology that must be overcome. The most important is the reproducibility of column fabrication, which is a problem with monoliths in general. Organo-silica monolithic columns in the traditional sizes (~4.6 mm ID) are not commercially available. Furthermore, as with the inorganic-based monolith, the fabrication of columns in the 1–2 mm diameter has been elusive as there are several technical difficulties that must be overcome. Nevertheless, the organo-silica approach has demonstrated such a great applicability with particle-packed columns that there is enough momentum to see the monolithic columns moving in that direction; the potential is clear. One area that we anticipate to continue growing relatively fast is in the use of organ-silica monolith for affinity chromatography, particularly with the entrapment approaches. Therefore, we expect to see the use of organo-silica monolithic columns to grow with time.

ACKNOWLEDGMENT

We acknowledge financial support from The National Science Foundation (CHE-0554677), U.S.A.

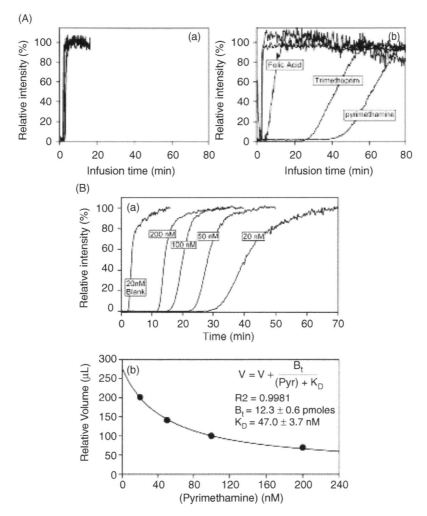

FIGURE 8.20 Binding of various inhibitors of DHFR using frontal affinity chromatography; the column is an organo-silica monolith entrapping DHFR. (**A**) Panel "a" shows control frontal chromatograms with no binding and panel "b" shows retention of the two DHFR inhibitors, trimethoprim and pyrimethamine, and for the weakly binding substrate folic acid. (**B**) Determination of K_d and B_t values for DHFR columns. Panel "a": frontal affinity chromatograms at four different ligand concentrations relative to a blank column; Panel "b": plot of elution volume vs. concentration of pyrimethamine. (Reprinted from R. J. Hodgson et al., *Anal. Chem.*, 76: 2780 (2004). With permission. Copyright American Chemical Society 2004.)

REFERENCES

1. J. E. MacNair, K. C. Lewis, J. W. Jorgenson, *Anal. Chem.*, 69: 983 (1997).
2. J. E. MacNair, K. D. Patel, J. W. Jorgenson, *Anal. Chem.*, 71: 700 (1999).
3. N. Wu, D. C. Collins, J. A. Lippert, Y. Xiang, M. L. Lee, *J. Microcol. Sep.*, 12: 462 (2000).

4. N. Wu, J. A. Lippert, M. L. Lee, *J. Chromatogr. A*, 911: 1 (2001).
5. J. M. Cintrón, L. A. Colón, *Analyst*, 127: 701 (2002).
6. J. R. Mazzeo, U. D. Neue, M. Kele, R. S. Plumb, *Anal. Chem.*, 77: 460A (2005).
7. H. J. Cortes, C. D. Pfeiffer, B. E. Richter, T. S. Stevens, *J. High Resolut. Chromatogr. Commun.*, 10: 446 (1987).
8. S. Hjerten, J. L. Liao, R. Zhang, *J. Chromatogr.*, 473: 273 (1989).
9. F. Svec, J. M. J. Frechet, *Anal. Chem.*, 64: 820 (1992).
10. J.-L. Liao, R. Zhang, S. Hjerten, *J. Chromatogr.*, 586: 21 (1991).
11. Q. C. Wang, F. Svec, J. M. J. Frechet, *J. Chromatogr. A*, 669: 230 (1994).
12. S. Hjerten, Y.-M. Li, J.-L. Liao, J. Mohammad, K. Nakazato, G. Pettersson, *Nature*, 356: 810 (1992).
13. H. Minakuchi, K. Nakanishi, N. Soga, N. Ishizuka, N. Tanaka, *Anal. Chem.*, 68: 3498 (1996).
14. T. B. Tennikova, B. G. Belenkii, F. Svec, *J. Liq. Chromatogr.*, 13: 63 (1990).
15. N. Tanaka, H. Kobayashi, K. Nakanishi, H. Minakuchi, N. Ishizuka, *Anal. Chem.*, 73: 420A (2001).
16. S. Eeltink, W. M. C. Decrop, G. P. Rozing, P. J. Schoenmakers, W. T. Kok, *J. Sep. Sci.*, 27: 1431 (2004).
17. F. Svec, T. B. Tennikova, Z. Deyl, Eds., *Monolithic Materials: Preparation, Properties, and Applications*, Elsevier, Amsterdam: 2003.
18. K. Miyabe, G. Guiochon, *J. Sep. Sci.*, 27: 853 (2004).
19. W. Li, D. P. Fries, A. Malik, *J. Chromatogr. A*, 1044: 23 (2004).
20. F. Svec, C. G. Huber, *Anal. Chem.*, 78: 2101 (2006).
21. K. K. Unger, *Porous Silica*, Elsevier, Amsterdam, 1979, p. 169.
22. H. Kobayashi, T. Ikegami, H. Kimura, T. Hara, D. Tokuda, N. Tanaka, *Anal. Sci.*, 22: 491 (2006).
23. P. Hatsis, C. A. Lucy, *Analyst*, 127: 451 (2002).
24. P. Hatsis, C. A. Lucy, *Anal. Chem.*, 75: 995 (2003).
25. A. Bugey, C. Staub, *J. Pharmaceut. Biomed. Anal.*, 35: 555 (2004).
26. N. Ishizuka, H. Kobayashi, H. Minakuchi, K. Nakanishi, K. Hirao, K. Hosoya, T. Ikegami, N. Tanaka, *J. Chromatogr. A*, 960: 85 (2002).
27. D. Allen, Z. El Rassi, *Analyst*, 128: 1249 (2003).
28. F. Gerber, M. Krummen, H. Potgeter, A. Roth, C. Siffrin, C. Spoendlin, *J. Chromatogr. A*, 1036: 127 (2004).
29. Q. Xu, M. Mori, K. Tanaka, M. Ikedo, W. Hu, *J. Chromatogr. A*, 1026: 191 (2004).
30. L. Xiong, R. Zhang, F. E. Regnier, *J. Chromatogr. A*, 1030: 187 (2004).
31. K. Cabrera, *J. Sep. Sci.*, 27: 843 (2004).
32. B. Chankvetadze, T. Ikai, C. Yamamoto, Y. Okamoto, *J. Chromatogr. A*, 1042: 55 (2004).
33. D. Victory, P. Nesterenko, B. Paull, *Analyst*, 129: 700 (2004).
34. D. Connolly, D. Victory, B. Paull, *J. Sep. Sci.*, 27: 912 (2004).
35. L. A. Colón, T. D. Maloney, A. M. Fermier, *J. Chromatogr. A*, 887: 43 (2000).
36. L. A. Colón, T. D. Maloney, A. M. Fermier, in *Capillary Electrochromatography*, Z. Deyl, F. Svec, Eds., Elsevier, New York, 2001, Chapter 4, pp. 111–164.
37. F. Svec, E. C. Peters, D. Sykora, C. Yu, J. M. J. Frechet, *J. High Resolut. Chromatogr.*, 23: 3 (2000).
38. C. Legido-Quigley, N. D. Marlin, V. Melin, A. Manz, N. W. Smith, *Electrophoresis*, 24: 917 (2003).
39. S. Xie, R. W. Allington, J. M. J. Fréchet, F. Svec, *Adv. Biochem. Eng./Biotechnol.*, 79: 87 (2002).

40. A. Tan, S. Benetton, J. D. Henion, *Anal Chem.*, 75: 5504 (2003).
41. F. Wei, Y. Fan, M. Zhang, Y.-Q. Feng, *Electrophoresis*, 26: 3141 (2005).
42. R. E. Majors, *LC-GC*, 23: 988 (2005).
43. K. Pflegerl, A. Podgornik, E. Berger, A. Jungbauer, *Biotechnol. Bioeng.*, 79: 733 (2002).
44. M. Motokawa, M. Ohira, H. Minakuchi, K. Nakanishi, N. Tanaka, *J. Sep. Sci.*, 29: 2471 (2006).
45. M. Kato, K. Sakai-Kato, T. Toyo'oka, *J. Sep. Sci.*, 28: 1893 (2005).
46. E. F. Hilder, F. Svec, J. M. J. Fréchet, *J. Chromatogr. A*, 1044: 3 (2004).
47. N. Tanaka, H. Nagayama, H. Kobayashi, T. Ikegami, K. Hosoya, N. Ishizuka, H. Minakuchi, K. Nakanishi, K. Cabrera, D. Lubda, *J. High Resolut. Chromatogr.*, 23: 111 (2000).
48. Y. Guo, L. A. Colón, *Anal. Chem.*, 67: 2511 (1995).
49. Y. Guo, L. A. Colón, *J. Microcolumn Sep.*, 7: 485 (1995).
50. L. A. Colón, in *Science and Technology of Polymers and Advanced Materials: Emerging Technologies and Business Opportunities*, Eds., P. N. Prasad, J. E. Mark, S. H. Kandil, Z. H. Kafafi, Plenum, 1998, p. 835.
51. D. Wang, S. L. Chong, A. Malik, *Anal. Chem.*, 69: 4566 (1997).
52. S. A. Rodríguez, L. A. Colón, *Chem. Mater.*, 11: 754 (1999).
53. J. D. Hayes and A. Malik, *Anal. Chem.*, 73: 987 (2001).
54. K. K. Unger, N. Becker, P. Roumeliotis, *J. Chromatogr.*, 125: 115 (1976).
55. K. J. Reynolds, L. A. Colón, *J. Liq. Chromatogr. Relat. Tech.*, 23: 161 (2000).
56. Y.-F. Cheng, T. H. Walter, Z. Lu, P. Iraneta, B. A. Alden, C. Gendreau, U. D. Neue, J. M. Grassi, J. L. Carmody, J. E. O'Gara, R. P. Fisk, *LC-GC*, 18: 1162 (2000).
57. U. D. Neue, T. H. Walter, B. A. Alden, Z. Jiang, R. P. Fisk, J. T. Cook, K. H. Glose, J. L. Carmody, J. M. Grassi, Y.-F. Cheng, Z. Lu, R. Crowley, *Am. Lab.*, 31: 36 (1999).
58. J. M. Cintron, L. A. Colón, *Analyst*, 127: 701 (2002).
59. K. D. Wyndham, J. E. O'Gara, T. H. Walter, K. H. Glose, N. L. Lawrence, B. A. Alden, G. S. Izzo, C. J. Hudalla, P. C. Iraneta, *Anal. Chem.*, 75: 6781 (2003).
60. C. Brinker, G. Scherer, *Sol-Gel Science: The Physics and Chemistry of Sol-Gel Processing*, Academic Press, New York, 1990.
61. L. A. Colon, D. Maloney Todd, J. Anspach, H. Colon, in *Advances in Chromatography*, 2003, vol. 42, pp. 43.
62. K. Nakanishi, *Bull. Chem. Soc. Jpn.*, 79: 673 (2006).
63. K. Nakanishi, N. Soga, *J. Non-Cryst. Solids*, 139: 1 (1992).
64. K. Nakanishi, K. Kanamori, *J. Mater. Chem.*, 15: 3776 (2005).
65. H. Minakuchi, K. Nakanishi, N. Soga, N. Ishizuka, N. Tanaka, *J. Chromatogr. A*, 762: 135 (1997).
66. H. Minakuchi, K. Nakanishi, N. Soga, N. Ishizuka, N. Tanaka, *J. Chromatogr. A*, 797: 121 (1998).
67. K. K. Unger, Ed., *Packings and Stationary Phases in Chromatographic Techniques*, Marcel Dekker, New York, 1990.
68. S. Pawlenko, *Organosilicon Chemistry*; Walter de Groyter, New York, 1986.
69. R. E. Majors, *LC-GC*, 18: 1214 (2000).
70. D. V. McCalley, *J. Chromatogr. A*, 902: 311 (2000).
71. L. R. Snyder, J. L. Glajch, J. J. Kirkland, *Practical HPLC Method Development*, 2nd ed. Wiley-Interscience, New York, 1997.
72. B. Trammell, L. Ma, H. Luo, D. Jin, M. A. Hillmayer, P. W. Carr, *Anal. Chem.*, 74: 4634 (2003).

73. A. Wehrli, J. C. Hildebrand, H. P. Keller, R. Stampfli, R. Frei, *J. Chromatogr.*, 149: 199 (1978).
74. J. J. Kirkland, J. C. Glajch, R. D. Farlee, *Anal. Chem.*, 61: 2 (1989).
75. A. B. Scholten, J. W. de Hann, H. A. Claessens, L. J. M. van de Ven, C. A. Cramers, *J. Chromatogr.*, 688: 25 (1994).
76. J. J. Kirkland, J. B. Adams, M. A. van Straten, H. A. Claessens, *Anal. Chem.*, 70: 4344 (1998).
77. M. J. Wirth, H. O. Fatunmbi, *Anal. Chem.*, 64: 2783 (1992).
78. M. J. Wirth, H. O. Fatunmbi, *Anal. Chem.*, 65: 822 (1993).
79. J. E. Sandoval, J. J. Pesek, *Anal. Chem.*, 63: 2634 (1991).
80. J. J. Pesek, M. T. Matyska, J. E. Sandoval, E. J. Williamsen, *J. Liq. Chromatogr. Rel. Technol.*, 19: 2843 (1996).
81. J. J. Pesek, M. T. Matyska, M. Oliva, M. Evanchic, *J. Chromatogr. A*, 818: 145 (1998).
82. Y. Guo, G. A. Imahori, L. A. Colón, *J. Chromatogr. A*, 744: 17 (1996).
83. M. Motokawa, H. Kobayashi, N. Ishizuka, H. Minakuchi, K. Nakanishi, H. Jinnai, K. Hosoya, T. Ikegami, N. Tanaka, *J. Chromatogr. A*, 961: 53 (2002).
84. S. Constantin, R. Freitag, *J. Sol-Gel Sci. Technol.*, 28: 71 (2003).
85. Y. Chen, Z. Zhang, X. Sui, J. D. Brennan, M. A. Brook, *J. Mater. Chem.*, 15: 3132 (2005).
86. K. A. Cobb, V. Dolknik, M. Novotny, *Anal. Chem.*, 62: 2478 (1990).
87. H. Dong, M. Lee, R. D. Thomas, Z. Zhang, R. F. Reidy, D. W. Mueller, *J. Sol-Gel Sci. Technol.*, 28: 5 (2003).
88. S.A. Rodríguez, L. A. Colón, *Appl. Spectrosc.*, 55: 472 (2001).
89. D. A. Loy, B. M. Baugher, C. R. Baugher, D. A. Schneider, K. Rahimian, *Chem. Mater.*, 12: 3624 (2000).
90. A. Itagaki, K. Nakanishi, K. Hirao, *J. Sol-Gel Sci. Technol.*, 26: 153 (2003).
91. D. A. Loy, A. Straumanis, D. A. Schneider, B. Mather, A. Sanchez, C. R. Baugher, K. J. Shea, *Polym. Prepr. (Am. Chem. Soc., Div. Polym. Chem.)*, 45: 591 (2004).
92. Y. Chujo, T. Saegusa, *Adv. Polym. Sci.*, 100: 11 (1992).
93. E. R. Pohl, F. O. Osterholz, in Silances, Surfaces and Interfaces, Ed., D.E. Leyden, Gordon Breach Science, Amsterdam, 1986, vol. 1, p. 481.
94. J. D. Jordan, R. A. Dunbar, D. J. Hook, H. Zhuang, J. A. Gardella, Jr., L. A. Colon, F. V. Bright, *Chem. Mater.*, 10: 1041 (1998).
95. H. Colon, X. Zhang, J. K. Murphy, J. G. Rivera, L. A. Colon, *Chem. Comm.*, 2826 (2005).
96. J. D. Hayes, A. Malik, *Anal. Chem.*, 72: 4090 (2000).
97. L. Yan, Q. Zhang, J. Zhang, L. Zhang, T. Li, Y. Feng, L. Zhang, W. Zhang, Y. Zhang, *J. Chromatogr. A*, 1046: 255 (2004).
98. B. Riegel, W. Kiefer, S. Hofacker, G. Schottner, *J. Sol-Gel Sci. Technol.*, 13: 385 (1998).
99. N. Huesing, U. Schubert, R. Mezei, P. Fratzl, B. Riegel, W. Kiefer, D. Kohler, W. Mader, *Chem. Mater.*, 11: 451 (1999).
100. L. Yan, Q. Zhang, W. Zhang, Y. Feng, L. Zhang, T. Li, Y. Zhang, *Electrophoresis*, 26: 2935 (2005).
101. L.-J. Yan, Q.-H. Zhang, Y.-Q. Feng, W.-B. Zhang, T. Li, L.-H. Zhang, Y.-K. Zhang, *J. Chromatogr. A*, 1121: 92 (2006).
102. K. Nakanishi, H. Shikata, N. Ishizuka, N. Koheiya, N. Soga, *J. High Resolut. Chromatogr.* 23: 106 (2000).
103. J. J. Pesek, M. T. Matyska, *J. Sep. Sci.*, 28: 1845 (2005).

104. J. J. Pesek, M. T. Matyska, E. J. Williamsen, M. Evanchic, V. Hazari, K. Konjuh, S. Takhar, R. Tranchina, *J. Chromatogr. A*, 786: 219 (1997).
105. D. M. Smith, R. Deshpande, C. J. Brinker, *Mater. Res. Soc. Symp. Proc.*, 271: 567 (1992).
106. S. S. Prakash, C. J. Brinker, A. J. Hurd, S. M. Rao, *Nature*, 374: 439 (1995).
107. N. Ishizuka, H. Minakuchi, K. Nakanishi, N. Soga, H. Nagayama, K. Hosoya, N. Tanaka, *Anal. Chem.*, 72: 1275 (2000).
108. R. J. Hodgson, Y. Chen, Z. Zhang, D. Tleugabulova, H. Long, X. Zhao, M. Organ, M. A. Brook, J. D. Brennan, *Anal. Chem.*, 76: 2780 (2004).
109. R. J. Hodgson, M. A. Brook, J. D. Brennan, *Anal. Chem.*, 77: 4404 (2005).
110. R. Mallik, D. S. Hage, *J. Sep. Sci.*, 29: 1686 (2006).
111. L.M. Ellerby, C.R. Nishida, F. Nishida, S.A. Yamanaka, B. Dunn, J.S. Valentine, J.I. Zink, *Science*, 255: 1113 (1992).
112. B.C. Dave, B. Dunn, J.S. Valentine, J.I. Zink, *Anal. Chem.*, 66: 1120A (1994).
113. C. M. Ingersol, F. V. Bright, *Anal. Chem.*, 69: 403A (1997).
114. I. Gill, *Chem. Mater.*, 13: 3404 (2001).
115. P. L. Edmiston, C. L. Wambolt, M. K. Smith, S. S. Saavedra, *J. Colloid Interface Sci.*, 163: 395 (1994).
116. S. Braun, S. Rappoport, R. Zusman, D. Avnir, M. Ottolenghi, *Mater. Lett.*, 10: 1 (1990).
117. K. Flora, J. D. Brennan, *Anal. Chem.*, 70: 4505 (1998).
118. U. Narang, M. H. Rahman, J. H. Wang, P. N. Prasad, F. V. Bright, *Anal. Chem.*, 67: 1935 (1995).
119. M. Cichna, D. Knopp, R. Niessner, *Anal. Chim. Acta*, 339: 241 (1997).
120. M. Kato, K. Sakai-Kato, T. Toyo'oka, *J. Sep. Sci.*, 28: 1893 (2005).
121. W. Jin, J. D. Brennan, *Anal. Chim. Acta*, 461: 1 (2002).
122. M. Kato, K. Sakai-Kato, H. Jin, K. Kubota, H. Miyano, T. Toyo'oka, M. T. Dulay, R. N. Zare, *Anal. Chem.*, 76: 1896 (2004).
123. M. Kato, K. Sakai-Kato, N. Matsumoto, T. Toyo'oka, *Anal. Chem.*, 74: 1915 (2002).
124. M. Kato, N. Matsumoto, K. Sakai-Kato, T. Toyo'oka, *J. Pharm. Biomed. Anal.*, 30: 1845 (2003).
125. K. Sakai-Kato, M. Kato, H. Nakakuki, T. Toyo'oka, *J. Pharm. Biomed. Anal.*, 31: 299 (2003).
126. K. Sakai-Kato, M. Kato, T. Toyo'oka, *Anal. Chem.*, 74: 2943 (2002).
127. M. A. Brook, Y. Chen, K. Guo, Z. Zhang, W. Jin, A. Deisingh, J. Cruz-Aguado, J. D. Brennan, *J. Sol-Gel Sci. Technol.*, 31: 343 (2004).
128. P. Kovarik, R. J. Hodgson, T. Covey, M. A. Brook, J. D. Brennan, *Anal. Chem.*, 77: 3340 (2005).
129. T. R. Besanger, R. J. Hodgson, D. Guillon, J. D. Brennan, *Anal. Chim. Acta*, 561: 107 (2006).
130. G. Ding, Z. Da, R. Yuan, J. J. Bao, *Electrophoresis*, 27: 3363 (2006).

9 Enhanced-Fluidity Liquid Mixtures: Fundamental Properties and Chromatography

Susan V. Olesik

CONTENTS

9.1 INTRODUCTION

An important goal in numerous chromatographic applications is to improve the separation power, P (Equation 9.1) or the speed of the separation [1]:

$$P(s^{-1}) = \frac{dN}{dt}. \tag{9.1}$$

This improvement is especially important in routine analyses and in large-scale evaluation studies, such as the combinatorial characterization of drug candidates. High temperature (HT), HT high-performance liquid chromatography (HT-HPLC), and supercritical fluid chromatography (SFC) are common methods considered when separation speed is essential due the increased mass transport and decreased viscosity obtained with both techniques [2]. However, both SFC and HT-HPLC have deficiencies that limit their applicability [2,3]. For example, carbon dioxide, the most commonly-used solvent in SFC, has limited solvent strength even with the addition of small quantities of polar modifiers. The increased temperatures used in HT-HPLC can cause on-column reactions of the analytes [4]. Herein, we focus on enhanced-fluidity liquid chromatography (EFLC) that shares many of the positive attributes of each of these techniques without the negative attributes.

Enhanced-fluidity liquids (EFLs) are mixtures that contain high proportions of liquefied gases, such as carbon dioxide [5]. Fluidity, f, is defined as the inverse of viscosity. EFL mixtures combine the positive attributes of commonly-used liquids, such as high solvent strength, with the positive attributes of supercritical fluids, such as low viscosity or high fluidity, low surface tension, high diffusivity. These attributes allow EFLC to contribute to the quest for increased separation power.

Recently the term, gas-expanded liquids (GXLs) [6–8] has been used to describe these unique mixtures while others use the term subcritical mixtures to describe the phase of matter. No matter what term is used to describe these mixtures one point should be clear; all of these mixtures are liquids not supercritical fluids. Furthermore, there is no discontinuity observed in moving from the supercritical condition to a liquid. However, EFL mixtures and supercritical fluids are two different phases of matter.

The review is organized in the following sections. The chromatographic theory relevant to EFLC and HT-HPLC is first described. Next a detailed description of the physicochemical properties of EFL mixtures is included. This is followed by a survey of the scope of liquid chromatography (LC) techniques that are presently using the attributes of EFLs. Finally, a discussion of future applications of EF-HPLC is included.

9.2 CHROMATOGRAPHIC THEORY

To allow the comparison of chromatographic performance with different phases, reduced (dimensionless) plate height, h, and reduced linear velocity, v, are typically used as defined in Equations 9.2 and 9.3, respectively [9,10]

$$h = \frac{H}{u} \tag{9.2}$$

$$v = \frac{u d_{\mathrm{p}}}{D_{\mathrm{m}}} \tag{9.3}$$

H is the plate height (cm); u is linear velocity (cm/s); d_{p} is particle diameter, and D_{m} is the diffusion coefficient of analyte (cm^2/s). By combining the relationships between retention time, t_{r}, and retention factor, k: $t_{\mathrm{r}} = t_0(1 + k)$, the definition of dead time, t_0, $t_0 = L/u$ where L is the length of the column, and $H = L/N$ where N is chromatographic efficiency with Equations 9.2 and 9.3, a relationship (Equation 9.4) for retention time, t_{r}, in terms of diffusion coefficient, efficiency, particle size, and reduced variables (h and v) and retention factor results. Equation 9.4 illustrates that mobile phases with large diffusion coefficients are preferred if short retention times are desired.

By rearranging this equation, the important parameters for increasing N/t_{r} are determined [11].

$$t_{\mathrm{r}} = \frac{N d_{\mathrm{p}}^2}{D_{\mathrm{m}}} \left(\frac{h}{v}\right) (1 + k) \tag{9.4}$$

$$\frac{N}{t_{\mathrm{r}}} = \frac{D_{\mathrm{m}}}{d_{\mathrm{p}}^2} \left(\frac{v}{h}\right) \left(\frac{1}{1 + k}\right) \tag{9.5}$$

Equation 9.5 shows that for an acceptable retention factor, the diffusion coefficient and particle size of the packing are the primary variables that can be varied to affect the N/t_{r}. An increase in the diffusion coefficient or a decrease in the particle size causes an increase in N/t_{r}. If the diffusion coefficient can be increased without impacting the retention factor significantly, then significant increases in separation speed will result. To minimally impact the retention factor the solvent strength of the mobile phase must be maintained.

The fastest HPLC separations are achieved using the maximum available pressure drop. Using reduced variables, Equation 9.6 illustrates a linear relationship between retention time and mobile phase viscosity for packed columns and fixed values of ΔP (pressure drop), N_{req} (required efficiency for a given separation) and ψ (a constant that describes the permeability of the packed bed) [4]

$$t_{\mathrm{r}} = \frac{\psi h^2 N_{\mathrm{req}}^2 \eta}{\Delta P} (1 + k). \tag{9.6}$$

A method that can decrease the viscosity of the mobile phase without impacting the mobile phase solvent strength (i.e., maintaining k) would therefore decrease the analysis time linearly. The next section illustrates the diffusion coefficients and viscosities, the unique relationship between them for EFL mixtures, their solvent-strength and other important properties.

9.3 PHYSICOCHEMICAL PROPERTIES OF ENHANCED-FLUIDITY LIQUID MIXTURES

9.3.1 MOLECULAR-LEVEL STRUCTURE OF ENHANCED-FLUIDITY LIQUID MIXTURES COMPARED TO THOSE OF SUPERCRITICAL FLUIDS

In supercritical fluids, significant intermolecular association results in large solvation spheres with hundreds of solvent molecules in the solvation sphere and extensive "free volume" encompassing the solvation spheres. The local solvent density is therefore much higher than the bulk density [12]. Local solvent density also varies with pressure and temperature. The significant local ordering is greatest near the critical condition of the supercritical fluid. As the bulk density of the solvent is increased by increasing the pressure, the local solvent density around an analyte decreases to values that are similar to the bulk value. Recently, high pressure NMR studies of supercritical CO_2 have shown that the local solvent density is quite sensitive to solute structure [13].

Are there similarities between the molecular-level properties of supercritical fluids and EFLs? The emerging data on EFL mixtures indicate the answer is yes. Experimental and theoretical studies of solvation structures within EFLs show significantly ordered solvation spheres around dissolved solutes. For liquid mixtures of methanol/CO_2 containing 10–40 mol% methanol, Souvignet et al. [14] measured large and negative infinite-dilution partial molar volumes for naphthalene at 299 K that were approximately an order of magnitude less negative than in supercritical CO_2 [15]. Therefore, the solvation sphere around naphthalene is extensive but not as large as in supercritical fluids. Recent molecular dynamical simulations of methanol/CO_2 EFLs illustrate that the addition of CO_2 to methanol up to 0.60 mole fraction CO_2 had minimal impact on the H-bond network of methanol [16]. However, with further addition of CO_2 the integrity of methanol's hydrogen-bond network decreased. Li et al. [17] illustrated preferential association occurring between organic molecules in EFL mixtures of cyclohexane/CO_2, acetonitrile/CO_2 and methanol/CO_2. As the proportion of CO_2 increased in these mixtures, clustering of the organic molecules increased. Therefore, like supercritical fluids, significant local density augmentation occurs in EFLs but the magnitude of the augmentation is smaller than in that in supercritical fluids. This augmentation varies with pressure and temperature [14].

9.3.2 DIFFUSION COEFFICIENTS OF SOLUTES IN ENHANCED-FLUIDITY LIQUID MIXTURES

The diffusion coefficients of small molecular weight solutes in supercritical fluids and liquids are typically in the range of 1×10^{-4} and 10^{-6} cm^2/s [18,19], respectively, while those of EFLs are typically intermediate [4] between these two values (10^{-5} cm^2/s).

The impact on the diffusion coefficient of benzene by adding liquid CO_2 to methanol at 298 K and 17.2 MPa [5,20] is shown in Figure 9.1. The addition of 40, 60, and 80 mol% CO_2 causes increases in the diffusion coefficient of benzene of 60%, 100%, and 182%, respectively. Figure 9.2 shows the variation in diffusion coefficient

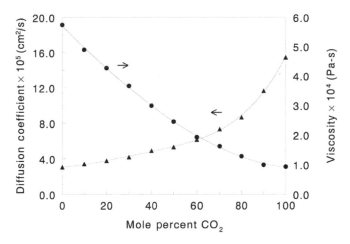

FIGURE 9.1 Variation of the diffusion coefficient (▲) of benzene and the solvent viscosity (●) for methanol/CO_2 mixtures at 298 K and 17.2 MPa. (Adapted from Y. Cui, S. V. Olesik, *Anal. Chem.*, 63: 1812 (1991); P. R. Sassiat et al., *Anal. Chem.*, 59: 1164 (1987).)

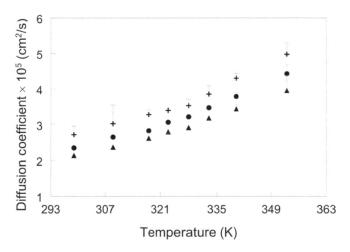

FIGURE 9.2 Variation of the diffusion coefficient of styrene with temperature for THF/CO_2 mixtures: (▲) pure THF, (●) 80/20 mol% THF/CO_2, (+) 60/40 mol% THF/CO_2 at 19.3 MPa. (Adapted from H. Yuan et al., *J. Chromatogr. Sci.*,35: 409 (1997).)

of styrene in THF/CO_2 mixtures at 19.3 MPa [21]. The addition of 40 mol% CO_2 to THF causes an increase in the diffusion coefficient of styrene of only 27% at 299 K. However, Figure 9.2 also illustrates that the combination of added CO_2 and slight increases in temperature provides substantial increases in the diffusion coefficient of styrene. By increasing the temperature to 353 K and adding 40 mol% CO_2, the styrene's diffusion coefficient increased 133%. Interestingly, for this EFL mixture, the addition of 40 mol% CO_2 to THF caused approximately a 27% increase in the

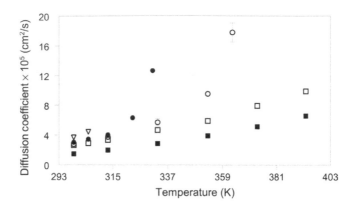

FIGURE 9.3 Variation of the diffusion coefficient of benzene as a function of temperature in (■) 0.70/0.30 mole fraction methanol/H_2O, (□) 0.56/0.24/0.20 mole fraction methanol/H_2O/CO_2, (○) 0.52/0.23/0.25 mole fraction methanol/H_2O/CO_2, (●) 0.49/0.21/0.30 mole fraction methanol/H_2O/CO_2, (▽) 0.42/0.18/0.40 mole fraction methanol/H_2O/CO_2 at 13.8 MPa. (Adapted from I. Bako et al., *J. Mol. Liq.*, 87: 243 (2000).)

diffusion coefficient for all temperatures studied (299–353 K). In comparing the variation of diffusion coefficient in the methanol/CO_2 mixtures and that in the THF/CO_2 mixture, greater enhancement in the diffusion coefficients are observed through addition of CO_2 in methanol than in THF. The observed difference in the impact of added CO_2 is likely due to the difference in intermolecular association in the original organic solvent. Liquid methanol has extensive hydrogen-bonded networks with as much as 80% of the molecules being involved in two hydrogen bonds, while tetrahydrofuran molecules are not expected to be highly associated [22].

The diffusion coefficients in EFLs with alcohol/H_2O mixtures were also studied [23,24]. Figure 9.3 shows the variation of the diffusion coefficient of benzene as a function of temperature (299–393 K) for EFL mixtures where the mole ratio of methanol/H_2O was maintained at 0.70/0.30 and the amount of CO_2 was increased from 0 to 0.40 mole fraction [23]. At 313 K, the addition of 40 mol% CO_2 caused a 100% increase in the diffusion coefficient of benzene. However, increasing the temperature and the proportion of CO_2 caused the largest increase in the diffusion coefficient. Over a 500% increase in the diffusion coefficient of benzene is observed when the temperature is increased to 363 K and 0.30 mole fraction CO_2 is combined with the 0.70/0.30 mole ratio methanol/H_2O mixture.

Souvignet et al. [24] studied the variation of the diffusion coefficients of nonpolar compounds, such as benzene and anthracene, and polar compounds, such as *m*-cresol and nitrophenol, in ethanol/H_2O/CO_2 mixtures. Figure 9.4 shows the variation of benzene's diffusion coefficient in 0.61/0.39 mole ratio ethanol/H_2O mixtures as a function of added CO_2 (0–40 mol%) and temperature (299–333 K). For the ethanol/H_2O mixture increasing the temperature from 298 to 333 K caused a 95% increase in the diffusion coefficient of benzene while adding 40 mol% CO_2 to the ethanol/H_2O mixtures increased the diffusion coefficient by 213%. However, the combination of both the addition of 40 mol% CO_2 and increasing the temperature to 333 K provided a

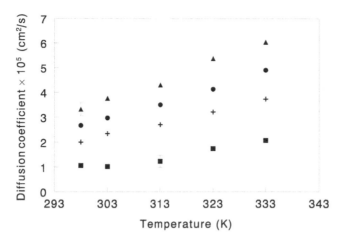

FIGURE 9.4 Effect of temperature and mole fraction CO_2 on the diffusion coefficient of benzene in ethanol/H_2O (0.61/0.39 mole ratio) at 17.2 MPa:($\blacksquare$) 0, (+) 20,($\bullet$)30, ($\blacktriangle$) 40 mol% CO_2. (Adapted from S. T. Lee, S. V. Olesik, *Anal. Chem.*, 66: 4498 (1994).)

470% increase in the diffusion coefficient. For a 50/50 mole ratio ethanol/H_2O mixture, the addition of 20 mol% CO_2 combined with a temperature increase of 333 K caused increases in the diffusion coefficients of all studied solutes by 325–380%, indicating that small additions of CO_2 with relatively small increases in temperature affect the diffusion coefficients of the solutes substantially [24].

9.3.3 Viscosity

Recent theoretical studies on the molecular structure of organic/CO_2 EFL mixtures show that the fraction of the total volume occupied by the van der Waals volumes of the mixture constituents decreases monotonically with addition of CO_2 to conditions that correspond to 60% of those of the original liquids. Therefore, as mentioned earlier, significant "free volume" is present in these mixtures [17]. Large free volumes result in significantly decreased viscosities as compared to those of common liquids. For example, experimental (Figure 9.1) [8] and theoretical studies [17] on methanol/CO_2 mixtures show an approximate monotonic drop in viscosity with addition of CO_2. For example, when 60 mol% CO_2 is added to methanol, the viscosity drops by 67%. The low viscosities are expected to be favorable for fast kinetics.

9.3.4 Relationship between Diffusivity and Viscosity

The Stokes–Einstein equation (Equation 9.7) is often used to describe the relationship between the diffusion coefficient of a solute and the viscosity of the solution

$$D = \frac{kT}{6\pi a\eta}$$

(9.7)

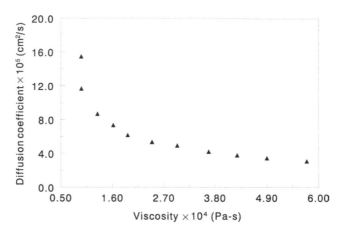

FIGURE 9.5 Variation of diffusion coefficient as a function of viscosity for methanol/CO$_2$ mixtures at 298 K and 17.2 MPa. (Adapted from Y. Cui, S. V. Olesik, *Anal. Chem.*, 63: 1812 (1991); P. R. Sassiat et al., *Anal. Chem.*, 59: 1164 (1987).)

where D is the diffusion coefficient, k is the Boltzmann constant, T is absolute temperature, a is the solvated radius, and η is viscosity. This equation predicts a linear relationship between D and η. Figure 9.5 shows the relationship between the diffusion coefficient of benzene in methanol/CO$_2$ mixtures and the solution viscosity at 298 K and 17.2 MPa. Clearly, the Stokes–Einstein equation does not describe the relationship between diffusion coefficients and solvent viscosity for this EFL mixture. The diffusion coefficient increases at a faster rate than the viscosity decreases. Other EFL mixtures show similar nonlinear correlations between D and $1/\eta$ [24]. Similarly, Debenedctti et al. [19] showed that the Stokes–Einstein equation only described the relationship between D and η for supercritical fluids at relatively high viscosities.

9.3.5 Dielectric Constants and Ionic Conductivities

Enhanced-fluidity liquid mixtures can exhibit moderately high dielectric constants. For example, Figure 9.6 shows the variation of the dielectric constant, ε, of methanol/CO$_2$ mixtures at 323 K and 11 MPa with increasing mole fractions of methanol [25]. For mixtures containing ≥ 0.42 mole fraction methanol, the dielectric constant increases approximately linearly with mole fraction of methanol [25].

The ionic conductivity of a solution depends on the viscosity, diffusivity, and dielectric constant of the solvent, and the dissociation constant of the molecule. EFL mixtures can carry charge. The conductivity of perfluoroacetate salts in EFL mixtures of carbon dioxide and methanol is large (10^{-5} to 10^{-4} S/cm for salt concentrations of 0.05–5 mM) and increases with salt concentration. The ionic conductivity of tetra-methylammonium bicarbonate (TMAHCO$_3$) in methanol/CO$_2$ mixtures has specific conductivities in the range of 9–14 mS/cm for pure methanol at pressures varying from 5.8 to 14.1 MPa, which decreases with added CO$_2$ to a value of 1–2 mS/cm for 0.50 mole fraction CO$_2$ for all pressures studied. When as much as 0.70 mole fraction

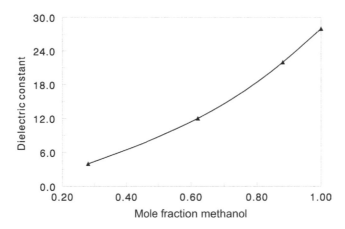

FIGURE 9.6 Variation of dielectric constant of methanol/CO_2 mixtures at 323 K and 11 MPa. (Adapted from R. L. Smith, Jr. et al., *Fluid Phase Equil.*, 194–197: 869 (2002).)

TABLE 9.1
pH Measurement of Methanol/H_2O/CO_2 at 20.7 MPa

Methanol/H_2O/CO_2 Mole Ratio	pH
68.2/30.6/1.2	4.54 ± 0.01
65.1/29.3/5.6	4.22 ± 0.03
61.7/27.7/10.6	4.38 ± 0.02
55.7/25/1/19.2	4.73 ± 0.03

Source: D. Wen, S. V. Olesik, *Anal. Chem.*, 72: 475 (2000).

CO_2 is in the mixture, the conductivity is negligible [26]. The decrease in conductivity with added CO_2 is believed to be caused by the diminishing dielectric constant. Changing the pressure of fluid did not impact the conductivity significantly [26]. However, the low viscosities of EFLs provide increased ionic conductance relative to that of a common liquid of comparable dielectric constant [27].

9.3.6 pH Control in Enhanced-Fluidity Liquid Mixtures

The pH of methanol/H_2O/CO_2 mixtures was measured spectrophotometrically with CO_2 proportions as high as 19.2 mol% using pH indicators [28]. Table 9.1 shows the variation of pH of methanol/H_2O/CO_2 mixtures as a function of added CO_2. Clearly, the impact of added CO_2 on the measured pH was not large. As the proportion of CO_2 increases, the dielectric constant of the solution decreases and the there is more dissolved CO_2 that could dissociate to carbonic acid. The data clearly show that the dissociation of carbonic acid is suppressed by the decreasing dielectric constant of the solution. When the pressure of these mixtures was varied from 12.06 to 20.7 MPa, a minimal change in the pH was observed.

This same study [28] illustrated the use of buffers to control pH in methanol/H_2O/CO_2 solutions. Acetate buffers had pH values from 4.89 to 6.4 and phosphate buffers had pH values from 5.2 to 6.9 depending on the proportion of CO_2 in the mixture.

9.3.7 SOLVENT STRENGTH

Kamlet-Taft solvatochromic parameters, α, β, π^*, provide information on hydrogen-bond acidity, hydrogen-bond basicity, and dipolarity/polarizability [5,29], respectively. These parameters were determined using spectrophotometry for a range of EFL mixtures. Figure 9.7 shows the variation of the Kamlet-Taft solvatochromic parameters, α (H-bond acidity), β (H-bond) basicity, and the π^* parameter for methanol/CO_2 mixtures at 17.2 MPa and 298 K. The hydrogen-bond acidity and basicity of mixture decreases minimally with large proportions (up to 0.80 mole fraction) of added CO_2. However, the π^* parameter decreases to approximately 50% of its original value with the addition of 0.65 mole fraction CO_2.

Yuan et al. [21] measured the Kamlet-Taft β and π^* parameters for THF/CO_2 mixtures (Figure 9.8). They illustrated that THF/CO_2 mixtures had Kamlet-Taft β parameters that were similar to that of THF even when as much as 0.6 mole fraction CO_2 is added, and π^* of the same mixture decreases to about half the value of that of THF when 0.50 mole fraction CO_2 is added to the mixture.

Figure 9.9 shows [30] similar data for the 0.70/0.30 mole ratio methanol/H_2O mixtures at 17.2 MPa and 298 K as a function of added CO_2. Interestingly, the

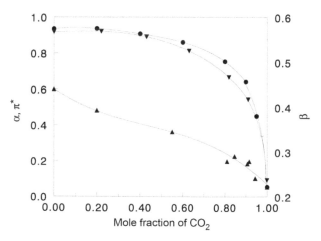

FIGURE 9.7 Variation of Kamlet-Taft solvatochromic parameters for methanol/CO_2 mixtures as a function of added CO_2 at 298 K and 17.2 MPa: ($\blacktriangle$)π^* (dipolarity/polarizability), ($\bullet$)α (H-bond acidity), and ($\blacktriangledown$)β (H-bond basicity). (Adapted from Y. Cui, S. V. Olesik, *Anal. Chem.*, 63: 1812 (1991).)

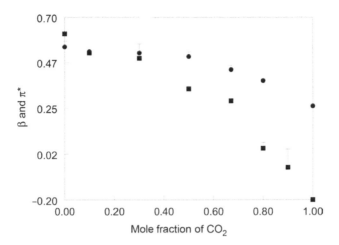

FIGURE 9.8 Variation of Kamlet-Taft solvatochromic parameters for THF/CO_2 mixtures as a function of added CO_2 at 13.74 MPa and 298 K: (■) π^* (dipolarity/polarizability) and (●)β (H-bond basicity). (Adapted from H. Yuan et al., *J. Chromatogr. Sci.*,35: 409 (1997).)

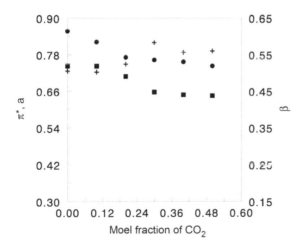

FIGURE 9.9 Variation of Kamlet-Taft solvatochromic parameters for methanol/H_2O/CO_2 mixtures as a function of added CO_2 with the methanol/H_2O mole ratio held at 2.3 at 298 K and 17.2 MPa: (■)π^* (dipolarity/polarizability), (●)α (H-bond acidity), and [+β (H-bond basicity). (Adapted from Y. Cui, S. V. Olesik, *J. Chromatogr. A*, 691: 151 (1995).)

β parameter actually increases with added CO_2, which must be a result of changes in the H-bond network between methanol and water because CO_2 has a negligibly small and negative β. However, the most important attribute is that the mixtures basically have high solvent strength even when high proportions of CO_2 are added to the mixtures.

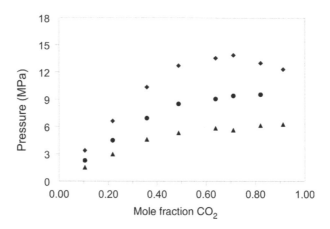

FIGURE 9.10   Methanol/CO₂ phase diagrams at (▲) 298 ● 323 (◆) 353 K. (Adapted from T. S. Reighard et al., *Fluid Phase Equil.*, 123: 215 (1996).)

9.3.8 PHASE DIAGRAMS

To take advantage of the properties of EFL mixtures in chromatographic applications, the phase behavior of the mixtures must be known because miscible mixtures are essential as mobile phases. The phase diagrams of a broad range of EFL mixtures have been characterized. The conditions (pressure and temperature) needed to maintain EFL mixtures are typically mild. For example, at 313 K the pressure necessary to maintain single phase mixtures from 0% to 100% CO_2 is below 7.8 MPa for binary mixtures of CO_2 with methanol [31], ethanol [32], 2-methyloxyethanol [33], THF [34], acetonitrile [31], acetone [35], *N*-methylpyrolidone [36], dichloromethane [36], and perfluorohexane [36]. Figure 9.10 shows phase boundary conditions for methanol/CO₂ at 298, 323, and 353 K and Figure 9.11 shows phase boundary conditions for CO_2 mixed with 2-propanol, 2,2,2-trifluoroethanol, acetonitrile, dichloromethane, tetrahydrofuran, and perfluorohexane at 313 K [36]. The conditions above the data points are single-phase liquid conditions, while the conditions below include liquid and gas phases. Lazzaroni et al. [36] noted that the solubility of CO_2 at 5.0 MPa increased from most soluble to least soluble in the order of perfluorohexane, tetrahydrofuran, dichloromethane, acetonitrile, N-methyl-2-pyrrolidone, nitromethanol, 2,2,2-trifluoroethanol, and 2-propanol. High miscibility with perfluorohexane is expected because it has similar solvent strength to CO_2. However, the other organic molecules are more polar and less miscible. Lewis acid–base interactions or polar interactions were thought to cause the high miscibility of the other organic solvents with CO_2.

Experimental phase diagrams of CO_2 with ethyl acetate [37] and alkyl carbonates [38] are also available. Molecular simulations that describe the binary phases diagrams of many organic solvents with CO_2 were also recently published [39]. Binary phase diagrams of methanol with fluoroform are also available [40]. At any given pressure and temperature, fluoroform is more miscible with methanol than CO_2. Ternary phase diagrams for a number of systems are also available, such as methanol/H_2O/CO_2 [41] acetonitrile/H_2O/CO_2, and methanol/H_2O/fluoroform [40].

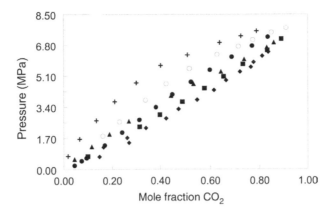

FIGURE 9.11 Phase diagrams of CO_2 combined with (♦) perfluorohexane, (■) tetrahydrofuran, (●) acetonitrile, (▲) dichloromethane, (○) 2,2,2,-trifluoroethanol, and (+) 2-propanol at 313 K. (Adapted from M. J. Lazzaroni et al., *J. Chem. Eng. Data*, 50: 6 (2005).)

9.4 ENHANCED-FLUIDITY LIQUID'S PROPERTIES AND CHROMATOGRAPHIC THEORY

The physicochemical properties described above clearly illustrate that EFLs have molecular level order similar to that of supercritical fluids, as well as, higher diffusion coefficients (typically doubling the original value for a addition of 60 mol% CO_2) and lower viscosities (typically decreasing by more than half the original value) than commonly used liquids. However, EFLs maintain the solvent strengths that are comparable to those of the commonly used liquids even when as much as 60 mol% CO_2. As shown in the applications below, the addition of approximately 40 mol% CO_2 to methanol often corresponds to chromatographic conditions comparable to those when the mobile phase temperature is raised to 333 K. The data also show that polar EFLs, can support charge transport and buffered solutions of varying pH can be readily produced as well. Many of these properties vary with pressure, which will be illustrated in the next section, and allows precise control of the chromatographic performance.

9.5 SCOPE OF ENHANCED-FLUIDITY LIQUID CHROMATOGRAPHY

9.5.1 NORMAL-PHASE CHROMATOGRAPHY

Normal-phase chromatography in both chiral and achiral modes is the most commonly used mode of SFC and EFLC. High throughput purification for drug discovery often starts with low proportions of a modifier, such as methanol, and then uses either a fast gradient (50–75 v/v%/min) [42] and/or fast flow rates (50 mL/min) with moderate gradients (10 v/v%/min) [43] to reach modifier levels of 60–70 v/v%. Berger showed that this type of separation using methanol/CO_2 mobile phases and cyano-functionalized silica supports is a more attractive alternative to the commonly used reversed-phase HPLC. (*Note*: Normal-phase HPLC is not commonly considered for these applications owing to the flammability of the commonly used mobile phases). Semipreparative separations using methanol/CO_2 mobile phases can be achieved in 6–8 min with small volumes collected, 3–5 mL, compared to 30–100 mL for the reversed-phase separation. In addition, methanol/CO_2 is readily evaporated, which allows complete sample purification in hours rather than in day when using water-based mobile phases. Furthermore, peak purity as high as 95% or greater were readily achieved with the CO_2/methanol mobile phase.

9.5.2 CHIRAL CHROMATOGRAPHY

Supercritical fluid chromatography is used extensively in the pharmaceutical industry for chiral separations. This is due to the fast equilibration times, high efficiency, and fast speed of analysis. However, for many applications, EFLC provides the best chiral separations in terms of increased resolution and efficiency [44]. Sun et al. [44] studied the use of EFLC for chiral separations using vancomycin as the stationary phase under normal- and reversed-phase conditions. For eight different types of drug compounds, EFL mixtures were optimal for the separation of all the racemate compared to SFC and HPLC with commonly used solvents. Figure 9.12 shows an example (for 1,1′ bi-2-naphthol) of the data collected in terms of resolution (R), retention factors (k), R/k, and reduced plate height (h) for normal-phase chiral separations. The minimum of the plate height was found at 40% CO_2; the resolution increased continuously with added CO_2; the average retention factors for the enantiomer pairs remained at values below 5 until >60% mol% CO_2. This created a maximum in resolution/retention factor at 40 mol% CO_2 as illustrated in Figure 9.12. This represents typical data for the chiral normal-phase separations, in which the optimal separation conditions were found between 40 and 60 wt% CO_2 in the mobile phase.

Liu et al. [45] studied the separation of 111 chiral compounds (heterocyclic compounds, chiral acids, β-blocker compounds, chiral sulfoxides, N-protected amino acids and underivatized amino acids) using three macrocyclic glycopeptide columns (teicoplanin, Chirobiotic-T; aglycone, Chirobiotic-TAG; and ristocetin, Chirobiotic R) using CO_2/methanol mixtures as the mobile phases. The optimal solvent conditions were 40–70 v/v% methanol combined with CO_2 for the β-blocker compounds, 40–60 v/v% for the heterocyclic compounds, 48–68 v/v% for the native amino acids with trifluoroacetic acid and/or triethylamine commonly added to improve the chromatographic peak shape. Clearly, for all three macrocyclic column types and for a large number (>60%) of the chiral compounds, EFL mixtures provided the higher resolution and the shorter amount of time compared to standard HPLC.

White recently illustrated the use of fast supercritical fluid and EFLC for drug discovery and purification [46]. The optimized isocratic separations used to scale up to preparative-scale separations were often EFL mixtures. For example, Figure 9.13 shows the optimized conditions for the separation of a drug candidate included 30% methanol (with 0.2% isopropyl amine)/CO_2 on a Chiralcel OJ-H column at 5 mL/min [46]. His work also illustrates by using gradients that start in supercritical conditions and then move into EFL mixture conditions provides efficient and fast separations.

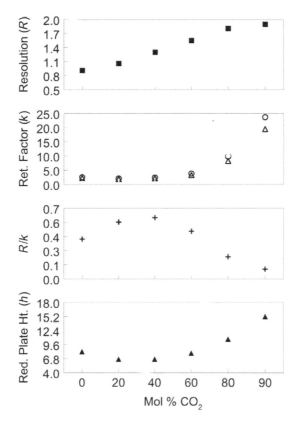

FIGURE 9.12 β-naphthol resolution (top ■), retention factors (○, △), resolution/retention factor (+), and reduced plate height (bottom ▲) as a function of mol% CO_2 combined with a 36/64 mol% ethanol/hexane at 17.2 MPa 298 K. (Adapted from Q. Sun, S. V. Olesik, *Anal. Chem.*, 71: 2139 (1999).)

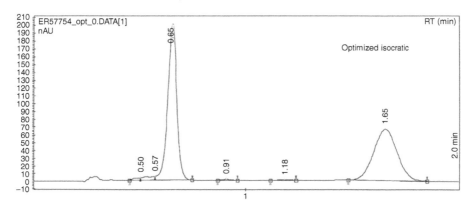

FIGURE 9.13 Optimized ioscratic separation of the enantiomer pair for a drug candidate. Mobile phases: 30% methanol containing 0.2% isopropyl amine/70% CO_2 on a Chiralcel OJ-H column, flow rate: 5 mL/min, outlet pressure 12.0 MPa and temperature 313 K. (Reprinted from C. White, *J. Chromatogr. A*, 1079: 163 (2005). With permission.)

Hoke et al. [47] recently did a detailed comparison of SFC-MS-MS, EFLC-MS-MS, and HPLC-MS-MS (hexane/2-propanol/trifluoroacetic acid) conditions for the bioanalytical determination of R and S ketoprofen in human plasma. The optimum chromatographic conditions included 55% methanol/45% CO_2 (EFL conditions) with a Chiralpak AD column. The performance parameters (specificity, linearity, sensitivity, accuracy, precision, and ruggedness) for SFC, EFLC, and HPLC were found to be comparable. However, the optimized EFLC conditions provided the analysis in one-third the amount of time for the LC-MS-MS conditions, which is 10-fold faster than an LC-UV method [48,49].

Geiser et al. [50,51] illustrated the screening of different chiral stationary phases and the separation of highly polar amine hydrochlorides using EFL methanol/CO_2 mixtures and the columns, Chiralpak-AD-H, Chiralpak-AS. This method is advantageous because no acid or base additive was required to achieve base line separation of the racemates and conversion to free base form for enantiomer separation was not required. Preparative-scale separations of the amine-hydrochloride were accomplished using similar mobile phase conditions [51]. Furthermore, this is believed to be the first chiral separation of highly polar solutes without the addition of acid or base additive to effect the separation.

9.5.3 Size Exclusion Chromatography

Yuan et al. [21] studied the addition of CO_2 to THF in an effort to improve size exclusion chromatography of polystyrene standards (M.W. range: 3,600–382,100) by using EFLs as mobile phases. Higher efficiency and shorter analysis times were obtained with the addition of CO_2 to the mobile phase. The average number of theoretical plates for the polystyrene standards increased by 62–76% by adding 40% CO_2 to the THF and the pressure drop across the column decreased by 38%, which resulted in a faster analysis time of nearly half the average retention time of that using pure THF.

9.5.4 Liquid Chromatography at the Critical Chromatography

Liquid chromatography at the critical condition (LCCC) is a technique that allows the heterogeneity of polymers to be separated. For example, the functionality distribution of hydroxylated polystyrene is determined by adjusting the combination of mobile phases and stationary phase conditions so that the Gibbs free energy of transfer polystyrene is zero (i.e., the backbone of the polymer is made "chromatographically invisible"). Under these LCCC conditions, when a functionalized polystyrene is injected into the system, the only species that will differentially interact with the surface are the functionalities. The resulting chromatogram consists of peaks for polymers with differing numbers of hydroxyl group per repeat unit and one peak at the critical condition of polystyrene. This is one of the very best ways of determining functionality distributions for polymers. The molecular weight distributions of polymers contained in copolymers are also determined by identifying the critical condition for

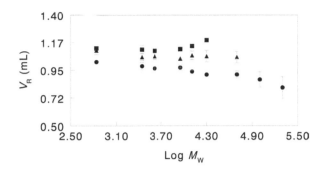

FIGURE 9.14 Variation of retention volume of polystyrene standards as a function of pressure for a 53.7% CO_2 in THF using a Jordi-Gel divinylbenzene column with 500 Å pore size 5 μm diameter particles: (●) 13.8 MPa, (▲) 9.31 MPa, (■) 6.89 MPa at 299 K. (Adapted from I. Souvignet, S. V. Olesik, *Anal. Chem.*, 69: 66 (1997).)

one of the polymers within the copolymer; then at those conditions the separation is based on the species that are not at their critical conditions, the other polymer(s).

For many polymer systems, it is difficult if at times impossible to find the critical condition of a component of polymer systems because the exact combination of solvent system and stationary phase is difficult to determine. Souvignet et al. [52] illustrated that EFL solvents are highly useful for LCCC because like supercritical fluid their solvent strength of the fluid can be varied by small changes in the pressure and temperature of the fluid. Figure 9.14 shows the calibration curve for polystyrene standards using an EFL THF/CO_2 mobile phase. By making small changes in the pressure, three different retention mechanisms were sampled (size exclusion, LCCC, and adsorption chromatography) [52]. Yun et al. [53] illustrated that highly efficient LCCC separations of functionalized polymers could be achieved with long columns under EF-LCCC conditions. Others showed that efficient separations of the components of copolymers could also be achieved [54]. Interestingly, the precise control of solvent strength with EFLs also allows polymers with a broader range of molecular weight to reach the critical condition before precipitation compared to that possible with conventional liquid mixtures [52,54].

9.5.5 GRADIENT POLYMER ELUTION CHROMATOGRAPHY

Another method of studying the chemical composition of copolymers is gradient polymer elution chromatography (GPEC) [55], which is also described as high-performance precipitation liquid chromatography (HPPLC) or liquid adsorption chromatography (LAC). With this technique, polymers are injected onto the column into a weak eluent. Owing to the limited solubility of the polymer in the eluent, precipitation onto the stationary phase occurs. The components of the polymers are then separated by applying a gradient with increasing solvent strength.

The addition of CO_2 to mobile phases in normal phase chromatography using silica gel stationary phases was used as an adsorption-promoting solvent [56]. Tetrahydrofuran or chloroform with 3.5% ethanol was the organic components in

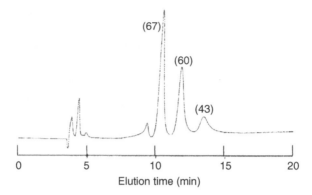

FIGURE 9.15   Separation of poly(styrene-comethyl methacrylate)s on a silica gel column with 5 µm particles using a solvent gradient including using chloroform with 3.5% ethanol added as the desorption promoting solvent and CO_2. Flow rates: CO_2, 0.5 mL/min; chloroform with ethanol additive (0.25–2.5 mL/min in 30 min) at 333 K and back pressure 20 MPa. (Reprinted from E. Kawai et al., *J. Chromatogr. A*, 991: 197 (2003). With permission.)

the EFL mixtures. Figure 9.15 shows the resulting chromatogram for statistical poly(styrene-comethyl methacrylate) with a solvent gradient that included CO_2 at a constant flow rate of 0.5 mL/min combining with a linear gradient of chloroform with 3.5% ethanol from 0.25 to 2.5 mL/min for 30 min. For both the THF and chloroform-based separations, the copolymers with the highest proportions of styrene had the least retention which indicated a normal-phase type retention mechanism. The separation that used chloroform (with 3.5% ethanol) and CO_2 had the highest resolution.

9.5.6 REVERSED-PHASE CHROMATOGRAPHY

Enhanced-fluidity liquid reversed-phase chromatography has numerous applications including the separation of nonpolar and polar compounds. For example, EFLC and nonaqueous reversed-phase HPLC are the common means of achieving effective separations of high molecular weight homologous compounds.

Gurdale et al. [57] studied the chromatographic characteristics of carbon dioxide/methanol and carbon dioxide/acetonitrile liquid mixtures under reversed-phase conditions for the purpose of determining which high fluidity mixtures would work best for the separation of high molecular weight compounds. Homologous series of alkylbenzenes with carbon numbers from 1 to 19 were used to determine methylene selectivity using 5–45% methanol or acetonitrile mixed with liquid CO_2. The acetonitrile/CO_2 provided a greater range in methylene selectivity than methanol/CO_2 mixtures (Figure 9.16). Increased proportions of acetonitrile caused enhanced methylene selectivity. A general trend for methylene selectivity as a function of methanol proportion for the methanol/CO_2 mobile phases. However, the 45% v/v methanol/CO_2 showed the greatest methylene selectivity but 5% methanol/CO_2 mixtures showed the next highest methylene selectivity. On the basis of variation of composition, temperature, and pressure of the mixtures, the polarity of the mixtures primarily controlled the selectivity variation and not density variations.

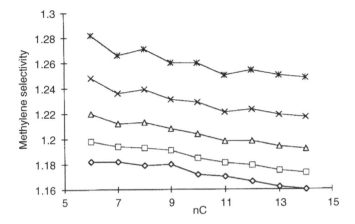

FIGURE 9.16   Variation of methylene selectivity (α_{CH_2}) as a function of carbon atom number (n_c) for the alkylbenzene series using acetonitrile as the modifier, pressure:10 MPa, temperature 283 K. (Reprinted from K. Gurdule et al., *Anal. Chem.*, 71: 2164 (1999). With permission.)

Temperature variation was more influential on methylene selectivity for the mixtures with high percentages of acetonitrile, while the reverse was true for methanol/CO_2 mixtures. Like in nonaqueous reversed-phase HPLC, a temperature increase lowered the methylene selectivity. The main conclusion from this work is that acetonitrile/CO_2 mixtures would be preferred over methanol/CO_2 mixtures as a first attempt to separate homologs.

Lee et al. [58] compared the reversed-phase EFLC to HPLC and elevated temperature HPLC for the separation of polyaromatic hydrocarbons using a Hypersil octadecyl polysiloxane column. The mole ratio of methanol/H_2O was held constant at 0.70/0.30 for the studied separations. Figure 9.17 shows that at 20.67 MPa, 299 K, and 0.12 cm/s linear velocity, a 0.49/0.21/0.30 mole fraction methanol/H_2O/CO_2 mixture achieved a separation of 16 polyaromatic hydrocarbons, PAHs, in 18 min as compared with 42 min when using 0.70/0.30 methanol/H_2O. By increasing the temperature of the 0.70/0.30 methanol/H_2O mobile phase mixture to 333 K and holding the other conditions constant, the separation was achieved in 16 min. However, by increasing the 0.49/0.21/0.30 mole fraction methanol/H_2O/CO_2 mobile phase to 333 K the separation of the 16 PAHs was achieved in less that 8 min with substantial improvement in chromatographic efficiency observed in comparison to all the other conditions studied. Cui et al. [30] also showed that using the 0.49/0.21/0.30 mole fraction methanol/H_2O/CO_2 mobile phase that the separation speed of separation, N/s, increased from 10 to 43 through the addition of 0.30 mole fraction CO_2 at 298 K and 20.67 MPa. Even greater increases in N/s were achieved by combining EFL with increased temperature. The rate constant for desorption of solutes from the reversed-phase stationary phase also increased by almost an order of magnitude with the addition of 0.30 mole fraction CO_2.

Lee et al. [59] later illustrated the use of four-coupled Hypersil octadecyl polysiloxane, reversed-phase HPLC columns, with the 0.49/0.21/0.30 mole fraction

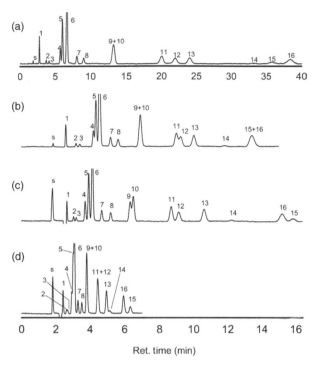

FIGURE 9.17 Separation of 16 polyaromatic hydrocarbons: s solvent; 1, benzene; 2, naphthalene; 3, acenaphthalene; 4, fluorene; 5, phenanthrene; 6, anthracene; 7, fluoranthene; 8, pyrene; 9, benzo[*a*]anthracene; 10, chrysene; 11, benzo[*b*]fluoranthene; 12, benzo[*k*]fluoranthene; 13, benzo[*a*]pyrene; 14, dibenzo[*a,h*]anthracene; 15, benzo[*ghi*]perylene; 16, indeno[1,2,3,-cd]- pyrene on a octadecylpolysiloxane column with 5 μm diameter particles using (A) 0.70/0.30 mole ratio methanol/H_2O mobile phase at 299 K; (B) same mobile phase as in A at 333 K; (C) 0.49/0.21/0.30 methanol/H_2O/CO_2 mixture at 299 K as the mobile phase; (D) same mixture as in C at 333 K as the mobile phase. (Adapted from S. T. Lee, S. V. Olesik, *Anal. Chem.*, 66: 4498 (1994).)

methanol/H_2O/CO_2 mobile phase and at 333 K produced efficiencies as high as 80,000 plates/m for 16 PAH priority pollutant standards as well as fat soluble vitamins and pharmaceuticals.

Phillips et al. [60] studied the use of reversed-phase EFLC for impurity/degradant profiling of a formulated drug candidate. Separations of drug candidate BMS-X with HPLC using typical reversed-phase stationary phases resulted in on-column hydroxylation. However, using a gradient elution separation (Figure 9.18), the drug candidates of interest, formulation components, and degradants are well separated in sixty minutes. This separation also did not require the addition of buffer to the mobile phases, which allows the monitoring of low molecular weight, weak chromophore process impurities by creating conditions favorable to greater retention, and detection at low wavelengths. The use of an unbuffered EFL also allows for the collection of high purity drug candidates and quick removal of the mobile phase.

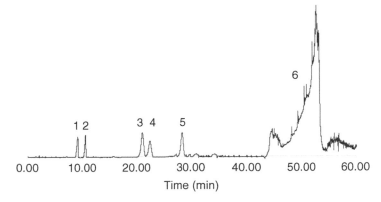

FIGURE 9.18 Separation of 1, methyl benzyl amine; 2, tartaric acid; 3, diastereomer; 4, drug candidate: BMS-X; 5, oxidation product; and 6, vehicle polymers. Using a ES-PFP column (250 × 4. 6 mm with 5 μm particles) with a mobile phase gradient starting with 95% modifier/5% CO_2 held for 25 minutes and ramped to 15% CO_2 for 5 minutes at 1.5 mL/min, 311 K and 18.0 MPa; modifier: 43% H_2O/56% methanol). (Unpublished data from S. L. Phillips et al., unpublished data. With permission.)

9.5.7 SHAPE SELECTIVITY CHROMATOGRAPHY

The separation of molecules of similar size but different shapes is a challenging problem. Coym and Dorsey [61] studied the chromatographic shape selectivity of monomer and polymeric ODS columns when liquid CO_2/acetonitrile mixtures with 0–100 v/v% acetonitrile, temperatures of 313–353 K, and a backpressure of 20.0 MPa were used. Three different methods of monitoring selectivity were used. The standard reference material (SRM) 869a developed by Sanders and Wise at the National Institute of Standards and Testing (NIST) [62], which involves determining the selectivity ratio, $\alpha_{TBN/BaP}$, for tetrabenzonaphthalene (TBN)/benzo[a]pyrene (BaP) pair was used as a numerical means of monitoring shape selectivity. A low value indicates enhanced shape selectivity. The selectivity of triphenylene/o-terphenyl pair was used to monitor planarity selectivity. This ratio monitors planarity because the two molecules have the same length/breath ratio but triphenylene is planar and o-terphenyl not [63]. Finally, the selectivity of the chrysene/benzo[a]anthracene pair was used to monitor length/breadth, L/B, selectivity. For the studied polyaromatic hydrocarbons on both the monomeric and polymeric stationary phases, increasing proportions of acetonitrile caused decreased retention which is behavior typical of a normal phase process.

For the monomeric stationary phase, Figure 9.19 shows the SRM 869a $\alpha_{TBN/BaP}$ decreasing with increasing proportions of aceonitrile in the mobile phase, which should indicate increased shape selectivity with added acetonitrile. This is the same observed trend when organic solvents are added to water in a reversed-phase separation. However, the selectivity measured from the triphenylene/o-terphenyl pair and the chrysene/benzo[a]anthracene pair showed exactly opposite trends in that the planarity and L/B selectivity was increased with decreasing proportions of acetonitrile.

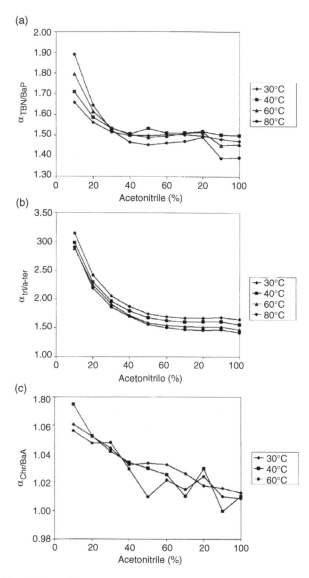

FIGURE 9.19 Effect of mobile phase composition on shape selectivity with a monomeric octadecylpolysiloxane stationary phase. column using (a) SRM 869a; (b) triphenylene/*o*−terphenyl; (c) chrysene/benzo[*a*]anthracene with column outlet pressure 20.0 MPa and flow rate 1 mL/min at pump head. (Reprinted from J. W. Coym, J. G. Dorsey, *J. Chromatogr. A*, 971: 61 (2002). With permission.)

Furthermore, increasing the temperature of the separation decreases the selectivity, which is similar to the trend observed in hydroorganic separations using ODS stationary phases.

For the polymeric stationary phase, the shape selectivity, as measured with SRM 869a $\alpha_{TBN/BaP}$, initially decreased with increasing acetonitrile in the mobile phase (Figure 9.20). However, for mobile phase compositions with >50% acetonitrile the SRM 869a $\alpha_{TBN/BaP}$ selectivity increased to values as large as for mobile phase mixtures with low proportions of acetonitrile. The shape of the curves for the planarity selectivity and the L/B selectivity were the same as those found for the monomeric stationary phases. However, the absolute values of the selectivities with the polymeric phases were markedly higher. The authors chose to accept the planarity and L/B selectivity indicators as the prime measures of the variation of selectivity with mobile phase composition in that they describe the shape selectivity generally increasing with decreasing proportions of acetonitrile.

These data are very interesting in that the measured shape selectivity seems dependent on the monitoring methods. Interestingly, the general shape of the curves for SRM 869a $\alpha_{TBN/BaP}$ selectivity and planarity selectivity as a function of acetonitrile concentration form the monomeric phase tracks well with the polarity variation of the mobile phases [17]. However, for the SRM 869a $\alpha_{TBN/BaP}$ with the polymeric phase the shape of the curve was much different which may indicate a mixed mode retention mechanism for those conditions. The authors suggested that solvation of the stationary phase by the mixed mobile phase was the cause of the observed variation in SRM 869a $\alpha_{TBN/BaP}$. Mores studies will be needed to better understand this phenomenon. However, the polymeric phase showed high planarity and L/B selectivity for <40% added acetonitrile.

9.5.8 ION-EXCHANGE OR ION-PAIR CHROMATOGRAPHY

Ion-exchanged and/or ion-pair chromatography using EFLC was recently evaluated. White and Burnett [64] recently illustrated that an ethyl pyridine silica-based column and a fast scan in a liquid CO_2/methanol mobile phase was more effective for drug discovery than reversed-phase HPLC. A 2 min fast gradient with five "calibration compounds" allowed ready scaling of the fast gradient to a preparative scale method that was 8.5 min long. The fast gradient consisted of starting with a mobile phase of 5% methanol/95% CO_2 then using a two-step ramp (first ramp was 25% v/v methanol/min) up to 40% methanol and then the second ramp was 75 v/v% methanol staring at 40% and ending at 60% methanol with a hold of 0.3 min). The methanol cosolvent contained a 0.2% isopropylamine, IPAm, to improve chromatographic peak shape. The preparative scale separation has a similar two step gradient but with more shallow slopes with a final mobile phase composition of 50 v/v% methanol (including 2% IPAm)/CO_2. The average resolution for 17 standard compounds was double that found for an optimized reversed-phase HPLC separation of the same standards. Furthermore, the evaporation time for the mobile phase in the preparative scale separation was reduced considerably.

Pinkston et al. [65] showed that the addition of low levels of ammonium salts to the modifiers allows the separation of polar and even ionic solutes such as sulfonate salts,

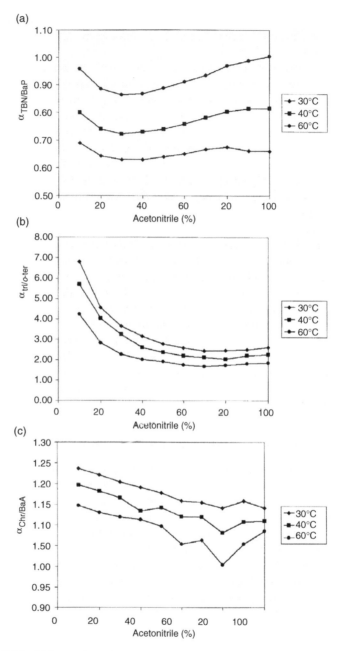

FIGURE 9.20 Effect of mobile phase composition on shape selectivity with a polymeric octadecyl-polysiloxane stationary phase. column using (a) SRM 869a; (b) triphenylene/*o*-terphenyl; (c) chrysene/benzo[*a*]anthracene with column outlet pressure 20.0 MPa and flow rate 1 mL/min at pump head. (Reprinted from J. W. Coym, J. G. Dorsey, *J. Chromatogr. A*, 971: 61 (2002). With permission.)

carboxylate salts, polyamines, and quaternary ammonium salts. As a direct result of this discovery, Zheng et al. [66] studied the interaction of sodium aryl sulfonates on bare silica and a cyano-bonded stationary phases using methanol/CO_2 mixtures that contained 2 mM of ammonium salts. These chromatographic studies combined with solid-state NMR provided evidence that the ammonium salts were adsorbing to the exposed silanols on the silica. As the amount of the ammonium ion in solution was increased, ammonium ion adsorbed to the stationary phase surface that generated a positively charge surface. Therefore, a significant fraction of the retention mechanism for the separation of the sulfonates was believed to be due to ion-exchange interaction at the surface.

As a direct result of the studies on the separation of sulfonates, Zheng et al. [67] showed that secondary amines and quaternary ammonium salts could be separated using a cyanophase column using sodium alkyl sulfonate additives in methanol/CO_2 mixtures with 30–40% methanol in the mixture. Ion-pairing was thought to be the retention mechanism responsible for the separation.

The separation of the same charged compounds were also accomplished on an ethyl-pyridine bonded silica surface and 30–40% methanol/CO_2 mobile phases without the need of added sulfonate modifier. Anionic compounds did not elute from the ethyl-pyridinium surface that lead the authors to hypothesize that the surface was positively charged. To further test this hypothesis, the separation of the same compounds on a strong anion exchange column, silica-based propyltri-methylammonium cationic surface, which exhibits are permanent positive charge was attempted. The same retention order was observed on the strong cation exchange surface.

9.6 SUMMARY AND FUTURE OF EFLC

In eight different LC techniques, EFLs were shown to provide advantages over commonly used liquids or supercritical fluids. High-speed gradient EFLC separations are becoming increasingly used for drug profiling. EFLC is one of the most commonly used techniques for chiral separations both at the analytical scale and preparative scale. LCCC is markedly more useful for organic polymer separations using EFLC because the solvent strength of EFLs can be precisely controlled with pressure or temperature variation. Similarly, GPEC using EFLs has high efficiencies and precise selectivity due to the precise control of the polymer precipitation under EFL solvent strength variation. Reversed-phase separations of a broad range of compounds are markedly more efficient and faster using EFLC than using an organic/H_2O mobile phase with markedly enhanced-shape selectivity possible with EFLC compared to hydroorganic mobile phase.

The expected areas of future expansion of EFLC are in the separation of highly polar solutes. Mixtures of 61.7/27.7/10.6 and 55.7/25/1/19.2 mol ratio methanol/H_2O/CO_2 were predicted to have dielectric constants of 38 and 34 [28], respectively. Ion-exchange EFLC should be viable in these and other higher polar EFL mixtures, such as acetontrile/H_2O/CO_2 and THF/H_2O/CO_2 mixtures. Ultra high-speed gradient separations may be possible in EFLC as a result of the fast desorption rate constants for solutes under EFLC conditions.

ACKNOWLEDGMENT

We are grateful for the support of this work through funding from the National Science Foundation.

REFERENCES

1. F. Erni, *J. Chromatogr.*, 282: 371 (1991).
2. H. Poppe, *J. Chromatogr. A*, 778: 3 (1997).
3. F. D. Anita, C. Horvath, *J. Chromatogr.*, 435: 1 (1988).
4. H. Chen, C. Horvath, *J. Chromatogr. A*, 705: 3 (1995).
5. Y. Cui, S. V. Olesik, *Anal. Chem.*, 63: 1812 (1991).
6. C. A. Eckert, C. L. Liotta, *J. Phys. Chem. B*, 110: 21189 (2006).
7. G. Musie, M. Wei, B. Subramaniam, D. H. Busch, *Coord. Chem. Rev.*, 219–221: 789 (2001).
8. M. Wei, G. Musie, D. H. Busch, *J. Am. Chem. Soc.*, 124: 2513.
9. J. C. Giddings, *J. Chromatogr.*, 13: 301 (1964).
10. J. C. Giddings, *Unified Separation Science*, John Wiley and Sons, New York (1991).
11. G. Guichon, *Optimization in Liquid Chromatography in High Performance Liquid Chromatography: Advances and Perspectives Vol 2*, C. Horvath, ed., Academic Press, New York (1980).
12. E. J. Beckman, *J. Supercrit. Fluids*, 28: 121 (2004).
13. M. Kanakubo, T. Umecky, C. C. Liew, T. Aizawa, K. Hatakeda, Y. Ikushima, *Fluid Phase Equil.*, 194–197: 859 (2002).
14. I. Souvignet, S. V. Olesik, *J. Phys. Chem.*, 99: 16800 (1995).
15. C. A. Eckert, D. H. Ziger, *J. Phys. Chem.*, 90: 2738 (1986).
16. T. Aida, H. Inomata, *Mol. Simul.*, 30: 407 (2004).
17. H. Li, M. Maroncelli, *J. Phys. Chem. B*, 110: 21189 (2006).
18. S. V. Olesik, J. L. Woodruff, *Anal. Chem.*, 63: 670 (1991).
19. P. G. Debenedetti, R. C. Reid, *AIChE J.*, 32: 2034 (1986).
20. P. R. Sassiat, P. Mourier, M. H. Caude, R. H. Rosset, *Anal. Chem.*, 59: 1164 (1987).
21. H. Yuan, I. Souvignet, S. V. Olesik, *J. Chromatogr. Sci.*, 35: 409 (1997).
22. I. Bako, P. Jediovszky, G. Palinkas, *J. Mol. Liq.*, 87: 243 (2000).
23. S. T. Lee, S. V. Olesik, *Anal. Chem.*, 66: 4498 (1994).
24. I. Souvignet, S. V. Olesik, *Anal. Chem.*, 70: 2783 (1998).
25. R. L. Smith, Jr., C. Saito, S. Suzuki, S. B. Lee, H. Inomata, K. Arai, *Fluid Phase Equil.*, 194–197: 869 (2002).
26. G. Levitin, D. W. Hess, *Electrochem. Solid-State Lett.*, 8: G23 (2005).
27. A. P. Abbott, J. C. Harper, *J. Chem. Soc. Faraday Trans.*, 92: 3895 (1996).
28. D. Wen, S. V. Olesik, *Anal. Chem.*, 72: 475 (2000).
29. V. T. Wyatt, D. Bush, J. Lu, J. P. Hallett, C. Liotta, C. A. Eckert, *J. Supercritical Fluids*, 36: 16 (2005).
30. Y. Cui, S. V. Olesik, *J. Chromatogr. A*, 691: 151 (1995).
31. T. S. Reighard, S. T. Lee, S. V. Olesik, *Fluid Phase Equil.*, 123: 215 (1996).
32. D. W. Jennings, R.-J. Lee, A. S. Tejak, *J. Chem. Eng. Data*, 36: 303 (1991).
33. S. N. Joung, C. W. Yoo, H. Y. Shin, S. Y. Kim, K.-P. Yoo, C. S. Lee, W. S. Huh, *Fluid Phase Equil.*, 185: 219 (2001).
34. J. Im, J. Lee, H. Kim, *J. Chem. Eng. Data*, 49: 35 (2004).
35. T. Adrian, G. Maurer, *J. Chem. Eng. Data*, 42: 668 (1997).
36. M. J. Lazzaroni, D. Bush, J. S. Brown, C. A. Eckert, *J. Chem. Eng. Data*, 50: 6 (2005).

37. M. Kato, D. Kodama, M. Sato, K. Sugiyama, *J. Chem Eng. Data*, 51: 1031 (2006).
38. J. Im, M. Kim, J. Lee, H. Kim, *J. Chem. Eng. Data*,49: 243 (2004).
39. Y. Houndonougbo, H. Jin, B. Rajagopalan, K.Wong, K. Kuczera, B. Subramaniam, B. Laird, *J. Phys. Chem. B*,110: 13195 (2006).
40. J. Zhao, S. V. Olesik, *Fluid Phase Equil.*, 154: 261 (1999).
41. S. T. Lee, T. S. Reighard, *Fluid Phase Equil.*, 122: 223 (1996).
42. T. A. Berger, W. H. Wilson, *J. Biochem. Biophys. Methods*,43: 77 (2000).
43. T. A. Berger, K. Fogelman, T. Staats, P. Bente, I. Crocket, W. Farrell, M. Osonubi, *J. Biochem. Biophys. Methods*, 43: 87 (2000).
44. Q. Sun, S. V. Olesik, *Anal. Chem.*, 71: 2139 (1999).
45. Y. Liu, C. R. Mitchell, T. L. Xiao, B. Zhang, D. Armstrong, *J. Chromatogr. A*, 978: 185 (2002).
46. C. White, *J. Chromatogr. A*, 1079: 163 (2005).
47. S. H. Hoke, II, J. D. Pinkston, R. E. Bailey, S. L. Tanguay, T. H. Eichhold, *Anal. Chem.*, 72: 4235 (2000).
48. J. Boisvert, G. Caille, I. J. McGilveray, S. A.Qureshi, *J. Chromatogr. B Biomed. Sci. Appl.*, 678: 237 (1995).
49. N. Rifai, M. Lafi, M. Sakamoto, T. Law, *Ther. Drug, Monit.*, 19: 175 (1997).
50. F. Geiser, R. Shah, *Chirality*, 16: 263 (2004).
51. F. Geiser, M. Schultz, L. Betz, M. Shaimi, J. Lee, W. Champion, Jr., *J. Chromatogr. A*, 865: 227 (1999).
52. I. Souvignet, S. V. Olesik, *Anal. Chem.*, 69: 66 (1997).
53. H. Yun, S. V. Olesik, E. Marti, *J. MicroColumn Sep.*, 11: 53 (1999).
54. S. Phillips, S. V. Olesik, *Anal. Chem.*, 74: 799 (2002).
55. H. J. A. Philipsen, *J. Chromatogr. A*, 1037: 329 (2004).
56. E. Kawai, K. Shimoyama, K. Ogino, H. Sato, *J. Chromatogr. A*, 991: 197 (2003).
57. K. Gurdale, E. Lesellier, A. Tchapla, *Anal. Chem.*, 71: 2164 (1999).
58. S. T. Lee, S. V. Olesik, *Anal. Chem.*, 66: 4498 (1994).
59. S. T. Lee, S. V. Olesik, S. Fields, *J. MicroColumn Sep.*, 7: 477 (1995).
60. S. L. Phillips, B. Kleintop, N. Khaselev, unpublished data.
61. J. W. Coym, J. G. Dorsey, *J. Chromatogr. A*, 971: 61 (2002).
62. L. C. Sanders, S. A. Wise, *Anal. Chem.*, 59: 2309 (1987).
63. K. Jinno, T. Nagoshi, N. Tanaka, M. Okamoto, J. C. Fetzer, W. R. Biggs, *J. Chromatogr.*, 392: 75 (1987).
64. C. W. White, J. Burnett, *J. Chromatogr. A*, 1074: 175 (2005).
65. J. D. Pinkston, D. T. Stanton, D. Wen, *J. Sep. Sci.*, 27: 115 (2004).
66. J. Zheng, T. Glass, L. T. Taylor, J. D. Pinkston, *J. Chromatogr. A*, 1090: 155 (2005).
67. J. Zheng, L. T. Taylor, J. D. Pinkston, *Chromatographia*, 63: 267 (2006).

10 Comprehensive Two-Dimensional Gas Chromatography

Danielle Ryan and Philip Marriott

CONTENTS

This chapter will present an overview of the comprehensive two-dimensional gas chromatography (GC×GC) technique, and in particular, in the context of considering the second dimension column separation as a monitor or analysis step subsequent to the first dimension. This allows the conceptual comparison of the hyphenated system GC×GC with the more familiar technique of gas chromatography–mass spectrometry (GC–MS). In hyphenated systems that offer truly greater information content, the two dimensions should be orthogonal (see later), and generally, in the chromatography realm, spectroscopic detection provides the most recognizable orthogonal second dimension to the chromatography separation step.

Thus, comparing GC–MS with GC×GC, in the former instance, we have

Stage 1 = GC separation: Stage 2 = mass spectral analysis

and in the latter instance, we have

Stage 1 = GC separation: Stage 2 = gas chromatography analysis.

Figure 10.1 presents a diagrammatic comparison of the two. Whereas GC/MS will present an undifferentiated mass spectrum of all ions that arise from overlapping

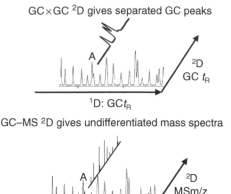

GC×GC ^{2}D gives separated GC peaks

GC–MS ^{2}D gives undifferentiated mass spectra

FIGURE 10.1 Contrast between GC×GC and GC–MS. In the former, solutes are continuously analyzed in the second dimension, to ideally provide separation on ^{2}D. In the latter, the MS will simply record all ions that arise from ionization processes in the source, regardless of peak overlap.

GC peaks (e.g., see peaks marked A in Figure 10.1), GC×GC offers the opportunity to resolve peaks in the second dimension stage. The undifferentiated mass spectra can only provide useful information where precise deconvolution of mass spectra will permit the individual overlapping compounds to be identified or measured. A more relevant comparison of the two systems would be the case of a mass spectrometer that generates only (or predominantly) molecular ions (e.g., such as using a chemical ionization method without fragmentation) so that one ion for each overlapping compound is obtained. In this case, we can then deduce how many peaks coeluted at that time. We can distinguish two fundamental cases that contrast the two systems. The first is where compounds elute from the first stage as well separated peaks. The second is where compounds are overlapping at the end of the first stage.

With GC–MS, the role of the MS stage is to provide identification. In a well-resolved chemical system, this should be a relatively reliable task (refer to any chapter on GC–MS), and for most analyses depends only on the specificity of the MS library search process. In an application where there is more than one (and this can mean multiple) compound eluting from the first stage at a given time, the mass spectrometer scan process will produce an undifferentiated mass spectrum of all the compounds that are present at that time. This is generally a complication that the analyst will not prefer. Taken in isolation, a mass spectrum of multiple coeluting compounds is of little value of its own. There are steps that can be taken to improve the specificity or the purity of the mass spectrum, such as background subtraction, but this is not useful where the subtraction protocol fails, such as might be the case for minor compounds interfered with by major components, or where many peaks coelute.

For GC×GC, if a peak from the first stage is fully resolved, we might believe that the second separating dimension offers little advantage. And again, in isolation, many would agree; if there are no overlapping peaks, what purpose does the second separation step play? However, as we see later, GC×GC offers a structured retention property, in a way that is not available to single column analysis. And in an orthogonal system, there is chemical information and a degree of specificity of retention that gives molecular identification possibilities. But we should also accept that GC×GC has its greatest value in the second case, where overlapping peaks elute from the end of the first stage. Here, the second GC column provides the opportunity for quantitative separation of these compounds (depending on column-phase choice and dimensions). This will be immediately translated into improved chemical analysis—quantitatively and qualitatively. Beyond this, there is now also the possibility of hyphenation of GC×GC with MS (see later) to provide an extraordinarily powerful identification system for chemical analysis of volatile compounds.

This chapter will begin by introducing GC×GC with focus on its general advantages of separation power, sensitivity, selectivity, and two-dimensional chromatogram structure, the technology of modulation, detector considerations including MS, and finally, present selected applications, which highlight various performance attributes of GC×GC.

10.1 INTRODUCTION

Comprehensive two-dimensional gas chromatography (GC×GC) was pioneered by Liu and Phillips [1] in 1991. In their seminal work, a metal-coated column providing two-stage elevated temperature (thermal) modulation was used to interface two serially coupled and somewhat orthogonal GC columns, ensuring the comprehensive analysis of a hydrocarbon standard mixture and a coal liquids sample. Despite the relative infancy of GC×GC, its implementation has been growing steadily, in recognition of the technique's proven superiority as compared to single-dimension gas chromatography, with respect to (i) peak capacity and resolving power, (ii) sensitivity, and (iii) the generation of structured chromatograms. It is also evident from conference programs that the GC×GC technique is attracting much attention from the GC community, since it is one of the main developments in GC over the past decade.

10.1.1 PEAK CAPACITY

The term peak capacity was coined by Giddings [2] in 1967 to describe the resolving power of a chromatographic separation. By definition, peak capacity represents the number of compounds that can be placed side by side, and evenly spaced, in a separation channel with a given minimum resolution performance (e.g., $R_s = 1.5$) for neighboring components. In a one-dimensional analysis, peak capacity is essentially determined by the resolution power of the single chromatographic column. Thus, any improvements in peak capacity can only be achieved at the expense of analysis time, for example, by using a longer column to improve resolution, or by using a narrower bore column, which has the effect of reducing sample capacity and increase inlet pressure. Conversely, in GC×GC, system peak capacity is the multiplicative peak capacity of each of the (two) dimensions, and because "fast" separation is performed on the short second dimension column, total analysis times with GC×GC are comparable to those of GC. Opportunities for solute resolution are by far superior using GC×GC, and highest resolution power is achieved when there is no correlation between column separation mechanisms, and orthogonal conditions exist.

Orthogonality in GC×GC may be achieved when the mechanisms controlling component separation in the two dimensions are very much different, and the characteristic chemical–physical parameter(s) leading to retention in one dimension is (are) independent of the parameter(s) in the second. Any correlation across the dimensions reduces the maximum theoretical peak capacity. In fact, a high degree of retention correlation can reduce a multidimensional separation to an essentially one-dimensional separation, with peaks distributed along a diagonal [3]. Nevertheless, an orthogonal column set does not necessarily result in optimum sample resolution, but rather offers the best opportunity for maximum use of the available separation space. It is only through correct instrumental tuning that column orthogonality can be fully exploited. In GC×GC, improved orthogonality may be achieved by temperature tuning, as demonstrated by Phillips and Beens [4]. By correctly raising the temperature, retention in the second column decreases to compensate (exactly) for the decreasing volatility of substances eluting from the first column. Recently, concepts of orthogonality have become increasingly important for the prediction of two-dimensional separations, and subsequent utilization of the two-dimensional separation space [5].

10.1.2 Sensitivity

The sensitivity enhancement capabilities of GC×GC compared to GC remains a contentious issue [6,7], however, it is clear that signal enhancement, in terms of peak height (but not peak area), is achieved using GC×GC compared to single dimension GC. This increase in peak signal arises due to the peak focusing that occurs during the modulation process; however, the extent of signal enhancement will depend on the modulation period (P_M) which is used. A smaller modulation period ensures greater sampling frequency of the first dimension peak, however, at the expense of signal enhancement. The modulation process also affects the extent to which resolution achieved in the first dimension is preserved. The apparent first dimension resolution is degraded as P_M increases, although the potential for signal enhancement increases, since the primary dimension peaks are collected in fewer and larger sample amount fractions [8]. Murphy et al. [9] investigated the effect of sampling rate on resolution in comprehensive two-dimensional liquid chromatography (LC×LC), and found that at least three or four samples per first-dimension peak should be sampled into the second dimension, depending on whether the sampling is in phase, or maximally out of phase, respectively. This will ensure high two-dimensional resolution. As a general rule, the recommendations of Murphy et al. [9] are abided by most GC×GC users.

However, signal also depends on detector data acquisition rate; noise increases with increased detection acquisition rate. Since GC×GC peaks are narrow, necessitating higher detection frequencies, S/N ratio improvements are not as great as might be initially anticipated.

10.1.3 STRUCTURED CHROMATOGRAMS

All chromatograms (even for single column analysis) have some degree of structure—for instance, homologous compounds elute with a predictable retention difference. As complexity increases, any evident pattern or structure is lost. Under orthogonal GC×GC conditions, the elution of particular compound classes will occur at defined positions within the two-dimensional separation space. Such structured retention, which is not so readily apparent using single column GC, is particularly beneficial in group-type analyses, and aids in the provisional classification of unknown analytes, for instance, by allowing their chemical–physical property to be estimated. Within a particular retention band, homologues will elute in a particular order, and structure can be assigned to the extent of analyte branching; for example, branched alkanes will elute before that of the normal alkane (of the same carbon number) in both first and second dimensions when a nonpolar first dimension is used together with a polar second dimension. This leads to the so-called "roof tile" effect, where branched alkanes are positioned to the lower left of their respective normal alkane (i.e., same C number) in the two-dimensional separation space [10]. Similar roof tile patterns relating to the separation of normal and branched members of a series are observed for other classes of compounds, with roof tiles becoming steeper as retention in the second dimension (and hence analyte polarity) increases. This is well demonstrated by alkyl naphthalenes, as shown in Figure 10.2, which compares the analysis of a diesel sample by GC–MS and GC×GC. Figure 10.2 clearly demonstrates the predictive power of GC×GC derived from structured retention and roof tile patterns compared to single-dimension analysis. Such capabilities are particularly useful for the analysis of complex petroleum type samples, where saturated, cyclic saturated, olefinic, heteroatomic, and aromatic elution occurs in defined retention bands.

10.2 MODULATORS

The modulator is the heart of the GC×GC system, and is positioned at the confluence of the coupled chromatography columns. The role of the modulator is to trap or isolate compounds present in a given time fraction eluting from the first-dimension column and reinject these components rapidly into the second column. This essentially yields a time-sampled chromatogram, from the first dimension (^{1}D) to the second dimension (^{2}D). It is critical that the modulator is capable of representatively and faithfully sampling peaks eluting from ^{1}D onto ^{2}D. This can be achieved by either complete or partial transfer of the first-column eluent, however, both techniques are considered comprehensive.

Essentially, modulators can be classified as either thermal or valve based. All thermal modulators described thus far are mass conservative. They are mass conservative because in almost all cases, the ^{1}D and ^{2}D columns are directly coupled, so that all solutes pass from ^{1}D to ^{2}D. These operate using thermal gradients, either through heating or cryogenic cooling. Heated thermal modulators operate through rapid heating of a short section of thin film (i.e., solute retarding) "modulating" column at the end of the ^{1}D column to facilitate solute acceleration into a narrow band for

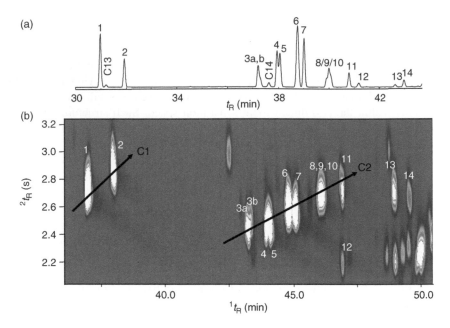

FIGURE 10.2 Characterization of alkyl naphthalene compounds in diesel using GC–MS (A) and GC×GC-FID (B). A first dimension column of 5% phenyl methyl polysilphenylene siloxane (BPX5; 30 m × 0.25 mm ID × 0.25 μm d_f) was used in both cases, and a ^{2}D of polyethylene glycol (BP20; 1 m × 0.1 mm ID × 0.1 μm d_f) was used in the latter. Figures A and B are approximately vertically aligned. Figure A shows the limited resolving power of single dimension analysis, whilst Figure B illustrates the structured retention capabilities of GC×GC for the C1– and C2-naphthalenes. Within each alkyl-naphthalene isomer group, retention in both dimensions decreases as the degree of branching increases, thus resulting in the observed roof tile effect. Roof tiles become steeper as retention in the second dimension increases. Compounds 1 and 2 refer to 2-methyl-naphthalene and 1-methyl-naphthalene, respectively, whilst compounds 3–11 represent the C2-napthalenes.

second-column (^{2}D) analysis. This was the concept that led to the design known as the thermal sweeper [11]. Cryogenic modulators operate by application of a cryogen at the end of the ^{1}D column or at the start of the ^{2}D column to retard and focus solute peaks. Rapid reheating of the column segment then enables "pulsing" of the time-sampled solute peaks between ^{1}D and ^{2}D. Cryogenic modulators are advantageous compared to thermal modulators in that the modulator does not need to be heated to temperatures greater than the GC oven temperature [8]; this enables higher GC oven temperatures to be used than those used with thermal modulators, and at present, elevated temperature thermal modulators are rarely used, having been replaced by cryogenic approaches. Examples of cryogenic modulators include that which was first described for GC×GC—longitudinally modulated cryogenic system (LMCS) [12], and subsequent variations on this theme including dual cryogenic jet system [13,14], the single-jet and single-stage cryogenic modulator [15], and the four-jet cryogenic modulator [16], and the loop modulator.

Valve-based modulators rely on flow switching mechanisms, and traditionally discrete (sub)sampling of ^{1}D effluent has been performed for subsequent separation on ^{2}D [17,18]. Valve switching may be performed at high frequencies, which is consistent with and enables very fast second dimension separations of ≤ 1 s [19]. Additionally, breakthrough problems, which may arise with cryogenic modulators as a result of inefficient trapping of very volatile compounds, are eliminated. However, problems have arisen where valve-based interfaces with limited temperature capacities are positioned within the chromatographic flow path [8], necessitating reduced maximum GC oven temperatures. This, of course, is detrimental for the analysis of high boiling point solutes. This problem has recently been addressed by Seeley and coworkers [20] who have developed a novel valve-based modulator, which is positioned outside the GC oven. This modulator incorporates two sample loops and a pneumatic switching system, and represents the only valve-based modulator that samples the entire ^{1}D effluent. It is therefore expected that implementation of this modulator will become popular; however, reliable operation of the modulator requires careful balancing of the pneumatics [8] and system flow rates.

10.3 DETECTORS

Peaks eluting from the second column of the GC×GC system typically have a base width of 150 ms or smaller; peaks as narrow as 50 ms base width have been reported. By necessity then, high data acquisition speed detectors must be used in GC×GC to properly monitor these fast eluting peaks, and detectors must also have very small internal (dead) volumes [21] to prevent band broadening, and preserve the integrity of the peak shape. Data acquisition rates of 50–200 Hz are generally required to ensure in the order of 10–15 data measurements per peak, and thus reliable representation of the chromatographic data. This corresponds to about four data points per peak standard deviation (4 points/σ). Contrary to this requirement, detectors with limited data acquisition rates of less than 50 Hz, such as quadrupole MS (qMS) and the atomic emission detector (AED) have been used favorably with GC×GC. In some instances, instrument manipulation has been required in order to ensure the collection of an adequate number of peak data points (see later).

The choice of detector used in GC×GC, as with single dimension GC, will also depend on the required chemical analysis, and the sensitivity requirements for the detection of these analytes. Detector selectivity is another important consideration. A selective detector is capable of isolating response signals for improved analysis (i.e., response vs. no response for target and matrix, respectively). Analogously, GC×GC provides selectivity of separation through enhanced temporal separation of analytes, facilitated by enhanced peak capacity. It may therefore be stated that separation selectivity in the second column in GC×GC is somewhat equivalent to signal selectivity in specific detection in conventional GC [22], although the timescale of these two measurements may be quite different. The inherent selectivity of GC×GC combined with selective detection therefore offers unrivalled potential in terms of sensitivity and detection capabilities, particularly when compared to one-dimensional counterparts and the more universal GC×GC-Flame ionization detector (FID).

Nevertheless, GC×GC-FID represents the most widely used form of GC×GC to date. This is because FID is a universal detector and is highly suited to the analysis of volatile organic compounds, plus it is an excellent choice for method development. Significantly, the detector is fast and can acquire data at rates of up to 200 Hz. Furthermore, it has a negligible dead volume, which ensures limited band broadening and thus maintains the column-based resolution. Recently however, attention has turned to GC×GC with selective detection in order to both simplify the analysis of highly complex samples (by responding to only target molecules) and push the boundaries of detection and sensitivity thresholds (as described earlier). Kristenson et al. 2003 [23] have evaluated the suitability of three commercially available electron capture detectors (ECDs) for use with GC×GC. The narrowest peaks were achieved using the Agilent μ-ECD operated at 50 Hz, which has an internal volume of 150 μL; a 10-fold reduction in internal volume compared to the standard Agilent ECD. However, band broadening was still observed when using the μ-ECD compared to that of an FID, which can be considered to be the best performing GC×GC detector in terms of maintaining peak profile. This was attributed to the comparatively minor dead volume of the FID of 10–20 μL (cited in Reference 23). Improvements in peak width using the μ-ECD could be achieved by using large make-up flow rates combined with high detection temperatures, yielding acceptable GC×GC results with respect to peak shape and resolution (although not as good as that with the FID under equivalent conditions). It must be realized however that greatly increasing make-up flow rates can result in compromised sensitivity due to dilution effects.

The sulfur chemiluminescence detector (SCD) enables highly sensitive, equimolar detection of sulfur-containing compounds, with an associated sensitivity of <0.5 pg sulfur s^{-1} [24]. Hua et al. [24] have used GC×GC-SCD for the differentiation and characterization of sulfur-containing compounds in diesel oil fractions. The SCD, operated at 100 Hz, produced second dimension peaks of 0.8–1.2 s wide, indicating that the detector was not sufficiently fast for GC×GC analyses. Hua et al. [24] attributed the broadness of these peaks to the larger cell volume of the SCD (compared to that of the more commonly used FID). Blomberg et al. [25] however refute this claim, and demonstrated that the lack of speed of the SCD is largely attributed to the speed of the electronics used, rather than the physical dimensions of the cell. Blomberg et al. [25] connected an electrometer intended for use with a flame-photometric detector (FPD) directly to the photomultiplier tube (PMT) of the SCD. The SCD with modified FPD electrometer, operated at 50 Hz, showed significantly improved ^{2}D peak shape compared to that of the original, unmodified SCD. Furthermore, peak widths decreased by approximately 50% to yield peaks comparable in width to those obtained using GC×GC-FID. Improvements in the electronic speed of the standard SCD should therefore make GC×GC-SCD a highly sensitive and powerful tool for the analysis of complex sulfur-containing samples.

AED combined with GC×GC was first reported by van Stee et al. [21] for element-selective detection in petrochemical samples. The AED used in this study had a maximum data acquisition rate of 10 Hz, which when applied to the detection of GC×GC peaks of base width of 100–300 ms, will not achieve the required minimum of 5–10 data points per peak. In order to meet this criterion, peak widths need to be broader, with necessary widths of 400+ ms required. Deliberate broadening of second dimension peaks was therefore facilitated by increasing the internal diameter of the transfer line connecting the GC and AED from 0.1 to 0.25 mm, and "optimization" of gas flow rates. Such modifications enabled sufficient band broadening and the generation of 5–6 data points per second-dimension peak.

Most recently, GC×GC with nitrogen phosphorus detection (GC×GC-NPD) has been reported, with the NPD operating reliably at 100 Hz [22]. However, it was observed that peak asymmetry and peak magnitude were greatly affected by detector gas flows. Peak magnitude (height) was found to be dependent on detector hydrogen and airflows, and maximum peak heights were observed when both gases were maximized. Peak asymmetry, however, was primarily dependent on detector airflow, and best peak symmetry was achieved when air flow was maximized. Significantly, second dimension peak asymmetry (^2A_s) ranged from a minimum of 1.6 to a maximum of 8.0, respectively, with the latter indicative of "source tailing" [26]. Optimization of the detector with respect to gas flows is therefore critical in GC×GC-NPD, since second-dimension peak tailing is detrimental to separation power. In terms of sensitivity, GC×GC-NPD was found to be approximately 20 times more sensitive (based on signal-to-noise ratios) for the detection of nitrogen containing methoxypyrazines analytes as compared to GC×GC-FID, and GC×GC-NPD had a larger linear detection range compared to GC×GC-FID. Such data reiterates the benefits of GC×GC combined with selective detection.

In all cases for the described detectors (μ-ECD, SCD, AED, NPD), modifications to either chromatographic separations (peak broadening) or detector parameters (flow rates, electronic speed) were necessary in order to produce satisfactory GC×GC results. It is therefore apparent that detector manufacturers need to develop new optimized detector geometries and evaluate detector performance in light of the requirements of GC×GC, in order for the routine implementation of GC×GC with selective detection. Such developments will impact significantly on the field of GC×GC by greatly simplifying the use of selective detectors, effectively eliminating the need for time-consuming manipulations and optimization before GC×GC analyses.

10.4 MASS SPECTROMETRY CONSIDERATIONS FOR GC×GC

The need for mass spectral detection as a routine tool in single dimension GC is much stronger than that for GC×GC analysis. Since the resolving power of GC is restricted to the single chromatographic column, where this is inadequate, MS detection provides capability for compound identification on the basis of spectral resolution. GC–MS, like GC×GC, is in fact an orthogonal two-dimensional analytical technique, which utilizes independent dimensions of retention time and mass spectra for sample characterization. In some cases, GC×GC is more powerful than GC–MS as a direct result of peak capacity and structured retention capabilities (outlined earlier). This is certainly the case for the analysis of isomers, where resolution and identification is entirely determined by the GC separation stage. In a comparative study of GC×GC and GC–MS for the analysis of complex hydrocarbon mixtures [10], the former was

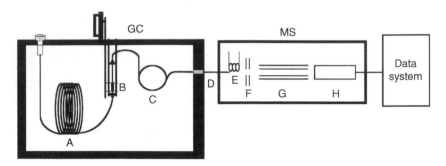

FIGURE 10.3 Schematic diagram of a GC×GC–MS system. A: ^{1}D column; B: LMCS cryogenic modulator; C: ^{2}D column; D: transfer line; E: ion source region; F: focusing lenses; G: analyzer region; H: electron multiplier.

labeled as powerful, as GC–MS is *potentially* powerful. The authors ascertained that GC–MS has not yet reached its potential separation power because a universal soft ionization method does not exist, and that quantification by GC×GC-FID is much more robust, reliable, and reproducible as compared to GC–MS. Note that MS does not provide peak resolution in time in the same manner as that of GC, but it does provide ion separation. GC–MS can be considered a two-dimensional separation method essentially when the MS gives a single ion per peak (i.e., as would be the case for a universal soft ionization process), provided then that peaks can be adequately resolved using GC×GC, there is no need for mass spectral detection to provide unique discrimination. Where resolution capabilities of GC×GC are insufficient, however, the three-dimensional technique of GC×GC–MS (Figure 10.3) offers a highly powerful information-rich alternative.

The MS of choice for GC×GC analyses is the time-of-flight mass spectrometer (TOFMS). This is a consequence of the fast spectral acquisition rate of the instrument, which is able to present spectra at rates up to 500 Hz. Using an acquisition rate of 50–100 spectra s^{-1}, therefore, equates to around 10–20 spectra per peak for peaks of 0.2 s width at baseline [27], ensuring accurate reconstruction of the two-dimensional chromatogram. Fast time-of-flight mass acquisition capabilities also prevents the occurrence of concentration skewing as the rapid chromatographic peak mass flux changes in the detector, thus preserving spectral continuity and enabling spectral deconvolution of coeluting chromatographic peaks with differing fragmentation patterns [28]. TOFMS deconvolution therefore represents an "additional tool" [29] to aid separation when combined with GC×GC analysis. This has been demonstrated by Dallüge et al. in Reference 29, wherein pesticide peaks that were separated by only 0.1 s in the second dimension were deconvoluted on the basis of unique ions, yielding suitably high similarity factors, and so reliable identification. It is therefore not surprising that GC×GC-TOFMS has been described as providing separation and identification power, which is unequalled in the field of gas chromatography [30].

Additional benefits of GC×GC-TOFMS include the generation of extracted ion chromatograms (EICs) postanalysis. This is particularly useful for group type analysis, or simplifying the identification of target analytes. In both cases, specific ions are chosen for the particular compound class or individual target with reference to the library reference spectra; searching for unique ions greatly improves chances for compound identification. Generation of EICs has been used by numerous groups, particularly for the analysis of highly complex samples such as coffee bean head-space and cigarette smoke. In the former case [31], alkyl substituted pyrazines were isolated using summed abundant spectral ions together with the unique ion. Summa-tion of these ions was found to yield significantly increased responses as compared to EICs based on a single ion, thereby simplifying the identification process. Similarly, in the analysis of cigarette smoke [32], alkyl substituted pyridines, pyrazines, and quinolines/isoquinolines were identified using EICs based on summed ions. EICs yiel-ded roof-tile structures of different alkyl substituted pyridines as a result of structured retention in GC×GC, which further validated TOFMS identifications.

Despite the advantages of TOFMS detection in GC×GC analyses, there is a gen-eral consensus that processing times for GC×GC-TOFMS analyses are arduously long, particularly for the analysis of highly complex samples. This is a direct result of the very large data files that are generated and the peak processing protocols that are employed in data reduction. Significant improvements have been made in the TOFMS software, however, data processing of GC×GC-TOFMS data still remains challenging and time consuming. It is therefore not surprising that there has been a recent increase in the application of GC×GC-qMS, which is quite simple with respect to data interpretation and manipulation compared to the more sophisticated GC×GC-TOFMS. It must be acknowledged that at present no GC×GC software is available to automatically process and present GC×GC-qMS results. In addition, GC×GC-qMS is relatively inexpensive compared to the latter and is a standard tech-nology in many laboratories; the cost or lack of immediate availability of the TOFMS instrument precludes its use in many laboratories for GC×GC purposes. Certainly, the quantitative capabilities of TOFMS are superior to those of qMS as a result of mass spectral skewing and limited peak data density in the qMS instrument. Nev-ertheless, GC×GC-qMS is still able to provide valuable qualitative information, and recently, semiquantitative [33] and quantitative [34] data have been reported in some studies.

The data acquisition rate of qMS detectors is quite slow, with a usual operation speed of approximately 4 Hz commensurate with standard GC-qMS. Obviously, this speed is inadequate for fast GC×GC analysis, and so specific techniques have been implemented to overcome these limitations. The first report of GC×GC-qMS was by Frysinger and Gaines [35]. In this work, chromatographic peaks were intentionally broadened by increasing the length of the thick film first dimension from 4 to 13 m (0.100 mm ID; 3.5 μm d_f). As a direct result, first dimension separation time increased from 60 to 440 min, and corresponding second dimension peak widths increased from 0.2 to almost 1 s, respectively, enabling about three full-scan mass spectra per peak (45–350 u scan; 2.43 scan s^{-1}). Although this is still insufficient for accurate peak shape and quantitation, the resulting mass spectra could be matched to library spectra for identification purposes. This, together with structured GC×GC retention, was particularly useful for the characterization of substituted benzenes in the petroleum sample, since chemical standards for such compounds (e.g., C5 and C6 substituted benzenes) are generally unavailable.

Shellie et al. [36] have used an alternative approach to GC×GC-qMS, whereby a reduced mass range of 41–228.5 United atomic mass units (u), equivalent to 20 scan s^{-1} (Hz), was used for the analysis of ginseng samples. This represents a significant increase in scan speed when compared to that of Frysinger and Gaines [35]. Improved resolution of the sample components and isolation of the analyte response from the chemical background response facilitated by GC×GC, combined with higher qMS scanning rates produced high quality mass spectra. In fact, spectral match qualities of up to 99% were observed for many components. On the basis of these findings, Song et al. [37] used a reduced mass range approach (42–235 u, equivalent to a scanning frequency of 19.36 Hz) for the analysis of 77 underivatized drug standards. A unique "truncated library" was created using mass spectra for the drugs on the basis of this reduced mass range, yielding spectral match qualities of at least 90% for 75% of the drugs analyzed. This approach of Song et al. [37] particularly demonstrates how lateral thinking and manipulation of instrumental/data parameters can extend the feasibility and capabilities of the GC×GC-qMS technique.

An alternative approach to increasing the data acquisition rate of the qMS detector is to employ selected ion monitoring (SIM). Using SIM, the mass spectrometer is set to scan over a very small mass range, typically one mass unit, or a few selected mass ions; hence, only analytes with the particular ion(s) will be detected. This generally necessitates that the identity of the analyte(s) be known before analysis in order to correctly choose the particular ions to monitor. The use of SIM therefore represents a selective approach to GC×GC qMS. Korytár et al. [38] achieved a data acquisition rate of 90 Hz using SIM (dwell time 0.01 s; interscan delay 0.001 s), compared to that of 23 Hz for a mass range of 300 u, for the analysis of organohalogen compounds. This represents an approximate fourfold increase in acquisition rate, translating to an equivalent increase in the number of peak data points. It should be noted that the qMS for this study was operated in the electron-capture negative ion (ECNI) mode (chemical ionization), as opposed to the electron ionization (EI) mode, which is generally used. The former represents a soft ionization technique, which is highly desirable for the sensitive and selective analysis of organohalide type compounds. Importantly, GC×GC with ECNI qMS caused no additional ^{2}D peak broadening as compared to simple EI qMS and was deemed by the authors to be highly suitable for the trace-level analysis of a variety of organohalogen compounds.

SIM was also used in the first semiquantitative report of GC×GC-qMS by Debonneville and Chaintreau [33] for the detection of 24 known allergens. The duration of analysis was divided into windows corresponding to the elution zone of each allergen in the first dimension. Target allergens were monitored by choosing a single and characteristic ion in each window, yielding a data acquisition rate of 30.7 Hz (dwell time 0.01 s). This is a significantly reduced scan rate compared to the results of Korytár et al. [38]. Nevertheless, 30.7 Hz was sufficient for measuring peaks of ^{2}D peak half widths as low as 54 ms (for limonene). The authors quoted a total of 28 data points over the region of elution of the limonene peak; however, the peak had a measurable response only for six data points. This represented about 92% of the total peak area, and was deemed adequate for the semiquantitative study reported. Additionally, "quantitation" of a mixture of five allergens was greatly affected by the nature of the second dimension column that was used, and it was found that the greater the orthogonality of the column set, the greater the peak broadening of the alcoholic components, and the worse their quantitation [33].

Adahchour et al. [34] found that GC×GC-qMS with a data acquisition rate of 33 Hz and a mass range of up to 200 amu ensured at least seven data points per peak and fully satisfactory quantitation of their target analytes. This is in accordance with the findings of Debonneville and Chaintreau [33], who found that 92% of the total area of the aforementioned limonene peak could be represented by only six data points. Adahchour et al. observed the generation of high-quality mass spectra, which enabled correct linearity and quantification of the test analytes, with Limit of detections (LODs) generally in the low picogram range. Thus "rapid scanning" qMS serves as a most useful alternative to TOFMS for applications of limited mass range.

The ultimate choice between TOFMS and qMS with respect to detection in GC×GC will depend on a variety of factors, including expense, and whether full quantitative analysis is necessary. Certainly, there are particular tradeoffs between using one technique over the other. For example, the spectral skewing associated with qMS impinges on the qualitative and quantitative capabilities of the GC×GC-qMS technique. Furthermore, GC×GC-qMS data requires external data conversion and presentation software [37] compared to that of GC×GC-TOFMS data, which has dedicated (two-dimensional) ChromaTOF™software for direct data presentation, provided through the only commercial manufacturer of GC×GC-TOFMS (LECO Corp, St. Joseph, MI). Alternatively, the greater scan range and raw scan numbers of the TOFMS compared to qMS necessitates data files of significantly increased size, requiring much longer postrun processing when subjected to proprietary processing protocols. Clearly, it is most important, based upon collective literature data [33,36,37] that GC×GC-qMS not be dismissed, and should be recognized as a powerful tool for qualitative characterization. At present, semiquantitative–quantitative analysis can be achieved [33,34], and it is expected that future developments will ensure the full quantitative capabilities of this technique as detector improvements are made.

10.5 SELECTED APPLICATIONS

The traditional application area of GC×GC is petroleum analysis, in direct recognition of the limitations of one-dimensional GC when applied to this task. Petroleum samples are highly complex, containing hundreds of thousands of components. Any attempt to separate these components obviously necessitates a technique with incredibly high peak capacity and was the motivation for the development of GC×GC [1]. The inherent benefit of structured retention in GC×GC is beautifully suited to the characterization of such samples as demonstrated by Beens et al. [39], and today, petroleum analysis represents one of the most popular application areas of GC×GC. Recent notable investigations of GC×GC applied to petroleum analysis include the application of the opposite column-phase arrangement in GC×GC, whereby a polar ^{1}D was coupled to a nonpolar ^{2}D [40]. Despite the reduced orthogonality compared to the regular arrangement used in GC×GC (nonpolar ^{1}D/polar ^{2}D), as a result of the ^{1}D separation being governed by both volatility *and* molecular specific interactions, group-type separation (saturates; monoaromatics; diaromatics; and triaromatics) could still be achieved. Compared to nonpolar–polar GC×GC, polar–nonpolar GC×GC is able to extend the separation space for some sample components (notably the nonpolar components that tend to elute with little ^{2}D retention on a nonpolar–polar column set), indicating that the peak capacity in the latter was better suited to the analytical task. Reversing the order of column set phases in GC×GC therefore offers alternative opportunities for the analysis of complex samples, and has also been applied to the analysis of compounds of interest in food samples [31,41].

The analysis of essential oils represents a significant and well-established application area for GC×GC, largely due to the work of Marriott and Shellie [27,42–45]. Similar to petroleum samples, essential oils may also be rather complex and contain a variety of compound classes including monoterpene and sesquiterpene hydrocarbons and their oxygenated analogs [46]. Structured retention ensures that the diversity of these analytes can be accommodated, and the resulting two-dimensional chromatograms can be used as sample fingerprints. Visual comparison of such fingerprints then facilitates rapid and facile sample discrimination, particularly between varietal samples. Work undertaken in the essential oils area serves as the foundation for related flavor and fragrance research, particularly relating to perfume and food samples, and has lead to implementation of GC×GC by key players in the fragrance industry, for example, Firmenich [33,43]. To date, food samples including garlic powder [47], coffee [31,41], wine [48,49], ginger [50], olive oil [34], and strawberry [51] have been investigated.

Analysis of environmental type samples including atmospherics, water pollutants, and pesticide residues represents a large GC×GC application area. This area continues to expand owing to increased pressure for ultra-trace and multiresidue analyses, which are certainly achievable using GC×GC. Furthermore, the need for time-consuming sample preparation and clean up procedures are effectively reduced since target components can be separated from coextracted impurities and interferences as a result of improved sensitivity and resolution capabilities. The benefits of GC×GC compared to single dimension analysis was most significantly demonstrated by Lewis et al. [52] for the analysis of urban atmospheres, whereby 10–20 peaks were produced using GC compared to 100–200 using GC×GC. Many of the (potentially hazardous) compounds detected using GC×GC were not known to exist in urban atmospheres before this investigation, as a result of the inadequacies of single dimension analysis arising from very low analyte concentrations and sample complexity. A similar situation is believed to exist for cigarette smoke, with the number of unidentified components estimated to be as high as 100,000 (cited in [30]). Clearly, single dimension GC cannot facilitate resolution of such an enormity of analytes; Dallüge [30] has thus demonstrated the power of GC×GC-TOFMS applied to the analysis of cigarette smoke. Figure 10.4 shows that essentially the entire two-dimensional separation space has been consumed by the sample components, reiterating the need for system peak capacity to be matched to sample complexity to ensure successful characterization.

One of the more recent application areas for GC×GC is forensic science, and it is expected that this area will grow because of the diversity of matrices that fall under the forensics science banner. Studies on fire debris and accelerants have been undertaken

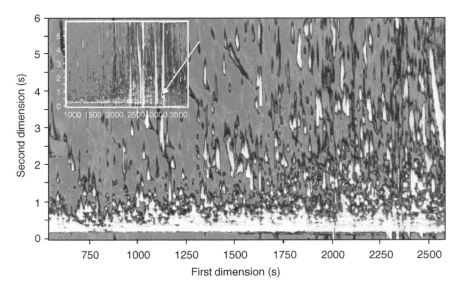

FIGURE 10.4 GC×GC–TOFMS contour plot (total ion chromatogram, TIC) of cigarette smoke showing the first-dimension range of 500–2600 s [30]. A column set of 5% phenyl methyl polysiloxane (DB-5; 30 m × 0.25 mm ID × 1 μm d_f)- polyethylene glycol (Carbowax; 1 m × 0.1 mm ID × 0.1 μm d_f) was used for ^{1}D and ^{2}D, respectively. (Reprinted from J. Dallüge, et al., *J. Chromatogr. A*, 974, 2002,169–184. Copyright 2002. With permission from Elsevier.)

using GC×GC, thus facilitating classification and differentiation of similar ignitable liquids, characterization of fire debris pyrolysates, and the detection of weathered ignitable liquids in such pyrolysates [53]. It is expected that the developed methods will play a major role in future prosecutions of arsonists. Legal implications are also associated with drug screening and antidoping control, which represents a very new and promising application area in GC×GC [37].

The interested reader is referred to recent reviews by Dallüge et al. [54] and Górecki et al. [8] for further details on GC×GC application areas.

REFERENCES

1. Z. Liu and J. B. Phillips, *J. Chromatogr. Sci.*, 29 (1991) 227.
2. J. C. Giddings, *Anal. Chem.*, 39 (1967) 1027.
3. C. J. Venkatramani, J. Xu, and J. B. Phillips, *Anal. Chem.*, 68 (1996) 1486.
4. J. B. Phillips and J. Beens, *J. Chromatogr. A*, 856 (1999) 331.
5. D. Ryan, P. Morrison and P. Marriott, *J. Chromatogr. A*, 1071 (2005) 47.
6. L. M. Blumberg, *J. Chromatogr. A*, 985 (2003) 29.
7. C. M. Harris, *Anal. Chem.*, 74 (2002) 410A.
8. T. Górecki, J. Harynuk, and O. Panić, *J. Sep. Sci.*, 27 (2004) 359.
9. R. E. Murphy, M. R. Schure, and J. P. Foley, *Anal. Chem.*, 70 (1998) 1585.
10. P. J. Schoenmakers, J. L. M. M. Oomen, J. Blomberg, W. Genuit, and G. van Velzen, *J. Chromatogr. A*, 892 (2000) 29.

11. J. B. Phillips, R. B. Gaines, J. Blomberg, et al., *J. High Resol. Chromatogr.*, 22 (1999) 3.
12. P. J. Marriott and R. M. Kinghorn, *Anal. Chem.*, 69 (1997) 2582.
13. J. Beens, M. Adahchour, R. J. J. Vreuls, K. van Altena, and U. A. Th. Brinkman, *J. Chromatogr. A*, 919 (2001) 127.
14. E. B. Ledford Jr. and C. Billesbach, *J. High Resol. Chromatogr.*, 23 (2000) 202.
15. M. Adahchour, J. Beens and U. A. Th. Brinkman, *Analyst*, 128 (2003) 213.
16. E. B. Ledford Jr., C. A. Billesbach, and J. R. Termaat, PTC/US01/01065.
17. C. A. Bruckner, B. J. Prazen, and R. E. Synovec, *Anal. Chem.*, 70 (1998) 2796.
18. J. V. Seeley, F. Kramp, and C. J. Hicks, *Anal. Chem.*, 72 (2000) 4346.
19. R. C. Y. Ong and P. J. Marriott, *J. Chromatogr. Sci.*, 40 (2002) 276.
20. R. W. LaClair, P. A. Bueno Jr., and J. V. Seeley, *J. Sep. Sci.*, 27 (2004) 389.
21. L. L. P. van Stee, J. Beens, R. J. J. Vreuls, and U. A. Th. Brinkman, *J. Chromatogr. A*, 1019 (2003) 89.
22. D. Ryan and P. Marriott, *J. Sep. Sci.*, 29 (2006) 2375.
23. E. M. Kristenson, P. Korytár, C. Danielsson, M. Kallio, M. Brandt, J. Mäkelä, R. J. J. Vreuls, J. Beens and U. A. Th. Brinkman, *J. Chromatogr. A*, 1019 (2003) 65.
24. R. Hua, Y. Li, W. Liu, J. Zheng, H. Wei, J. Wang, X. Lu, H. Kong and G. Xu, *J. Chromatogr. A*, 1019 (2003) 101.
25. J. Blomberg, T. Riemersma, M. van Zuijlen and H. Chaabani, *J. Chromatogr. A*, 1050 (2004) 77.
26. W. M. Draper, *J. Agric. Food Chem.*, 43 (1995) 2077.
27. R. Shellie, P. Marriott, and P. Morrison, *Anal. Chem.*, 73 (2001) 1336.
28. J.-F. Focant, A. Sjödin, and D. G. Patterson Jr., *J. Chromatogr. A*, 1040 (2004) 227.
29. J. Dallüge, M. van Rijn, J. Beens, R. J. J. Vreuls, and U. A. Th. Brinkman, *J. Chromatogr. A*, 965 (2002) 207.
30. J. Dallüge, L. L. P. van Stee, X. Xu, J. Williams, J. Beens, R. J. J. Vreuls, and U. A. Th. Brinkman, *J. Chromatogr. A*, 974 (2002) 169.
31. D. Ryan, R. Shellie, P. Tranchida, A. Casilli, L. Mondello, and P. Marriott, *J. Chromatogr. A*, 1054 (2004) 57.
32. X. Lu, M. Zhao, H. Kong, J. Cai, J. Wu, M. Wu, R. Hua, J. Liu, and G. Xu, *J. Sep. Sci.*, 27 (2004) 101.
33. C. Debonneville and A. Chaintreau, *J. Chromatogr. A*, 1027 (2004) 109.
34. M. Adahchour, M. Brandt, H.-U. Baier, R. J. J. Vreuls, A. M. Batenburg, and U. A. Th. Brinkman, *J. Chromatogr. A*, 1067 (2005) 245.
35. G. S. Frysinger and R. B. Gaines, *J. High Resol. Chromatogr.*, 22 (1999) 251.
36. R. A. Shellie, P. J. Marriott, and C. W. Huie, *J. Sep. Sci.*, 26 (2003) 1185.
37. S. M. Song, P. Marriott, and P. Wynne, *J. Chromatogr. A*, 1058 (2004) 223.
38. P. Korytár, J. Parera, P. E. G. Leonards, J. de Boer, and U. A. Th. Brinkman, *J. Chromatogr. A*, 1067 (2005) 255.
39. J. Beens, J. Blomberg, and P. J. Schoenmakers, *J. High Resol. Chromatogr.*, 23 (2000) 182.
40. C. Vendeuvre, R. Ruiz-Guerrero, F. Bertoncini, L. Duval, D. Thiébaut, and M.-C. Hennion, *J. Chromatogr. A*, 1086 (2005) 21.
41. L. Mondello, A. Casilli, P. Q. Tranchida, P. Dugo, S. Festa, and G. Dugo, *J. Sep. Sci.*, 27 (2004) 442.
42. R. Shellie and P. Marriott, *Flavour Frag. J.*, 18 (2003) 179.
43. R. Shellie, P. Marriott, and A. Chaintreau, *Flavour Frag. J.*, 19 (2004) 91.
44. R. Shellie, P. Marriott, and C. Cornwell, *J. Sep. Sci.*, 24 (2001) 823.

45. R. Shellie, L. Mondello, P. Marriott, and G. Dugo, *J. Chromatogr. A*, 970 (2002) 225.
46. P. Marriott and R. Shellie, *Trends Anal. Chem.*, 21 (2002) 573.
47. M. Adahchour, J. Beens, R. J. J. Vreuls, A. M. Batenburg, E. A. E. Rosing, and U. A. Th. Brinkman, *Chromatographia*, 55 (2002) 361.
48. D. Ryan, P. Watkins, J. Smith, M. Allen and P. Marriott, *J. Sep. Sci.*, 28 (2005) 1075.
49. Y. Shao, P. Marriott, and H. Hügel, *Chromatographia*, 57 (2003) S-349–S-353.
50. Y. Shao, P. Marriott, R. Shellie, and H. Hügel, *Flavour Frag. J.*, 18 (2003) 5.
51. A. Williams, D. Ryan, A. O. Guasca, P. Marriott, and E. Pang, *J. Chromatogr. B*, 817 (2004) 97.
52. A. C. Lewis, N. Carslaw, P. J. Marriott, R. M. Kinghorn, P. M., A. L. Lee, K. D. Bartle, and P. Michael J, *Nature*, 405 (2000) 778.
53. G. S. Frysinger and R. B. Gaines, *J. Forensic Sci.*, 47 (2002) 471.
54. J. Dallüge, J. Beens, and U. A. Th. Brinkman, *J. Chromatogr. A.*, 1000 (2003) 69.

Index

A

A35512B, vancomycin analog, 114
A-40, 926 (MDL 62,476) glycopeptide,
 118–119
A-40, 926 CSP, 152
A82846, vancomycin analog, 114
 A82846A (Eremomycin), 114
 chemical structure, 115
Ab-initio methods
 SO–SA complexes structure determination,
 60
Acetonitrile, 14
Acetyl carnitine
 enantioseparation of, 148
O-Acylcarnitine derivatives, enantioseparation
 of, 145
Additives effect in SFC, 225–228
ADHD (attention deficit hyperactivity
 disorder), 150
β-Adrenoreceptors antagonists, *see* β-blockers
Adsorption isotherm measurements, 44
AED (adsorption energy distributions), 278
Affinity ligand encapsulation (entrapment),
 411–416
AFM (atomic force microscopy), 267–268
Aglycones
 enantiorecognition studies by, 157–161
Alkyl-stationary phases, 252–256
Allethrin, 83
Amides, enantioseparation of, 144
Amino acids
 amino sulfonic acids, chromatographic
 applicability spectrum, 76–78
 and derivatives, analysis, 66–76
 enantioseparation of, 139–143
 N-blocked α-amino acids, 141
 nonproteinic amino acids, 141
 secondary amino acids, 143
 underivatized α-amino acids, 140
 unusual amino acids, 141–143
Aminopropylated silica with bifunctional
 spacers, 126–127
Amino-terminated organosilanes
 for glycopeptide containing CSPs synthesis,
 125–128
 aminopropylated silica with bifunctional
 spacers, 126–127

Amphetamine derivatives, enantioseparation
 of, 150–151
Analytical shape computation techniques, 282
Anion-exchange retention process, 7
ANP (aqueous normal-phase
 chromatography), 344
Ansamycins, 111–112, *see also* Rifamycin B;
 Rifamycin SV
APCI (atmospheric pressure chemical
 ionization), 136, 191
APPI (atmospheric pressure photoionization),
 136
APTES (3-aminopropyltriethoxysilane),
 402
ATR IR spectroscopy, of SO–SA complexes,
 55
Avoparcin, 116–118
 avoparcin α
 chemical structure, 117
 avoparcin β
 chemical structure, 117

B

Barbiturates enantioseparation of, 151–152
Basic compounds
 liquid chromatographic separations of,
 305–348
 RP chromatography of, 307–344
 influence of the solute, 330–331
 mobile phase, influence of, 331–338
 overloading, 310–319
 RP column based on silica, 319–329,
 see also RP columns
 silanols, 307–310, *see also separate entry*
 temperature, 338–340
 testing methods for RP columns,
 340–344
Beer's law, 354
Benzodiazepines
 enantioseparation of, 151–152
BGE (background electrolyte), 39
Bifunctional spacers
 aminopropylated silica with, 126–127
Bi-Langmuir model, 44